Volume 2

Ceramic Transactions

SILICON CARBIDE '87

Edited by James D. Cawley, The Ohio State University
Charles E. Semler, Semler Materials Services

The American Ceramic Society, Inc.
Westerville, Ohio

Proceedings of the Silicon Carbide 1987 Symposium held in Columbus, OH, on August 2-5, 1987.

LIBRARY OF CONGRESS
Library of Congress Cataloging-in-Publication Data

Silicon carbide '87.

    Includes index.
    Contents: v. 2. Ceramic transactions
    1. Silicon carbide--Congresses.
I. Cawley, James D. II. Semler, Charles E.
TP245.S5S365 1989   666   88-35004
ISBN 0-944904-06-8 (v. 2)

ISBN 0-944904-06-8

Copyright © 1989, The American Ceramic Society, Inc. All rights reserved.

No part of this book may be reproduced, stored in a retrieval system, or transmitted in any form or by any means, electronic, mechanical, photocopying, microfilming, recording, or otherwise, without written permission from the publisher.

Printed in the United States of America

1 2 3 4 5—94 93 92 91 90 89

Foreword

Silicon carbide might aptly be termed the original advanced ceramic material because of its commercial uses in multiple applications for decades. However, because it does not exist in significant quantities in nature, commercial use relies entirely on synthetic material. The development of the Acheson process in 1891 brought silicon carbide production to a commercial scale in the abrasives industry. It continues to play a major role in that industry; however, it was widely recognized that silicon carbide offers a unique set of electrical, optical, thermal, and mechanical properties which are appropriate to a very wide range of engineering applications. In many cases, significant developments in processing technology were necessary before these properties could be successfully exploited. One example is the use of silicon carbide as a high temperature semiconductor; although the electronic structure of silicon carbide was recognized in the 1950s, it took the very recent development of a process for the fabrication of reproducible, low defect concentration, large area single crystals to build electronic devices. The development of this process is discussed within this volume by workers from the NASA Lewis Research Center. Many other examples exist, including (1) the development of a process for sintering submicron silicon carbide powder to near theoretical density by Prochazka and others, which led to the fabrication of silicon carbide monoliths with excellent high temperature strength, and (2) the development of polymer precursors which has been instrumental in the development of fibers.

It was the intent of this symposium, held in conjunction with the ACerS Central Ohio Section, the Department of Ceramic Engineering at the Ohio State University, NASA Lewis Research Center, and the Standard Oil Engineered Materials Company, to assess and discuss the state-of-the-art in silicon carbide processing and application. The symposium was lively, characterized by a high level of technical discussion both within and outside of the sessions.

The efforts of the staff of the American Ceramic Society in organizing and conducting the meeting as well as in the preparation of these proceedings is gratefully acknowledged. Thanks are extended to our colleagues who contributed manuscripts, served as reviewers and/or session chairs, and in particular to James A. DiCarlo, NASA Lewis Research Center, and Bryan D. Foster, Norton Co., for their stimulating talks during the symposium luncheon.

Charles E. Semler   James D. Cawley

Ceramic Transactions is a new proceedings series designed to meet two needs: high quality content and rapid publication. Volumes in the series come from meetings, symposia, and forums. Each paper is reviewed by two peers, and final manuscripts are prepared by authors in a "camera-ready" format. The volumes in this series would not be possible without the hard work, dedication, and cooperation of editors, reviewers, and authors, who all deserve a great deal of thanks.

This volume in the *Ceramic Transactions* series was generated electronically from disks prepared by authors. Text was converted to files readable by Word Perfect, Verson 5.0, and was output in a consistent format on a Hewlett-Packard Laser Jet Series II printer.

Your comments, questions, and suggestions for future *Ceramic Transactions* volumes are welcomed and should be addressed to the Director of Publications, The American Ceramic Society, Inc.

Contents

Section I. Synthesis and Characterization

Pressureless Sintering and Properties of Plasma Synthesized
SiC Powder 3
H.R. Baumgartner and B.R. Rossing

Synthesis of Submicron Silicon Carbide Powder 17
K.M. Rigtrup and R.A. Cutler

Silicon Carbide Powders by Gaseous Pyrolysis of
Tetramethysilane 35
H.-D. Wu and D.W. Readey

Thermal Carburization of Single-Crystal Silicon 47
J.G. Sheek and J.D. Cawley

Alloying of Silicon Carbide with Other Ceramic Compounds
—A Review 63
A. Zangvil and R. Ruh

Effect of BeO on the Microstructure and Phase Stability of SiC ... 83
A. Zangvil, Y.-W. Chang, and R. Ruh

Synthesis and Characterization of HSC Silicon Carbide 93
W.M. Goldberger, A.K. Reed, and R. Morse

New Method for the Quantitative Analysis of Free Carbon in
Silicon Carbide 105
H. Knoch, K.A. Schwetz, and W.D. Long

Enhanced Formation of 4H Polytype in Silicon Carbide
Materials 113
S.S. Shinozaki, J. Hangas, K. Maeda, and A. Soeta

Section II. Green State Processing

Ultrafine SiC Powder Produced by Turbomilling 125
D.E. Wittmer, Sr.

Turbomilling Parameters Affecting the Ultrafine Grinding of

Alpha-SiC 137
J.L. Hoyer
Ultrasonic Impact Grinding 149
D.O. Moore
Suspension Processing of Beta-SiC Powders 157
B.A. Bishop and H.K. Bowen
Slip Casting and Sintering of Silicon Carbide 175
E. Carlstrom, M. Persson, E. Bostedt, A. Kristoffersson, and R. Carlsson
Nuclear Magnetic Resonance Imaging for Detecting Binder/
Plasticizers in Green-State Structural Ceramics 187
W.A. Ellingson, J.L. Ackerman, L. Garrido, and S. Gronemeyer

Section III. Mechanical Properties and Environmental Stability

Effects of Various Consolidation Techniques on
Microstructure, Strength, and Reliability of Alpha-SiC 201
S. Dutta
Microstructure and Mechanical Properties of Pressureless
Sintered Alpha-SiC 215
S.G. Seshadri, M. Srinivasan, and K.Y. Chia
SiC-Based Ceramics with Improved Strength 227
T.B. Jackson, A.C. Hurford, S.L. Bruner, and R.A. Cutler
The Fracture Resistance of a Sintered Silicon Carbide Using
the Chevron-Notch Bend Specimen 241
M.G. Jenkins, A. Ghosh, K.W. White, A.S. Kobayashi, and R.C. Bradt
Deformation Behavior of Reaction-Bonded, Chemically Vapor
Deposited and Sintered Silicon Carbides at Elevated
Temperatures 253
R.F. Davis, C.H. Carter, Jr., and J.E. Lane
The Behavior of SiC and Si_3N_4 Ceramics in Mixed Oxidation/
Chlorination Environments 275
J.E. Marra, E.R. Kreidler, N.S. Jacobson, and D.S. Fox
Oxidation of SiC Ceramic Heat Exchanger Materials in the
Presence of Chlorine at 1300° C 289
S.Y. Ip, M.J. McNallan, and M.E. Schreiner

Active Oxidation of SiC in Low Dew-Point Hydrogen Above 1400°C 301
H.-E. Kim and D.W. Readey

Thermal Expansion and Elastic Anisotropies of SiC as Related to Polytype Structure 313
Z. Li and R.C. Bradt

Section IV. Applications

Improved Silicon Carbide for Advanced Heat Engines. Part I: Process Development for Injection Molding 343
T.J. Whalen and W. Trela

Improved Silicon Carbide for Advanced Heat Engines. Part II: Pressureless Sintering and Mechanical Properties of Injection Molded Silicon Carbide 355
T.J. Whalen and J.R. Baer

Fabrication of Sintered Alpha-SiC Turbine Engine Components 367
R.W. Ohnsorg and M.O. Ten Eyck

Large Silicon Carbide Radiant Tube Production Processes 387
M.C. Kasprzyk

Section V. Fibers and Composites

SiC Whiskers and Platelets 397
S.C. Weaver, R.D. Nixdorf, and G. Vaughan

Single Phase Alpha-SiC Reinforcements for Composites 407
W.D.G. Boecker, S. Chwastiak, F. Frechette, and S.-K. Lau

Silicon Carbide Fibers from Methylpolysilane Polymers 421
J. Lipowitz, G. LeGrow, T. Lim, and N. Langley

Characterization of Si, C, N, O Fibers by Analytical STEM and Scanning Auger Techniques 435
Y.-W. Chang, A. Zangvil, and J. Lipowitz

Section VI. Electronic Applications

Crystal Growth of SiC for Electronic Applications 447
L.G. Matus and J.A. Powell

Chemical Vapor Deposition, In Situ Doping, and MESFET Performance of Beta-SiC Thin Films 457
H.J. Kim, H.-S. Kong, J.A. Edmond, J.T. Glass, and R.F. Davis

Structural Characterization of Ion Implanted Beta-SiC Thin Films 479
J.A. Edmond, R.F. Davis, and S.P. Withrow

Effects of Cathode Materials and Gas Species on the Surface Characteristics of Dry Etched Monocrystalline Beta-SiC Thin Films 491
J.W. Palmour, R.F. Davis, P. Astell-Burt, and P. Blackborow

Microstructure and Thermoelectric Energy Conversion in Porous SiC Ceramics 501
K. Koumoto, M. Shimohigoshi, S. Takeda, and H. Yanagida

Index 511

Section I

Synthesis and Characterization

Pressureless Sintering and Properties of Plasma Synthesized SiC Powder

H. Robert Baumgartner

Aluminum Company of America
Alcoa Center, PA 15069

Barry R. Rossing

Lanxide Corporation
Newark, DE 19711

Abstract

Beta-phase silicon carbide powders produced in an arc-plasma reactor are highly sinterable due to their small, unagglomerated particle size, high purity and homogeneous distribution of co-formed sintering aids. Minimum levels for each of the boron and free carbon additives required for effective densification have been defined. The densification and microstructural development of plasma-SiC powder have been studied as a function of sintering temperature. Powders of <0.3 wt% B can be pressureless-sintered to 98% of theoretical density at 2100°C. Property characterizations on densified material include wet and dry abrasion resistance, 4-point bend strength to 1400°C, dynamic fatigue resistance at 1400°C, Young's modulus, fracture toughness, hardness and thermal diffusivity.

Introduction

Silicon carbide is probably the most widely used nonoxide ceramic. It exhibits, in a pronounced manner, most of the positive attributes associated with ceramic materials. Its high hardness and chemical inertness make it useful as a wear material, while its strength retention and resistance to oxidation and thermal shock make it attractive in structural applications at elevated temperatures.

Many methods have been used to synthesize silicon carbide. Among these are the classical carbothermic reduction of silica, the pyrolysis of natural[1] and synthetic polymers[2] and the siliconizing of carbonaceous matter[3]. Recently, a number of processes have been developed which directly produce submicron-sized powders of silicon carbide from gaseous reactants. Included in these methods are the use of laser radiation to initiate the reaction[4] and the use of plasmas, either of the radio frequency (r.f.)[5] or electrical arc[6] type, to provide the thermal activation. The nature of silicon carbide powder produced by the d.c. arc process and the properties of bodies sintered from this powder form the subject of this paper.

Powder Production and Characterization

A schematic diagram of the d.c. arc process is shown in Figure 1. Hydrogen is used as the plasma gas and also serves as a reductant. The most common silicon feedstock is silicon tetrachloride. A variety of feedstocks may be used for the carbon source; pure hydrocarbons or halogenated hydrocarbons with 1 to 12 carbon atoms have been used[6]. Reactants are entrained in a carrier gas and fed into the tail flame of the plasma within the reactor. The sequence of the reaction steps is believed to be the reduction of the silicon tetrachloride, followed by the carburization of molten silicon droplets[7]. The particulate product is transported pneumatically to a suitable filter where it is collected. Process off-gases, containing hydrogen chloride, are passed through a caustic scrubber. The reactor is capable of producing approximately 4 kg of silicon carbide powder per hour.

In order to sinter silicon carbide to near theoretical density without the application of pressure, sintering aids, most frequently boron and carbon, must be used. In the plasma process, these aids are incorporated into the powder within the reactor. Boron trichloride is used for the boron dopant and an excess of the carbon feedstock provides the free carbon. The co-formation of the sintering additives and the silicon carbide powder produces a homogeneous distribution of the additives within the powder. Boron is present as a solid solution in the silicon carbide. The homogeneity increases the effectiveness of the additives, thereby requiring lower levels for effective densification.

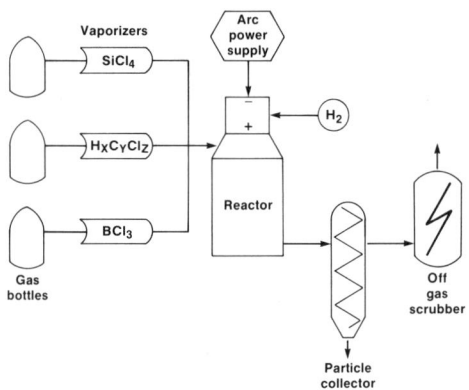

Fig. 1. Schematic of Plasma-Arc Process for SiC Powder Production.

Characteristics of the powder are shown in Table 1. The powder is predominately of the beta (cubic) silicon carbide polytype, as determined by X-ray diffraction. A mean particle size of 0.5 micron diameter has been found to sinter well. The particle size distribution, as characterized by the D_{90}/D_{10} ratio, shows that the powder is not monodispersed. The principal impurities in the powder are low levels of oxygen and adsorbed chlorine. The level of oxygen is unusually low for a submicron silicon carbide powder.

Table 1. Characteristics of SiC Powder Made by Plasma-Arc Process.

Average (mass) diameter	0.5×10^{-6} m
D_{90}/D_{10}	12
Surface area	10 m^2/g
XRD phase	beta (cubic) SiC

Chemistry

Element	Weight (%)
B	0.2-0.3
Co	0.6
O	0.1
Cl	0.15
Ag, Al, As, Au, Ba, Be, Bi, Ca, Cd Ce, Co, Cr, Cu, Fe, Ga, Ge, Hg, In K, La, Li, Mg, Mn, Mo, Na, Nb, Ni P, Pb, Pt, Rb, Sb, Sn, Sr, Ta, Te Ti, V, W, Yb, Zn, Zr	Below limit of detectability ($<$ 0.0003-$<$ 0.2) by a.e.s

A transmission electron micrograph of the powder is shown in Figure 2. The powder is composed of largely unagglomerated particles, a characteristic which is essential for good sinterability. A small fraction of the powder, <2 wt%, is in the form of partially sintered agglomerates. These detrimental particles are removed during downstream powder processing. In general, the particles are spheroidal without well developed facets. Some of the larger particles appear to be polycrystalline. It has been reported[7] that the morphology of plasma-produced silicon carbide particles is dependent on their temperature of formation with a low temperature of formation favoring faceted crystals. This suggests that the current particles form at a relatively high temperature. In contrast to particles formed in an r.f. plasma under carbon-rich conditions[8], there is no indication of a condensed, thick shell of carbon on the current particles.

The nature of the powder surface has been examined by ESCA[9]. Detected were carbon-carbon bonding, associated with free carbon, carbidic carbon linkages, associated with silicon carbide, silica-like silicon-oxygen bonds and a weak, diffuse peak which possibly may be associated with boron.

SINTERING BEHAVIOR

Powder was dispersed in a water-based suspension and spray dried to produce granules for pressing. Powder compacts were pressed mechanically at 55 MPa, followed by isostatic pressing at 150 MPa to yield a green density of 1.89 g/cm^3. The compacts were pressureless-sintered under an inert atmosphere (either helium or argon) at 2000-2200°C for 0.5-1 h. Sintered density as a function of sintering temperature is shown in Figure 3. Maximum density occurred within the 2100-2175°C interval.

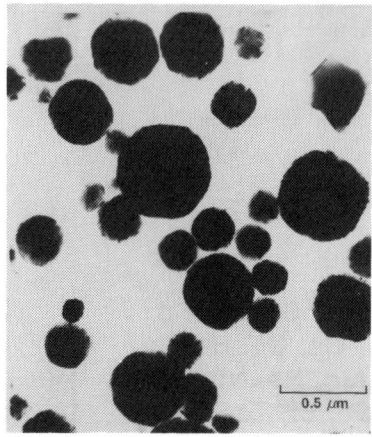

Fig. 2. Transmission Electron Micrograph of Plasma-SiC Powder.

As previously mentioned, the use of sintering aids is necessary to achieve high sintered densities in silicon carbide. Figure 4 shows the effect of boron level on sintered density. There is a critical boron concentration, near 0.15 wt%, below which high densities are not achieved. Also shown on the plot are data for two commercially available beta powders. Boron in the form of boron carbide was added to the latter powders. For equivalent amounts of boron, higher sintered densities are obtained with the plasma powder. The effect of free carbon level upon sintered density is shown in Figure 5. A minimum amount of free carbon is necessary for good densification and the plasma powder requires lower additive levels. Less additives are needed for the plasma powder because of their better distribution and, for the carbon additive, because of the lower oxygen content of the plasma powder.

The development of the microstructure of plasma silicon carbide with sintering temperature is shown in the series of micrographs of Figure 6. The microstructures were obtained from the samples used for Figure 3 and are representative of the bulk of the powders. A molten salt etch of $KOH-KNO_3$ has been used to differentiate the beta and alpha phases. The alpha phase appears and rapidly grows to a 200-300 micron grain size over a narrow temperature range. The beta-to-alpha transformation begins below 2075°C and alpha growth proceeds rapidly with increasing temperature. Both platelet and "feather" forms[10] of the alpha phase are visible in the microstructure with "feathers" predominant.

For incompletely understood reasons, the exact temperature at which rapid alpha growth occurs and the predominant morphology of the alpha phase varies with powder lot. In a small number of powder cases, the extent of the transfor-

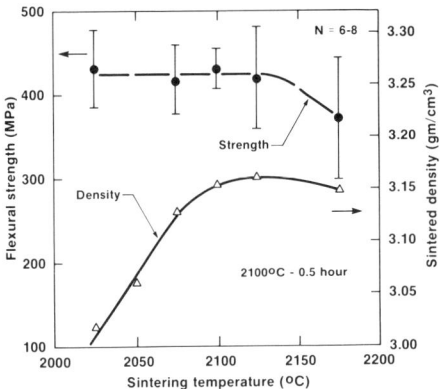

Fig. 3. Four-Point Flexural Strength (25°C) and Density vs. Sintering Temperature for Plasma SiC.

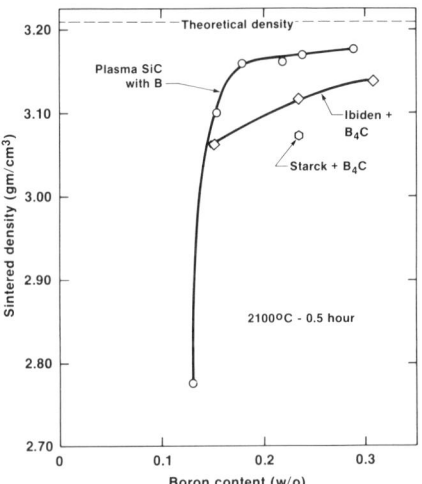

Fig. 4. Sintered Densities of SiC Compacts Using Differing SiC Powders and Boron Contents.

mation was retarded by about 25-50°C and the platelet morphology was more prevalent. Since the strength of high density material is controlled by the size of the alpha grains, the illustrated case (Figure 6) is the worst case for strength.

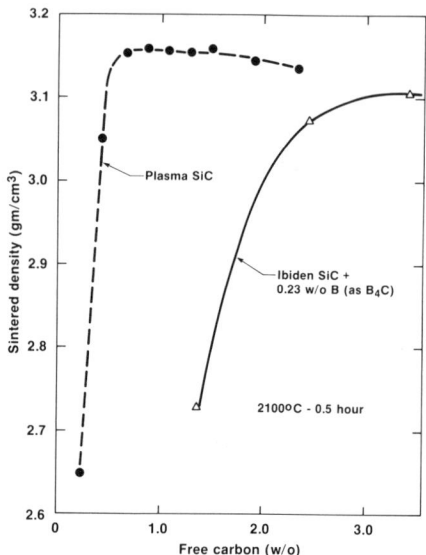

Fig. 5. Sintered Density vs. Free Carbon Content.

PROPERTY CHARACTERIZATION

Fast Fracture Strength

Strength bars were prepared in accordance with the procedures of Military Specification MIL-STD-1942(MR) in the Type B specimen size (bend spans of 20 mm over 40 mm). Bars were broken in 4-point bending in the as-ground condition. Room temperature bend strengths for bars sintered over a range of temperatures are shown in Figure 3. Each datum point represents the mean of 10 tests and ± 1 standard deviation is represented by the vertical bars. Despite the significant increase in density over the sintering range, the strength is unusually insensitive to sintering temperature. The strength plateau, approximately 430 MPa (62 ksi), is similar to the strength levels reported for other strong, sintered silicon carbides[11]. At the highest sintering temperature, the strength decreases slightly. This decrease is due to the growth of large alpha grains which are critical flaws, such as is illustrated in Figure 7.

As is well known, the fracture strength of brittle materials is dependent upon the nature and severity of the critical flaws, which in turn are conditioned by processing and finishing procedures. In addition, reported strengths are influenced by bar size and method of bend testing. Recent room temperature 4-point bend

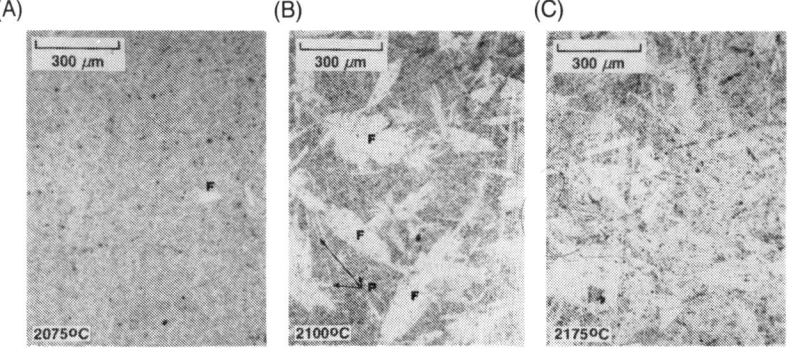

Fig. 6. Development of Plasma–SiC Microstructure After 0.5 h at: (A) 2075°C, (B) 2100°C and (C) 2175°C. Note Platelet (P) and Feather (F) Morphologies of Alpha Phase.

Fig. 7. Fracture Origin of Alpha Feather, Bar = 22 microns.

strengths of machined plasma silicon carbide have been in the 475–540 MPa (69–78 ksi) range. Three-point bend strengths of 650 and 456 MPa have been reported[12] for as-fired and machined bars, respectively, made from beta-SiC powder produced in a similar reactor but processed by somewhat different procedures.

The 4-point bend, fast fracture strength of plasma-SiC as a function of temperature of measurement is shown in Figure 8. Each datum point is the mean of 7 bars broken in air in the as-ground condition at a testing machine crosshead speed of 0.051 cm/min. The vertical bars represent ± 1 standard deviation. The mean strengths increase from a value of 369 MPa (53.5 ksi) at 25°C to 488 MPa (70.8 ksi) at 1400°C. The increase in strength with temperature is attributed to the

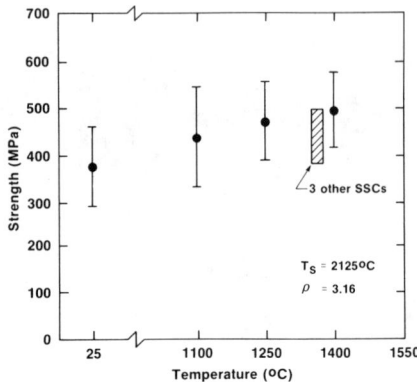

Fig. 8. Four-Point Bend, Fast Fracture Strength of Plasma SBSC.

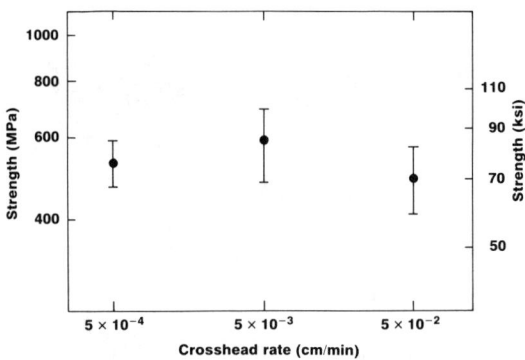

Fig. 9. Dynamic Fatigue of Beta SiC at 1400°C in Air.

presence of relatively severe grinding damage and to the fact that the bars were tested in the as-ground condition. When the material is tested at elevated temperature in air, surface oxidation tends to "heal" the flaws which enables the material to exhibit a higher strength. The cross-hatched region in Figure 8 represents the range of mean strengths at 1371°C of three other well-regarded silicon carbide materials[11].

Dynamic Fatigue Testing

A dynamic fatigue test, wherein strength is measured at different stressing rates, was conducted on bars of material identical to those used to generate Figure 8. This test is often used to evaluate the occurrence of slow crack growth in materials[13]. Seven bars were broken at each of the crosshead rates of 0.051, 0.0051 and 0.00051 cm/min in air at 1400°C.

The dynamic fatigue data are reported in Figure 9. It is seen that the strength goes through a maximum at the intermediate stressing rate. This behavior suggests the operation of two opposing mechanisms which influence the strength. The increase in strength at the intermediate rate is attributed to additional flaw healing by oxidation as a lower stressing rate prolongs the oxidation time. The mean strength at the intermediate rate is 587 MPa (85.2 ksi).

The decline in strength at the slowest stressing rate suggests the operation of a weakening mechanism, such as thermally activated slow crack growth. The existence of slow crack growth has been observed[14] in other silicon carbide materials at lower stress levels and temperatures and, therefore, may also be the mechanism responsible for the current weakening.

Miscellaneous Properties

Other mechanical and thermal properties of plasma-SiC are shown in Table 2. The values are consistent with what are expected of a dense, high purity silicon carbide. Young's modulus was determined by measurements of ultrasonic wave velocities, the thermal diffusivity by the xenon flash-lamp method and the fracture toughness by an indentation method[15]. The thermal conductivity was calculated from the diffusivity, measured specific heat and density.

Abrasion and Corrosion Resistance

Silicon carbide is of interest as a wear material because of its high hardness and excellent corrosion resistance[16]. The abrasion resistance of plasma-SiC was evaluated by the ASTM B-611 wet abrasion test and the ASTM G-65 dry abrasion test. The tests were originally developed to evaluate the wear resistance of tungsten carbide cermets. A schematic of the wet test is shown in Figure 10. In the test, the specimen is pressed against a rotating steel disk while an abrasive slurry of fused alumina in water is fed between the sliding couple. After 100 wheel revolutions, the abraded weight loss of the specimen is measured and converted into a volumetric loss. The latter is the standard by which this test compares the wear resistance of various materials. The dry abrasion test is similar in concept, but uses a different disk material, abrasive, load and number of disk revolutions.

The abrasion resistances of the various materials that were tested are shown in Table 3. The wear results are the mean values of triplicate tests. Two types of plasma silicon carbide are compared with 2 grades of tungsten carbide, a commercially available sintered alpha silicon carbide, an experimental SiAlON material and 3 types of dense, high purity alumina from 3 manufacturers with a range of grain size. Relative wear comparisons between different materials are valid only if made within a given test type. For example, finer grained aluminas are more wear resistant in the wet test, but the ranking order reverses in the dry test. The 2 sintered beta silicon carbides (SBSC) from the plasma process differ only in their sintering temperatures. Although the higher sintering temperature yields a higher density, it also increases the extent of the beta-to-alpha transformation and produces a weaker material with somewhat poorer abrasion resistance. With the exception of the highly wear resistant tungsten carbide with cobalt binder, the less

Table 2. Mechanical and Thermal Properties of Plasma SiC.

Young's Modulus	410 GPa
Hardness (V_{500})	2800 kg/mm^2
K_{IC} (Indentation)	3.6 MPa·m$^{1/2}$
Thermal diffusivity (25°C)	0.54 cm^2/s
Thermal conductivity (25°C)	114 W/m·K
Thermal expansion coefficient	
25° - 500°C	4.0 x 10^{-6}/°C
25° - 1000°C	4.8 x 10^{-6}/°C

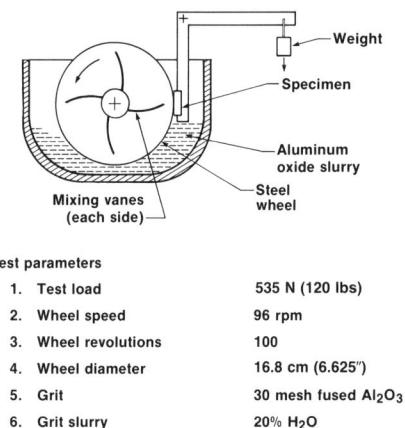

Test parameters
1. Test load — 535 N (120 lbs)
2. Wheel speed — 96 rpm
3. Wheel revolutions — 100
4. Wheel diameter — 16.8 cm (6.625")
5. Grit — 30 mesh fused Al_2O_3
6. Grit slurry — 20% H_2O

Fig. 10. Wet Abrasion Test ASTM B-611 (Modified).

transformed plasma material is superior to the other materials. In the dry test, the WC-Co composition is superior, followed by the closely bunched silicon carbides and the WC-Ni.

Abrasion resistance is not the sole criterion for ceramic wear performance. Other factors, such as corrosion resistance, can influence the acceptability of a material for a given application. Table 4 compares corrosion rates of the materials in corrosive aqueous media. Material recession rates were calculated from weight

Table 3. Abrasion Wear Results Under Wet (ASTM B-611) and Dry (ASTM G-65) Test Conditions.

Material	Wet Wear Volume ± Std. Dev. (mm^3/100 rev)	Dry Wear Volume ± Std. Dev. (mm^3/6000 rev)
WC-6% Co	11.2 ± 0.1	2.62 ± 1.39
WC-10% Ni	55.1 ± 2.5	5.25 ± 0.24
Plasma SBSC (TS = 2150°C; 96.9%)	18.6 ± 1.3	6.54 ± 0.88
Plasma SBSC (TS = 2200°C; 98.4%)	40.5 ± 2.1	7.27 ± 1.54
Commercial SASC	47.8 ± 1.1	6.39 ± 0.28
Si$_{6-x}$Al$_x$O$_x$N$_{8-x}$ (x = 0.5)	28.0 ± 1.8	14.5 ± 0.36
99.8% Al$_2$O$_3$ (1.5 μm)	97.2 ± 14.4	66.8 ± 2.63
99.8% Al$_2$O$_3$ (7.4 μm)	354.0 ± 1.5	35.2 ± 4.05
99.5% Al$_2$O$_3$ (11.7 μm)	579.0 ± 13.2	10.7 ± 0.30

Table 4. Corrosion Rates (MPY) of Wear Materials in Corrosive Media.

Test Solution*	SBSC	99.8% Al_2O_3	WC-6% Co	WC-10% Ni	Sialon; $x = 0.5$; 6% Y_2O_3
40 w/o NaOH	<0.05	0.39	0.58	0.16	27.
98 w/o H_2SO_4	<0.05	0.42	7.1	11.5	0.7
70 w/o HNO_3	<0.05	0.15	65.	79.	42.
37 w/o HCl	<0.05	0.13	103.	158.	63.

*168 hrs, 100°C, continuously stirred

losses sustained during 168 h of exposure and extrapolated to an annual loss. The corrosion resistance of all three silicon carbides was comparable and significantly superior to the other materials. The combination of good abrasion and corrosion resistance makes the plasma-SBSC material a candidate for severe wear applications.

Summary

The d.c. plasma-arc process is capable of producing beta-phase silicon carbide powders of high purity from gaseous reactants. The powders are submicron in size, largely unagglomerated, and contain low levels of co-formed sintering aids of boron and free carbon in a homogeneous distribution. These attributes enable the powder to be pressureless-sintered to near theoretical density at around 2100°C in inert gas atmospheres. The minimum amounts of boron and free carbon required to promote good densification have been defined. The powders are readily dispersible in aqueous suspensions without degradation of powder characteristics necessary for easy sintering. This feature permits the advantageous application of suspension-based processing techniques.

Sintered bodies produced from the plasma powder have exceptional mechanical, thermal and wear resistant properties. These properties may be modified by controlling the sintering conditions to induce microstructural variations, for example, the degree of the beta-to-alpha transformation, in the sintered body. Four-point bend strengths in excess of 585 MPa (85 ksi) have been achieved at 1400°C in SBSC.

References

[1] J. Lee and I. B. Cutler, "Formation of Silicon Carbide from Rice Hulls," *Am. Ceram. Soc. Bull.*, **54** [2], 195-98 (1975).

[2] S. Yajima et al., "Synthesis of Continuous SiC Fibers with High Strength," *J. Amer. Ceram. Soc.*, **59** [7-8], 324-27 (1976).

[3] C. W. Forrest, P. Kennedy and J. V. Shennan, in Special Ceramics, Vol. 5, P. Popper, ed., Brit. Ceram. Res. Assoc., Stoke-on-Trent, England, 99-123 (1972).

[4] Y. Suyama, R. D. Marra, J. S. Haggerty and H. K. Bowen, "Synthesis of Ultrafine SiC Powders by Laser-Driven Gas Phase Reactions," *Am. Ceram. Soc. Bull.*, **64** [10], 1356-59 (1985).

[5] G. J. Vogt, C. M. Hollabaugh, D. E. Hull, L. R. Newkirk and J. J. Petrovic, "Novel RF Plasma System for the Synthesis of Ultrafine, Ultrapure SiC and Si_3N_4," pp. 283-89 in Matl. Res. Soc. Symp. Proc., Vol. 30, J. H. Crawford et al. eds., North Holland, New York, 1984.

[6] U.S. Patent Nos. 4,133,689 and 4,295,890.

[7] G. Saiki and J. Koudo, "Synthesis and Sintering of B-Doped SiC Powders by Plasma Arc Method," Poster 16-BP-86, Am. Ceram. Soc. Annual Meeting, Chicago, Illinois, 1986 April 29.

[8] D. S. Phillips, T. N. Taylor and G. J. Vogt, "Ultrafine SiC Powders2—Surface Microstructure and Dispersibility," Poster 12-BP-86, Amer. Ceram. Soc. Annual Meeting, Chicago, Illinois, 1986 April 29.

[9] D. M. Wilhelmy and S. A. Jankosky, "Surface Characterization of Beta-SiC Powders," Poster 20-BP-86, Am. Ceram. Soc. Annual Meeting, Chicago, Illinois, 1986 April 29.

[10] C. A. Johnson and S. Prochazka, "Microstructures of Sintered SiC," in Ceramic Microstructures '76, R. M. Fulrath and J. A. Pask, eds., Westview Press, Boulder, Colorado, 1977.

[11] L. J. Lindberg, "Durability Testing of Ceramic Materials for Turbine Engine Applications," at 24th Automotive Technology Development Contractors' Coordination Meeting, Dearborn, Michigan, 1986.

[12] S. Prochazka, C. A. Johnson and R. A. Giddings, "Investigation of Ceramics for High Temperature Turbine Components," Report prepared and Contract N62269-75-C-0122 for Naval Air Systems Command, 65 (1975).

[13] A. G. Evans and H. Johnson, "The Fracture Stress and Its Dependence on Slow Crack Growth," *J. Mater. Sci.*, **10** [2], 214-22 (1975).

[14] K. D. McHenry and R. E. Tressler, "High Temperature Dynamic Fatigue of Hot-Pressed SiC and Sintered Alpha-SiC," *Am. Ceram. Soc. Bull.*, **59** [4], 459-61 (1980).

[15] A. G. Evans and E. A. Charles, "Fracture Toughness Determinations by Indentation," *J. Am. Ceram. Soc.*, **59** [7-8], 371-2 (1976).

[16] R. W. Lashway, "Silicon Carbide for Mechanical Shaft Seals," Paper A4 at 9th International Conference on Fluid Sealing, 1981 April, Noordwijkerhout, Netherlands.

Synthesis of Submicron Silicon Carbide Powder

Kevin M. Rigtrup and Raymond A. Cutler

Ceramatec, Inc.
2425 South 900 West
Salt Lake City, UT 84119

Abstract

An exothermic reaction which produces submicron silicon carbide by metallothermically reducing silica in the presence of carbon was studied. The effect of reactants (magnesium, silica, and carbon powders) on the size of silicon carbide synthesized was investigated. Mg starting size only affected the initiation temperature. Starting silica powders had surface areas ranging between 1 and 400 m^2/g, while carbon (or graphite) reactants ranged from 10 to 1000 m^2/g. The size of the silicon carbide, which ranged from 0.02 to 0.2 micrometers, was controlled by the starting size of the carbon reactant. A plausible reaction mechanism based on the experimental results is discussed. The above reaction was compared to the synthesis of SiC from Si and C. The sinterability of the powders was investigated and compared with commercially available powders.

Introduction

High hardness, high thermal conductivity, excellent oxidation resistance, and good creep resistance of dense SiC make it a useful material for a wide variety of applications. The sintering of SiC with B and C additions[1,2], and more recent work using Al and C (or Al, B, and C) sintering aids[3,4] have shown that densification occurs only when submicron SiC powder is used.

The different methods which have been used to produce submicron SiC powders can be classified into two categories: 1) carbothermal reduction of silica or compounds containing silica, and 2) pyrolysis of silane compounds. The grinding and purification of alpha SiC made by the Acheson process[2], the synthesis of beta SiC from silicates[5], and the carbothermal reduction of silica[6-8] are all examples of the first category. The second category involves the synthesis of SiC from silane compounds using a variety of vapor phase reactions and methods[9].

The investigation of the carbothermal reduction of silica at the University of Utah[5,6,10] has shown that transport occurs via gaseous intermediates, namely SiO and CO, as shown below:

$$SiO_{2(s)} + CO_{(g)} \rightarrow SiO_{(g)} + CO_{2(g)} \qquad (1)$$

$$CO_{2(g)} + C_{(s)} \rightarrow 2CO_{(g)} \qquad (2)$$

$$SiO_{(g)} + 2C_{(s)} \rightarrow SiC_{(s)} + CO_{(g)} \tag{3}$$

$$SiO_{2(s)} + 3C_{(s)} \rightarrow SiC_{(s)} + 2CO_{(g)} \tag{4}$$

The overall reaction is in agreement with the earlier work of Pultz and Hertl[11]. The kinetics of the overall reaction support the formation of SiO as a vapor phase instead of Si as an intermediate species[5,6,10-13]. Lee[6] showed that reaction kinetics are dependent on both silica and carbon size and that Reactions (1) and (2) are rate limiting.

The synthesis of submicron SiC via the carbothermal reduction of SiO_2 in the presence of Mg permits the rapid production of fine powders[14]. It is necessary to understand the reaction mechanism in the overall reaction

$$SiO_2 + C + 2Mg \rightarrow SiC + 2MgO \tag{5}$$

in order to control the particle size of the synthesized silicon carbide. Particle size control is desired since ultrafine powders have high oxygen contents due to their surface silica layer[15] and low green densities result due to their high surface area. Powder surface areas in the 5-20 m^2/g range are desired. Such powders are active enough to sinter, yet can be consolidated to green densities greater than 60% of theoretical. The purpose of this paper is to report a plausible reaction mechanism for the carbothermal reduction of silica in the presence of magnesium and to compare this reaction method with the direct synthesis of SiC from the elements.

EXPERIMENTAL PROCEDURES

The main experimental procedures used in this study have been reported elsewhere[14]. The reactants were dispersed in hexane and stirred for 5 minutes in a high speed mixer and dried. The loose powder was wrapped in graphite foil and reacted inside a fused quartz tube in flowing Ar by heating at a rate of approximately 15°C/min. Silica reactants had surface areas ranging between 1 and 400 m^2/g, while carbon (or graphite) reactants ranged from 10 to 1000 m^2/g. Magnesium was either -60 +100 mesh or -325 mesh[14]. After the reaction, the product was leached first with HNO_3 to remove MgO, Mg, and Mg_2Si, followed by NaOH to remove Si and SiO_2. The leaching process involved stirring the powder in an excess amount of dilute (20%) boiling HNO_3 until no further reaction was noted (typically 5-10 minutes), filtering off the HNO_3, repeating the leaching with excess dilute (20%) boiling NaOH, filtering off the NaOH, leaching with dilute HNO_3 as before, and water washing three times, followed by an acetone rinse.

RESULTS AND DISCUSSION

Effect of Reactant Particle Size

Although the work at the University of Utah[6] showed the effect of reactant particle size on reaction kinetics for synthesizing submicron silicon carbide, there was no investigation of the critical variables that control the size of the silicon carbide. This is understandable since very fine, well-dispersed carbon (i.e., starch or carbon blacks) was needed in order to get the reaction to go to completion[6].

Harris et al.[16] found that SiC-C-SiO$_2$ agglomerates retained the same shape as the SiO$_2$-C precursors, but they did not investigate the effect of the precursor size on the final SiC particle size. The work of Klinger et al.[13] is the only reference which showed that the SiC particle size is controlled by the carbon size, as would be expected by Reactions (1)-(4). Klinger et al. used 105-149 micron SiO$_2$ particles mixed with graphite particles less than 44 microns in diameter. They reported that the silicon carbide-graphite reaction product was approximately the size of the graphite reactant particles, although they could not distinguish between SiC and C particles.

In order to determine the effect of reactant particle size on the size of SiC synthesized, the size of Mg, SiO$_2$, and C was independently varied in Reaction (5). In an effort to minimize residual free carbon in the product, all powder mixtures were purposely made carbon deficient (0.9 moles C in Reaction (5)).

Magnesium Reactant Size: The size of Mg had no effect on the surface area of the reacted powder. This is to be expected since Mg melts shortly after the reaction initiates and transport occurs either in the liquid or vapor phase. The Mg particle size affects the initiation temperature, with -325 mesh magnesium resulting in reaction initiation at 550-590°C and identical compositions containing -60,+100 mesh Mg initiating at 660-680°C. Since the vapor pressure of Mg is high (100 Pa at 580°C) it is not surprising that the finer Mg initiated the reaction below the melting point of Mg.

Silica Reactant Size: Silica reactants ranging in surface area between 1 and 500 m^2/g were each mixed with the same carbon (acetylene black) and magnesium source. The reaction initiation temperature, which varied between 550 and 590°C, and surface area of the reacted product, which ranged between 5 and 9 m^2/g, showed no clear trends with the size of the silica reactant used. The reacted product consisted mainly of MgO and beta SiC, although Mg, Si, and Mg$_2$Si were also detected by x-ray diffraction[17]. Acid leaching with dilute HNO$_3$ removed MgO, Mg, and Mg$_2$Si, and treatment with a strong base (dilute NaOH) removed Si and SiO$_2$. The surface area of the powder increased by an order of magnitude, from 5-9 m^2/g to 60-130 m^2/g, upon leaching, probably due to liquid phases (i.e., Si, Mg, Mg$_2$Si, and Mg) filling interstices in agglomerates consisting of submicron SiC. After leaching, the surface areas of SiC synthesized from larger SiO$_2$ reactants (1-50 m^2/g) were comparable to the surface areas of SiC produced from smaller silica reactants (200-500 m^2/g), strongly suggesting that SiO$_2$ starting size does not affect the particle size of the SiC formed by the exothermic reaction. Scanning electron micrographs confirmed that silica starting size does not affect the size of SiC synthesized[17]. A vapor phase transport mechanism involving SiO, as discussed in the next section, explains the above results. Although SiO$_2$ particle size does not affect the resultant SiC particle size, previous work on the carbothermal reduction of silica has shown that both silica and carbon particle size greatly influence the reaction kinetics[5,6,10].

Carbon Reactant Size: In order to determine the effect of carbon size on the size of SiC synthesized, five furnace carbon blacks (surface areas ranging between 24 and 1000 m/g), one acetylene black (75 m^2/g)), and two graphite powders (9 and 13 m^2/g) were used as starting powders in Reaction (5) while holding silica (50 m^2/g) and magnesium (1-10 microns) size constant. The reactions initiated between 554 and 598°C and no trend in initiation temperature was observed as a function of

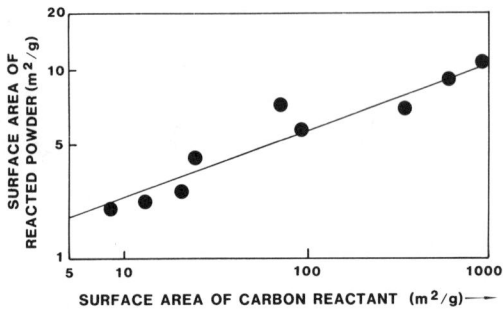

Fig. 1. Surface area of Reaction (5) powder as a function of carbon reactant surface area. Carbon content was substoichiometric (0.9 moles in Reaction (5)) to avoid free carbon formation.

carbon particle size. This is consistent with Mg-SiO$_2$ interaction as the initiation step in the reaction sequence. As shown in Figure 1, the surface area of the reacted powder increased with increasing carbon reactant surface area (decreasing carbon particle size). The amount of free Si increased with increasing C particle size[17], indicating that when the carbon reactant size is too large, the reaction to form SiC is incomplete. This is consistent with reaction kinetics causing the size of SiO$_2$ and C reactants to have an effect on reaction rate. Graphite peaks were always observed in XRD patterns of reacted powders when using graphite reactants. It is therefore unclear whether SiC was free of uncombined carbon when amorphous carbons were used in place of graphite in Reaction (5). Figure 2 shows surface area data for leached powders as a function of C reactant size. The surface area of SiC increases with increasing surface area of carbon reactant. Scanning electron microscopy (see Figure 3) confirmed that the ultimate particle size of synthesized SiC did indeed decrease with increasing C reactant surface area[17].

While it was possible to make SiC powders having average ultimate particle sizes as small as 0.02 micrometers, the powders were always agglomerated (see Figure 3). Agglomerates as large as 30 microns were formed and sufficient sintering of individual particles occurred that milling was required to break them down to near 1 micron in size. The largest SiC particles synthesized (see Figure 3(c)) were made from 9 m^2/g graphite powder. The XRD pattern of this SiC powder showed significant amounts of unreacted graphite, as compared with the SiC from the amorphous carbon and acetylene blacks which showed only SiC peaks (primarily beta) in the leached powders. Energy dispersive spectroscopy (EDS) of individual particles showed higher Si count rates for the larger SiC (Figure 3(c)) than the finer SiC powders (Figure 3(a) and 3(b)) verifying that the larger particles were not unreacted graphite. Since x-rays used with the diffractometer penetrate through the entire particles (penetration is approximately 5-25 microns) and x-rays in EDS are emitted from shorter depths, it is probable that the coarser graphite does not fully react and unreacted graphite is contained within a SiC shell. A comparison between leached powders and reactants[17] strongly indicates that carbon morphology controls the morphology of SiC synthesized by this exothermic reaction.

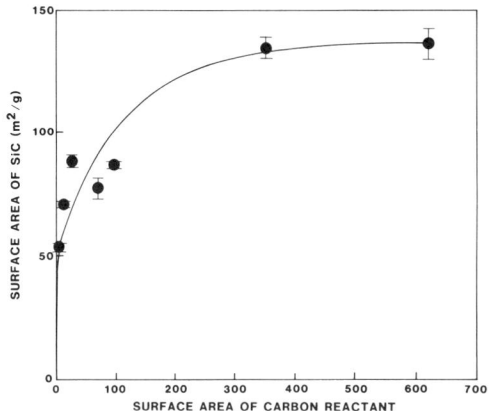

Fig. 2. Surface area of leached Reaction (5) powder as a function of carbon reactant surface area.

The objective of these experiments, however, was to show that it was possible to synthesize SiC with surface areas below 30 m^2/g by increasing the size of the carbon reactant. Experiments using Reaction (5) were unsuccessful in obtaining single phase SiC (by XRD) with low surface areas. The powders could be coarsened, as expected, by heating to elevated temperatures. A leached SiC powder which had a surface area of 145 m^2/g was heated to 1500°C and held for one hour. The surface area after heating was 60 m^2/g. Heating the powders to control surface area is not effective since coarsening of ultimate particles hardens the agglomerates, making agglomerate size reduction via milling difficult. Although the carbon size study was ineffective in making powders with desirable surface areas, it was instructive in learning about the morphology of the synthesized SiC.

Reaction Mechanism

Based on the carbothermal reduction of silica to form silicon carbide[5,6,10], it was presumed that Reaction (5) proceeds through an SiO intermediate as follows:

$$SiO_{2(s)} + Mg_{(l,g)} \rightarrow SiO_{(g)} + MgO_{(s)} \tag{6}$$

$$SiO_{(g)} + 2C_{(s)} \rightarrow SiC_{(s)} + CO_{(g)} \tag{3}$$

$$Mg_{(l,g)} + CO_{(g)} \rightarrow MgO_{(s)} + C_{(s)} \tag{7}$$

$$SiO_{2(s)} + C_{(s)} + 2Mg_{(l,g)} \rightarrow SiC_{(s)} + 2MgO_{(s)} \tag{5}$$

Since SiO$_2$ is reduced to SiO and transported through the vapor phase, the size of carbon would control the size of SiC, in agreement with the experimental results

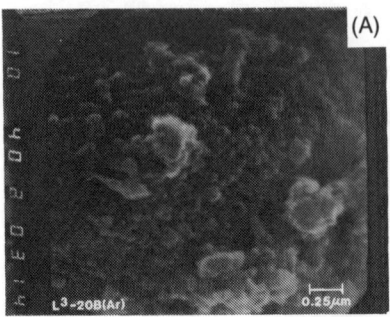
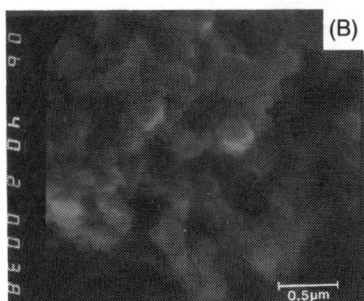

Fig. 3. Scanning electron micrographs of SiC synthesized by Reaction (5). (A) Carbon starting size was 650 m^2/g), (B) Carbon starting size was 75 m^2/g, and (C) Graphite starting size was 9 m^2/g.

above. Experimental verification of SiO and CO as the intermediate gas phases are needed to support the above observations. The obvious advantage of an exothermic reaction, such as Reaction (5) above, is that the fast kinetics of the reaction permit control of the size of SiC synthesized since particle coarsening is limited. Rice and McDonough[18] have discussed the difficulties of in-situ densification during the synthesis of self-propagating reactions. The alternative of synthesizing powders, followed by densification[14], is quite attractive.

Carbon Stoichiometry: In order to verify the reaction mechanism proposed above it was desired to determine the effect of carbon stoichiometry on surface area. Six batches of powders were made using 50 m^2/g fumed SiO$_2$*, 1-10 micron Mg†, and acetylene black carbon‡ to study the effect of free carbon and free silicon on the surface area of reacted powders. The amount of carbon was therefore varied according to the reaction

$$SiO_2 + 2Mg + xC \rightarrow xSiC + (1-x)Si + 2MgO \qquad (8)$$

The carbon contents in the six powders were 0. 0.7, 0.8, 0.9, 1.0, and 1.1 moles (relative to 1.0 mole of SiO_2). As discussed above, the resulting product also contains free Si, Mg_2Si, and Mg, as determined by XRD, in addition to SiC and MgO. Since the carbons above were amorphous, free carbon could also be present and therefore affect the surface area. Table 1 gives the initiation temperature of the reaction, surface area of the reacted powder, and surface area of the powder after leaching. As shown in Figure 4, the surface area of the unleached product decreased rapidly for carbon deficient specimens, consistent with the formation of free Si, as discussed above. Upon leaching, the surface areas increased by an order of magnitude, consistent with results obtained above. The surface area of the leached SiC powder was relatively insensitive to carbon content (see Figure 4). The results do not conclusively show that free carbon is absent in stoichiometrically reacted compositions, however, since the carbon used had a surface area of approximately 75 m^2/g. A high surface area carbon (i.e., greater than 300 m^2/g) substituted for acetylene black would allow free carbon to be detected by surface area measurements. As expected, XRD of the reacted powders showed that Si and Mg_2Si form in preference to SiC when C is substoichiometric (see Equation (8))[17]. However, in the presence of excess carbon, free Si, Mg, and Mg_2Si were still formed. The silicon-to-silicon carbide ratio was a strong function of carbon content[17], as expected.

Table 1. Effect of Carbon Content on Initiation Temperature and Surface Area.

Carbon (moles)	Initiation Temperature (°C)	Surface Area (m^2/g) As-reacted	Surface Area (m^2/g) Leached
0.0	547	1.1	dissolved
0.7	563	5.0	74.4
0.8	570	5.5	72.0
0.9	587	7.1	70.5
1.0	573	8.6	76.3
1.1	598	9.3	77.5

The increasing reaction initiation temperature with surface area (see Table 1) is explained based on an initial reaction between SiO_2 and Mg (see Equation (6)). Carbon, therefore, acts as a filler material and decreases the rate of heat build up in a diluted exothermic reaction. These data show that carbon is not involved in the initiation reaction and support the reaction mechanism proposed earlier.

Reaction Temperature: Using heat capacity data for MgO and SiC[19], an adiabatic reaction temperature of 2695K was calculated for Reaction (5) if initiating the reaction at 298K. Using the same heat capacity data and preheating to 800K to initiate the reaction increases the adiabatic temperature to 3080K. SiC sublimes and dissociates at elevated temperatures and atmospheric pressure (vapor pressures of 0.1 and 1 atm at 2783K and 3311K, respectively, are reported[20]). It is therefore possible that free Si observed in XRD patterns is the result of decomposition of SiC.

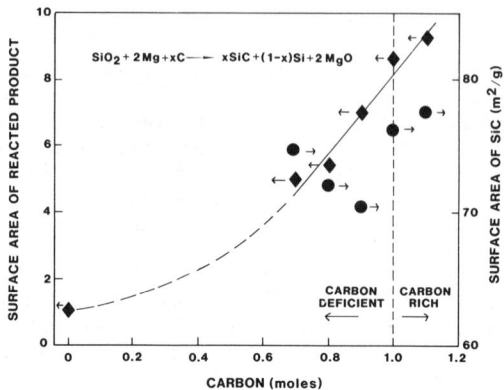

Fig. 4. Effect of carbon content on surface area of reacted (MgO-SiC) and leached (SiC) powders made by Reaction (8).

Free Si would therefore be present after decomposition to react with Mg to form Mg_2Si.

One method for controlling the reaction temperature is to add an inert filler to the reactants. The end product, SiC, would be a natural choice of a filler. In order to avoid influencing the surface area of the product, MgO (which can be removed by leaching) was chosen as the filler material for the experiments described below. Calculated adiabatic temperatures for the reaction

$$SiO_2 + 2Mg + C + xMgO \rightarrow SiC + (2+x)MgO \qquad (9)$$

initiating at 800K decrease from 3080K to 2180K as the MgO filler content increases from 0 to 70 wt. %[17]. The above calculations do not take into account the presence of Mg, Mg_2Si and Si in the reacted powder. In addition, there are heat losses that occur as the loose powder is reacted.

Magnesium oxide[§] was added as a filler material to silica, magnesium, and carbon, keeping the carbon substoichiometric (0.9 moles of C to each mole of SiO_2) to minimize free carbon in the product. The surface area of the MgO filler was 26 m^2/g. The reaction initiation temperature increased from 587 to 626°C as the filler content increased from 0 to 60 wt. % MgO. Magnesia, like carbon, increases the mean free path between silica and magnesium, and thereby requires higher temperatures for the reaction to initiate.

While no reaction was observed visually when the MgO filler content reached 70 wt. %, x-ray diffraction showed that the reaction had still occurred. Peak intensities from XRD patterns clearly showed that free Mg, Mg_2Si, and Si contents decreased with increasing filler content (i.e., lower reaction temperature)[17]. By adding 70 wt. % MgO filler to Reaction (8) reactants (x=0.9), it was possible to obtain the predicted reaction products. Not only does the addition of inert material give the desired reaction products, it also controls the reaction. With no filler, the powder is spewed throughout the tube by the expanding gas. As the filler content increases, the reaction becomes less and less violent, until no reaction is visually observed. The yield after leaching increased from 83% with no filler to 100% with greater than 30% MgO addition.

The observation that Si, Mg, and Mg_2Si all decrease, relative to SiC, as the reaction temperature is lowered suggests that decomposition of SiC is occurring as the reaction temperature increases. Self-diffusion coefficients of Si and C in SiC single crystals are reported to be approximately 10^{-13} and 10^{-11} cm^2/s, respectively, at 2000°C[21,22]. Assuming C diffusion to be rate limiting in Reaction (5), the maximum diffusion distance for C to be half of the diameter of the carbon particles (i.e., approximately 200 Å), and the time at temperature to be 1 second, the required diffusion coefficient for C diffusion through the SiC case surrounding the C particle is on the order of 10^{-12} cm^2/s. The reaction temperature must therefore approach 2000°C. As indicated above, the predicted adiabatic temperature is approximately 2000°C by the time the filler content is increased to 60 wt. %. Actual temperature measurements indicate that the reaction temperature may be below 1700°C[17] suggesting that a faster diffusion path (i.e., possibly grain boundary diffusion) may be operational. If diffusion of C or Si is rate limiting, and the sequence for Reaction (5) is

$$SiO_2 + Mg \rightarrow SiO + MgO \qquad (6)$$
$$SiO + 2C \rightarrow SiC + CO \qquad (3)$$
$$Mg + CO \rightarrow MgO + C \qquad (7)$$

as proposed above, or

$$SiO_2 + 2Mg \rightarrow Si + 2MgO \qquad (10)$$
$$Si + C \rightarrow SiC \qquad (11)$$

where metallic Si is formed in preference to SiO gas, then one would predict increased amounts of unreacted Mg or Si as the temperature decreases. It therefore appears that lower reaction temperatures limit SiC decomposition and Mg volatilization, thereby allowing a more complete reaction to the desired end product.

Contrary to expectation, alpha (2H) SiC peaks were readily apparent after leaching high MgO filler powders, whereas alpha SiC peaks were difficult to detect in XRD patterns of SiC reacted at higher temperature[17]. The beta to alpha transformation in SiC is dependent on both temperature and sintering aids[23-25]. Transition elements (such as Fe) are known to promote beta SiC formation during the carbothermal reduction of silica[5,6], whereas AlN formation stabilizes the isomorphic 2H SiC polytype[26]. It appears that Mg, which has lattice parameters similar to 2H SiC, promotes the formation of 2H SiC polytype or lowers the transformation temperature. Although the exact reason for the beta → alpha transformation is not known, it is very interesting and surprising that it occurs more readily with decreasing reaction temperature (i.e., increasing MgO filler content).

The surface areas of the "as-reacted" and "leached" powders are shown in Figure 5. The surface area of the leached product increases with increasing filler content (decreasing reaction temperature), most likely due to minimization of localized particle bonding and sintering. SEM micrographs at different filler contents do not show substantial differences in bonding, although there is indication of finer

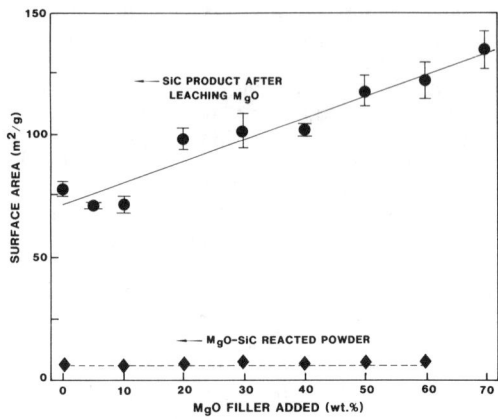

Fig. 5. Surface area of "as-sintered" and "leached" Reaction (9) powders showing the effect of MgO filler additions.

ultimate particle size with increasing MgO filler[17]. The agglomerate size appears to increase with filler content. Lower reaction temperatures allow the powder to remain together (as opposed to distributing it around the tube due to the rapid expansion of the Ar gas) and permit larger agglomerate formation. The surface area of the as-reacted powder is unchanged with increasing filler content even though the MgO filler has twice the surface area of the reacted powder. This could be due to liquid forming phases filling interstices and lowering the surface area of the reacted agglomerates. This explanation is not satisfying, however, since the liquid forming species (i.e., Si, Mg, and Mg_2Si) all decrease with increasing filler content, as discussed previously. It is not plausible that free carbon is affecting the surface area of the leached powder since the surface area of the acetylene black is too low to account for the observed trend.

The addition of inert filler lowers the reaction temperature, causing a reduction in undesired side products (i.e., Mg, Si, and Mg_2Si) and an increase in surface area and reaction yield. The alpha to beta ratio of SiC increases with increasing filler content. The use of filler materials is a practical method for controlling reaction rates and allowing larger batch sizes or continuous synthesis processes to be utilized.

All powders showed only SiC in XRD patterns after leaching. Chemical analysis, however, determined that there was substantial Mg (2.8 to 9.6% depending on C surface area) and O (6.3 to 15.1%) suggesting that MgO was trapped inside agglomerates and SiO_2 was present on the high surface area leached powders[17]. In order to see if the MgO was amorphous, a leached "SiC" powder was heated to 1950°C for 30 minutes. XRD showed coarsening of the SiC peaks but no MgO, Mg, or Mg_2Si peaks. It is readily apparent that the lower the surface area of the carbon reactant, the lower the tendency to trap MgO. It is also possible that some Mg is in solution with wurtzite (2H) SiC since both have hexagonal structures with similar lattice parameters, although adding a filler did not substantially alter the chemistry of the product[17]. High quality commercial silicon carbide powders have C contents of 30.0-30.5% and O less than 0.5%. The chemical analyses show that ultimate particle size must be increased in order to control oxygen content and that a more effective method of removing Mg must be used if this process is to be effective in making high purity SiC. Work at LLNL has shown that high purity B_4C can be made when reducing B_2O_3 with Mg in the presence of carbon[27]. This suggests that similar reactions or different leaching methods can result in high purity ceramic powders. As in the present work, the boron carbide produced by the thermite-type reaction had high surface area and was submicron in particle size[27].

Direct Reaction of Si and C

Background: The direct reaction of Si and C is a possible reaction step (see Reaction (11)) in the reduction of SiO_2 and subsequent carburization to form SiC in Reaction (5). The direct reaction of elements is often avoided when synthesizing carbides (or nitrides and borides) due to the cost of the metal powders and the fact that they were obtained by the reduction of oxides. Silicon metal, however, is relatively inexpensive due to the large amount of ferrosilicon produced. Silicon metal has been used to make silicon nitride via controlled nitridation.

Prochazka[1] synthesized submicron SiC using the direct reaction of elements but saw little densification with B and C additions, suggesting that free Si is detrimental to densification. Prochazka chose to synthesize submicron SiC by the more expensive chloromethyl silane synthesis route. Prochazka did not publish details of his synthesis of submicron beta SiC using Si and C. Hosaka et al.[28] synthesized beta SiC by reacting Si and C in an atmosphere having an oxygen content of 0.3 to 35%, with initiation occurring between 800 and 1400°C. The introduction of oxygen into the reaction evidently created enough SiO or CO transport to initiate the reaction below the melting point of Si (1410°C). Since the exothermic reaction melted the silicon it was found that the Si size was not critical in forming submicron SiC. Si less than 200 microns in diameter was acceptable and fine carbon size was desired (preferably in the 0.005 to 5 micron size range). The average SiC particle size was 4 microns. By adding SiO_2 and C (to form SiC via Reaction (4) which is endothermic) Sasaki et al.[29] were able to control the reaction temperature and end up with an average particle size of 0.5 microns. Their processing was identical to the previous work of Hosaka et al.[28] and neither group reported being able to sinter the powders.

Although a significant amount of work has been performed using the direct reaction of elements to form a wide variety of ceramics[30], this combustion synthesis work has neglected SiC formation from Si and C, since Reaction (11) is only mildly exothermic and is difficult to initiate at room temperature using igniters. The adiabatic temperature for Reaction (11) is 1825K using heat capacity data[19] for SiC between 298 and 3260K. Preheating to 1500K increases the adiabatic temperature to 2820K. It is obvious that temperatures generated by the reaction can be high enough to form SiC even when filler has been added to the reaction. The use of MgO filler in experiments described above suggested that SiC could be synthesized even in mildly exothermic reactions and it was desired to compare Reaction (11) with Reaction (5) in order to determine if control of carbon size could control the size of SiC synthesized.

Synthesis of Submicron Powders: High purity Si[¶] with an average particle size of approximately 5 microns was used to speed reaction kinetics and carbon was varied as described above. The silicon and carbon were milled for 3 hours in a stainless steel ball mill (10:1 ball (WC-6Co) to charge ratio) in hexane to mix the two components and the slurry was air dried. The effect of carbon size on SiC surface area was studied using batch sizes of 55–70 grams (typical of batch sizes used in the earlier study of Reaction (5)) and five C reactants having surface areas of 650, 350, 75, 25, and 13 m^2/g. The powders were reacted by wrapping them in graphite foil and injecting them (under Ar) into the hot zone of a furnace preheated to 1950°C. The thermal mass of the furnace was such that the heat of the reaction had only a slight effect on the temperature control of the furnace. The residence time of the powder in the furnace was approximately 2 minutes.

All of the powders contained only SiC (primarily beta) as identified by XRD, except for the powder made using graphite (13 m^2/g) instead of carbon. Surprisingly, only trace amounts of Si and C were observed in the XRD pattern of the SiC synthesized from the graphite reactant, in contrast to the large amount of unreacted graphite which occurred when the same graphite reactant was used in Reaction (5)[17]. All synthesized powders were submicron in size with average ultimate particle sizes ranging from 0.02 microns for high surface area carbons to 0.6 microns for SiC made using the graphite reactant. Figure 6 shows 0.1–0.3 micron SiC formed using acetylene black in Reaction (11). The particles are equiaxed and agglomerated. Chemical analysis showed that oxygen contents were as low as 0.3% and the ((C-O)/Si) atomic ratios were near unity, as desired.

Fig. 6. SiC synthesized by Reaction (11) using 75 m^2/g acetylene black as the carbon source.

Surface areas, which ranged between 1.6 and 18 m^2/g, showed a trend of increasing surface area with increasing carbon reactant size[17]. While this trend suggests a diffusion controlled reaction, both 75 m^2/g acetylene black carbon and 13 m^2/g graphite resulted in SiC with ultimate particle size of approximately 0.6 microns. This suggests that solution-precipitation may be important in SiC formation. The solubility limit for C in liquid Si at 2300°C is approximately 10%[31], suggesting that solution-precipitation could have a significant role in Reaction (11). The most convincing argument for solution-precipitation is the conversion of graphite to SiC in Reaction (11) and only partial conversion in Reaction (5). It appears that SiC synthesis by Reaction (5) is based on gaseous reactants (i.e., SiO and CO as described by Reactions (6), (3), and (7), as discussed above), while Reaction (11) allows for solution of C in liquid Si followed by precipitation of SiC, and subsequent C or Si diffusion to the Si-SiC or C-SiC interface. It should be noted that the reaction process is poorly understood at present and more work is needed before firm conclusions can be made.

Reaction Initiation Temperature: Another series of Reaction (11) powders were made to further examine the tendency of increasing surface area with decreasing carbon particle size and to verify that the reaction initiates with the melting of Si. The powders were reacted by wrapping the loose powders in graphite foil and injecting them into the pre-heated furnace and holding them at temperature for 5 minutes. Table 2 gives surface area data and XRD data for reacted powders. After heating Reaction (11) powders for 5 minutes at 1700°C the surface area is still a function of carbon starting size, although possible free carbon in these specimens complicates the clear interpretation of these data. The SiC surface area, as reported in Table 2, was 2-5 m^2/g due to the longer time at temperature (after the reaction has occurred). This particle coarsening was also observed by heating identical powders to different temperatures. Above 1700°C, the surface area decreases with increasing temperature. These same data show that the reaction initiates with the melting of Si in highly reducing environments, as expected.

Table 2. Effect of Carbon Reactant Size and Reaction Temperature on SiC Formation From the Elements (Reaction (11)).

Carbon Type	Surface Area (m^2/g) Carbon[†]	Powder[‡]	Temp. (°C)	Phase Analysis by XRD[*] Si	SiC
7000[§]	649	3.3	1700	0	100
3500[§]	352	5.2	1700	0	100
1020[§]	93	3.9	1700	0	100
410[§]	24	1.6	1700	0	100
A.B.[¶]	76	2.9	1700	0	100
1020	93	33.0	1000	100	0
1020	93	31.6	1100	100	0
1020	93	33.2	1200	100	0
1020	93	32.3	1300	>99	<1
1020	93	30.9	1400	95	5
1020	93	3.7	1500	5	95
1020	93	4.5	1600	2	98
1020	93	3.8	1700	<1	>99
1020	93	2.7	1800	0	100
1020	93	0.9	1900	0	100

[*]Comparison of Si and SiC (111) planes.
[†]Surface area of unreacted carbon.
[‡]Surface area of reacted (unreacted if temperature is less than 1410°C) powder.
[§]Raven series carbon powders (Columbian Chemicals, Co.).
[¶]Acetylene black carbon (Gulf Chemical Co.).

Surface area and electron microscopy show that powders produced by Reaction (5) have higher surface areas and smaller ultimate particle sizes than powders made by Reaction (11). This gives strong support for SiO and CO gas formation in Reaction (5). The ability to more fully react 1-5 micron graphite particles in Reaction (11), as compared to Reaction (5), suggests that solution of carbon in liquid Si is occurring in Reaction (11) and not in Reaction (5). It would therefore appear that Reaction (5) proceeds through Reactions (6), (3) and (7) and not through Reactions (10) and (11). Furthermore, the observed dependence of SiC size on carbon reactant size, suggests that carbon or silicon diffusion through SiC is important in Reaction (11) as well as Reaction (5).

Sinterability of SiC Powders

Efforts to densify powders from Reactions (5) and (11) by conventional B and C additions[1,2] were not successful, whereas both types of powders could be densified by liquid phase sintering[32]. Due to the inability of leaching to rid Reaction (5) SiC powders of Mg or MgO, sintering studies were not conducted with this powder. Sintering of Reaction (11) powder (93 m^2/g furnace black** as the C source) was investigated using a liquid phase sintering approach[32]. The powder was isostatically pressed at 200 MPa and pressureless sintered at 2000°C for 5 minutes in Ar. Shrinkage was 16.0% and % open porosity was 0.8. The four point bend strength was 490 MPa, hardness (measured with 75 N load on 136° Vicker's diamond indenter) was 25.2 GPa, and fracture toughness[33] was 3.0 MPa·$m^{1/2}$ at room temperature. These data are comparable or superior to commercially available alpha SiC and show that Reaction (11) deserves further investigation as a method for producing sinterable submicron SiC. Further work is needed to show that minimization of agglomeration can be used to make powders which are sinterable with B and C additions.

CONCLUSIONS

Both carbon stoichiometry and MgO filler addition studies showed that Mg-SiO_2 reactions initiate Reaction (5) and that the reaction initiation temperature increases with increasing carbon and/or filler content. Surface area of reacted MgO-SiC powders decreases with increasing free silicon. Lowering of the reaction temperature by the addition of inert filler, such as MgO powder, is an effective method for decreasing undesirable products (Si, Mg, and Mg_2Si) of the reaction. Lowering of the reaction temperature increased the ratio of alpha/beta SiC formed and increased the yield of the reaction. Prereacted SiC would be an effective method for controlling reaction temperature.

It was possible to make submicron SiC by either reduction of the oxide (Reaction (5)) or the direct reaction of the elements (Reaction (11)). High surface area (> 50 m^2/g) resulted from Reaction (5) while lower surface area (2-20 m^2/g) powders generally resulted from Reaction (11). The ultimate SiC particle size synthesized from either reaction was controlled by the size of the carbon (or graphite) reactant. Reaction (11) was effective in converting graphite powders to SiC, whereas Reaction (5) resulted in significant unreacted graphite suggesting that solution-precipitation is occurring during Reaction (11). It is postulated that Reaction (5) occurs via the following reactions:

$$SiO_{2(s)} + Mg_{(l,g)} \rightarrow SiO_{(g)} + MgO_{(s)} \quad (6)$$

$$SiO_{(g)} + 2C_{(s)} \rightarrow SiC_{(s)} + CO_{(g)} \quad (3)$$

$$Mg_{(l,g)} + CO_{(g)} \rightarrow MgO_{(s)} + C_{(s)} \quad (7)$$

$$SiO_{2(s)} + C_{(s)} + 2Mg_{(l,g)} \rightarrow SiC_{(s)} + 2MgO_{(s)} \quad (5)$$

Leaching of Reaction (5) powders was ineffective in completely removing Mg from SiC, despite x-ray diffraction patterns showing no MgO peaks, since Mg greater than 2% was always detected by wet chemistry. Further experiments are planned to investigate solid solution formation between SiC and Mg. Oxygen contents were high (6-15%) due to the high surface area of the SiC powders and possibly due to the inability to remove all of the MgO.

The direct reaction of Si and C (Reaction 11) was an effective method for making submicron SiC having low oxygen (0.3%). Reaction (11) powders could be sintered, aided by a liquid phase, to near theoretical density with high hardness and strength and toughness comparable to commercially available sintered SiC. The use of fillers to minimize the formation of hard agglomerates during the synthesis step is needed in order to make submicron SiC which is sinterable using boron and carbon additions. Further efforts are needed to show the relative importance of solution-precipitation and diffusion in Reaction (11). The abundant supply of relatively inexpensive silicon and the high rate of reaction make the direct reaction of the elements an attractive alternative to the carbothermal reduction of silica (Acheson Process) for the production of submicron silicon carbide.

ACKNOWLEDGEMENTS

This work was supported by Lawrence Livermore National Laboratory (LLNL) under subcontract 5839605. Discussions with Dr. J. Birch Holt of LLNL, Professor Anil V. Virkar of the University of Utah and Professor Zuhair A. Munir of the University of California at Davis are appreciated.

---

[*] OX-50, Degussa Corp. (Teterboro, NJ).
[†] Hart Metals, Inc. (Tamaqua, PA).
[‡] Shawinigan, Gulf Chemical Co. (Englewood Cliffs, NJ).
[§] Grade MG330, Atlantic Equipment Engineers (Bergenfield, NJ).
[¶] HQ-Silgrain, Elkem Metals Co. (Niagara Falls, NY).
[**] Raven 1020, Columbian Chemical Co. (Tulsa, OK).

References

[1] S. Prochazka, "Sintering of SiC," *Proc. of the Conference on Ceramics for High Performance Applications*, Hyannis, MA., 1973, edited by J. J. Burke, A. E. Gorum and R. M. Katz (Brook Hill Publ. Co., 239-252 1974).

[2] J. A. Coppola, L. N. Hailey and C. N. McMurtry, "Process for Producing a Sinterable Silicon Carbide Ceramic Body," U.S. Patent No. 4,124,667, November 7, 1978.

[3] W. Boecker, H. Landfermann, and H. Hausner, "Sintering Alpha Silicon Carbide with Additions of Aluminum," *Pow. Met. Int.*, **11** [2] 83-85 (1979).

[4] H. Tanaka, Y. Inomata, K. Hara, and H. Hasegawa, "Normal Sintering of Al-doped Beta SiC," *J. Mater. Sci. Letters*, **4**, 315-317 (1985).

[5] J. G. Lee and I. B. Cutler, "Formation of Silicon Carbide from Rice Hulls," *Bull. Am. Ceram. Soc.*, **54** [2] 195-198 (1975).

[6] J. G. Lee, "Carbide and Nitride Ceramics by Carbothermal Reduction of Silica," Ph.D. Dissertation, University of Utah (1976).

[7] B. W. Jong, "Formation of Silicon Carbide from Silica Residues and Carbon," *Bull. Am. Ceram. Soc.*, **58** [8] 788-789 (1979).

[8] G. C. Wei, "Beta SiC Powders Produced by Carbothermic Reduction of Silica in a High-Temperature Rotary Furnace," *J. Am. Ceram. Soc.*, **66** [7], C-111-C-113 (1983).

[9] Y. Suyama, R. M. Marra, J. S. Haggerty and H. K. Bowen, "Synthesis of Ultrafine SiC Powders by Laser-Driven Gas Phase Reactions," *Bull. Am. Ceram. Soc.*, **64** [10] 1356-1359 (1985).

[10] P. D. Miller, J. G. Lee and I. B. Cutler, "The Reduction of Silica with Carbon and Silicon Carbide," *J. Am. Ceram. Soc.*, **62** [3-4] 147-149 (1979).

[11] W. W. Pultz and W. Hertl, "SiO_2 + SiC Reaction at Elevated Temperatures: I," *Trans. Faraday Soc.*, **62** [9] 2499-2504 (1966).

[12] B. C. Bechtold and I. B. Cutler, "Reaction of Clay and Carbon to Form and Separate Al_2O_3 and SiC," *J. Am. Ceram. Soc.*, **63** [5-6] 271-275 (1980).

[13] N. Klinger, E. L. Strauss and K. L. Komarek, "Reactions Between Silica and Carbon," *J. Am. Ceram. Soc.*, **49** [7] 369-375 (1966).

[14] R. A. Cutler, A. V. Virkar and J. B. Holt, "Synthesis and Densification of Oxide-Carbide Composites," *Ceram. Eng. and Sci. Proc.*, (Am. Ceram. Soc., Columbus, OH., 715-728, 1985).

[15] J. Hojo, K. Miyachi, Y. Okabe and A. Kato, "Effect of Chemical Composition on the Sinterability of Ultrafine SiC Powders," *J. Am. Ceram. Soc.*, **66** [7] C-114-C-115 (1983).

[16] L. A. Harris, C. R. Kennedy, G. C. Wei and F. P. Jeffers, "Microscopy of SiC Powders Synthesized by Reacting Colloidal Silica and Pitch," *J. Am. Ceram. Soc.*, **67** [6] C-121-C-124 (1984).

[17] R. A. Cutler and K. R. Rigtrup, "Synthesis of Submicron Powders via Exothermic Reactions and Characterization of Densified Ceramics," Ceramatec, Inc. report Prog: 87012601 (March 1987).

[18] R. W. Rice and W. J. McDonough, "Intrinsic volume Changes of Self-Propagating Synthesis," *J. Am. Ceram. Soc.*, **68** [5] C-122-C-123 (1985).

[19] O. Kubaschewski and C. B. Alcock, *Metallurgical Thermochemistry*, 5th Edition (Pergamon Press, Oxford, 1979).

[20] W. F. Knippenberg, "Growth Phenomena in Silicon Carbide," Thesis, University of Leiden, Netherlands, Philips Research Reports, 18 [3], 161-274 (June 1963).

[21] J. D. Hong, M. H. Hon and R. F. Davis, "Self-Diffusion in Alpha and Beta Silicon Carbide," *Ceramurgia Int.*, **5** [4] 155-160 (1979).

[22] J. D. Hong and R. F. Davis, "Self-Diffusion of Carbon-14 in High Purity and N-Doped Alpha SiC Single Crystals," *J. Am. Ceram. Soc.*, **63** [9-10] 546-552 (1980).

[23] D. H. Stutz, S. Prochazka and J. Lorenz, "Sintering and Microstructure Formation in Beta-Silicon Carbide," *J. Am. Ceram. Soc.*, **68** [9] 479-82 (1985).

[24] R. M. Williams, B. N. Juterbock, S. S. Shinozaki, C. R. Peters, and T. J. Whalan, "Effects of Sintering Temperatures on the Physical and Crystallographic Properties of Beta-SiC," *Bull. Am. Ceram. Soc.*, **64** [10] 1385-89 (1985).

[25] S. Shinozaki, R. M. Williams, B. N. Juterbock, W. T. Donlon, J. Hangas and C. R. Peters," Microstructural Developments in Pressureless-Sintered Beta-SiC Materials with Al, B, and C Additions," *Bull. Am. Ceram. Soc.*, **64** [10] 1389-93 (1985).

[26] I. B. Cutler, P. D. Miller, W. Rafaniello, H. K. Park, D. P. Thompson and K. H. Jack," New Materials in the Si-C-Al-O-N and Related Systems," *Nature*, 275, 434-435 (1978).

[27] J. B. Holt, private communication (1986).

[28] T. Hosaka, T. Sasaki, and H. Suzuki, "Process for Producing Powder of Beta Type Silicon Carbide," U.S. Patent 4,117,096 (Sept. 26, 1978).

[29] T. Sasaki, I. Komaru, and R. Yoshioka, "Process for Producing Beta Silicon Carbide Fine Powder," U.S. Patent 4,217,335 (August 12, 1980).

[30] J. F. Crider, "Self-Propagating High Temperature Synthesis—A Soviet Method for Producing Ceramic Materials," *Ceram. Eng. Sci. Proc.*, **3** [9-10] 519-528 (1982).

[31] R. Elliot, *Constitution of Binary Alloys*, 1st Supplement, McGraw-Hill New York, NY 1965).

[32] J. L. Huang, A. C. Hurford, R. A. Cutler and A. V. Virkar, "Sintering Behavior and Properties of SiCAlON Ceramics," *J. Mater. Sci.*, **21**, 1448-56 (1986).

[33] G. R. Anstis, P. Chantikul, B. R. Lawn and D. B. Marshall, "A Critical Evaluation of Indentation Techniques for Measuring Fracture Toughness: I, Direct Crack Measurements," *J. Am. Ceram. Soc.*, **64** [9] 533-538 (1981).

SILICON CARBIDE POWDERS BY GASEOUS PYROLYSIS OF TETRAMETHYLSILANE

HUANN-DER WU AND DENNIS W. READEY

The Ohio State University
Columbus, OH 43210

ABSTRACT

Silicon carbide powders were prepared by gaseous thermal decomposition of $Si(CH_3)_4$ in a flow-through reactor between 800°C and 1500°C. β-SiC powder similar to that produced with other heating techniques was obtained. The particle size can be controlled by the deposition temperature and the $Si(CH_3)_4$ concentration in the gas phase. The degree of crystallinity of the powders depends strongly on the synthesis temperature. A core-shell structure is frequently found in the particles due to differences in reaction conditions along the length of the reactor. This core-shell structure can be easily removed by subsequent heat treatment.

INTRODUCTION

The preparation of ceramic powders by gas phase reactions offers the advantages of high purity, loose aggregation, small particle size, and a narrow particle size distribution.[1] Silicon carbide had been produced by gas phase reactions as early as 1909.[2] However, the preparation of SiC powders by vapor phase reactions has received renewed interest in the last ten years due to the need for controlled starting materials for the production of high performance structural ceramics. Both plasma[3,4] and laser[5-10] heating have been used to prepare SiC powder by gas phase reactions. The rationale for use of these heating techniques is that the size of the heated reaction zone can be easily controlled producing a powder with a relatively narrow particle size distribution. Nevertheless, SiC powders have also been prepared with more conventional flow-through furnace reactors[11,12] and the resulting powders have been shown to give good sinterability.[13,14] The purpose of this research was to prepare and characterize powders prepared in a conventional flow-through furnace reactor and to evaluate tetramethylsilane (TMS), $Si(CH_3)_4$, as a precursor material for the preparation of silicon carbide powders.

EXPERIMENTAL

The flow-through reactor system consisted of a furnace* capable of reaching 1700°C lined with a high alumina (99.8%) tube (ID = 45 mm). Prepurified argon (oxygen content < 10 ppm) was the carrier gas. NMR-grade, tetramethylsilane† with a purity of 99.9% was used as the reactant. The partial pressure of TMS in the gas was controlled by bubbling some of the argon gas stream through the liquid TMS. The gas containing TMS vapor was mixed with the remaining gas stream in

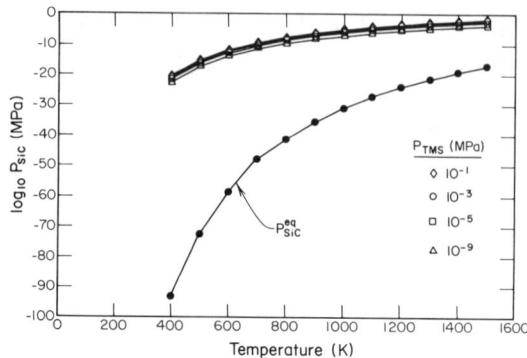

Fig. 1. Equilibrium partial pressure of silicon carbide vapor from the decomposition of TMS and the equilibrium vapor pressure over solid SiC as a function of temperature.

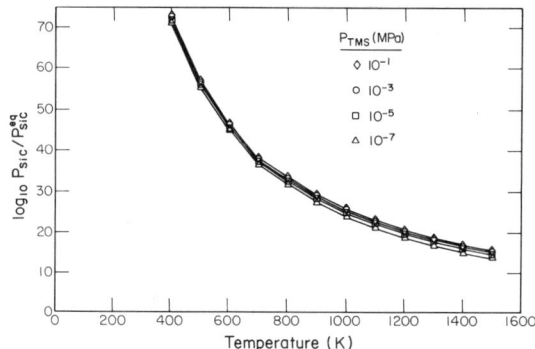

Fig. 2. Supersaturation ratio of silicon carbide vapor for the decomposition of TMS as a function of temperature.

a mixing chamber containing glass beads before entering the furnace. The amount of TMS in the gas stream could be calculated by the amount of liquid used during an experiment, the gas flow rate, and the length of the run. From tabulated thermodynamic data[15], the equilibrium partial pressures and the supersaturations were calculated from the following reactions:

$$Si(CH_3)_{4(g)} = 3\ CH_{4(g)} + SiC_{(g)}$$
$$SiC_{(g)} = SiC_{(s)}$$

$$\text{Supersaturation ratio} = \frac{P(SiC_{(g)})}{P_e(SiC_{(g)})}$$

where $P_e(SiC_{(g)})$ is the equilibrium partial pressure of SiC gas from the decomposition of TMS.[9] Admittedly, the reaction from gaseous TMS to solid SiC may not proceed in exactly these steps. Nevertheless, the thermodynamic data presented in Figures 1 and 2 clearly show that decomposition is highly favorable over the range of temperatures used in this research.

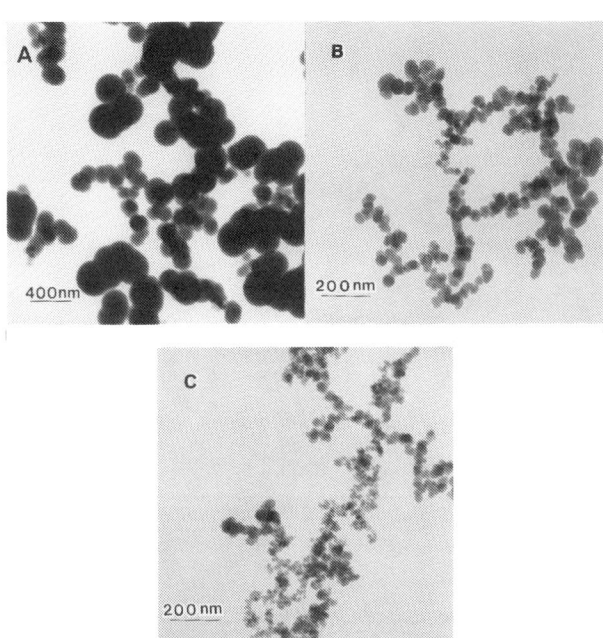

Fig. 3. Microstructure of SiC powder synthesized at various temperatures with $P(TMS) = 2.5 \times 10^{-3}$ MPa (2.5×10^{-2} atm). (*A*) 800°C. (*B*) 1000°C. (*C*) 1200°C.

Powders were collected at the end of the furnace in a flask by natural sedimentation from the gas stream and exit gases were burned in a natural gas flame. Microstructures were examined with a transmission electron microscope[†]. Particle size distributions were determined from TEM images with an image analyzer[§]. Between 130 and 500 particles were used for a distribution. Free carbon was determined from weight loss (TGA, Perkin Elmer) during low temperature oxidation in air. Crystallization and other thermal effects were determined by differential thermal analysis, DTA[¶], in nitrogen at a heating rate of 20°C/min.

Results

Ultrafine SiC powders were produced by pyrolysis of TMS as shown in Figure 3. The particle size decreased with increasing reaction temperature and increased with increasing TMS partial pressure as demonstrated in Figure 4. Figure 5 displays a typical particle size distribution. X-ray diffraction[**] data showed that powders produced below 1100°C were amorphous but could be crystallized to β-SiC by heating one hour above 1100°C as evidenced in Figure 6. DTA analysis also indicated crystallization about 1100°C, Figure 7.

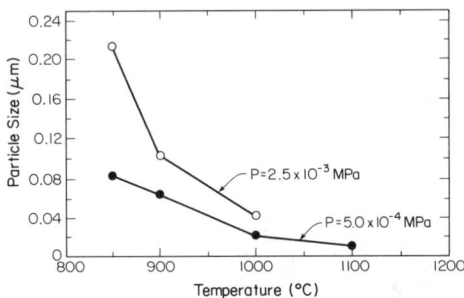

Fig. 4. The average particle size of SiC powder prepared at various temperatures.

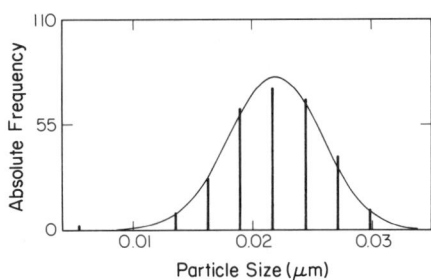

Fig. 5. Particle size distribution of SiC powder prepared at 1000°C with P(TMS) = 5.0×10^{-4} MPa (5.0×10^{-3} atm).

X-ray diffraction patterns of a series of powders prepared with a TMS pressure of 1.8×10^{-2} MPa (0.18 atm) between 850°C and 1500°C, at a flow rate of 1000 cc/min (corresponding to a hot zone dwell time of about 4 sec), are displayed in Figure 8. Again, below 1100°C, the powders are amorphous but crystalline SiC at high temperatures. Graphite peaks are found above 1100°C. Both below and above about 1000°C, a "core-shell" structure was observed as seen in Figure 9. No

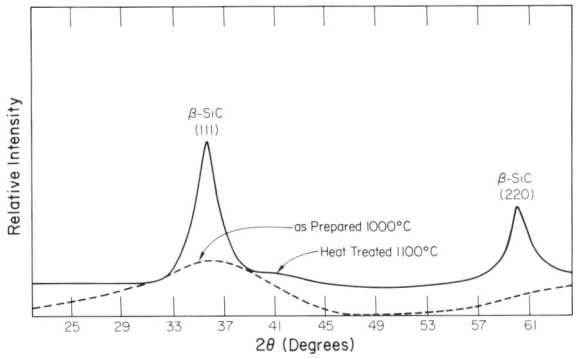

Fig. 6. X-ray diffraction patterns of as-prepared SiC at 1000°C and after heat treatment at 1100°C.

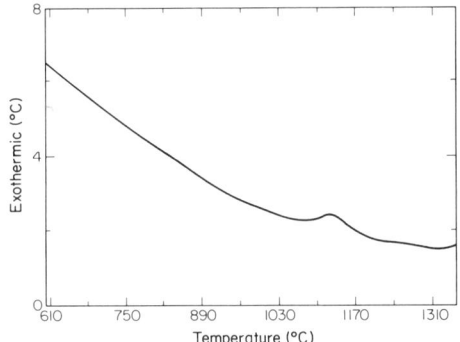

Fig. 7. DTA heating curve of SiC powder prepared at 1000°C showing crystallization just above 1100°C.

Silicon Carbide

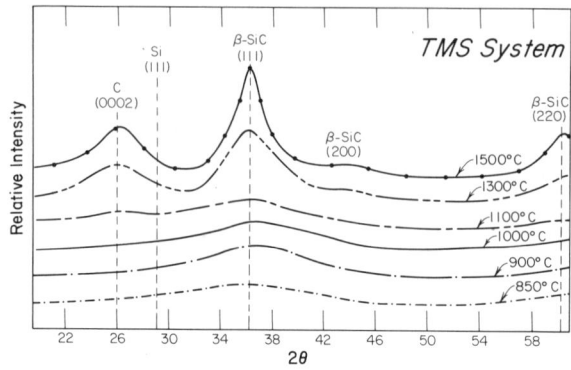

Fig. 8. X-ray diffraction patterns for SiC powder prepared at various temperatures showing the presence of carbon at high reaction temperatures.

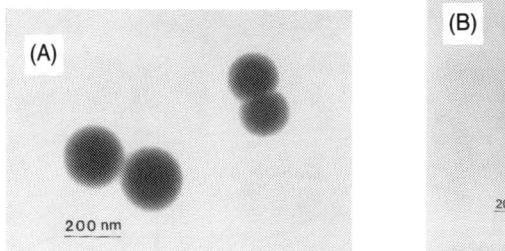 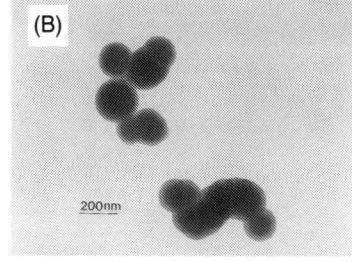

Fig. 9. Microstructures of powders prepared with P(TMS) = 1.8×10^{-2} MPa (0.18 atm) showing core-shell structure. (A) 850°C. (B) 1300°C.

core-shell structure was discernable in powders prepared at 1000°C. The results of oxidation experiments on these powders is shown in Figure 10. Powders prepared below 1000°C exhibited weight gain while those prepared at higher temperatures lost weight. For comparison, the weight lost by a pure carbon powder is included. Assuming this shell to be carbon, then the amount of free carbon as a function of deposition temperature determined by TGA is shown in Figure 11. Figure 12 displays the core-shell structure of a powder before and after air heat treatment at 800°C. The shell is clearly gone after heat treatment. X-ray diffraction data in Figure 13 corroborate that carbon is removed during the air heat treatment. In contrast, the same powder heated in argon at 1400°C shows a greater degree of crystallinity and the continued presence of carbon. The presence of the shell after heat treatment is clear in Figure 14. On the other hand, powders prepared below 1000°C and heat treated at 1400°C in argon showed no core-shell

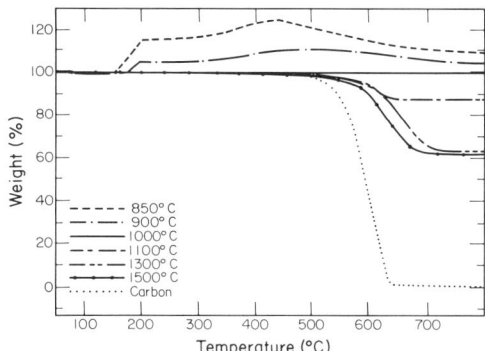

Fig. 10. TGA data of heating powders in air which were prepared at various temperatures. A weight loss curve for pure carbon is included for comparison.

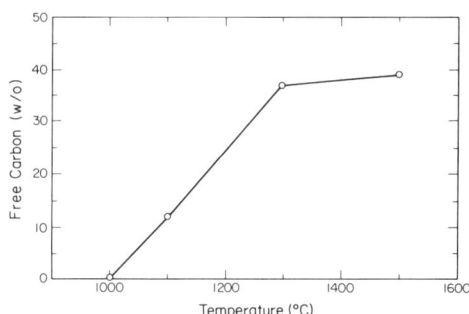

Fig. 11. The weight percent of free carbon in a core-shell powder as a function of preparation temperature as function of reaction temperature for P(TMS) = 1.8×10^{-2} MPa (0.18 atm).

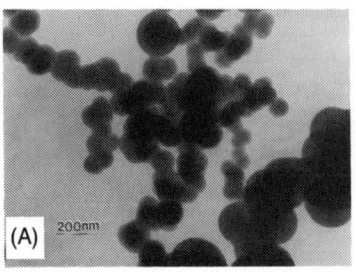
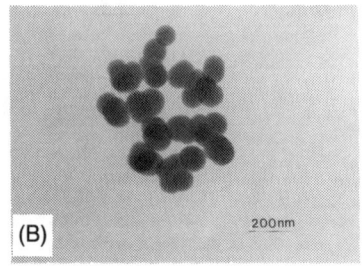

Fig. 12. Microstructures of: (A) powder as-prepared at 1300°C and (B) after heat treatment at 800°C in air.

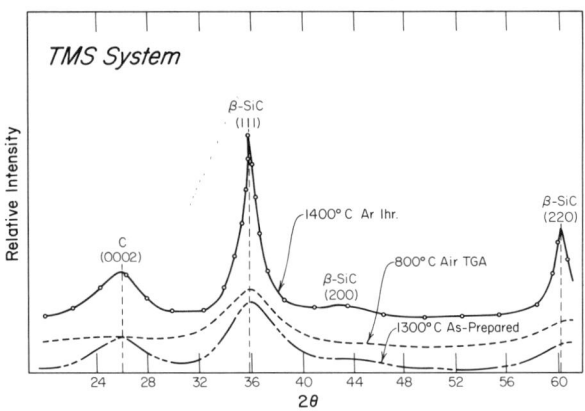

Fig. 13. X-ray diffraction patterns of the as-prepared 1300°C powder, after heat treatment in air at 800°C showing the disappearance of the carbon peak, and after 1400°C heat treatment in argon showing the continued presence of the carbon peak.

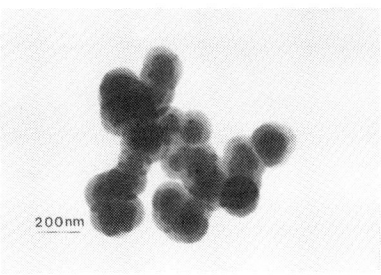

Fig. 14. Microstructure of SiC powder synthesized at 1300°C and subsequently heat treated at 1400°C in argon showing the continued presence of a now clearly more cystalline carbon layer.

structure present, Figure 15. In fact, heat treatment at temperatures as low as 800°C eliminated the shell on these powders, Figure 16.

DISCUSSION

Figures 3 and 4 confirm that powders made in a flow-through furnace reactor are essentially the same as those produced by other heating techniques. The particle size, degree of agglomeration, and the particle size distributions are all similar. Furthermore, considering the variation in particle size that can be obtained simply by varying the reaction temperature, the flow-through reactor system may offer more control over the resultant powder than other heating techniques. Certainly, TMS is a suitable precursor for the preparation of β-SiC powder. There are two possible mechanisms for the formation of SiC powders from gaseous TMS. The first is the formation of partially-pyrolyzed polymerized particles at low temperature which then decompose into solid SiC at higher temperatures. The second is the direct nucleation and growth of SiC powders from the vapor phase. The results of this study show that at low reaction temperatures, a core-shell structure is obtained. The core is SiC and the shell is some material which gains weight during oxidation and can be pyrolyzed to pure SiC during heat treatment in a non-oxidizing atmosphere. These results strongly suggest that the shell may be a partially-pyrolyzed, perhaps polymeric, organo-silicon material. The observation of organo-silicon polymers at low decomposition temperatures have lead others[11] to conclude that formation of SiC from vapor phase pyrolysis of TMS occurs by formation of an intermediate partially-polymerized organo-silicon compound.

However, a core-shell structure is observed both at high and low temperatures in these studies. At high temperatures the shell is carbon and at about 1000°C, no shell is formed. These results can be explained as follows. At low reaction temperatures, not all of the TMS is reacted. As a result, it partially decomposes on SiC particles in the cooler exit end of the furnace. At higher temperatures, the reaction goes essentially to completion leaving only methane as the gaseous product. At elevated temperatures, methane can decompose to carbon and does so on the already-formed SiC particles. Near 1000°C, for the conditions used in these studies, the degree of reaction and the residence time in the hot zone are in approximate

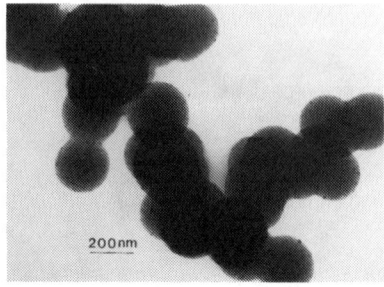

Fig. 15. Microstructure of SiC powder synthesized at 850°C and subsequently heat treated at 1400°C in argon showing the disappearance of the shell and a uniform SiC particle.

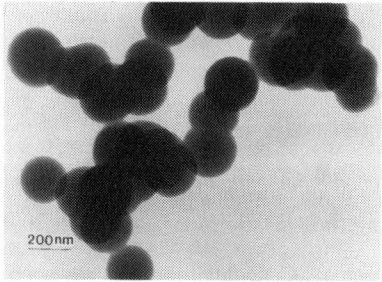

Fig. 16. Microstructure of SiC powder synthesized at 850°C and subsequently heat treated at 800°C in air.

balance to prevent the formation of either a partially-pyrolyzed or carbon layer. Presumably, under different experimental conditions, the optimum temperature would be different. That the shell formed under low temperature deposition conditions in this study indeed forms after the SiC particles are formed is further corroborated by the fact that this shell structure could be heat treated in argon at temperatures as low as 800°C to form SiC, Figure 16. Therefore, the results of these studies suggest that β-SiC powders are formed directly from the vapor phase pyrolysis of TMS over the temperature range studied, 850°C to 1500°C. The core-shell structure results from reactions in the cooler exit end of the furnace.

Conclusions

Tetramethylsilane can be used as a precursor for the vapor phase preparation of SiC powders. The particle size and particle size distribution produced in a flow-through furnace reactor under certain conditions are essentially the same as those of powders made by less conventional heating techniques. The degree of crystallinity of the powders increases with reaction temperature. For a given set of deposition conditions, single phase SiC powder is produced. Under less than optimum conditions, a core-shell structure results. At low temperatures the shell is partially-pyrolyzed TMS and at elevated temperatures it is carbon. Both can be easily removed by suitable heat treatment to give single phase SiC powders. Shells of partially-pyrolyzed TMS formed at low temperatures can be decomposed to SiC at elevated temperatures in argon. Similarly, particles coated with carbon during high temperature deposition can be heat treated in an oxidizing atmosphere at relatively low temperatures to remove the carbon.

---

[*]Rapid Temp, C M Corporation, Bloomfield, NJ
[†]Morton Thiokol, Chicago, IL
[‡]Model 200CX, JEOL, Tokyo, Japan
[§]Videoplan, Carl Zeiss, Inc., Thurmond, NY
[¶]Model XRG 3100, Phillips Electronic Inst., Mahwah, NJ
[**]Model 1700, Perkin Elmer Corp., Newark, NJ

References

[1] A. Kato, J. Hojo, and Y. Okabe, "Formation of Ultrafine Powders of Refractory Nitrides and Carbides by Vapor Phase Reaction," Memoirs of the Faculty of Engineering, Kyushu University, 41 [4] 319-334 (1981).

[2] J. N. Pring and W. Fielding, "The Preparation at High Temperatures of Some Refractory Metals from Their Chlorides," *J. Chem. Soc.*, 95, 1497-1506 (1909).

[3] G. J. Vogt, C. M. Hollabaugh, D. E. Hull, L. R. Newkirk, and J. J. Petrovic, "Novel RF-Plasma System for the Synthesis of Ultrafine, Ultrapure Silicon Carbide and Silicon Nitride," *Mater. Res. Soc. Symp. Proc.*, 30, 283-289 (1984).

[4] Peter Chuen Sun Kong, "Synthesis of Ultrafine β-SiC Powders in a Thermal Plasma," Ph.D. Dissertation, University of Minnesota, 1985.

[5] R. A. Marra and J. S. Haggerty, "Synthesis and Characteristics of Ceramic Powders Made from Laser-Heated Gases," *Ceram. Eng. and Sci. Proc.*, 3, [1-2] 3-19 (1982).

[6] S. C. Danforth and J. S. Haggerty, "Synthesis of Ceramic Powders by Laser-Driven Reactions," *Ceram. Eng. and Sci. Proc.*, 2 [7-8] 466-479 (1981).

[7] W. R. Cannon, S. C. Danforth, J. H. Flint, J. S. Haggerty, and R. A. Marra, "Sinterable Ceramic Powders from Laser-Driven Reactions: I, Process Description and Modeling," *J. Am. Ceram. Soc.*, 65, [7] 324-330 (1982).

[8] W. R. Cannon, S. C. Danforth, J. H. Flint, J. S. Haggerty, and R. A. Marra, "Sinterable Ceramic Powders From Laser-Driven Reactions: II, Powder Characteristics and Process Variables," *J. Am. Ceram. Soc.*, 65, [7] 330-335 (1982)

[9] Y. Suyama, J. S. Haggerty, and H. K. Bowen, "Synthesis of Ultrafine SiC Powders by Laser Driven Gas Phase Reactions," *Chem. Soc. of Japan*, 10, 1539-1541 (1984).

[10] Y. Suyama, R. A. Marra, J. S. Haggerty, and H. K. Bowen, "Synthesis of Ultrafine SiC Powders by Laser-Driven Gas Phase Reactions," *Bull. Am. Ceram. Soc.*, 64, [10] 1356-59 (1985).

[11] Y. Okabe, J. Hojo, and A. Kato, "Formation of Fine Silicon Carbide Powders by a Vapor Phase Method," *J. Less-Common Metals*, 68, 29-41 (1979).

[12] Y. Okabe, J. Hojo, and A. Kato, "Formation of Silicon Carbide Powders by the Vapor Phase Reaction of the SiH_4-CH_4-H_2 System," *Chem. Soc. of Japan (Japanese)*, No. 2, 188-93 (1980).

[13] Y. Okabe, K. Miyachi, J. Hojo, and A. Kato, "Sintering Behavior of Ultrafine Silicon Carbide Powders Obtained by a Vapor Phase Reactions," *Chem. Soc. of Japan (Japanese)*, No. 9, 1363-1370 (1981).

[14] K. Miyachi, Y. Okabe, J. Hojo, and A. Kato, "Sintering of Ultrafine Silicon Carbide Powders by Using Boron and Carbon as Sintering Aids," *Chem. Soc. of Japan (Japanese)*, No. 1, 28-33 (1983).

[15] *JANAF Thermochemical Tables*, (U.S. Government Printing Office, Washington, D.C.), 1971.

THERMAL CARBURIZATION OF SINGLE-CRYSTAL SILICON

J. G. SHEEK

Ultramet
12173 Montague St.
Pacoima, CA 91331

J. D. CAWLEY

Department of Ceramic Engineering
Ohio State University
2041 College Road
Columbus, OH 43210-1178

ABSTRACT

The thermal carburization of single-crystal silicon was investigated. Carbon sources included solid dense carbon, carbon black, and a CH_4/H_2 atmosphere. Analysis of scale microstructure, as well as growth direction kinetics, indicate that vapor phase transport plays an important role in the process. In the case of carburization using carbon black or dense carbon, the scales were observed to grow inward and the rate limiting step appears to be delivery of carbon to the Si/SiC interface. In the CH_4/H_2 atmosphere, scale growth is outward and the transport of silicon as a vapor leads to growth of the scale.

INTRODUCTION

The direct thermal carburization of silicon has been the subject of a number of studies[1-8]. The growth of a silicon carbide (SiC) scale around a silicon substrate involves mass transport in response to a chemical potential gradient(s) and efforts have been made to determine the relative magnitude of the carbon, and silicon diffusion coefficients from these experiments. This is an attractive experiment in light of the conflicting assessment of the relative rates of carbon and silicon bulk self-diffusion and because the silicon grain boundary self-diffusion has not been measured[9-12].

In the earlier work[1-3], it was inferred that the SiC displayed inward growth, i.e the reaction occurred at the Si/SiC interface. However, the later work, both tracer studies[5] and geometrical measurements[4,5], has indicated that the reaction takes place at the C/SiC interface. In carburizing experiments using filaments of silicon roughly 50 microns thick, carried out in the temperature range of 1200°– 1400°C and an atmosphere of 1% CH_4 and 99% H_2, it has been observed that hollow shells of SiC are formed and that the geometry of the internal void mimics that of the original silicon. From this it was inferred that the solid state diffusion of silicon was rapid compared to that of carbon (see Appendix 1 for an analysis of the relationship between scale growth and diffusion coefficients). However, it was not demonstrated in either work that no short circuit paths for diffusion, e.g. pores, were present. The generation of growth defects which may serve as short circuit diffusion paths has been observed in SiC scales grown in a CH_4 atmosphere[6].

Hydrocarbons were used as the carbon source in all but one[3] of these studies, and different atmospheres have been employed. Some experiments have employed an inert gas such as argon[1-3], while others have employed hydrogen/hydrocarbon mixtures[4,5,7,8] or simple hydrocarbon atmospheres[6]. Both types of experiments were repeated in this work. Large samples and long times were employed in order to amplify any differences in growth morphology. The results indicate that outward growth results from carburization in a CH_4/H_2 atmosphere, with inward growth when a solid source of carbon and a helium atmosphere are used.

Experimental Details

Sample Preparation

A boule of single crystal silicon* was wafered to a thickness of approximately 1.5 mm and cut into bars (not necessarily square) approximately 1.5 × 1.5 mm. Although some polished wafers were employed, the bulk of the samples were used "as-sawed." The samples were cleaned in an acetone wash. After cleaning, weights and dimensions were measured and recorded.

The carbon source was one of three different forms: 75-nm diameter particles of carbon black powder†, a dense high purity carbon rod roughly 0.5 cm in diameter‡, and a certified 1% methane/99% hydrogen mixture§. The particle size of the carbon black was verified by transmission electron microscopy (TEM) using an image analyzer¶.

Experimental Procedure

A schematic diagram of the furnace and apparatus used for the carburization studies is shown in Figure 1. During carburization runs, a very slight positive pressure was maintained, due to the gas exiting through a flask containing oil.

Several initial solid carbon experiments were done using a wafer dense carbon placed on a polished single-crystal silicon wafer. All subsequent solid carbon experiments employed carbon black. In these, carbon black powder was placed in a boron nitride (BN) crucible, with silicon bars placed on top of the carbon powder. The crucible was then placed in a graphite retort and covered with a graphite lid. Preliminary experiments showed that the crucible must be sealed in air in order for SiC scale growth to occur; the significance of this will be discussed. A 3.2 mm diameter alumina rod was attached to the graphite retort to facilitate loading and retrieval of the retort to and from the hot zone of the furnace. An atmosphere of flowing helium surrounded the retort in the furnace, with the gas flow rate maintained at approximately 100 ml/min. Helium was used rather than argon so that a liquid nitrogen cold trap could be used for extraction of water vapor. The gas was vented through a silicone oil-filled flask and finally exhausted into the room.

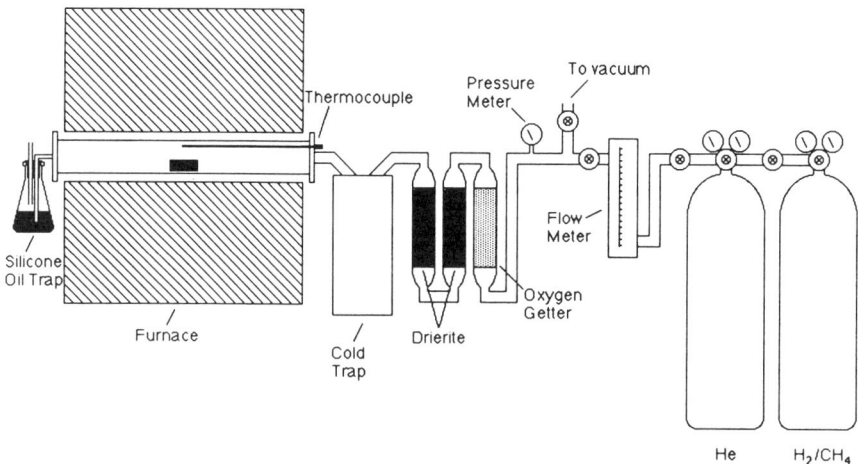

Fig. 1. Schematic of the carburization apparatus.

In the CH_4/H_2 carburization experiments, a silicon bar was placed upright in an alumina holder and inserted into the hot zone of the furnace by the alumina rod described previously. The sample was loaded at reaction temperature while under a helium purge. After the sample was in the hot zone, a vacuum of −686 mm Hg was pulled in order to remove residual air. At that time, the apparatus was backfilled with helium until ambient pressure was reached. The 1% CH_4/99% H_2 mixture was then admitted into the furnace, and the helium was shut off. The gas mixture was allowed to pass over the sample for specified lengths of time at 1390°C, with the gas flow rate maintained at 200 ml/min. The gas exited the furnace via the same route as did the helium in the carbon black experiment, except that the flammable gases were burned as they passed over a resistively heated platinum coil.

Upon completion of the carburization cycles, the samples were removed from the furnace, sectioned using a diamond-impregnated blade, and analyzed with optical and scanning electron microscopy (SEM). Some of the samples were etched with a 50/50 volume % nitric acid/hydrofluoric acid mixture to remove the silicon substrate and allow easier measurement of the SiC scale thickness. Several samples were also analyzed with X-ray diffraction (XRD) to determine the SiC crystal structure.

RESULTS

Carburization with Solid Carbon

Three different types of experiments were conducted. In the first, a wafer of high density carbon was placed on a polished single-crystal silicon wafer. After exposure to 1400°C for 10 hours, a yellow-green scale was observed on the silicon wafer which XRD revealed to be β-SiC. The reaction product was consistent with the observations of Tombs et al. on silicon carburized with graphite in an argon atmosphere[3]: the surface finish of the silicon was not degraded, and a discoloration was observed where the two materials were in contact. The graphite crucible was partially converted to SiC (although no reaction product was found on the dense carbon wafer). The results of Tombs et al.[3] also indicated that the formation of SiC is dependent on the reactivity of the solid carbon, and this was confirmed. It was found that the rate of SiC formation increased as the carbon source was switched from dense carbon to porous graphite to carbon black. All subsequent experiments used carbon black. These experiments were carried out at 1300°, 1350°, and 1400°C.

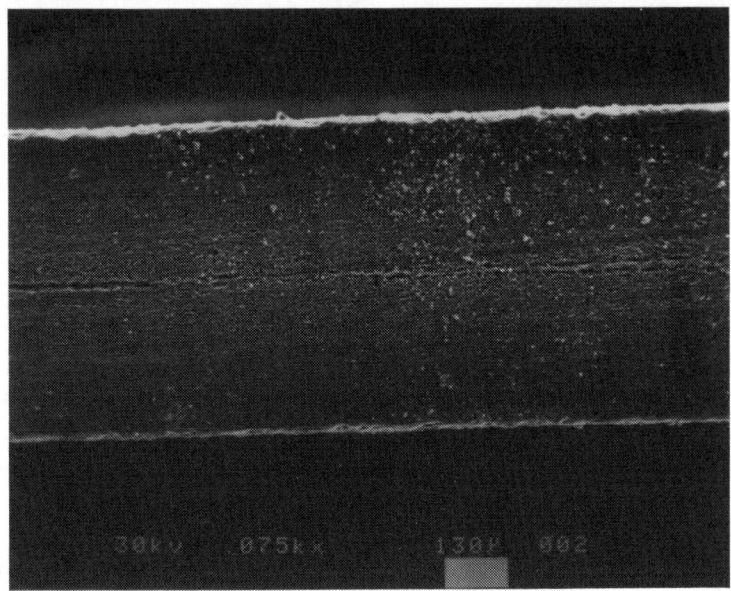

Fig. 2. SEM micrograph of a polished silicon wafer which was completely converted to SiC by carburization by carbon black in a helium atmosphere at 1400°C for 72 hours.

It was determined that no outward growth occurs when silicon is carburized with carbon black. This conclusion was reached after making measurements of the specimens with calipers before and after thermal carburization. No detectable difference in sample dimensions was observed, even in specimens completely converted to SiC by carburization for 72 hours at 1400°C (see Figure 2). In an attempt to detect outward growth too small to be measured with calipers, a small piece was removed from the ends of two specimens, which were then carburized at 1350°C for two hours. After carburization, SEM photos were taken to compare the unreacted pieces with the carburized pieces. The images were measured, and again, no dimensional changes were observed. Si/SiC composite maintained sharp corners during growth, indicative of inward growth where carbon diffusion has taken place.

At 1400°C, SiC scales ranged from 28.33 μm after 2 hours to 571.67 μm after 44 hours. The scales grown at 1350°C ranged in thickness from 42 μm after 5 hours to 335 μm after 71 hours. Plots of thickness vs. time were constructed for both the 1400° and 1350°C experiments and displayed a somewhat linear dependence of thickness on time (see Figure 3). The scatter in these results is attributed to variations in the amount and packing of the carbon powder. Only a qualitative measure of the amount of powder was made, since it was anticipated that the scale growth would be rate limited by diffusion across the scale. However, as will be discussed, it appears that the growth is controlled by the supply of carbon. Although runs were made at 1300°C, no significant scale growth was observed, even after 71 hours.

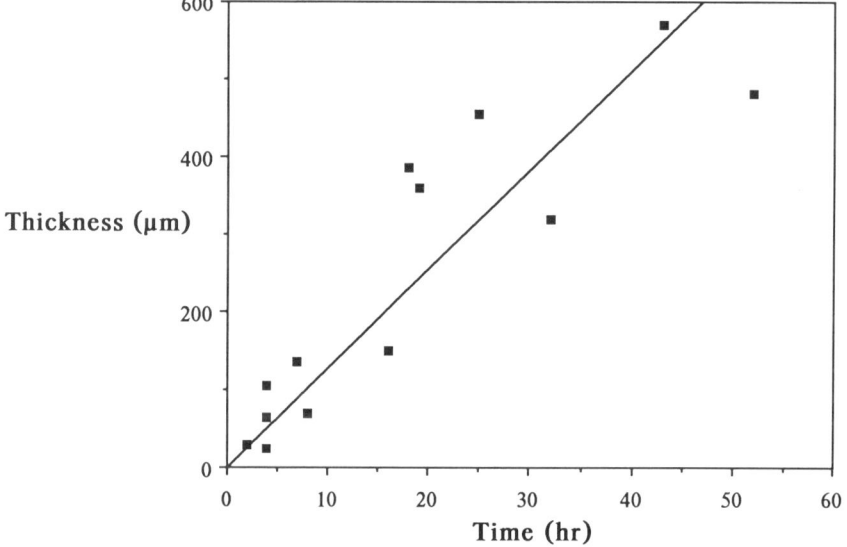

Fig. 3. Thickness of the SiC scale as a function of time for silicon carburized with carbon black in helium at 1400°C.

The microstructures of the scales grown at 1400° and 1350°C are very similar, although less porosity is noted in the scales grown at 1350°C. All scales displayed a density gradient, the density being greatest near the surface. An example is shown in figure 4. A well-defined line of porosity is formed at about 205 μm from the outer surface. Also, more porosity is noted in the corners than in the rest of the scale. The material consists of layered platelets. Observation of the surface of a sample reacted at 1400°C showed the porous nature of the scales produced in the

Fig. 4. SEM micrograph of a SiC scale produced by carburization with carbon black in helium at 1400°C for 19 hours.

carbon black experiments. All SiC scales were green in color, indicative of β-SiC, as confirmed by X-ray analysis.

THERMAL CARBURIZATION OF SILICON WITH METHANE

In contrast to the solid carbon experiments, the specimens carburized in the H_2/CH_4 atmosphere exhibited outward growth, which is consistent with the results of Kato and Okabe[7] and Freiderich and Coble[8]. The kinetics were also markedly different. No correlation between time and scale thickness could be made. Substantial variations in thickness and microstructure were observed (see Figures 5A and B). In general, the scales produced in these experiments were thinner and more dense than those produced using carbon black. Even with long reaction times, up to 117 hours, scales only 30 μm thick were produced, as contrasted with the complete conversion of a wafer using carbon black. Despite the differences in scale thickness and microstructure, a gap between the silicon and resultant SiC was always observed.

A solid phase of carbon was visually detected on the sample surface and confirmed by X-ray analysis. The X-ray pattern revealed crystalline graphite, β-SiC, and an amorphous phase believed to be carbon. SiC whisker formation on the external surface was noted in samples allowed to react for longer times.

(a)

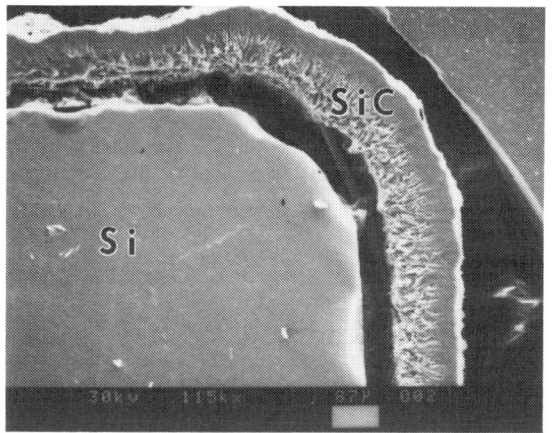

(b)

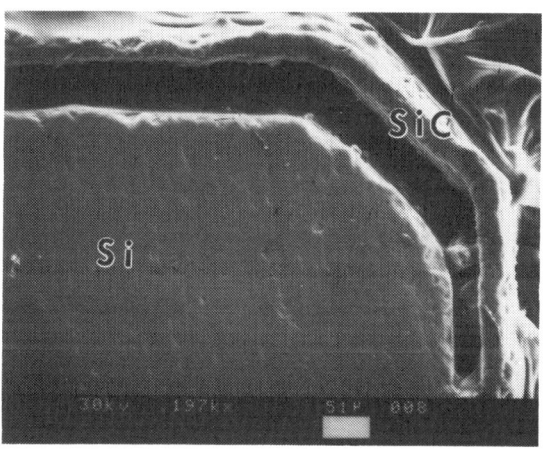

Fig. 5. SEM micrographs of single crystal silicon partially carburized in a 1% CH_4/99% H_2 atmosphere at 1390°C for (a) 2 hours and (b) 5 hours.

Thermal Carburization of Silicon with Carbon Black Followed by Methane

Several silicon specimens were partially carburized using carbon black and subsequently subjected to additional carburization using the CH_4/H_2 atmosphere. SEM examination of these specimens showed that the scale from the carbon black grew inwardly during the carbon black exposure, and an additional scale grew outwardly during the carburization with methane. Such a duplex scale is shown in Figure 6.

Fig. 6. SEM micrograph of a SiC scale produced by partial carburization using carbon black for 1350°C for 2 hours, followed by further carburization in a CH_4/H_2 atmosphere at 1390°C for 117 hours.

DISCUSSION

Carbon Black Experiments

The results of the carbon black experiments may be summarized as follows: 1) the scale was of constant thickness around the entire sample, despite the fact that only one side was in contact with the carbon black; 2) inward growth of the SiC was observed; 3) the outer dimensions of the SiC scale are defined by the original dimensions of the silicon substrate; 4) SiC was found in the carbon black. The first observation leads to the conclusion that the carbon arrives at the silicon surface via the gas phase. The second indicates an inward flux of carbon which reacts at the Si/SiC interface. Since SiC has a silicon molar density roughly one-half of that of single-crystal silicon, the third observation requires an appreciable outward flux of silicon which does not react at the gas/SiC interface (i.e., in order for the SiC to occupy the same volume as the parent silicon, half the silicon must be removed from that volume). The fourth observation confirms that an outward flux of silicon does exist.

In order to explain these observations, it is postulated that a finite pressure of oxygen is trapped in the crucible during sealing in air. Although the oxygen pressure inside the crucible was not measured during the anneal, no growth was observed in unsealed crucibles (or in crucibles with a loosely fitting lid), implying strongly that sealing in air plays an active role.

The following steps are proposed to describe the process:

1. Solid carbon reacts with oxygen to form the carbon-bearing molecule (CBM). The proposed reaction is $C + 1/2\ O_2 \rightarrow CO$. At 1400°C, little carbon dioxide will exist, and the physical evaporation of carbon is negligible[14].
2. Carbon is transported, as CO, to the sample surface.
3. CO then diffuses through the scale, via pores, to the Si/SiC interface. There it is reduced, and SiC is formed by the reaction $CO + 2Si \rightarrow SiC + SiO$.
4. The SiO diffuses through the scale in the same fashion as the CO.
5. The SiO gas travels to the solid carbon source and upon reaction, SiC is formed: $SiO + 2C \rightarrow SiC + CO$.

It is assumed that the mass transport across the scale occurs by gas phase diffusion for several reasons. First, the microstructures are observed to be porous. Second, the rate of scale growth is dependent on the form of the solid carbon and the kinetics appear linear, indicating that the rate limiting step is the supply of carbon. In addition, the absolute magnitude of the growth rate was much more rapid than would be expected for solid state diffusion.

Thus the inward growth occurs because transport of carbon across the SiC scale is possible, and the CO will react with the single-crystal silicon. Note that for every molecule of SiC formed a molecule of SiO is also formed, thereby removing the necessary half of the silicon. No outward growth is observed because the carbon activity in the CO atmosphere is insufficient for reaction with the CO. The SiO will diffuse until it comes in contact with the solid carbon and then reacts to form SiC.

CH_4/H_2 Experiments

The results of the CH_4/H_2 experiments may be summarized as follows: 1) a layer of carbon was deposited on the outer surface of the scale; 2) outward growth of the SiC was observed; 3) the scale was of constant thickness around the entire sample, despite the fact that a gap was formed between the silicon substrate and the scale; 4) scale growth was erratic, as it appears that the scales grow rapidly until they are on the order of 30 μm thick, after which time very little growth occurs. The first observation is consistent with thermodynamic calculations for the reaction $CH_4 \rightarrow C + 2H_2$, using the tabulated thermodynamic data[14] and assuming 1 atm of H_2. The second observation indicates that an outward flux of silicon exists which reacts with the carbon at the outer surface. Apparently, no inward flux of carbon occurs. The third observation suggests that the activity of silicon is constant around the interior surface of the scale, consistent with vapor phase transport of silicon. The fourth observation suggests that a transition in the transport mechanism occurs after the scale reaches a critical thickness.

The formation of SiC under these conditions is proposed as follows:

1. It is assumed that the methane deposits a film of solid carbon which is always present on the surface of the sample. If no initial SiC scale exists on the silicon substrate, the carbon reacts immediately to form SiC.

2. Once the SiC scale has formed, hydrogen diffuses through the scale, via pores, to the silicon surface and reacts to form SiH_4.

3. SiH_4 diffuses to the outer surface of the scale where it reacts with the carbon by the reaction $SiH_4 + C \rightarrow SiC + 2H_2$.

4. After some amount of time, the pores in the scale are closed due to local growth of SiC, and solid state diffusion will be necessary for growth to continue. As a result, the scale growth rate will decrease sharply.

The assumption that pores exist in the scale allowing gas phase transport, and that they eventually become closed, is based on the work of Mogab and Leamy[6]. These investigators observed irregular growth kinetics for SiC formed by carburization with CH_4 at pressures > 10^{-5} atm and temperatures of 1100°C (i.e., rapid growth followed by little or no growth). They were able to correlate this to the existence and closure of porous defect channels. The irregular morphologies of the scales produced under their conditions appear similar to those observed in the present work. The limiting scale thickness observed, approximately 30 μm, is significantly greater than the 40-80 nm observed by Mogab and Leamy. This may be due to the rough surface of "as-sawed" samples and/or the difference in carburization temperatures between the two studies.

The outward growth is due to the vapor phase transport of silicon. Thermodynamic calculations, assuming 1 atm H_2 and using tabulated data[14], suggest that the species is SiH_4. Elemental silicon vapor, which is calculated to have a vapor pressure within a factor of four of that of SiH_4, may also contribute to scale growth. The lack of any inward growth is the result of the absence of a gaseous carbon carrier.

Double Carburization Experiments

The results of partial carburization with carbon black followed by further carburization with CH_4/H_2 can be interpreted in light of the above discussion. The formation of the initial layer of the duplex scale results from the process outlined in the Carbon Black Experiments above. The result of this will be a porous scale of SiC on a silicon substrate. In the CH_4/H_2 atmosphere, therefore, the process will be that outlined in the CH_4/H_2 experiments above, and the growth direction will reverse.

CONCLUSIONS

Studies on the carburization of single-crystal silicon with either carbon black or a CH_4/H_2 atmosphere indicate that vapor phase transport plays an important role in the process. In the case of carbon black, the rate limiting step is the delivery of carbon to the Si/SiC interface. It is hypothesized that the carbon carrier is CO formed by the reaction of carbon with oxygen trapped inside the sealed crucible. In the CH_4/H_2 atmosphere, transport of silicon as a vapor (either SiH_4 or Si) leads to growth of the scale. In a system in which vapor phase transport must be considered, the relationship between the scale geometry and the transport processes is more complex than that derived in appendix 1 due to the role of the atmosphere.

In particular, it has been shown that thermal carburization of silicon can lead to either inward or outward scale growth, depending on the atmosphere. In the carbon black experiments the oxygen-containing atmosphere produced appreciable concentrations of both carbon and silicon gaseous species. Conversely, in the CH_4/H_2 atmosphere, the low concentration of a carbon carrier limited inward growth and allowed SiC outward growth until SiC formation limited growth kinetics.

An unambiguous demonstration that the scales formed in a thermal carburization experiment are gas-tight is therefore necessary before it can be construed as indicating the relative mobility of silicon and carbon. Interestingly, in the experiments reported here, the thickness at which the scales showed a transition in growth kinetics is larger than the sample dimensions employed in the studies used to address the issue of chemical diffusion.[7,8] It is therefore plausible that those scales were not gas-tight. Since no tracer data on the diffusion of silicon along grain boundaries exists, it remains an open question whether carbon or silicon is more mobile in silicon carbide.

---

*Crysteco Co.
‡Cabot Co.
†Sigri Corporation, Carbon and Graphite; Somerville NJ
§OSU Gas Stores
¶Zeiss Videoplan

References

[1] W. G. Spitzer, D. A. Kleiman, and C. J. Frosch, "Infrared Properties of Cubic Silicon Carbide Films," *Phys. Rev.* **113** [1], 133-36 (1959).

[2] H. Nakashima, T. Sugano, and H. Yanai, "Epitaxial Growth of SiC Film on Silicon Substrate and its Crystal Structure," *Japanese J. Appl. Phys.* **5** [10], 874-78 (1966).

[3] N. C. Tombs, J. J. Comer, and J. F. Fitzgerald, "Cubic Beta-Silicon Carbide Films on Silicon Substrates," *Solid-State Electron.* **8**, 839-42 (1965).

[4] P. Rai-Choudhury and N. P. Formigoni, "Beta-Silicon Carbide Films," *J. Electrochem. Soc.* **116** [10], 1440-43 (1969).

[5] J. Graul and E. Wagner, "Growth Mechanism of Polycrystalline Beta-SiC Layers on Silicon Substrate," *Appl. Phys. Lett.* **21** [2], 67-69 (1972).

[6] C. J. Mogab and H. J. Leamy, "Epitaxial Growth of Beta-SiC on Si: Kinetics and Growth Mechanism," and H. J. Leamy and C. J. Mogab, "Epitaxial Growth of Beta-SiC on Si: Growth Morphology and Perfection," in Silicon Carbide 1973, edited by R. C. Marshall, J. W. Faust, and C. E. Ryan, University of South Carolina Press (1974).

[7] A. Kato and Y. Okabe, "Formation of Micro SiC Tubes by the Carburization of Si Whiskers," *J. Am. Ceram. Soc.* **63** [3-4], 236 (1980).

[8] K. H. Freiderich and R. L. Coble, "Influence of Boron on Chemical Interdiffusion in SiC During the Conversion of Silicon Fibers to SiC," *J. Am. Ceram. Soc.* **66** [8], C-141-C-142 (1983).

[9] R. N. Ghostagore and R. L. Coble, "Self-Diffusion in Silicon Carbide," *Phys. Rev.* **143** [2], 623 (1966).

[10] M. H. Hon and R. F. Davis, "Self-Diffusion of Carbon-14 in Polycrystalline Beta-Silicon Carbide," *J. Mat. Sci.* **14**, 2411 (1979).

[11] M. H. Hon, R. F. Davis, and L. Newbury, "Self-Diffusion of Silicon-30 in Polycrystalline Beta-Silicon Carbide," *J. Mat. Sci.* **15**, 2073 (1980).

[12] J. D. Hong and R. F. Davis, "Self-Diffusion of Carbon-14 in High Purity and N-Doped Alpha-Silicon Carbide Single Crystals," *J. Am. Ceram. Soc.* **63** [9-10], 546-52 (1980).

[13] B. E. Deal and A. S. Grove, "General Relationship for the Thermal Oxidation of Silicon," *J. Appl. Phys.* **36** [12], 3770-78 (1965).

[14] JANNAF Thermochemical Tables (1971).

Appendix

Consider the geometry of partially carburized silicon, as illustrated in Figure A1. If the assumption is made that the growth front(s) is planar and that the scale is gas-tight, it is possible to analyze the relationship between mass transport and the scale geometry loosely following an analysis developed for thermal oxidation[13].

In general, both species may be mobile, and the increase in scale thickness with time, $d(X)/dt$, is related to the carbon and silicon flux densities across the scale, J_C and J_{Si}, according to

$$d(X)/dt = J_C/n_C - J_{Si}/n_{Si} \tag{1}$$

where n_i is the molar density of the ith species. Since carbon and silicon have equal molar densities, the subscript is dropped, and equation 1 may be rewritten

$$d(X)/dt = (J_C - J_{Si})/n \tag{2}$$

If it assumed that the two fluxes operate in parallel, i.e. do not interfere with each other, the flux densities may be written as

$$J_C = -(nD_C/RT)\, \mu_C/X \tag{3}$$

and

$$J_{Si} = -(nD_{Si}/RT)\, \mu_{Si}/X \tag{4}$$

where μ_i is the chemical potential of the ith species and D_i is the effective diffusion coefficient, which includes both volume and grain boundary contributions (its value is therefore dependent on microstructure). Note that if the scale is porous and mass transport is dominated by gas phase diffusion, the partial pressure of the relevant gas should replace n in this expression. The assumption of linear chemical potential allows the differences $\mu = \mu(X_2) - \mu(X_1)$ and $X = X_2 - X_1$ to be substituted for the differentials in equations 3 and 4.

The chemical potential differences for silicon and carbon are determined by the phase equilibria at each interface. The equilibrium constant for the reaction $Si + C \rightarrow SiC$ may be written

$$k = a_{SiC}/(a_C a_{Si}) = \exp(-G_r/RT) \tag{5}$$

Assuming unit activity for the solid silicon carbide at both X_1 and X_2 allows a unique relation between the activities of carbon and silicon

$$a_C a_{Si} = 1/k = \exp(+G_r/RT) \tag{6}$$

A consequence of this is that the chemical potential gradient for silicon is equal in magnitude and opposite in sign to that for carbon,

$$\mu_C = RT\ln\{a_c(X_2)a_{Si}(X_1)k\} \tag{7}$$

$$\mu_{Si} = RT\ln\{1/(a_c(X_2)a_{Si}(X_1)k)\} = -RT\ln\{a_c(X_2)a_{Si}(X_1)k\} \tag{8}$$

where $a_C(X_2)$ is the activity of carbon on the outside of the scale and $a_{Si}(X_1)$ is the activity of silicon on the inside of the scale. Both of these are fixed by the experimental conditions. Note that if $a_C(X_2)=a_{Si}(X_1)=1$, then $\mu_C = G_r$ and $\mu_{Si} = -G_r$. Substitution into equation 2 with equations 3,4,7, and 8 yields

$$d(X)/dt = -(D_C+D_{Si})\ln\{a_c(X_2)a_{Si}(X_1)k\}/X \tag{9}$$

Integration gives

$$X = [-(D_C+D_{Si})\ln\{a_c(X_2)a_{Si}(X_1)k\}t + A]^{1/2} \tag{10}$$

where A is the constant of integration. Applying a general initial condition of an arbitrary initial scale thickness, $X(t=0) = X_i$, and rearranging gives

$$X = [-(D_C+D_{Si})\ln\{a_c(X_2)a_{Si}(X_1)k\}]^{1/2}\, [t +]^{1/2} \tag{11}$$

where is the equivalent time to have grown the initial scale.

It is also possible to specify the velocities of the two interfaces relative to a reference frame fixed to an arbitrary lattice site in the SiC. Consider point "P" in Figure A1 as fixed to the lattice. The distance between P and X_1 increases only when there is a flux of carbon from X_2 to X_1 and reaction at the Si/SiC interface, thus

$$d(P-X_1)/dt = +D_C\ln\{a_c(X_2)a_{Si}(X_1)k\}/X \tag{12}$$

On the other hand, the distance between X_2 and P increases as a result of silicon diffusion and reaction at the C/SiC interface.

$$d(X_2-P)/dt = -D_{Si}\ln\{a_c(X_2)a_{Si}(X_1)k\}/X \tag{13}$$

The condition of outward scale growth may be stated

$$d(X_2-P)/dt \gg -d(P-X_1)/dt \tag{14}$$

and by simple substitution this requires $D_{Si} \gg D_C$. Of course, a necessary condition for inward growth is $D_C \gg D_{Si}$.

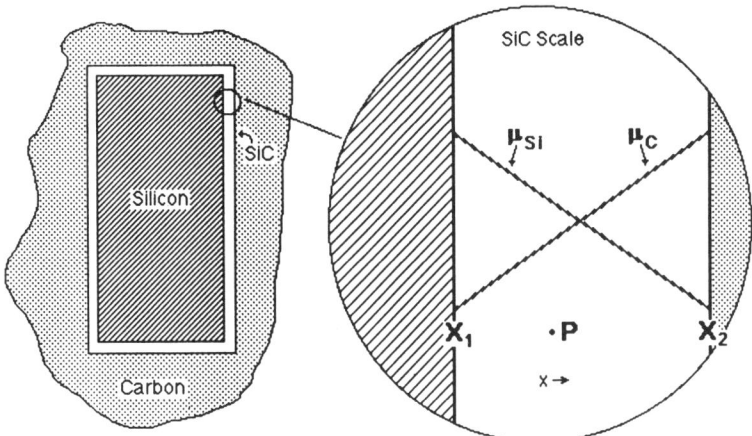

Fig. A1. Schematic illustration of silicon carburization. A single crystal of silicon is completely surrounded by a silicon carbide scale, which is in turn uniformly surrounded by a source of carbon with constant activity. Symmetric chemical potential gradients for carbon and silicon exist across the scale.

Alloying of Silicon Carbide with Other Ceramic Compounds—A Review

Avigdor Zangvil

Materials Research Laboratory and Department of Materials Science and Engineering
University of Illinois at Urbana-Champaign
Urbana, IL 61801

Robert Ruh

Air Force Wright Aeronautical Laboratories
Wright-Patterson AFB, OH 45433

Abstract

A comprehensive review of the literature on alloying of silicon carbide with other ceramic compounds is presented. Studies on processing, microstructure, phase relationships and properties are described, covering the SiC-AlN, SiC-Al$_2$OC-AlN, SiC-BN and SiC-BeO systems, with some mention of the SiC-GaN system. Particular emphasis is given to solid solution formation and polytype stabilization by the additives.

Introduction

In the past decade there has been increasing interest in possibilities for alloying of SiC with other ceramic compounds, in order to modify and control various mechanical and physical properties and to improve sinterability. A number of compounds, such as B4C, Al$_2$O$_3$, BN, and BeO have been added in small amounts as sintering aids, while other compounds, such as AlN and Al$_2$OC, could be added in large amounts to form extensive solid solutions with SiC and modify its properties. The present article focuses on additives that can potentially modify properties other than sinterability, by the formation of solid solutions, control of microstructure development, formation of compounds and, significantly, control of polytype stability. The latter point is important because the various polytypes of SiC have different electronic structures and possess different band gaps, and therefore have different physical properties at any given temperature.

The potential of SiC alloying has been demonstrated in several areas. Ervin[1] was the first to report the existence of SiC-AlN solid solutions, and found that they have superior oxidation resistance because their protective oxide (mullite or aluminosilicate glass) had a thermal expansion coefficient similar to that of the base SiC-AlN material. Sukhanek et al.[2] calculated such parameters as piezoelectric modulus and non-linear optical susceptibility of SiC-AlN solid solutions. They concluded that these materials "hold promise as materials for optoelectronics and acoustoelectronics." Variable-gap semiconductor structures as well as semiconductors with a pronounced gradient of the refractive index have been mentioned as potential applications of the SiC-AlN and SiC-GaN systems. Improved control of microstructure and mechanical properties of SiC by alloying with AlN and Al_2OC has also been demonstrated[3-6], as described in the following section.

Compounds which are potential alloying additives to SiC are listed in Table I, along with their lattice constants. They all have hexagonal wurtzite-type structures. Other compounds which have been mentioned as potential additives to SiC include $BeSiN_2$ and $MgSiN_2$[7], both forming extensive solid solutions with AlN, and a list of compounds with similar structures, such as $LiSiN_3$, LiSiON and $MgSiAlN_3$[8]. Extensive studies have been done in the SiC-AlN system, and considerable work has also been reported in the SiC-Al_2OC-AlN, SiC-BN and SiC-BeO systems. The present paper intends to review these studies, emphasizing processing methods, phase relationships, microstructures and properties.

Table I. Potential Alloying Additives To SiC.

Compound	a_0(Å)	c_0(Å)	c_0/a_0
BN_w	2.55	4.20	1.647
BeO	2.698	4.380	1.623
β-Be_3N_2	2.841	9.693	3.412
2H-SiC	3.076	5.048	1.641
AlN	3.111	4.979	1.600
Al_2OC	3.19	5.09	1.596
GaN	3.186	5.178	1.625

PROCESSING

Materials in the systems of interest have been prepared by a variety of methods. Cutler et al. and Rafaniello et al., assuming that SiC-AlN-Al_2OC solid solutions cannot be obtained by heating the powdered solid components due to the extremely low diffusion coefficients, attempted to produce the materials by in situ carbothermal reduction of silica and aluminum hydroxide in the presence of a carbon source (starch) and nitrogen gas[3,8]. One possible reaction would be[9]

$$3SiO_2 + Al(OH)_3 + 4.5C + N_2 \xrightarrow{1600°C} SiC \cdot AlN + 1.5\ CO + 1.5\ H_2O + 0.5N_2 \tag{1}$$

The intimate dispersion of the materials indeed resulted in the formation of SiC-AlN-Al_2OC solid solutions, although their homogeneity was subsequently questioned.

Another chemical route, described by Dobson[9], involved a carbon reduction of an M-SiAlON, or Alpha' SiAlON. For example,

$$CaSi_9Al_3ON_{15} + 10C \xrightarrow{1800°C} 3(3SiC \cdot AlN) + Ca + CO + 6N_2 \qquad (2)$$

SiC-AlN solid solutions were reported to form in this manner; they reportedly contained excess nitrogen.

Tsukuma et al.[10] reported that a new compound with the wurtzite structure could be obtained by the reaction

$$Si_3N_4 + Al_4C_3 \rightarrow Si_3Al_4N_4C_3 \qquad (3)$$

using a gas autoclave at 10 MPa argon pressure and 1800°C for 1 h. A $Si_3Al_5N_5C_3$ compound with the same structure was also obtained. It was noted later that these compounds should be considered as compositions in the continuous $(SiC)_{1-x}(AlN)_x$ solid solution series[11].

Ruh and Zangvil[4,12] have shown that, in spite of the low diffusivity, SiC-AlN solid solutions can be obtained by hot-pressing a mixture of the individual components at 2100°C, through a direct solid state reaction

$$(1-x)SiC + xAlN \rightarrow (SiC)_{1-x}(AlN)_x \qquad (4)$$

Hot-pressing methods were later applied to SiC-Al_2OC-AlN as well, resulting in a ternary solid solution. Al_2O_3 and Al_4C_3 were used instead of the unstable Al_2OC[13].

Schwetz and Lipp found recently that, by using 1 wt% metal oxide addition to mixtures of AlN and SiC powders, almost complete densification could be achieved through pressureless sintering at $\leq 1900°C$[14]. They reported that the process resulted in 2H AlN-SiC solid solution with a 27R SiAlON binder phase.

The fabrication of SiC alloy single crystals has also been demonstrated. Knippenberg et al. have grown mixed SiC-AlN crystals epitaxially on SiC platelets at temperatures in excess of 2100°C[15], and Rutz has grown epitaxial SiC-AlN crystals on an Al_2O_3 substrate with <0001> orientation, by heating SiC and AlN source materials to 1900-2020°C and using a $N_2 + H_2$ carrier gas[16]. Sorokin et al. have grown both SiC-AlN and SiC-GaN solid solution films on SiC substrates[17].

THE SiC-AlN SYSTEM

Microstructure and Phase Relationships

As mentioned above, Ervin was the first to report that "SiC-AlN solid solutions of varying composition, as well as other complex phases" formed when AlN was deposited from the vapor phase in the pores of a SiC body, heated to 1400 to 2200°C[1]. At about the same time, Shaffer reported that "SiC formed at 2500°C from the elements in the presence of excess Al and under 15-100 atm (1.5~10 MPa) of N_2 pressure has the 2H or wurtzite structure of AlN"[18]. The amount of aluminum and the nitrogen pressure were shown to control the phase stability of SiC, but solid solution formation was not mentioned. Yates, on the other hand, could not find any evidence of mutual solid solubility when SiC and AlN were produced by hot-pressing Al_4C_3 and Si_3N_4 at 1600 to 2200°C and 1.4 to 42 MPa[19]. Only SiC and AlN were formed; however, in light of later results, these may have actually been

SiC-rich and AlN-rich solid solutions, at least in the higher part of the temperature range. Later, Schneider et al.[20] could find no extensive solubility of SiC and AlN in a study of the SiC-Si$_3$N$_4$-AlN-Al$_4$C$_3$ system. Their samples were hot pressed at 1760–1860°C.

Cutler et al.[8,21] and Rafaniello et al.[3] were the first to employ a more direct approach towards the preparation and study of SiC-AlN and SiC-AlN-Al$_2$OC solid solutions. Based on principles of atom exchange, they obtained the solid solutions through reactions such as (1) and (2). A continuous solid solution with the 2H wurtzite structure was found to exist between 2–100% AlN in the SiC-AlN system and between 1–100% Al$_2$OC in the SiC-Al$_2$OC system. X-ray diffraction (XRD) showed a single phase in all these compositions. The compositional homogeneity of the solid solutions within the powders was not confirmed, but the general microstructure of hot-pressed samples obtained from these powders was found to be fine-grained and uniform[3].

The present authors found that, when SiC and AlN mixtures are hot-pressed at 2100–2300°C, wurtzite-type solid solutions are formed[4]. At 2100°C, 35% AlN* or more were needed in order to obtain a single phase solid solution. Lattice parameters of the solid solutions closely followed Vegard's law. Grain size of the solid solutions for a given firing temperature was significantly smaller than for the same SiC or AlN raw powders fired to the same temperature. For compositions with less than 35% AlN, multiphase assemblages were found, as shown in Table II. Multiphase assemblages were also found in all the samples which had been hot-pressed at 1700–2000°C. Rafaniello et al. suggested that a miscibility gap exists in the SiC-AlN system in the low temperature range[3,7], and that the previously obtained solid solution powders were actually mixtures of SiC-rich and AlN-rich solid solutions, mistakenly identified as a single homogeneous solid solution because of the limited angular resolution of the XRD methods. Ish-Shalom and Zangvil indeed found that the XRD patterns from SiCAlON powders with 20–80% AlN, produced by a carbothermal reduction process, exhibited peak splits which corresponded to SiC-rich and AlN-rich materials[22]. Rafaniello et al. showed, moreover, that annealing at 1700°C of hot-pressed samples resulted in precipitation of a minor phase from the solid solution. Precipitates containing about 93% AlN formed when a 47.5% AlN material was annealed, and precipitates with about 8% AlN appeared in a 75% AlN sample. The existence of a miscibility gap was also confirmed by Zangvil and Ruh[23] by analytical STEM in samples hot-pressed at 1850–1950°C. Although firing times were only ~ 1 h and the samples were far from reaching equilibrium, analysis of grains in contact indicated that SiC + ~ 93% AlN was in "local equilibrium" with SiC + ~ 3.5% AlN, the SiC having a faulted β structure, denoted β' (unfaulted β contains much less AlN). This was in general agreement with the above results.

Recently, Kuo et al. found modulated structures, indicative of spinodal decomposition, in hot-pressed and annealed SiC-AlN materials with 50 and 75% AlN[24]. Annealing was carried out at 1620 to 1900°C for 56 to 320 h in nitrogen atmosphere. One of the microstructures is shown in Fig. 1. Selected area electron diffraction indicated the 2H wurtzite structure in the whole area. Since N$_2$ was used, and in view of results by Patience et al.[25] and Dobson[9] (to be described below) one should consider the possibility that nitrogen solubility in 2H-SiC (and, perhaps, in 2H-SiC-AlN) had a role in the decomposition process. Sukhanek et al. estimated the boundaries of the miscibility gaps (assuming heterovalent solid solution

Table II. Phases Present in SiC-AlN Compositions[4].

Composition	Temp. (°C)	Time (h)	Phases Present
SiC	1950	0.5	3C,6H,4H,2H,21R
SiC	2100	1	6H, 4H, 2H, 3C?
SiC	2300	1	2H,4H,6H,21R,3C?
SiC+10.2%AlN	2100	1	4H,2H,6H,15R,3C
SiC+11.4% AlN	2300	0.5	2H,4H,15R,21R
SiC+24.8% AlN	1950	1	3C,2H
SiC+24.8% AlN	2100	1	2H,3C,4H,6H
SiC+34.0% AlN	2100	4	2H
SiC+36.9% AlN	2100	1	2H,15R,21R,3C?
SiC+45.9% AlN	2000	4	2H,4H,6H,3C
SiC+49.4% AlN	2100	1	2H
SiC+59.4% AlN	2200	1	2H
SiC+60.0% AlN	2100	1	2H,3C
SiC+65.4% AlN	2100	1	2H
SiC+75.4% AlN	2300	0.5	2H
SiC+80.0% AlN	2100	2	2H
AlN	1700	0.5	2H
AlN	2100	1	2H

approximation with $\eta=2$ rather than regular solution approximation) for the SiC-AlN and SiC-GaN systems[2]. The calculated miscibility gap for SiC-GaN was found to extend to 2200°K, compared to 1000°K for SiC-AlN.

The single wurtzite-type phase obtained at 2100°C was found, by analytical STEM, to contain steep compositional gradients even within individual grains[4]. This is shown in Fig. 2. However, there was no evidence of any miscibility gap, as a continuous range of compositions was present. By hot-pressing SiC-AlN powders obtained in situ by carbothermal reduction, better homogeneity was achieved at 2300°C than by hot-pressing a mixture of SiC and AlN (which were separately obtained by carbothermal reduction processes)[7]. This may be attributed to the fact that the SiC-AlN powder obtained in situ was, most likely, a mixture of SiC-rich and AlN-rich solid solutions (and not of pure SiC and AlN), which made the homogenization process much easier.

It should be added that two studies reported the formation of solid solutions even at $\leq$ 1900°C. One study is that of Schwetz and Lipp, mentioned above, in which a 20/80 SiC/AlN solid solution was obtained[14]. The other is that of Tsukuma et al.[10], in which solid solution powders were obtained through Reaction 3 at 1800°C. Hot-pressing these powders at 3.0 GPa and 1300-1900°C for 1 h was reported by Shimada et al. to result in solid solutions as well[5]. In all these cases there was no attempt to verify the homogeneity of the solid solutions, or the possible existence of a mixture of two solid solutions, by analytical TEM/STEM, and it is suggested that SiC-rich and AlN-rich solid solutions may have been present.

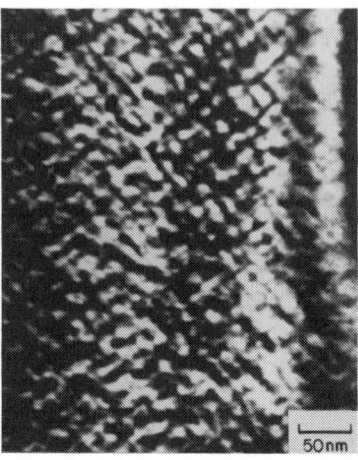

Fig. 1. Electron micrograph of a sample containing 50 mol% SiC and 50 mol% AlN annealed at 1900°C for 56 h showing a modulated structure. (Courtesy of Kuo, Virkar and Rafaniello[24]).

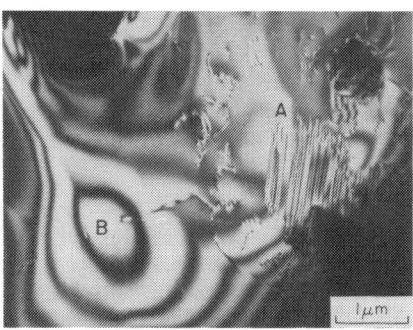

Fig. 2. Bright-field transmission electron micrograph showing part of a 2H solid-solution grain in a sample of SiC+65.4% AlN. The composition is ~ 5% SiC-95%AlN in "A" and 65% SiC-35%AlN in "B"[4].

The difficulties encountered in the attempts to obtain homogeneous SiC-AlN solid solutions are clearly related to the low diffusion coefficients. However, while the diffusion coefficients and solubilities of individual elements in SiC (and probably in AlN as well) are known to be low[26,27], it is not clear at all that interdiffusion of the SiC and AlN as compounds should also be low. The rate of lattice diffusion is related to the solubility limit, and the high mutual solubility at

Table III. Diffusion of SiC in AlN[28].

Temperature (°C)	Approximate Diffusion Distance After 1 h (μm)	$D(cm^2s^{-1})$
1950	0.2–0.5	$(0.5-1.5)\times10^{-12}$
2100	2	1×10^{-11}
2200	8	2×10^{-10}
2300	15	6×10^{-10}

high temperatures may mean that higher diffusion coefficients are possible. Zangvil and Ruh calculated approximate diffusion coefficients of SiC in AlN from diffusion distances in hot-pressed SiC-AlN samples and in SiC-AlN diffusion couples, all hot-pressed in vacuum. The results are shown in Table III[28]. The corresponding activation energy was roughly estimated to be as high as 900 kJ·mol^{-1} and the pre-exponential term was in the order of 10^8 cm^2 s^{-1}, an unusually high value. It was suggested that lattice diffusion of coupled Al-N and Si-C pairs was responsible for these high values. The strong temperature dependence of diffusion distances explains the fact that solid solutions have not been observed in XRD after short processing times at $\leq$ 2000°C.

A SiC-AlN phase diagram was recently proposed by Zangvil and Ruh[23] and is reproduced in Fig. 3. It is based upon the presently available data—mostly XRD and analytical STEM data obtained by the present authors for high temperatures ($\geq$ 2100°C)[4,23,29-31] and data obtained by Rafaniello et al. and by the present authors for the lower temperatures[7,23,32].

It should be noted that Dobson suggested an entirely different phase diagram for the SiC-AlN system[9], showing a 2H solid solution with up to ~ 40% AlN at the low temperature range and a cubic (3C) solid solution with up to 20%AlN at about 2100°C. However, this diagram was obtained from samples which were processed through nitrogen-evolving reactions (e.g. Reaction 2) and hot-pressed in nitrogen atmosphere; up to 12% nitrogen was reported to dissolve in 2H-SiC (and probably in 2H-SiC-AlN as well) thereby stabilizing the wurtzite structure[9,25]. It is suggested that the resulting phase relationships may apply only under such nitrogen-saturated processing conditions.

The question whether the phase diagram in Fig. 3 represents true pseudo-binary relationships has been addressed by the present authors. Ultra-thin-window energy dispersive X-ray spectroscopy and scanning Auger analysis were used to analyze local concentrations of all the elements of interest (Si, C, Al, N, O). The analyses showed that Si:C and Al:N ratios in the solid solutions were indeed unity or near unity for all phases in the vacuum-hot-pressed materials, including the diffusion couples, e.g. Fig. 4[23].

Phase transformations in the SiC-AlN system have also been studied. The precipitation and decomposition phenomena after annealing at 1600–1900°C[7,13,24] were described above. Transformations from various SiC structures into the 2H SiC-AlN solid solution have been studied by the present authors[23]. The 4H→2H and 15R→2H transformations were described as diffusion-controlled stacking rearrangements through a layer displacement mechanism, similar to the mechanism described by Pandey and Krishna for SiC[33], and the transformation zones were up

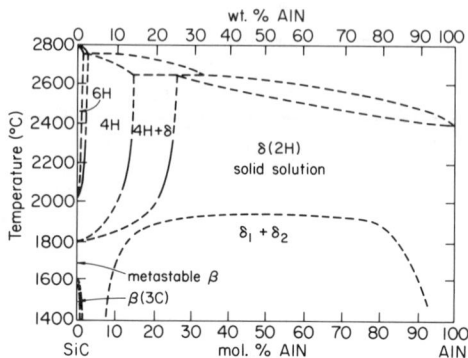

Fig. 3. A tentative SiC-AlN phase diagram[23].

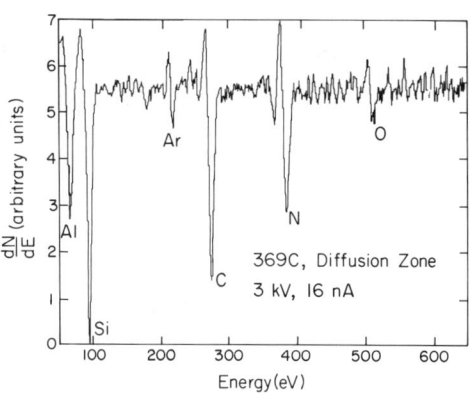

Fig. 4. Auger derivative spectrum from the solid solution in a 2200°C/2 h diffusion couple, showing approximately 40% AlN, 2% oxygen[23].

to 30 nm wide (Fig. 5). The 3C→2H transformation proceeded through a heavily faulted cubic structure (β'), which contained up to 4% AlN[23].

The reverse 2H→3C transformation was observed through lattice imaging by Dobson[9] in "nitrogen-2H" and in SiC-AlN solid solutions. It took place by a layer-displacement mechanism through a disordered intermediate stage.

A comprehensive review of polytypic transformations in silicon carbide by Jepps and Page has been published[34]. It includes a discussion of impurity effects on polytype stability and polytype transformations.

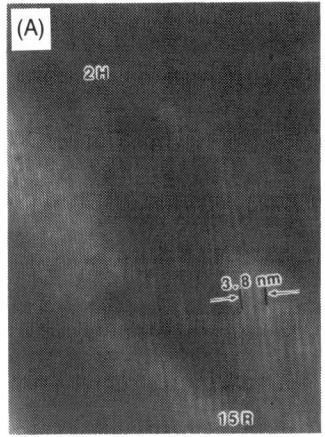 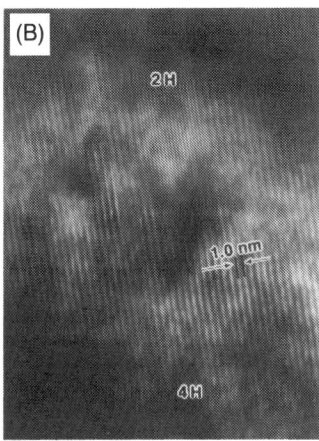

Fig. 5. The mechanisms of the (A) 15R→2H and (B) 4H→2H transformations. In both cases, a band of several unit cells in width is transformed through a diffusion controlled process of layer rearrangement[23].

Properties

Basic physical and mechanical properties have been studied in SiC-AlN solid solutions and, in some cases, compared with properties of SiC-AlN composites of corresponding compositions. Rafaniello et al. obtained a nearly linear relationship between Young's modulus and AlN content in samples hot-pressed at 1950-2030°C[5]. However, these samples are now believed to have been closer to two-phase mixtures than to homogeneous solid solutions. Ruh et al. have studied the behavior of room temperature Young's moduli in detail[35]. First, the effects of porosity on the moduli of SiC and AlN were determined. The Young's moduli of solid solutions and composites, corrected to zero porosity, were then determined, as shown in Fig. 6, and compared with composite mixture theory. The elastic moduli of two-phase particulate composites were lower than those of the corresponding solid solutions and also lower than values calculated for composite theory.

Microhardness measurements from three different studies are shown in Fig. 7. The differences probably result from the differences in the measurement techniques as well as differences in the processing conditions of the samples, which are described in the corresponding articles[7,5,4]. Still different values were obtained by using a 67N load Vickers test[5].

Flexural strength values were reported to be low, in the order of 200 MPa, for the inhomogeneous solid solution. The strength decrease seemed to be related to the maximum grain size which, in turn, increased with firing temperature[4].

The fracture toughness of SiC-AlN solid solutions was determined using the indentation technique. The K_{IC} values dropped from about 4.5 to about 3 MPa·m$^{1/2}$ by adding 20% AlN to SiC, then remained nearly unchanged up to 95% AlN[3]. Shimada et al. found similar results by using the same technique on their

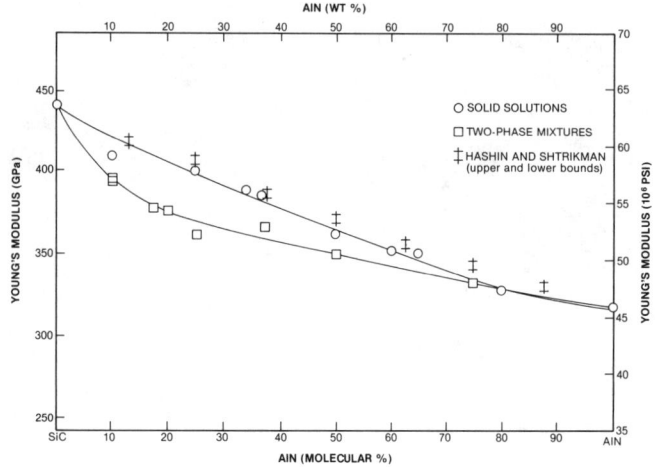

Fig. 6. Young's modulus of SiC-AlN compositions[35].

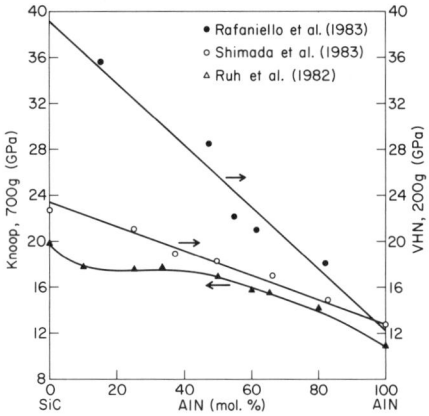

Fig. 7. Microhardness vs. composition of SiC-AlN according to three studies[7,5,4].

high-pressure hot-pressed samples[5]. They also studied the temperature dependence of K_{IC} and found a slow decrease of about 0.5 MPa·m$^{1/2}$ with increasing temperature from R.T. to 1200°C.

Creep resistance was measured for hot-pressed SiC-35wt% AlN by four point bending at 1400-1500°C under a stress of 1-3 MPa. Above about 1425°C the SiC-AlN samples showed better creep resistance than pure SiC (with 1 wt% B)[3]. The role of grain size in the creep of SiC-AlN materials was studied by Jou et al.

Both dislocation creep and diffusional creep were thought to be operating at the temperature range of the experiments, 1400 to 1520°C. The creep rate dependence on temperature and grain size is shown in Fig. 8[36].

The thermal expansion coefficient for R. T. to 800°C exhibited a nearly linear behavior vs. AlN content[3]. Thermal diffusivity or thermal conductivity were measured by several investigators[3,5,37]. Figure 9 presents these results. There are considerable differences in absolute values among the various studies, but one common feature is clearly the drop in thermal conductivity (and diffusivity) for the solid solutions relative to the end materials (SiC and AlN). Shimada et al. suggested that this is due to increased phonon scattering by the disordered potential field in the both cation- and anion-substituted system[5]. This is supported by the fact that samples hot-pressed at lower temperatures, in which complete solid solutions have not formed, exhibited a much smaller drop in thermal conductivity, compared to identical compositions hot-pressed at $\geq$ 2100°C[37].

Sukhanek et al. estimated the nonlinear optical susceptibility d_{333} of SiC-AlN solid solutions as function of composition. The values ranged from about 1.5 × 10^{-11} m/V for AlN to a high of over 6 × 10^{-11} m/V for the equimolar solid solution[2]. Rutz plotted the measured optical density vs. absorbed photon energy and determined the absorption edge to be at 4.6 eV for a SiC-56.5% AlN crystal[16].

THE SiC-Al$_2$OC-AlN SYSTEM

Extensive binary and ternary solid solutions have been found by Cutler et al.[8,21] for 2-100%AlN and 1-100% Al$_2$OC. These results were obtained by XRD, but optical microscopy and TEM indicated that two phases, presumably a SiCAlON solid solution and a SiC-rich phase, existed after hot-pressing at 2000°C. Moreover, splitting of the 2H peaks appeared in a sample annealed 50 h at 2050°C[13]. Reactive pressureless sintering at below 2100°C also resulted in two-phase mixtures[6]. This may imply that the miscibility gap in the SiC-AlN system extends into the ternary system[13]. Densification in the SiC-Al$_2$OC-AlN system occurs through a liquid phase sintering mechanism, and the addition of boron as sintering aid is not necessary[6].

The ternary solid solution can form due to the stabilization of Al$_2$OC by SiC and AlN, or by the presence of nitrogen[6]. The stabilization of Al$_2$OC by AlN has also been reported by Sandberg[38].

The degree of homogeneity of the samples depended upon the type and source of SiC powder used. It was suggested that "the kinetics of dissolution of β-SiC in AlN and Al$_2$OC to form a 2H solid solution are faster than if the starting SiC is α-SiC". This is based upon hot-pressing at 2000°C in N$_2$[13].

Further work is needed to establish the phase relationships in the system, including low temperature annealing to determine the misability gap and high temperature processing to determine the series of stable solid solutions at above 2100°C.

Four point bend strength, hardness and fracture toughness for the SiCAlON materials have been reported[6]. The bend strength of pressureless sintered samples increased slightly with increasing Al$_2$OC content in the SiC-AlN solution, and ranged between 310-330 MPa. The strength decreased with Al$_2$OC content in hot-pressed samples, from ~ 600 MPa at 10 wt% Al$_2$OC to around 250 MPa at ~ 50% Al$_2$OC. This was thought to be due to the free carbon in samples with > 10% Al$_2$OC (carbon and Al$_2$O$_3$ were found in these samples). Most significantly,

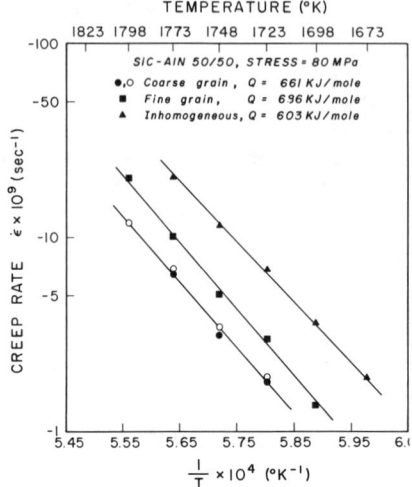

Fig. 8. Creep rate vs. temperature for fine grained, coarse grained and partially inhomogeneous specimens. (Courtesy of Jou, Kuo and Virkar[36]).

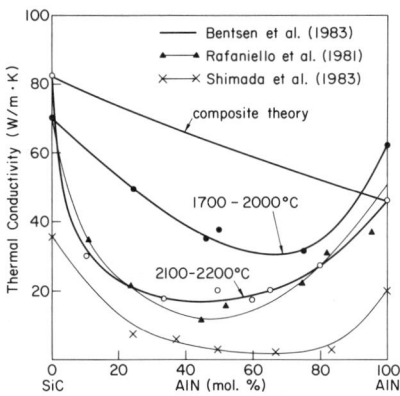

Fig. 9. Thermal conductivity vs. composition of SiC-AlN materials according to several investigators[3,5,37].

the fracture toughness of some SiCAlON compositions appeared to be higher than that of SiC (~ 4.2 compared with 3.0 MPa·m$^{1/2}$), using an indentation technique.

THE SiC-BN SYSTEM

Shaffer reported that "SiC formed at 2500°C from the elements in the presence of boron and under nitrogen pressure is cubic as the high pressure form of BN"[18]. Shaffer and Whitney observed the formation of β-SiC with decreased unit cell dimensions (from 4.359 to 4.348 Å) by reacting the elements (Si, C, B and N_2 gas) or a mixture of Si, C and BN at 2500°C and 3.2 MPa pressure for 5 min.[39]. They assumed that B and N entered into solid solution with SiC. Increased microhardness values, relative to SiC, were also observed with increasing BN content in the samples. It was claimed that the major phase was "β-SiC" (with decreased unit cell) even in samples with 80% BN. All this suggested that an extensive solid solution may exist in the SiC-BN system, in spite of the large difference in lattice constants of SiC and BN (BN is almost 20% smaller). Sokhor and Fel'dgun, on the other hand, could not find any evidence of deviation from the lattice constant of β-SiC in high pressure processing of cubic SiC and cubic BN powder mixtures at up to 1850°C[40]. Murata and Smoak measured the solubilities of BN, BP and B_4C in 6H-SiC, in order to correlate the densification behavior with the solubility limit of the sintering aid[41]. Sintering was done with various amounts of additives at 2200°C for 30 minutes in argon atmosphere. X-ray diffraction measurements of the lattice parameters vs. composition showed solubility limits of 1.61% BN, 1.92%BP and 0.36%B_4C, all in 6H-SiC.

The present authors have tried to clarify the discrepancies in the reported results and prepared hot-pressed SiC-50 wt.%BN at 35 MPa and up to 2250°C, using cubic SiC and graphite-type BN. X-ray diffraction and transmission electron microscopy showed that some of the SiC retained the cubic structure even at 2250°C. Moreover, cubic (β) SiC was usually adjacent to BN grains, while other SiC polytypes were found deeper into SiC areas. An example is shown in Fig. 10[28]. Hot isostatic pressing at 2000°C for 1 or 2 h under 186-200 MPa N_2 pressure was also done, in an attempt to further stabilize the cubic phase[42,43]. Cubic SiC was predominant in the resulting samples, and it again appeared adjacent to the BN particles. Other polytypes, such as 6H, 15R and 21R, also appeared. The lattice parameters of β-SiC decreased only slightly relative to pure SiC, to values which correspond to 0.5-1%BN solubility (assuming Vegard's law is applicable). It was clear, however, that β-SiC and SiC polytypes with low hexagonality are stabilized by the presence of BN, probably by the solution of BN up to saturation level near the BN and in smaller amounts (which may correspond to saturation in other polytypes) further away from the BN grain. This description is strongly supported by the appearance of a sequence of polytypes (3C-8H-6H-15R), whose hexagonality increases with increasing distance from the interface, as shown in Fig. 11[43].

Some mechanical and thermal properties of SiC-BN materials have been measured. These should be viewed as properties of SiC-BN particulate composites, since extensive solid solubility does not occur. Ruh et al. determined Young's moduli for SiC-BN composites, Fig. 12[43]. Valentine et al.[44,45] determined the strength and thermal shock resistance of SiC-BN hot-pressed composites, and found an increase in thermal shock resistance from 314°C for SiC to 526°C in SiC + 25%BN[45]. Ruh et al.[46] investigated the thermal diffusivity of the composites and found a high degree of anisotropy with respect to the hot-pressing direction as a result of preferred orientation of the BN phase.

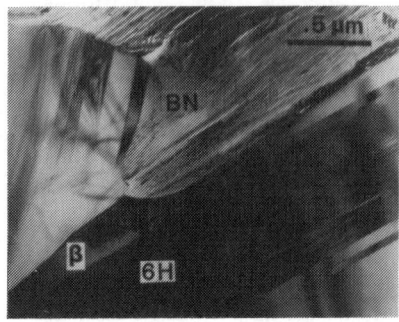

Fig. 10. SiC-50 wt.%BN which had been hot-pressed for 45 min. at 2100°C, showing stabilization of cubic SiC[28].

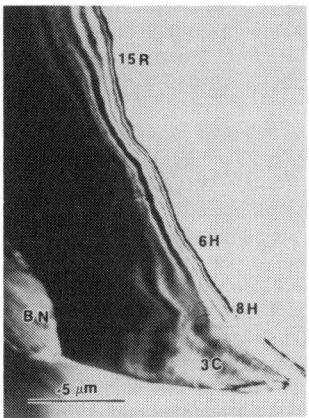

Fig. 11. A sequence of SiC polytypes increasing in hexagonality with increasing distance from the adjacent BN, in a SiC-62%BN sample hot-pressed 30 min, 2250°C. Hexagonality is 0%, 25%, 33% and 40% in the 3C, 8H, 6H and 15R structures, respectively[43].

The SiC-BeO System

One system in which an alloying additive has very strong effects on the physical properties of SiC is the SiC-BeO system. Takeda et al. added BeO in small amounts, 1.6 or 3.2%, to SiC, which was then hot-pressed to high density at 2050°C and 30 MPa for 1 h[47,48]. The resultant materials have exceptionally high thermal conductivity and electrical resistivity values (Table IV), which makes them

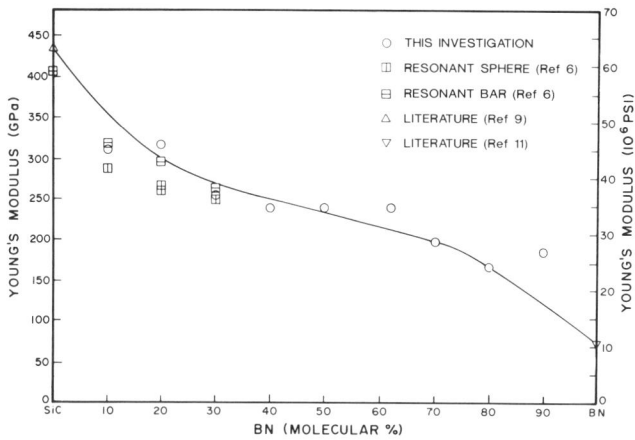

Fig. 12. Young's modulus of SiC-BN composites versus BN content[43].

Table IV. Properties of SiC-BeO Hot-Pressed Materials[47,48] Compared with Pure SiC, AlN and BeO.

Property\Material	SiC	SiC+ 1.6%BeO	SiC+ 3.2%BeO	BeO	AlN
R. T. Thermal Conductivity (w/m.K)	~ 40	270	270	240	60–240
R. T. Electrical Resistivity (Ω cm)	10^{-2}–10^6	>10^{13}	4×10^{13}	>10^{14}	10^{13}–10^{15}
Thermal Expansion ($10^{-6}\cdot K^{-1}$)	~ 4	3.7	3.7	8.0	4.6

attractive for applications such as for electronic substrates. Takeda et al.[48] studied the effects of various elemental and compound additives on the microstructure and strength of sintered SiC. The compounds BeO, B_4C, BN and AlN produced high density hot-pressed SiC ($\geq$ 97% of theoretical density). Hot-pressing with a BeO addition resulted in a narrower grain size distribution than with boron or boron carbide; no abnormal grain growth occurred in the case of BeO addition and this resulted in higher strength levels. For example, 6H-SiC with 3.2%BeO, hot-pressed at 2040°C for 1 h had a flexural strength of 450 MPa. However, AlN and Al_2O_3 additions resulted in finer grains and higher strengths than BeO additions.

The SiC-BeO materials had a thermal conductivity higher than that of Al and electrical resistivity > 10^{13} Ω cm (Table IV). Microstructural analysis has been performed in order to understand those unusual properties. Lattice images of SiC grains near SiC/SiC boundaries seemed to show local variations, which were attributed to beryllium segregation[47]. Beryllium also had other effects, such as the epitaxial formation of Be_2C on SiC interfaces into the BeO grains. More 4H, growing within the 6H-SiC matrix, was observed after hot-pressing with BeO, but the 4H was not necessarily adjacent to BeO. It has been suggested that Be at the grain boundaries could be the cause for the high thermal conductivity, while an impurity-depleted layer could explain the high electrical resistivity[50,51]. The resistivities of the grains themselves was estimated at less than 10 Ω cm.

Shinozaki et al. found, in a TEM study, an enhanced formation of the 4H polytype in the SiC-BeO materials[52]. Transformation into 4H was more pronounced when β-SiC, rather than 6H-SiC, was used as starting material. Impurities, such as Ti, also aided in the formation of the 4H structure.

Zangvil et al. found a seemingly opposite effect: 6H was stabilized relative to 4H in SiC-BeO powder mixtures with 20-75%BeO, hot-pressed at 2100-2280°C[53]. The apparent contradiction between the results of the two studies was thought to be due to the different impurity levels, firing temperature and firing atmosphere. The limiting of SiC grain growth by BeO grains in the high-BeO composites may also play a role in determining the amounts of phases present in the samples. Samples with lower BeO contents were also studied. Those with 1.6 and 3.2%BeO, identical with materials studied in[47,48], showed formation of Be_2C and contained some glassy grain boundaries[53]. A sample with 5.5%BeO, hot-pressed in N_2 at 2050°C, contained less Be_2C. Moreover, most BeO grains in this sample were found to be Be(O,N)x solid solutions with the wurtzite structure. Safaraliev et al.,[54] predicted high mutual solubilities of SiC and BeO, based on St. John and Bloch's[55] quantum-defect electronegativity vs. hybridization diagram. Such high solubility has not been detected to date even at 2200°C[53].

SUMMARY

Alloying of silicon carbide with various ceramic compounds holds a potential for powerful property engineering. Thermal, electronic, optical and mechanical properties are affected by various amounts of compound additives in solid solution, depending on the system. The SiC-AlN system has been studied with considerable detail, mostly in the past decade. Phase relationships, diffusivities, phase transformations and some basic physical properties have been investigated. More limited work has been reported on the SiC-Al_2OC-AlN, SiC-BN and SiC-BeO systems. Several more potential additives are awaiting investigation.

ACKNOWLEDGMENT

The authors thank Y. W. Chang for his kind assistance. This work was supported by the U.S. Department of Energy, Division of Materials Sciences, through the Materials Research Laboratory of the University of Illinois under Contract No. DE-AC02-76ER01198.

---

*Compositions are given in molecular percent, unless otherwise stated. Molecular and weight percents are almost identical in the SiC-AlN system.

References

[1] G. Ervin, Jr., "Silicon Carbide-Aluminum Nitride Refractory Composite," U.S. Pat. No. 3,492,153, North American Rockwell Corp., Jan. 27, 1970.

[2] G. Sukhanek, Yu. M. Tairov and V. F. Tsvetkov, "Estimating the Basic Electrophysical Parameters of Solid Solutions of Silicon Carbide and $A^{III}B^{V}$ Nitrides," *Sov. Tech. Phys. Lett.*, 9 [6] 317-8 (1983).

[3] W. Rafaniello, K. Cho and A. V. Virkar, "Fabrication and Characterization of SiC-AlN Alloys," *J. Mat. Sci.*, 16, 3479-88 (1981).

[4] R. Ruh and A. Zangvil, "Composition and Properties of Hot Pressed SiC-AlN Solid Solutions," *J. Am. Ceram. Soc.*, 65 [5] 260-5 (1982).

[5] M. Shimada, K. Sasaki and M. Koizumi, "Fabrication and Characterization of AlN-SiC Ceramics by High Pressure Hot-Pressing," pp. 466-72 in Proceeding of the First International Symposium on Ceramic Components for Engines, Oct. 17-19, 1983, Hakone, Japan. Edited by S. Somiya, E. Kanai and K. Ando. KTK Scientific Publishers, Tokyo, 1984.

[6] J.-L. Huang, A. C. Hurford, R. A. Cutler and A. V. Virkar, "Sintering Behavior and Properties of SiCAlN Ceramics," *J. Mat. Sci.*, 21 [4] 1448-56 (1986).

[7] W. Rafaniello, M. R. Plichta and A. V. Virkar, "Investigation of Phase Stability in the System SiC-AlN," *J. Am. Ceram. Soc.*, 66 [4] 272-76 (1983).

[8] I. B. Cutler, P. D. Miller, W. Rafaniello, H. K. Park, D. P. Thompson and K. H. Jack, "New Materials in the Si-C-Al-O-N and Related Systems," *Nature*, 275, 534-35 (1978).

[9] M. M. Dobson, "Silicon Carbide Alloys," Research Reports in Materials Science, Series One, P. E. Evans, ed., The Parthenon Press, Carnforth (England), 1986.

[10] K. Tsukuma, M. Shimada and M. Koizumi, "A New Compound $Si_3Al_4N_4C_3$ with the Wurtzite Structure in the System Si_3N_4-Al_4C_3," *J. Mat. Sci. Lett.*, 1, 9 (1982).

[11] A. Zangvil and R. Ruh, "The $Si_3Al_4N_4C_3$ and $Si_3Al_5C_5C_3$ Compounds as SiC-AlN Solid Solutions," *J. Mat. Sci. Lett.*, 3, 249-50 (1984).

[12] A. Zangvil and R. Ruh," Structure and Morphology of HP SiC-AlN Composites," paper presented at the American Ceramic Society 82nd Annual Meeting, Chicago, IL, April 27, 1982. For abstract see paper No. 77-B-80, *Am. Ceram. Soc. Bull.*, 59 [3] 359 (1980).

[13] S. Y. Kuo, Z. C. Jou, A. V. Virkar and W. Rafaniello, "Fabrication, Thermal Treatment and Microstructure Development in SiC-AlN-Al_2OC," *J. Mat. Sci.*, 21 3019-24 (1986).

[14] K. A. Schwetz and A. Lipp, "Sintering and Properties of AlN-Rich Aluminum Nitride-Silicon Carbide Mixed Crystals," Paper No. 49-BP-87, p. 30 in Am. Ceram. Soc. 89th Annual Meeting Abstracts, The American Ceramic Society, Westerville, OH, 1987.

[15] W. F. Knippenberg et al., U.S. Patent No. 3,634,149, 1972.

[16] R. F. Rutz, "Epitaxial Crystal Fabrication of SiC-AlN," U.S. Pat. No. 4,382,837, IBM Corp., May 1983.

[17] N. D. Sorokin, Yu. M. Tairov, and V. F. Tsvetkov, "Study of the Composition of Solid Solutions SiC-AlN and SiC-GaN by the Method of X-ray Spectral Analysis," p. 227 in Abstracts of Reports at the Third All-Union Symposium on Scanning Electron Microscopy and Analytic Methods for Studying Solids [in Russian], Zvenigorod. Nauka, Moscow, 1981.

[18] P. T. B. Shaffer, "Problem in Silicon Carbide Device Development," *Mat. Res. Bull.*, 4, 513-524 (1969).

[19] P. C. Yates "Dense, Submicron Grain AlN-SiC Bodies," U.S. Pat. No. 3,649,310, March, 1972.
[20] G. Schneider, L. J. Gauckler and G. Petzow, "Phase Equilibria in the Si,Al,Be/C,N System", pp. 399-408 in Mat. Sci. Monographs, Vol. 6., edited by P. Vincenzini, Elsevier Publishing Co., Amsterdam, New York and Oxford, 1980.
[21] I. B. Culter, "Solid Solution and Process for Producing a Solid Solution," U.S. Pat. No. 4,141,740, Feb. 27, 1979.
[22] M. Ish-Shalom and A. Zangvil, "In Situ SiCAlN's: Formation and Characteristics of Powders and Compacts," paper presented at the Am. Ceram. Soc. 83rd Annual Meeting, Washington, D.C., May 1981. For abstract see paper No. 74-B-81, *Am. Ceram. Soc. Bull.*, **60** [3] 378 (1981).
[23] A. Zangvil and R. Ruh, "Phase Relationships in the SiC-AlN System," submitted to the American Ceramic Society.
[24] S. Y. Kuo, A. V. Virkar, and W. Rafaniello, "Modulated Structures in SiC-AlN Ceramics," *J. Am. Ceram. Soc.*, **70** [6] C-125-C-128 (1987).
[25] M. M. Patience, P. England, D. P. Thompson and K. H. Jack, "Ceramic Alloys of Silicon Carbide with Aluminum Nitride and Nitrogen," presented at the Intl. Symp. on Ceramic Components for Engines, Oct. 17, 1983, Hakone, Japan.
[26] Y. A. Vodakov, and E. N. Mokhov, "Diffusion and Solubility of Impurities in Silicon Carbide," pp. 508-19 in Silicon Carbide-1973, edited by R. C. Marshall, J. W. Faust, Jr., and C. E. Ryan, University Press, S.C., 1974.
[27] Y. Tajima and W. D. Kingery, "Solid Solubility of Aluminum and Boron in Silicon Carbide," *J. Am. Ceram. Soc.*, **65** C 27-C 29 (1982).
[28] A. Zangvil and R. Ruh, "Solid Solutions and Composites in the SiC-AlN and SiC-BN Systems," *J. Mat. Sci. and Engineering*, **71** 159-164 (1985).
[29] R. Ruh and A. Zangvil, "Hot-Pressed SiC-AlN Mixtures: Composition and Properties of Solid Solutions," paper presented at the Am. Ceram. Soc. 83rd Annual Meeting, Washington, D.C., May 1981. For abstract see paper No. 115-B-81, *Am. Ceram. Soc. Bull.*, **60** [3] 381 (1981).
[30] A. Zangvil, R. Ruh and M. Ish-Shalom, "TEM and Analytical Microscopy of Materials in the System SiC-AlN," paper presented at the Am. Ceram. Soc. 83rd Annual Meeting, Washington D.C. May 1981. For abstract see paper No. 116-B-81, *Am. Ceram. Soc. Bull.*, **60** [3] 381 (1981).
[31] A. Zangvil and R. Ruh, "Analytical Electron Microscopy of SiC-AlN Diffusion Couples," paper presented at the Am. Ceram. Soc./TMS-AIME Joint Meeting, Louisville, KY, Oct. 1981. For abstract see paper No. 96-B-81F, *Am. Ceram. Soc. Bull.*, **60** [8] 855 (1981).
[32] W. Rafaniello, Ph.D. dissertation, University of Utah, Salt Lake City, UT, 1984.
[33] D. Pandey and P. Krishna, "Mechanism of Solid State Transformations in Silicon Carbide," pp. 198-206 in Silicon Carbide-1973, edited by R. C. Marshall, J. W. Faust, Jr., and C. E. Ryan, University Press, S.C., 1974.
[34] N. W. Jepps and T. F. Page, "Polytypic Transformations in Silicon Carbide," pp. 259-306 in Crystal Growth and Characterization of Polytype Structures, edited by P. Krishna, Pergamon Press, Oxford, 1983.
[35] R. Ruh, A. Zangvil and J. Barlowe, "Elastic Properties of SiC, AlN, and Their Solid Solutions and Particulate Composites," *Am. Ceram. Soc. Bull.*, **65** [10] 1368-74 (1985).
[36] Z. C. Jou, S.-Y. Kuo, and A. V. Virkar, "Elevated-Temperature Creep of Silicon Carbide-Aluminum Nitride Ceramics: Role of Grain Size," *J. Am. Ceram. Soc.*, **69** [11] C-279-C-281 (1986).

[37] L. D. Bentsen, D. P. H. Hasselman, and R. Ruh, "Effect of Hot-Pressing Temperature on the Thermal Diffusivity/Conductivity of SiC/AlN Composites," *J. Am. Ceram. Soc.*, 66 [3] C-40 (1983).

[38] B. Sandberg, Ph.D. dissertation, University of Trondheim, 1981.

[39] P. T. B. Shaffer and E. D. Whitney, "Silicon Carbide Containing Boron and Nitrogen in Solid Solution," U.S. Pat. No. 3,554,717, the Carborundum Co., Niagara Falls, NY, Jan. 12, 1971.

[40] M. I. Sokhor and L. I Fel'dgun, "Polymorphic Transitions in the Composite Material SiC-BN," Izvestiya Akademii Nauk SSSR, *Neorganicheskie Materialy*, 12 [10] 1877-78 (1976).

[41] Y. Murata and R. H. Smoak, "Densification of Silicon Carbide by the Addition of BN, BP and B4C, and Correlation to their Solid Solubilities," pp. 382-399 in Proc. of International Symposium on Factors in Densification and Sintering of Oxides and Non-Oxide Ceramics, Oct. 3, 1978, Hakone, Japan.

[42] R. Ruh and A. Zangvil, "Phase Studies of SiC-BN Composites," paper presented of the Amer. Ceram. Soc. 87th Annual Meeting, Cincinnati, OH, May 6, 1985. For abstract see paper No. 30-BP-85, *Am. Ceram. Soc. Bull.*, 64 [3] (1985).

[43] R. Ruh, A. Zangvil and R. R. Wills, "Phase and Property Studies of SiC-BN Composites," *Advanced Ceramic Materials*, in press.

[44] P. G. Valentine, "Strength and Thermal Shock Resistance of SiC-BN Composites," Ph.D. dissertation, Air Force Institute of Technology Air University, 1983.

[45] P. G. Valentine, A. N. Dalazotto, R. Ruh and D. C. Larsen, "Thermal Shock Resistance of SiC-BN Composites," *Advanced Ceramic Materials*, 1 [1] 81-87 (1986).

[46] R. Ruh, L. D. Bentsen and D. P. H. Hasselman, "Thermal Diffusivity Anisotropy of SiC/BN Composites," *J. Am. Ceram. Soc.*, 67 [5] C-83-C-84 (1984).

[47] Y. Takeda, K. Usami, K. Nakamura, S. Ogihara, K. Maeda, T. Miyoshi, S. Shinozaki, and M. Ura, "Grain Boundary Structure of Highly Resistive SiC Ceramics with High Thermal Conductivity," pp. 253-9 in Additives and Interfaces in Electronic Ceramics (Advances in Ceramics, Vol. 7), edited by M. F. Yan and A. Heuer, The American Ceramic Society, Columbus, OH, 1983.

[48] Y. Takeda and K. Maeda, "Mechanical Properties of SiC Ceramic with Addition of BeO," Paper No. 60-BP-87, p. 33 in Am. Ceram. Soc. 89th Annual Meeting Abstracts, Westerville, OH, 1987.

[49] Y. Takeda and K. Nakamura, "Effects of Additives on Microstructure and Strength of Dense Silicon Carbide," pp. 215-9 in Proc. 23rd Japan Congress on Materials Research, The Society of Materials Science, Kyoto, Japan, 1980.

[50] K. Maeda, T. Miyoshi, Y. Takeda, K. Nakamura, S. Ogihara, and M. Ura, "Grain-Boundary Effect in Highly Resistive SiC Ceramics with High Thermal Conductivity," pp. 260-8 in Additives and Interfaces in Electronic Ceramics (Advances in Ceramics, Vol. 7), edited by M. F. Yan and A. Heuer, The American Ceramic Society, Columbus, OH, 1983.

[51] S. Ogihara, K. Maeda, Y. Takeda, and K. Nakamura, "Effect of Impurity and Carrier Concentrations on Electrical Resistivity and Thermal Conductivity of SiC Ceramics Containing BeO", *J. Am. Ceram. Soc.*, 68 [1] C-16-C-18 (1985).

[52] S. Shinozaki, J. Hangas, K. Maeda, and A. Soeta, "Enhanced Formation of 4H Polytype in SiC Materials," these proceedings.

[53] A. Zangvil, Y. W. Chang and R. Ruh, "Effect of BeO on the Microstructure and Phase Stability of SiC," these proceedings.

[54]G. K. Safaraliev, G. K. Sukhanek, Yu. M. Tairov and V. F. Tsvetkov, "Criterion for the Formation of Solid Solutions Based on Silicon Carbide," *Inorganic Materials*, **22**, [11] 1610–12 (1986).

[55]J. St. John and A. N. Block, "Quantum-Defect Electronegativity Scale for Nontransition Elements," *Phys. Rev. Lett.*, **33**, [18] 1095–96 (1987).

Effect of BeO on the Microstructure and Phase Stability of SiC

Avigdor Zangvil and Yeu-Wen Chang

Materials Research Laboratory and
Department of Materials Science and Engineering
University of Illinois at Urbana-Champaign
Urbana, IL 61801

Robert Ruh

Air Force Wright Aeronautical Laboratories
Wright-Patterson AFB, OH 45433

Abstract

SiC-BeO materials containing 1.6 to 75 mole % BeO were hot pressed under various conditions, and the resultant microstructures and phase compositions were analyzed by X-ray diffraction and analytical electron microscopy techniques. Extensive SiC-BeO mutual solid solubility was not detected, but some solubility of BeO in SiC at 2000°C and above was indicated by an apparent stabilization of the 6H polytype, relative to the 4H and 15R polytypes, adjacent to BeO grains. BeO was found to transform locally into Be_2C, and thin glassy layers were found at some SiC/SiC grain boundaries. Hot pressing under nitrogen atmosphere resulted in solution of considerable amounts of nitrogen in the BeO structure.

Introduction

Beryllium oxide has been proposed as a sintering aid to silicon carbide[1]. Silicon carbide with 3.2% BeO* was densified to 97% of theoretical density by hot-pressing at 2040°C for 1 hour, and had a flexural strength of 450 MPa. However, the fact that BeO is a toxic material dictates that only applications in which it is indispensable be considered, and that minimum amounts be used. One material which answers these requirements is the recently developed hot-pressed α-SiC with 3.2% BeO, which, under appropriate processing conditions, possesses extremely high electrical resistivity (> 4×10^{13} Ω cm) and a thermal conductivity higher than that of Al or even pure BeO[2]. The various thermal and electrical properties of the materials have been studied, along with sophisticated measurements of current-voltage characteristics[3]. It has been suggested that low impurity concentration within the grains causes the high electrical resistivity, while Be segregation at grain boundaries is responsible for the high thermal conductivity[2,3,4]. Microstructural effects of the BeO were also reported[2], and a material with similar properties but lower BeO content (1.6%) was recently reported by the same group[5]. Safaraliev et al.[6] predicted high mutual solubilities of SiC and BeO, based on St. John and Bloch's[7] electronegativity vs. hybridization diagram.

We have been interested in the SiC-BeO system as part of our studies of the alloying of SiC with various ceramic compounds. These studies are reviewed elsewhere[8]. The purpose of the present study was to investigate mutual solid solubilities of SiC and BeO at high temperatures (2050 to 2280°C), interphase reactions, and the effect of BeO on the microstructure and phase assemblages of a variety of SiC-BeO materials.

Experimental Procedures

Samples used in this study are listed in Table I. The raw powders for the H1 and H2 material were a high purity α-SiC powder and a >99.5% BeO powder, both having a grain size of about 2 μm^2. Powders for the PML samples were H.C. Starck SiC and A.O. Mackay BeO, both >99% pure and <1 μm in size. Powders for the NBD1 sample were 99% pure SiC, 3μm average particle size, and Brush Wellman Grade UOX BeO, <1μm. Samples were hot-pressed at 2050° to 2280°C for up to 1 h under about 30 MPa. Hot pressing was usually done in vacuum, except that argon was used at the highest temperature. One sample was hot-pressed under nitrogen atmosphere. A considerable weight loss, believed to be mostly due to the evaporation of BeO, was observed in samples hot-pressed at 2200°C and above.

Table I. Hot-Pressed SiC-BeO Compositions.

Sample No.	Source*	%BeO (mol.%)	SiC Type	Conditions	SiC
H1	1	1.6	α(6H)	2050°C/1h/30MPa/Vac.	N.A.
H2	1	3.2	α(6H)	2050°C/1h/30MPz/Vac.	N.A.
PML1	2	20	β	2100°C/1h/28MPa/Vac.	4H,6H,(3C)
PML2	2	50	β	2100°C/2h/28MPa/Vac.	6H,4H,(3C)
PML3	2	50	β	2200°C/.6h/28MPa/Vac.	3C,4H,6H
PML4	2	20	β	2200°C/1h/28MPa/Vac.	4H,6H,3C
PML5	2	50	β	2220°C/1h/28MPa/Vac.	N.A.
PML6	2	75	β	2280°C/.7h/28MPa/Vac.	4H,6H,3C
PML7	2	75	β	2100°C/1h/28MPa/Vac.	3C,4H,6H
NBD1	3	5.5	β	2050°C/1h/33MPa/N2	N.A.

*1Hitachi Ltd., Hitachi, Japan
2Powder Metallurgical Laboratory, Max-Planck-Institute, Stuttgart, FRG
3National Beryllia Division, General Ceramics, Inc., Haskell, NJ.

Samples were sliced and prepared for X-ray diffraction (XRD) and transmission electron microscopy (TEM) by conventional techniques. Some TEM samples exhibited charging due to the high electrical resistivity, but coating with a conducting layer was usually avoided. XRD was done at room temperature using Ni-filtered CuKα radiation on commercial equipment.[†] Scanning transmission electron microscopy (STEM) and TEM studies were done on two instruments[‡] equipped with electron energy loss (EELS) detectors[§] and energy dispersive X-ray spectrometers (EDS).[¶] The EDS of the EM420 had an ultra-thin window capability, which enabled analysis of carbon, nitrogen and oxygen.

RESULTS AND DISCUSSION

Figure 1 shows an area in a SiC-5.5% BeO, hot-pressed in nitrogen (NBD1). The microstructure is generally uniform. The material contained residual BeO grains, which could not be distinguished from the SiC without further analysis (below), and some impurity inclusions. Thin layers of a glassy phase were also observed at some SiC/SiC grain boundaries of the sample, using dark field imaging. Another phenomenon was the appearance of Be_2C within some BeO grains adjacent to the SiC. This is shown in Fig. 2. Energy loss spectroscopy showed that the intermediate area (#2) contained Be, O and C, and its appearance suggested that Be_2C particles formed within the BeO by diffusion of carbon into it. The source of the carbon, whether the SiC itself or some excess carbon in the material, could not be verified.

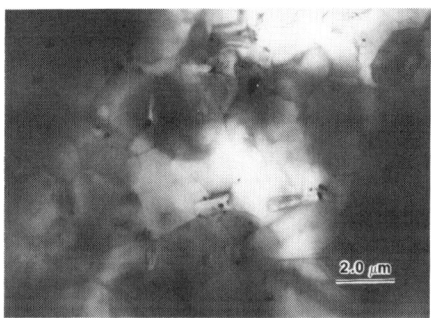

Fig. 1. A typical uniform microstructure of SiC-5.5% BeO, hot-pressed in nitrogen, 2050°C, 1 h, 33 MPa.

Thermodynamically, however, the reaction

$$3SiC + 2BeO \rightarrow 3Si + Be_2C + 2CO$$

is not favorable[9], so it is likely that excess carbon is indeed responsible for the formation of Be_2C. This SiC -3.2% BeO, hot-pressed in vacuum (H2), exhibited more Be_2C areas. The reason for the difference may be the different atmosphere used in the processing. In fact, most of the BeO grains in sample NBD1, hot-pressed in N_2, were found to contain considerable amounts of nitrogen,

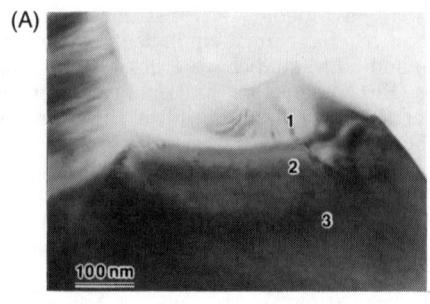

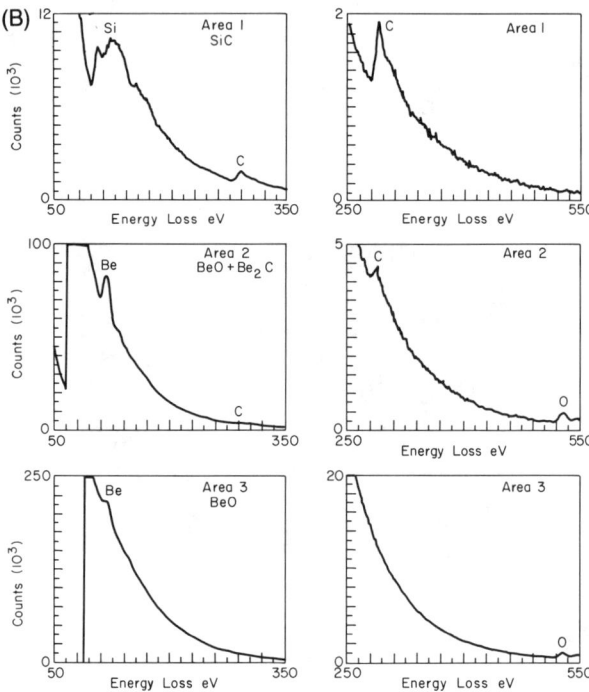

Fig. 2. A BeO + Be_2C area formed within the BeO grain adjacent to SiC, in the same sample as in Fig. 1. (A) is the bright field TEM micrograph; (B) shows EELS spectra in the three areas, as indicated.

assumably as $Be(O,N)_x$ solid solutions. These grains preserved the BeO wurtzite structure. An example is shown in Fig. 3, where the analysis was accomplished by ultra-thin-window EDS. It is interesting that Be_2C was not observed in BeO grains which contained nitrogen, indicating that the diffusion and reaction of carbon with

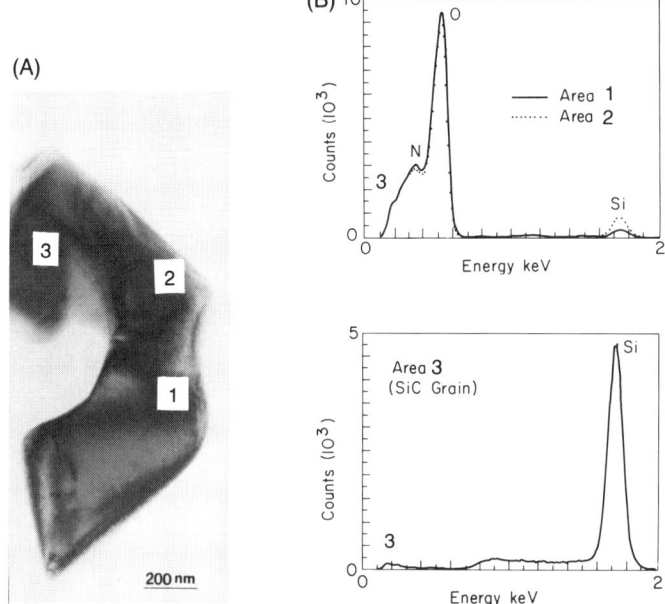

Fig. 3. A nitrogen-enriched BeO grain adjacent to a SiC grain (3) in the same sample as above. (A) is the bright field micrograph; (B) shows ultra-thin window EDS spectra, indicating that the area 2, which is closer to SiC, is richer in Si and somewhat lower in nitrogen compared to area 1.

BeO may be inhibited by the nitrogen. The occurrence of $Be(O,N)_x$ may be compared with the recent report on the formation of $Si_x(C,N)$ (x smaller than unity in both cases) with the wurtzite structure, essentially through a reaction of silicon nitride and carbon[10].

XRD analysis of the PML samples provided approximate polytype assemblages by order of abundance, as shown in Table I. Comparing PML 1,2 and 7, all hot-pressed at 2100°C in vacuum, one observes that the 6H and the cubic β-SiC (3C) becomes more abundant relative to the 4H structure with increasing BeO content. The 3C structure becomes more abundant with increasing BeO also at 2200°C, when one compares PML4 and PML3, but this tendency seems to be reversed at 2280°C-PML6 contained 75% BeO but showed 4H, 6H and 3C as the SiC structures, in this order.

TEM/STEM analyses provided some support to the XRD results. In Fig. 4, from a SiC-20% BeO sample (PML1), one observes that the 6H polytype of SiC is adjacent to the BeO, while 4H is further away, with a mixed 6H + 4H structure in the intermediate area. Fig. 4B shows windowless EDS spectra from the two structures which show that oxygen may be somewhat higher near the BeO, in the 6H-SiC structure. Fig. 5 shows a 15R-SiC adjacent to BeO, with 4H further away.

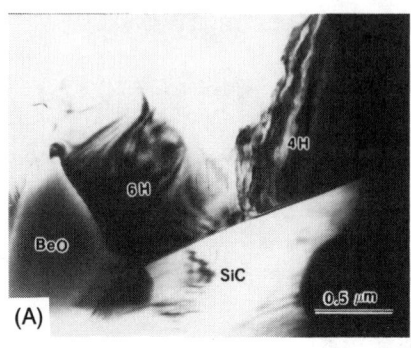

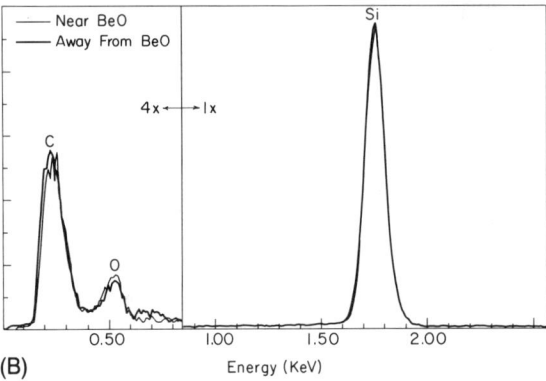

Fig. 4. 6H and 4H polytypes in a SiC grain of a SiC-20% BeO sample, hot-pressed at 2100°C and 28 MPa for 1 h, in vacuum (A). The windowless EDS spectra are shown in (B).

The structures are identified by their corresponding selected area electron diffractions (B,C). Since 15R is more cubic than 4H, this result seems in agreement with that of Fig. 4 and with the XRD results. In other words, polytypes with lower hexagonality seem to be stabilized by the BeO.**

It should be noted that this result is not in agreement with a concurrent report[11], which indicates that 4H formation is enhanced relative to 6H in materials similar to our H1 and H2 samples. Differences in impurity, atmosphere and firing temperature are likely to be responsible for the difference in results, as these parameters are known to affect the polytype structure. Figure 5D shows the lattice fringe image of the 4H→15R transformation zone. It is similar in appearance to the diffusion-controlled transformation zones observed in SiC-AlN and SiC-BN materials[12,13].

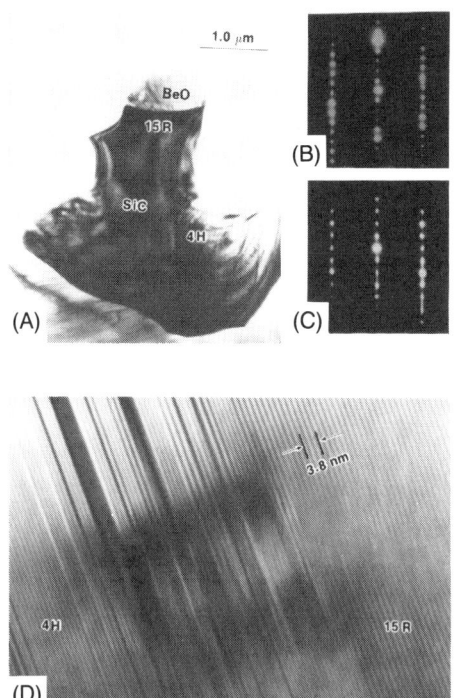

Fig. 5. 15R and 4H polytypes in another grain of the same sample as in Fig. 4, showing 15R is adjacent to BeO (A). The selected area electron diffractions for the 15R and 4H region are shown in (B) and (C), respectively, whereas (D) is a lattice fringe image of the 4H→15R transformation zone.

Summary

SiC-BeO samples with 1.6 to 75 mol.% BeO were hot-pressed under a variety of conditions. Mutual solid solubility is not extensive even above 2200°C. The effect on polytype stability seems to be in the direction of the cubic and 6H structures, judged from XRD information. TEM suggests that 6H and 15R are stabilized relative to 4H, i.e., lower hexagonality polytypes are stabilized.

Samples with 5.5% BeO or less showed Be_2C formation within the BeO, adjacent to the interface with SiC. Small amounts of grain boundary glassy phase were also observed in these samples, indicating possible liquid phase sintering. In a sample hot-pressed under nitrogen atmosphere, a $Be(O,N)_x$ solid solution forms, with the BeO wurtzite structure.

Acknowledgments

The authors wish to thank Drs. Y. Takeda and K. Maeda of the Hitachi Research Laboratory and Mr. J. Lynch of the National Beryllia Division for providing samples for this study. We also thank Dr. J. Lorenz and Professor G. Petzow of the Max-Planck Institute in Stuttgart for allowing us to use their facilities and providing us assistance in sample preparation.

This work is supported by the U.S. Department of Energy, Division of Materials Sciences, through the Materials Research Laboratory of the University of Illinois under Contract No. DE-AC02-76ER01198.

---

*Molecular percent is used throughout this work.
†Model XRG 3000, Philips Electronic Instruments, Mount Vernon, NY.
‡Models EM430 (300 Kv) and EM420 (120 Kv), Philips, Eindhoven, The Netherlands.
§Gatan, Inc., Warrendale, PA.
¶EDAX International, Prairie View, IL.
**3C, 6H, 15R and 4H are 0%, 33%, 40% and 50% hexagonal, respectively.

References

[1] Y. Takeda and K. Nakamura, "Effects of additives on microstructure and strength of dense silicon carbide," pp. 215-9 in Proc. 23rd Japan Congress on Materials Research, The Society of Materials Science, Kyoto, Japan, 1980.

[2] Y. Takeda et al., "Grain boundary structure of highly resistive SiC ceramics with high thermal conductivity," pp. 253-9 in Additives and Interfaces in Electronic ceramics (Advances in ceramics, Vol. 7), M. F. Yan and A. Heuer, eds., The American Ceramic Society, Columbus, Ohio, 1983.

[3] K. Maeda et al., "Grain-boundary effect in highly resistive SiC ceramics with high thermal conductivity," ibid, pp. 260-8.

[4] S. Ogihara, K. Maeda, Y. Takeda and K. Nakamura, "Effect of impurity and carrier concentrations on electrical resistivity and thermal conductivity of SiC ceramics containing BeO," *J. Am. Ceram. Soc.*, **68** [1] C-16-C-18 (1985).

[5] Y. Takeda and K. Maeda, "Mechanical properties of SiC ceramic with addition of BeO," Paper presented at the 89th Annual Meeting of the American Ceramic Society, Pittsburgh, Pennsylvania, April 1987.

[6] G. K. Safaraliev, G. K. Sukhanek, Y. M. Tairov and V. F. Tsvetkov, "Criterion for the formation of solid solutions based on silicon carbide," *Inorganic Materials*, **22** [11] 1610-12 (1986).

[7] J. St. John and A. N. Bloch, "Quantum-defect electronegativity scale for nontransition elements," *Phys. Rev. Lett.*, **33** [18] 1095-96 (1974).

[8] A. Zangvil and R. Ruh, "Alloying of SiC with other ceramic compounds (A Review)," these proceedings.

[9] K. Negita, "Effective sintering aids for SiC ceramics: Reactivities of SiC with various additives," *J. Am. Ceram. Soc.*, **69** [12] C-308-C-310 (1986).

[10] M. M. Dobson, "Silicon carbide alloys," Research Reports in Materials Sciences, Series One, P. E. Evans, ed., The Parthenon Press, Carnforth (England), 1986.

[11] S. Shinozaki, J. Hangas, K. Maeda and A. Soeta, "Enhanced formation of 4H polytype in SiC materials," these proceedings.

[12] A. Zangvil and R. Ruh, "Phase relationships in the SiC-AlN system", submitted to the American Ceramic Society.

[13] R. Ruh, A. Zangvil and R. R. Wills, "Phases and property studies of SiC-BN composites", *Advanced Ceramic Materials.*, in press.

Synthesis and Characterization of HSC Silicon Carbide

W. M. Goldberger, A. K. Reed and R. Morse

Superior Graphite Company
120 South Riverside Plaza
Chicago, IL 60606

Abstract

HSC Silicon Carbide is a beta crystalline form of silicon carbide made by a continuous electrothermal furnace process developed for large scale production by the Superior Graphite Company. Similar to the conventional Acheson process, the reactants are silica sand and solid carbon. However, being a continuous rather than batch process, the reaction conditions differ from those in the Acheson process and the HSC process yields a microcrystalline product especially suited for making fine microgrits and ultra-fine ceramic grade silicon carbide powders.

Background

Silicon carbide grain and grit for refractory materials and abrasives have been produced almost entirely by the Acheson electric furnace process. The Acheson process, used commercially since before the turn of the century, is a batch process involving the charge-resistance heating of a mixture of silica sand with solid carbon to a temperature above 2000°C.[1] Reaction times are long and the silicon carbide that is formed varies from very large crystals of high purity that form near the central core of the charge mixture, to a finer grained mixture of silicon carbide and unreacted carbon and silica. The increasing interest during the past ten years in high purity silicon carbide microgrits and ultra-fine powders for advanced materials has led to development of a number of new processes for continuous rather than batch production of silicon carbide. The HSC Silicon Carbide Process is one of these newer methods that has been developed by the Superior Graphite Company and operated commercially since 1981.

The HSC Process is a continuous version of the Acheson Process. Similar to the Acheson furnace method, sand and petroleum coke are the basic reactants. Each of these materials is fed continuously to the HSC furnace and heated electrothermally. The product silicon carbide is continuously removed. Being a fully continuous and steady state operation, the process conditions can be carefully controlled. The HSC process can be operated at high throughput for large scale commercial production.

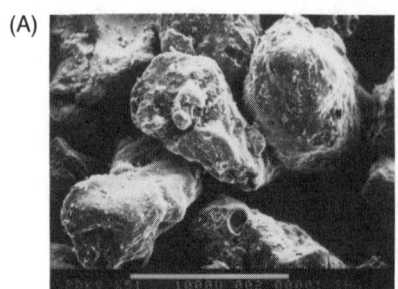
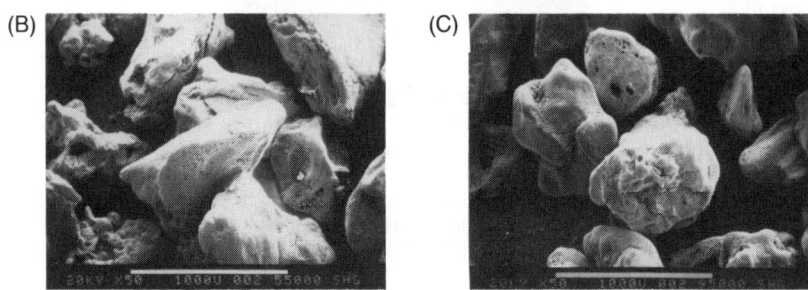

Fig. 1. The Development of Silicon Carbide Coating on Particles of Petroleum Coke—50× (A) Coke Particles Before Silicon Carbide Reaction; (B) Partially Reacted Coke—55% SiC; (C) Fully Reacted Coke—95% SiC.

SILICON CARBIDE FORMATION MECHANISM IN THE HSC PROCESS

It is generally accepted that reduction of silica with solid carbon to form silicon carbide occurs through a sequence of reactions that involve the formation of intermediate volatile forms of silicon[2-5]. The following reactions are an example of one proposed sequence:

$$C + SiO_2 \quad SiO + CO \tag{1}$$

$$2C + SiO \quad SiC + CO \tag{2}$$

giving the net overall reaction,

$$3C + SiO_2 \quad SiC + 2CO \tag{3}$$

Although silicon carbide does form at temperatures as low as 1400°C, the HSC Process is operated at 1900°C to achieve a rapid reaction rate and a high degree of conversion of the reactants. The process can be operated at higher temperatures;

however, silicon losses begin to occur as the temperature approaches the decomposition temperature of silicon carbide.

Clear evidence that the reaction proceeds via an intermediate volatile form of silicon is shown in the photomicrographs of Figures 1 and 2. The sequence of photographs in Figure 1 illustrates the development of a coating of silicon carbide on the external surfaces of the particles of petroleum coke as the coke is converted to silicon carbide.

Silicon carbide also forms within the pores of the coke particles as shown in the series of photomicrographs taken of polished sections of particles at different levels of conversion.

As the overall reaction of eq. (3) shows, two of every three reacting carbon atoms are gasified and leave the particle as carbon monoxide. The reaction product therefore has an inherently porous structure as evident from the microsections of Figure 2.

(A) 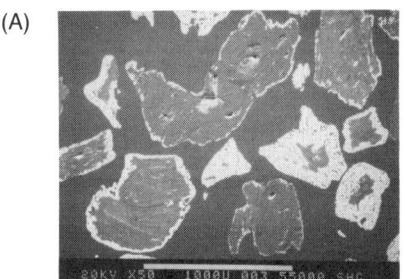 (B)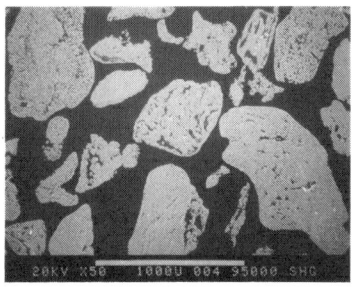

Fig. 2. The Formation of Silicon Carbide as a Coating and within the Pore Structure of Particles of Petroleum Coke—50× (A) Partially Reacted; (B) Fully Reacted—75% SiC.

A major factor influencing the rate of silicon carbide formation is the characteristic of the petroleum coke. A coke having a high degree of macroporosity is preferred. BET surface area is not a valid criteria in that a high BET surface coke having a high percentage of micropores can be less accessible to the silicon bearing gases than a lower surface area coke that has a larger pore structure.

The overall rate of reaction is, of course, proportional to the available carbon and as the conversion of the coke increases, the overall rate of silicon carbide formation decreases. Thus, there is a practical limit to the extent of conversion for a continuous process. For the HSC process, the practical upper limit is about 95 percent silicon carbide content. The balance is mainly free carbon. Production of low carbon content products is done by various post treatments using conventional methods of carbon removal, size reduction and chemical purification. The processing steps applied to produce the basic commercial grades of HSC silicon carbide is given in Figure 3.

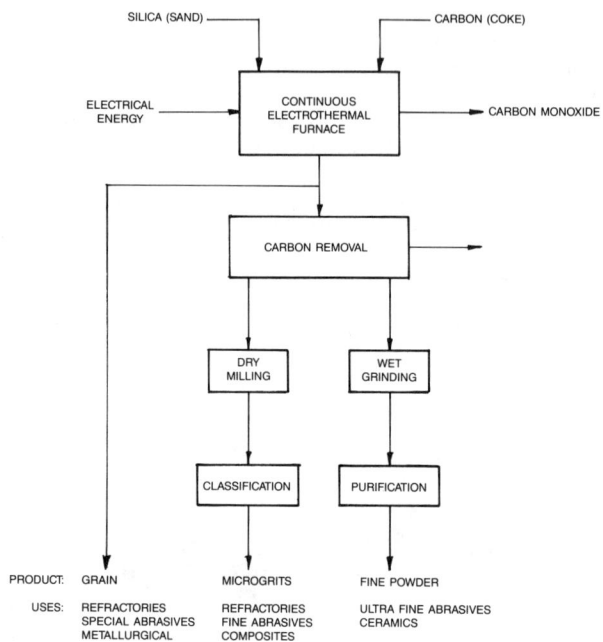

Fig. 3. General Flowsheet for the Production of HSC Silicon Carbide Products.

The run-of-furnace product is a free flowing grain generally minus 3×10^{-3}m (1/8-inch). It can be made to contain any silicon carbide content up to about 95 percent by adjusting the relative feed rates of silica sand and petroleum coke. Feed rate is also adjusted to provide adequate retention time in the reaction zone to achieve the desired degree of conversion.

The removal of free carbon from the run-of-furnace product can be done either by combustion or by physical separation. The choice of method depends on end use and the grain size of the desired product. Physical separation requires milling the run-of-furnace grain to liberate the unreacted free carbon from silicon carbide. Liberation is essentially complete when the grain is reduced to minus 270 Tyler mesh (0.037×10^{-3}m) particle size. Froth flotation is particularly effective in removal of free carbon because of the substantial difference in the wettability characteristics of free carbon and silicon carbide.

Because the HSC furnace is operated with free carbon present, the run-of-furnace product contains essentially no free silica. However, some surface oxidation of the silicon carbide can occur when carbon is removed by combustion. Production of powder for ceramic use requires comminution to sub-micron size and this can also cause some surface oxidation. Oxide levels can be reduced by chemical methods. Any contamination due to the comminution step is also removed by wet chemical purification methods.

CHARACTERIZATION OF HSC SILICON CARBIDES

Run-of-Furnace (ROF Grade)

The HSC products include the direct run-of-furnace grains and the various milled, classified and chemical purified grits and powders made by post processing of the furnace product. The run-of-furnace grain shown in Figures 1 and 2 comprises agglomerated very fine crystallites of the beta form silicon carbide. The beta crystal structure is verified by x-ray diffraction (Fig. 4). The ultimate crystallite size determined by line broadening is calculated to be in the range 350-700 Å units. A typical size and chemical analyses for the direct furnace run products are given in Tables 1 and 2.

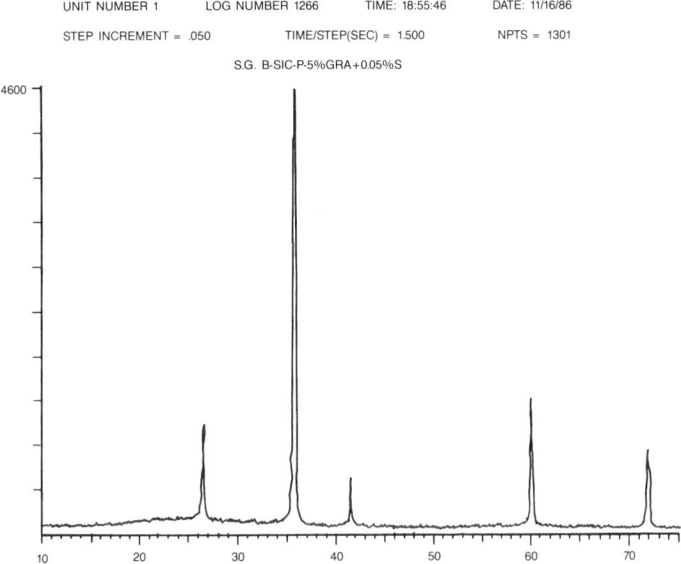

Fig. 4. X-Ray Diffraction of HSC Silicon Carbide Shows Beta Crystal Structure.

The bulk density and particle density of the ROF product increases with increasing silicon carbide content as would be expected. This is shown in Figure 5.

It should be noted that the particle density data shown in Figure 5 were obtained by liquid pycnometry and indicate particle densities approaching the theoretical density of silicon carbide (3.21 g/cc) at the higher silicon carbide contents. However, the actual particle is a lightly sintered agglomerate with substantial internal porosity as indicated by the relatively low bulk density for a silicon carbide grain.

Table 1. Typical Size Distribution of HSC Run-of-Furnace (ROF) Silicon Carbide Grain.

Tyler Mesh Size	Weight Percent Retained
Plus 20	35
20 × 35	45
35 × 48	12
48 × 65	5
Minus 65	3

Table 2. Specification and Typical Chemical Analyses of HSC Run-of-Furnace (ROF) Silicon Carbide Grain.

Component or Element	Specification Percent
Silicon Carbide	55–95
Iron	0.5 Max
Silica	0.5 Max
Silicon (Metallic)	0.1 Max
Sulfur	1.0 Max
Carbon (Free)	Balance from Above
	Parts Per Million
Aluminum	200
Vanadium	100
Titanium	50
Zirconium	50
Calcium	10
Magnesium	10

HSC Microgrits

The ROF grain is quite friable and is readily broken, for example by air milling, into smaller grains or particles of generally in the size range 5–25 micron size as shown in Figure 6.

It can be noted that these particulate are crystalline in nature with a somewhat platelet like character. The particle size distribution of the HSC microgrits depends on the methods used for comminution and classification. The range in particle size distribution is indicated in Figure 7.

Typical chemistry for these products is given in Table 2.

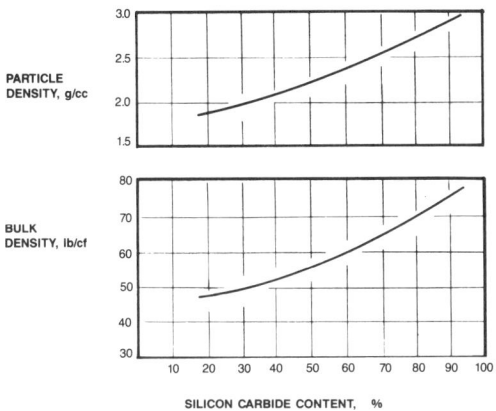

Fig. 5. The Variation of Density of HSC Run-of-Furnace (ROF) Grain with Silicon Carbide Content.

Fig. 6. Agglomerated Crystallites of HSC Silicon Carbide Obtained by Milling the Run-of-Furnace Grain—5000×.

Silicon Carbide

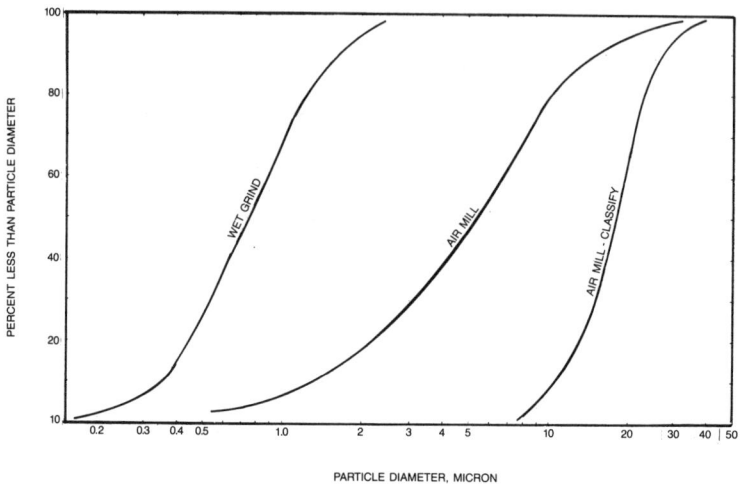

Fig. 7. Particle Size Distribution of Various Grades of HSC Silicon Carbide.

Ultra Fine HSC Powder

Production of shaped articles of silicon carbide by hot pressing and sintering methods requires powders of ultra-fine particle size and a high degree of chemical purity. Particle sizing with an average of less than 1 micron having a BET surface area greater than 10 square meters per gram are generally specified for sintering. Powder that meets these criteria is made by wet grinding the HSC run-of-furnace grain after carbon removal. Because of the hardness and highly abrasive characteristic of silicon carbide, wet milling to ultra-fine powder can cause contamination of the product with components from the milling media and mill walls. These are removed by a combination of chemical purification procedures and wet classification methods. The resulting properties of the ultra-fine HSC powder are listed in Table 3.

The HSC powder is quite stable with regard to oxidation with no significant oxidation occurring below 800°C. The weight gain shown in the TGA diagram of Figure 8 corresponds to 4.88 percent oxidation when the powder is heated in air to 1200°C and is typical.

High surface area silicon carbide powders do adsorb oxygen and possibly other gases from the atmosphere at ordinary temperature. The extent of adsorbed surface oxygen is directly related to the surface area of the powder. The oxygen content of different production lots of HSC silicon carbide powder is indicated in Figure 9.

Table 3. Typical Chemical Analyses of Processed Grades of HSC Silicon Carbide.

Element	HSC Grade		
	ROF*	Microgrit†	Ultra Fine Powder‡
	Analyses in Percent		
O	<0.1	0.15	0.75
N	0.2	0.2	0.21
C	0.6	0.6	0.85
	Analyses in Parts Per Million		
Al	60	160	180
Ca	10	170	nd
Cu	nd	nd	30
Fe	70	100	160
Ni	20	15	80
Ti	50	130	80
V	100	250	90
Zr	100	70	nd

Note: Elements not shown are below detection limit.
      nd = not determined
* Run-of-Furnace Grain After Removal of Free Carbon
† Milled Run-of-Furnace - All Minus 325 Mesh
‡ Wet Milled and Chemically Purified. Average Particle Diameter 0.7 Micron with 15.0 sq. meter/g BET Surface Area

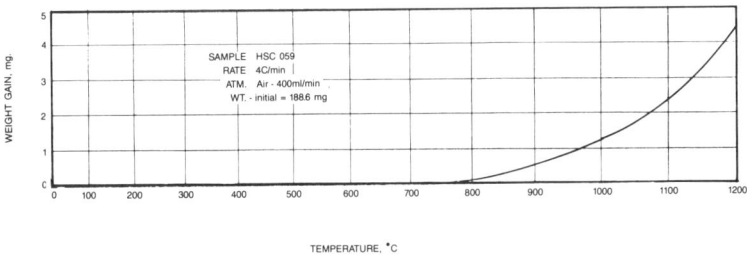

Fig. 8. Oxidation of HSC Ceramic Grade Powder Heated in Air in TGA Apparatus.

Silicon Carbide

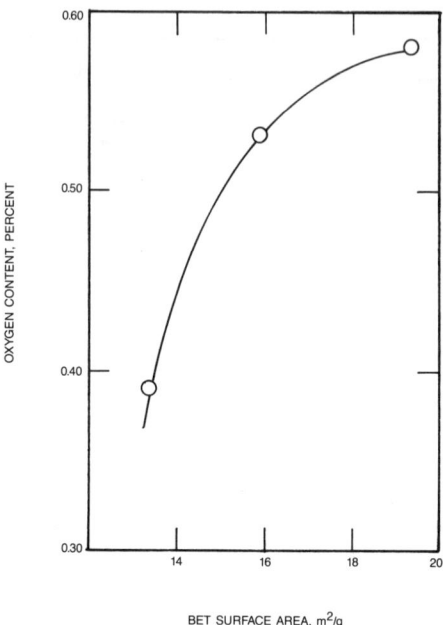

Fig. 9. The Oxygen Content of HSC Ceramic Grade Powder with BET Surface Area.

The loss in weight observed when heating the powder occurs mainly below 500°C and very little difference was noted in the measured weight losses when heated in hydrogen or nitrogen (Figure 10).

It is concluded that the oxygen contained in the HSC powder is mainly chemisorbed rather than chemically bound. Further evidence of room temperature adsorption is also seen in Figure 10 which shows the gain in weight of the heat treated and cooled sample after exposure to air at room temperature.

Summary

An electrothermal process has been developed to produce silicon carbide grain from silica sand and petroleum coke as a continuous version of the Acheson furnace process. The HSC process, however, is operated at conditions which yield silicon carbide in the beta microcrystalline form. The furnace product is a relatively coarse grain comprising agglomerates of silicon carbide crystallites and free carbon. This grain is friable and is readily processed to remove free carbon and to make sized microgrits and powders especially suited for composites and consolidated engineering ceramic materials.

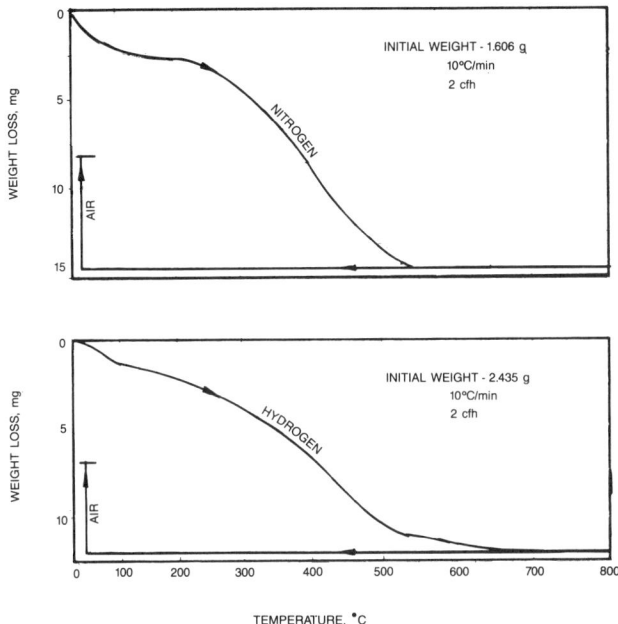

Fig. 10. Change in Weight of HSC Ceramic Grade Powder in Various Gas Atmospheres.

References

[1] E. G. Acheson, Production of Artificial Crystalline Carbonaceous Materials, 492,767 U.S. (February 28, 1893).

[2] J-G Lee and I. B. Cutler, Formation of Silicon Carbide from Rice Hulls. *Ceramic Bulletin*, **54**, 2, 195 (1975).

[3] F. Viscomi and L. Himmel, Kinetic and Mechanistic Study on the Formation of Silicon Carbide from Silicon Flour and Coke Breeze. *Journal Metal*, **30**, 6, 21 (1978).

[4] B. W. Jong, Formation of Silicon Carbide from Silica Residues and Carbon. *Ceramic Bulletin*, **58**, 8, 789 (1979).

[5] P. Kennedy and B. North, The Production of Fine Silicon Carbide Powder by the Reaction of Gaseous Silicon Monoxide with Particulate Carbon. *Proc. Brit. Cer. Soc.*, **33**, *No. Fabr. Sci.*, **3**, 1 (1983).

New Method for the Quantitative Analysis of Free Carbon in Silicon Carbide

H. Knoch and K. A. Schwetz

Elektroschmeizwerk Kempten
GmbH

W. D. Long

Wacker Chemicals (USA), Inc.

Introduction

It is important to know the exact content of free carbon in sintered silicon carbide (SSiC). The free, or residual, carbon is present in SSiC after the "solid state" pressureless sintering process as a secondary phase that influences the material properties, often in a negative way (Figure 1). This residual carbon results from the addition of defined quantities of carbon as a sintering additive (generally < 5 wt.% C). The carbon additive to α SiC sinterable powder has two important functions:

i. Chemical reduction of oxide layers on the surface of SiC particles that block the sintering process: The carbon that is used in this reduction is released as carbon monoxide gas ($SiO_2 + 3C = SiC + 2CO$).
ii. Sintering mechanism: Activation of shrinkage enhancing grain boundary and/or lattice diffusion, and retardation of surface diffusion evaporation/condensation processes, which inhibit shrinkage. The carbon used for this purpose stays in the sintered body as the carbon secondary phase.

When adjusting the desired carbon content in the final product via the quantity of the carbon containing additive (elemental carbon or novolak as precursor material) in the starting mixture, the free carbon of the silicon carbide powder (Figure 2) must be taken into account. The quantity of free carbon that is present in the doped powder mixture, or after removal of binders in the green bodies is critical for the sintering process. It also influences the final properties of the sintered bodies.

Procedure

The DIN standardized direct and indirect procedures for analysis of carbon (burning in oxygen[1,2] are not suitable for the quantitative determination of free carbon in doped SiC powder and the resulting sintered bodies. Silicon carbide powder that is fine enough to sinter oxidizes at 400°C, and above, together with the

Fig. 1. Homogeneously dispersed graphite particles in α-SiC, fracture surface, back scattered electrons.

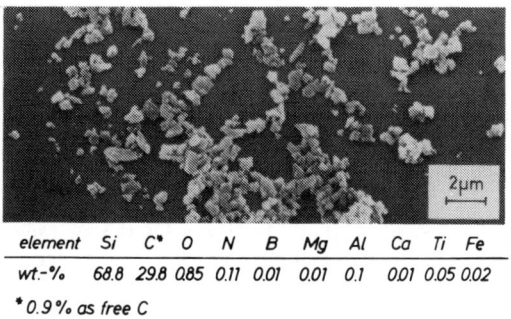

element	Si	C*	O	N	B	Mg	Al	Ca	Ti	Fe
wt.-%	68.8	29.8	0.85	0.11	0.01	0.01	0.1	0.01	0.05	0.02

*0.9% as free C

Fig. 2. Scanning electron micrograph of submicron α-SiC powder and associated chemical analysis.

free carbon. Therefore, a new method has been established,[3] which is based on the wet chemical carbon oxidation according to Meerson and Samsonov[4] in hot chromosulfuric acid as follows:

$$3C + 2K_2Cr_2O_7 + 8H_2SO_4 = 2K_2SO_4 + SO_4 + 2Cr_2(SO_4)_3 + 8H_2O + 3CO_2$$

All free carbon, together with a small amount of the carbon bonded to silicon carbide (CSiC) is oxidized to CO_2 in a reactor that is kept at a constant temperature. From there on it is transferred by an argon gas stream into an absorption cell (Coulomat 702) via a CuO-catalysis furnace. The CO_2 is coulometrically titrated as a function of time. The free carbon content can be calculated from the plotted curve of CO_2 against time; then the fraction of the combined carbon is determined by extrapolation. Figure 3 shows a typical analysis diagram for a sample weight of 25mg at an acid temperature of 120°C for 100 minutes.

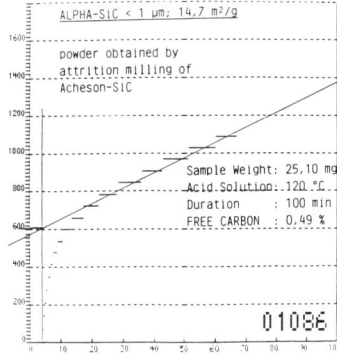

Fig. 3. Example of a free carbon evaluation plot.

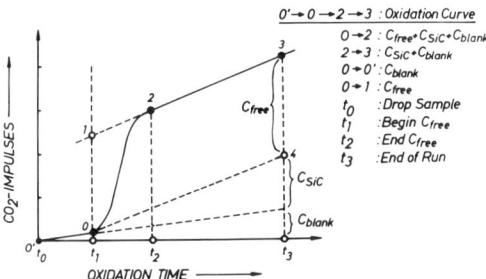

Fig. 4. Quantitative determination of free carbon in silicon carbide via wet oxidation (formed CO_2 versus time).

Figure 4 shows two distinct stages of reaction during oxidation. In the first one (steeper slope) all free carbon and a small amount of combined carbon react. The second stage (the curve straightens with "lower" slope) corresponds to the slow and incomplete oxidation of the silicon carbide. The percentage of free carbon is calculated by graphical extrapolation from the printed plot using the formula:

$$\frac{\text{Corrected impulses} \times 2}{\text{Sample weight} \times 100} = \% \ C_{free}$$

The graphical extrapolation (see Figure 4) is executed by the following steps:

i. the starting point t1 is determined from the first inflection (point 0) on the oxidation curve, which corresponds to the destruction of capsule and the start of reaction.
ii. a new ordinate is drawn through t1.

iii. the line between points 2 and 3 is extended to the left until it intersects the ordinate in point 1; the slope of the line 2-3 is a measure of the particle size of the carbide sample.
iv. the extrapolated impulses are converted to corrected impulses by subtracting the default impulses at t1.

Figures 5(A), 5(B), give a schematic description of the apparatus. The detailed measuring instructions are described elsewhere by Schwetz and Hassler.[3] The percussion mill used to reduce sintered material back into fine powder for analysis is shown in Figure 6.

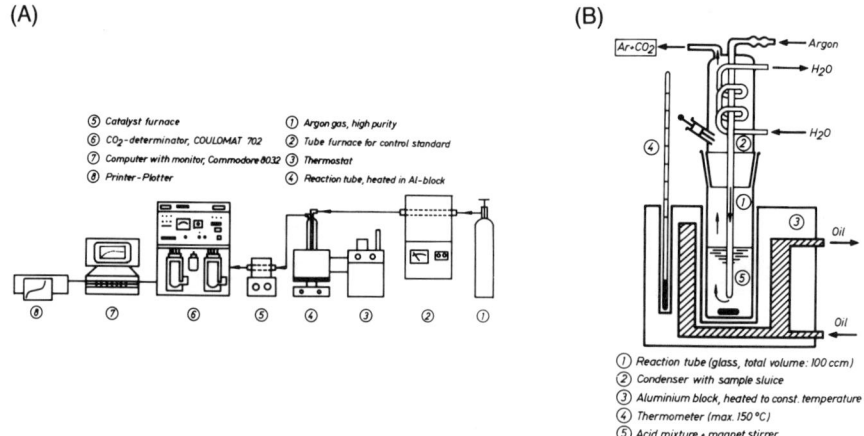

Fig. 5. (A) Schematic description of the apparatus for determination of free carbon in silicon carbide via wet oxidation in chromosulfuric acid. (B) Detailed picture of reaction apparatus (reaction tube assembly).

Accuracy, Detection Limit and Comparison with the Indirect Free Carbon Method

The relative standard deviation is:

max. 20% for 0.01 - 0.1 wt.% Cfree
 5 - 10% for 0.1 - 1.5 wt.% Cfree
 < 5% for more than 1.5 wt.% Cfree

For the detection limit a value of 10 micrograms of free carbon, with a concentration limit of 0.01% carbon for a 100mg sample has been determined. Table 1 shows the free carbon values of five SiC samples that have been found by wet chemical oxidation compared with values that have been obtained by the indirect DIN method modified per the techniques described by Kriegesmann and

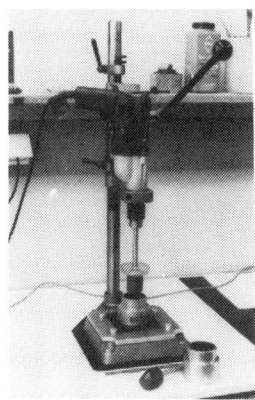

Fig. 6. Percussion mill.

Schwetz.[5] As can be seen from Table 1, the relative deviation of the methods was an average of 7% with a maximum of 12%. This grade of conformity of the two methods is regarded to be sufficient for the free carbon analysis of doped silicon carbide samples.

An example of the relationship of the free carbon data generated by this method to grit size is given in Figure 7. Note that the slope of the rate of reaction time is directly related to the particle size of the powder sample.

Table 1. Free Carbon in SiC (comparison of analyses).

Name	Sample Part. Size(μm)	Weight % C_{free}	
		New method	By combustion in oxygen[*]
SiC 813	$\leq$3	0.25	0.28
SiC 820	$\leq$1	0.98	1.04
"ultrafein"	<10	0.40	0.45
HIP SiC[†]	<30	1.16	1.17
HIP SiC	<30	1.08	1.04

[*]Modified DIN method: 900°C, 15min, 400mg weight sample.
[†]Undoped sintered bodies after milling to analyzable size in the TETRABOR® percussion drilling high speed mill.

Silicon Carbide

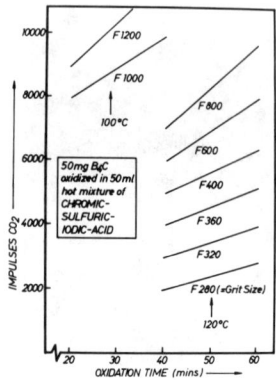

Fig. 7. Linear sections of B4C oxidation isotherms: slope is a function of grit size.

References

[1] DIN 51075, Teile 2, Deutsche Norm Okt. 1982, Beuth-Verlag, Berlin 30.
[2] DIN 51075, Teile 3+5, Deutsche Norm Okt. 1982, Beuth-Verlag, Berlin 30.
[3] K. A. Schwetz and J. Hassler, *Journal of the Less Common Metals*, 117 (1986), 7-15.
[4] G. A. Meerson and G. V. Samsonov, Zavod Lab. 16 (1950), 1423-8.
[5] J. Kriegesmann and K. A. Schwetz, BMFT - Forschungsbericht 1/1/776/30/77. "Entwichklung und Herstellung hockwarmfester Formkoerper aus SiC, insbesondere fuer den Gasturbinenbau," Foerderungskennzeichen: 01ZA 025-Z 13NTS 1003, 24 Seiten (1977).

Enhanced Formation of 4H Polytype in Silicon Carbide Materials

Samuel S. Shinozaki and Jon Hangas, Research Staff

Ford Motor Company
Scientific Research Laboratory
Dearborn, MI 48121-2053

Kunihiro Maeda and Atsuko Soeta

Hitachi Research Laboratory, Hitachi, Ltd.
Hitachi, Ibaraki, Japan

Abstract

Polytypic lamellar formation in silicon carbide (SiC) has been found to be influenced by types and level of additives and impurity elements. When β-SiC (FCC) powder is sintered using sintering aids such as aluminum (Al) + boron (B) + carbon (C), or BeO, the β-phase transforms predominantly to the 4H polytype. In comparison, polytypic transformation from 6H to 4H in α-SiC with BeO is found to occur at a much slower rate. Residual impurities such as titanium (Ti) also help to form thin 4H lamellae grown epitaxially on α-SiC basal plane. Microstructure of 4H formation was analyzed by analytical electron microscopy (AEM).

Introduction

It has become increasingly important to understand the polytypic lamellar formation in silicon carbide (SiC) materials, due to more extensive use of SiC materials for electronics and advanced structural applications. Polytypic distribution in these materials has been known to be dependent on consolidation conditions, such as phases (α and β) of starting powders, types of sintering aids, sintering temperature and environment[1,2,3]. Enhanced 4H polytype growth in SiC with Al, B and C additives with β-phase starting powder has already been reported, using x-ray diffraction (XRD) analysis as well as analytical electron microscopy (AEM)[4,5]. A similar beta to 4H phase transformation has been observed in SiC materials with other sintering aids such as BeO[6]. In these materials, several different polytypes are distributed within individual grains. However, when the starting powder is α-phase, the 6H to 4H polytypic transformation is extremely slow. No thick 4H lamellae have been observed in the 6H matrix, only a long period polytype with some combinations of 6H and 4H. Thus, the mechanism of the polytypic transformation may be different during β to α phase transformation, in comparison with polytypic transformation in α-SiC materials. In this report, some differences in mechanism of the enhanced 4H transformation in β-SiC and α-SiC materials will be reported.

Experimental Procedure and Results

In this investigation, one AEM sample was prepared out of each of the following materials:

Sample A

Boron (1.2wt%), carbon (4.0wt%) and aluminum (1.5wt%) were added to an ultrafine grade β-SiC powder known as Betarundum (made by Ibiden Co., Tokyo, Japan). This material was pressureless-sintered in a graphite furnace under a protective atmosphere of Ar and He. The temperature increase rate and holding time (30 min.) were computer-controlled and only the final holding temperature was changed for each sample within a range of 1200–2200°C[5]. As explained in detail in Ref. 5, XRD analysis was utilized to determine the amount of the various polytypes[8]. Figure 1(a) shows that 4H is the predominant high temperature polytype in this material. The AEM Sample (A) was prepared from this material sintered at 1950°C, where 15R-4H transformation started to occur.

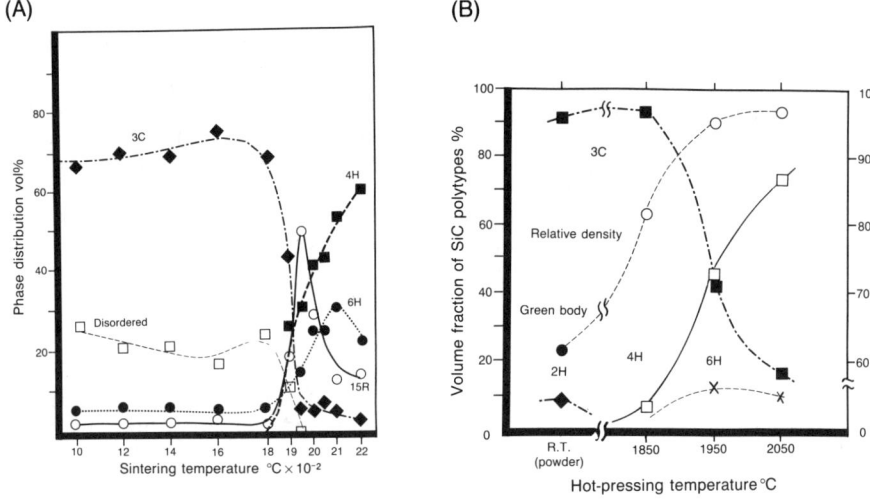

Fig. 1. (A) and (B): Quantitative x-ray diffraction analysis of polytype distributions in SiC materials with Al-B-C addition[5] and another SiC material with BeO addition[6], sintered at various temperatures, respectively. In (B), theoretical density of the SiC-BeO material is shown.

Sample B

Only BeO(2wt%) was added as a sintering aid in the β-SiC powder. The mixture was hot-pressed at various temperatures within a range of 1600–2050°C under a pressure of 30MPa for one hour[6]. As shown by XRD analysis for this material (Fig. 1(b)), the 4H polytype was found to be predominant at temperatures above 1950°C. For this investigation, the AEM Sample (B) was prepared from the material hot-pressed at 1950°C.

Sample C

The AEM Sample (C) was prepared from a material where BeO (2wt%) was added to α-SiC powder and hot-pressed at 2050°C under a pressure of 30MPa for one hour. This material is known as Hitaceram 101[7], produced by Hitachi Research Laboratory.

ANALYTICAL ELECTRON MICROSCOPY ANALYSIS

Polytype identification using the AEM was accomplished by selected area diffraction (SAD) pattern analysis and by direct resolution of structural lattice images. Most AEM studies were performed at or near the $\langle 110 \rangle$ zone axis in β-phase. Micro-chemical analyses of individual SiC grains and grain boundaries were performed utilizing energy dispersive x-ray analysis and electron energy loss spectroscopy. The instruments used in the analyses were JEM 2000FX and Siemens EM102 at Ford, and a VG HB501 STEM with a field emission gun at Michigan State University.

Significant AEM results are summarized as follows:

Sample A

Beta-15R-4H transformation in SiC with Al, B, and C addition: When Al, B and C are added to β-SiC starting powder, the polytype distribution in the sintered body is complicated by the transient inclusion of 15R polytype (Figure 1(a))[5]. An example of the 15R-4H interphase boundaries is shown in Figure 2. A small additional amount of Al was detected on the boundary where several ledges were analyzed simultaneously between 15R-4H. The 15R-4H transformation is possibly enhanced by rapid ledge advance on the basal plane at temperatures above 1950°C, due to an Al-B rich second phase.

Sample B

β-4H transformation in β-SiC with BeO addition: One example of 4H lamellar formation in a β-phase zone is shown in Figure 3. The 4H structure was identified by structural lattice image spacing of 1.0 nm and convergent beam electron diffraction (CBED) pattern analysis shown in Figure 4. The β to 4H transformation may be enhanced by the advancement of partial dislocations at the interface which may be rich with BeO, but BeO could not be detected by means of existing AEM techniques. The lattice parameters were determined from 4H lamellae in

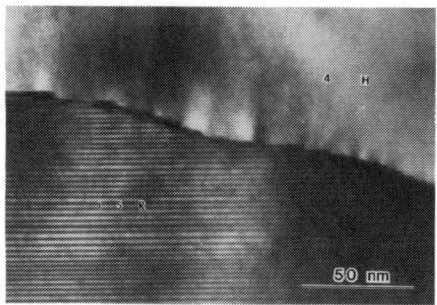

Fig. 2. High resolution structural lattice image shows a typical 15R-4H boundary with common basal plane.

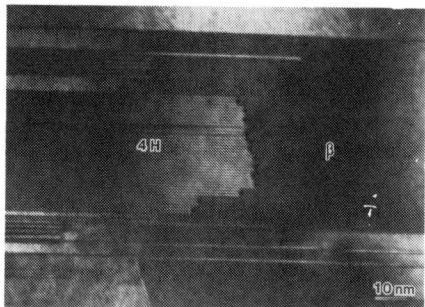

Fig. 3. Structural lattice image with 1.0 nm in spacing was identified as 4H polytype, grown into β-phase matrix. Dark spots in front of the 4H lamella indicate partial dislocation advancement.

(A) 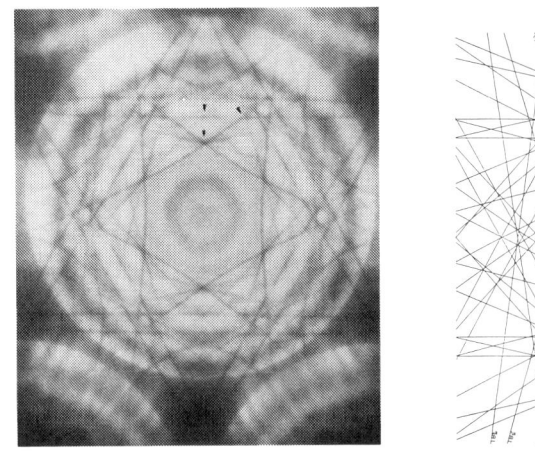 (B)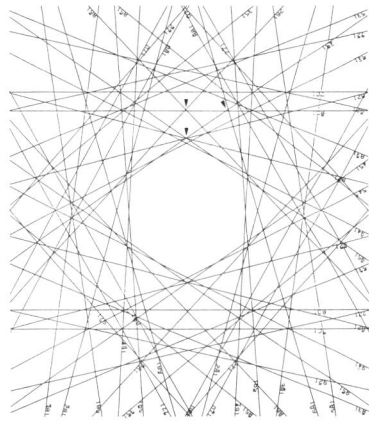

Fig. 4. (A) [0001] zone CBED pattern of 4H lamella in sample B) and (B) computer simulation using a=0.3080 and c=1.008 nm.

samples A) and B) to see if any changes occur with the addition of different elements. No changes were observed and computer simulation results showed a=0.3080±.0004 and c=1.008±.001 nm[9].

Sample C

Polytypic transformation in α-SiC materials with BeO: The effects of the BeO sintering aid on the polytype distribution are different in sample C), where the starting powder is α-SiC (mostly 6H). No 4H lamellae have been observed in the 6H zone. Several long period polytypes containing some combinations of 6H and 4H have been observed (Figure 5), which may be some intermediate structure in the transition from 6H to 4H.

More commonly, thin sheaths of various polytypes are situated on the outside (0001) surfaces of a 6H zone in a SiC grain. Figure 6 shows one such thin sheath which is in a form of disordered lamellae with mostly 4H and other polytypes (often thin β-phase lamellae are included). Occasionally some ordered 4H lamellae are observed, as shown in Figure 7. In this case, there are double layers of a Ti-rich 4H lamella, where low levels of Ti were detected by EDS and 4H SiC lamella, which were epitaxially grown on 6H SiC lamella. The Ti-rich 4H lamella is an extension of a Ti-rich impurity particle. Similarly, thin transition metal-rich lamellae have been observed with other elements.

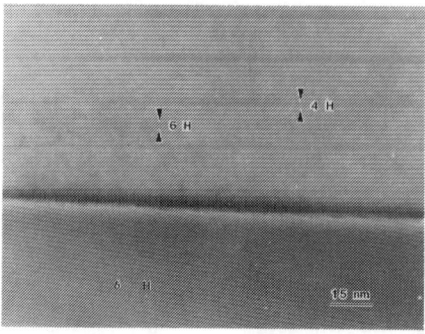

Fig. 5. Structural lattice image of two adjacent grains shows that the bottom grain has pure 6H structure and the upper grain is an example of long period polytype.

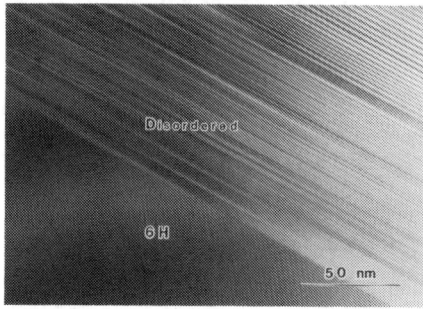

Fig. 6. Structural lattice image shows disordered sheaths formed on one side of 6H zone, indicating strong effect of sintering aid on polytype distribution in newly grown zone.

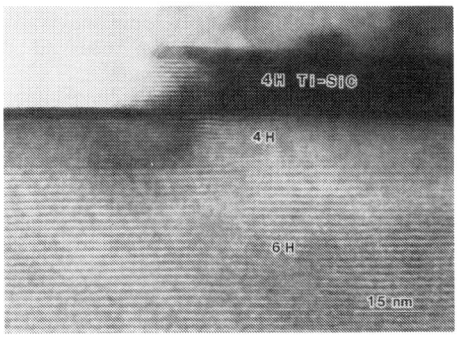

Fig. 7. Ti-rich 4H lamella adjacent to Ti-rich impurity grain. Structural lattice image shows that the Ti-rich 4H lamella was grown on pure SiC lamella with 4H structure.

DISCUSSION

The results of this investigation have shown that 4H lamellar formation in β-SiC materials may be strongly enhanced by some types of sintering aids. As shown in Figure 1, the densification and the phase transformation are nearly completed at a relatively low temperature of approximately 1950°C, which is at least 100°C lower than SiC materials with only B and C additions[5]. It is also observed that the combination of Al, B, and C as sintering aid enhances the β to α phase transformation compared to the hot pressed SiC with BeO addition. In the SiC-BeO materials, no 15R formation was observed. It may be speculated that the 15R polytype was transformed to 4H during the hot pressing process in the SiC-BeO materials or the formation of 15R need not be a precursor for 4H formation for SiC-BeO materials. It is noted that the effects of sintering aids may be altered, due to the fact that chemical compositions of second phases in SiC with Al-B-C additions changes as sintering temperature is increased from 1200 to 2100°C, while BeO is stable up to 2100°C. Changes in the polytype distribution in SiC materials may thus be related to the type of the second phase present.

When the α-SiC powder (mostly 6H polytype) was hot-pressed with BeO addition, 4H lamellar formation was observed only as thin sheaths outside the 6H zone and not within the 6H zone. It may be concluded that the effect of the additives on the polytype distribution is predominantly on β-SiC accompanying the phase transformation to α-phase and not directly on α-SiC[1,2]. The AEM analysis revealed that low levels of elements such as Al and Ti were detected in the 4H lamellae, while no such elements were detected in the 6H zone at the center of a grain. Thus, 4H lamellar formation may be strongly enhanced by these elements.

CONCLUSIONS

1. 4H lamellar formation in β-SiC materials is enhanced by some elements or oxides such as Al, Ti or BeO.

2. In the α-SiC materials, 6H to 4H transformation is extremely slow. Under normal sintering conditions only long period polytypes with some combinations of only 6H and 4H are observed. No thick 4H lamellae have been observed within 6H zones.

3. An Al-rich second phase advances on (0001) planes and enhances the β-15R-4H polytypic transformation, while in SiC-BeO materials the β-phase transforms directly to 4H.

4. Low levels of Ti and Al were detected by EDS in the 4H lamellae and interphase boundaries, respectively.

Acknowledgements

The authors wish to thank Ms. B. N. Juterbock and Dr. R. M. Williams (deceased) for preparing samples with Al, B and C additions, Dr. C. R. Peters for his quantitative x-ray diffraction analysis and Dr. W. T. Donlon for his continuous and invaluable discussions and a critical reading of this manuscript.

References

[1] S. Shinozaki and K. R. Kinsman, "Evolution of Microstructure in Polycrystalline Silicon Carbide," Proc. of Crystalline Ceramics; Edited by Hayne Palmer III, R. F. Davis and T. M. Hare, p. 641 (1978).

[2] K. R. Kinsman and S. Shinozaki, "Influence of Interfacial Structure on the Morphology of the α to β and β to α Transformation in Polycrystalline SiC," Proc. of Int. Conf. on Solid-state Phase Transformation; Edited by H. Arronson, p. 605 (August 1981).

[3] T. Hase, H. Suzuki and T. Iseki, "Sinterability of Submicron SiC Prepared from Siliconization of Carbon Black under Presence of Al Additives," Yogyo-Kyokai-Shi, 87 [11], 1979.

[4] Y. Tajima and W. D. Kingery, "Solid Solubility of Aluminum and Boron in Silicon Carbide," Comm. Am. Ceram. Soc., 65 [2] C-27 (1982).

[5] R. M. Williams, B. N. Juterbock, S. S. Shinozaki and C. R. Peters, and T. J. Whalen, "Effect of Sintering Temperatures on the Physical and Crystallographic Properties of beta-SiC," Am. Ceram. Soc. Bull., 64 [10] p. 1385 (1985).

[6] A. Soeta, K. Maeda and Y. Suzuki, "Polytypes in SiC Ceramics with BeO addition," Ceram. Soc. of Japan, Annual Meeting Bulletin, p. 519, (1985).

[7] Y. Takeda, K. Usami, K. Nakamura, S. Ogihara, K. Maeda. T. Miyoshi, S. Shinozaki, and M. Ura, "Grain-boundary Structure of Highly Resistive SiC Ceramics with High Thermal Conductivity," p. 253 in Advances in Ceramics, Vol. 7; Edited by M. F. Yan and A. H. Heuer. The American Ceramic Society, Columbus, OH, 1983.

[8] J. Ruska, L. Gauckler, J. Lorenz and H. V. Rexer, "The Quantitative Calculations of SiC Polytypes from Measurements of X-ray Diffraction Peak Intensities," J. Mat. Sci., 14, 2013 (1979).

[9] J. Hangas, S. Shinozaki and W. T. Donlon, "Lattice Parameter Measurements of SiC Materials Using CBED HOLZ Lines," Proc. of the 44th Ann. Meeting of Electron Microscopy Soc. of Am.; Edited by G. W. Bailey, p. 464 (1986).

Section II

Green State Processing

ULTRAFINE SiC POWDER PRODUCED BY TURBOMILLING

DALE E. WITTMER, SR.

Associate Professor
Dept. of Mechanical Engineering and Energy Processes
Southern Illinois University
Carbondale, IL 62901

ABSTRACT

Silicon carbide powder was produced with high surface area by autogenous turbomilling 200-mesh SiC with 30-mesh SiC of the same composition. Additions of boron and carbon were made during the turbomilling process to improve the homogeneous dispersion of the boron and carbon sintering aids in the ultrafine SiC powder. The processing methods and results of strength and fracture behavior of hot-pressed powders are presented in this paper.

INTRODUCTION

The economical production of ultrafine high-purity ceramic powders without appreciable contamination is a recognized obstacle in the processing of advanced ceramics. Standard milling techniques, although economical for materials with low hardness, generally require extended milling times and may produce significant levels of contamination.

Turbomilling is an attrition grinding process* that was originally developed and patented by Feld and Clemmons[1] for grinding and beneficiation of industrial minerals. Commercially delaminated kaolins and ultra-fine titanium pigments, in addition to several industrial minerals, are currently being produced by successful upscale of the original process and turbomilling apparatus. These techniques have been reported and reviewed by Stanczyk and Feld[2,3] and Sadler, Stanley, and Brooks[4]. Stanley et al.[5] were the first to investigate the production of submicron SiC powders by turbomilling, and more recently autogenous turbomilling of high-purity ceramic powders has been reported by Wittmer[6], Hoyer and Petty[7], and Prochazka[8].

The turbomilling process involves the intense milling and agitation of a slurry composed of a relatively coarse milling media, the material to be milled, and a suspending fluid. The turbomill consists of a cage-like rotor made of vertical bars connected to a drive shaft, a stationary stator composed of vertical bars fixed to rings at the top and bottom, and the mill body. For autogenous turbomilling the material to be ground and the coarse material are of identical composition to minimize contamination of the powders produced.

Several operating parameters have been found to control the effectiveness and efficiency of the turbomilling process. Stanczyk and Feld[2] found that the most important variables affecting particle size reduction and energy consumption were the type, size and shape of the grinding media, the grinding media to fines ratio, rotor speed, slurry density, degree of slurry dispersion and angular arrangement of the rotor and stator vertical bars. Sadler, Stanley and Brooks[4] reported that optimum milling medium concentrations for coarse particles can be approximated from mean free path theory to occur when the particle mean free path is equivalent to approximately 0.7 particle diameter. Milling rates were observed to increase with rotor velocity and milling efficiency was found to be constant for rotor velocities over the 800 to 1600 r.p.m. range.

The original turbomill was made of stainless steel, however relatively recent improvements have resulted in the development of a laboratory scale all polymer turbomill composed of ultra-high molecular weight polyethylene (UHMWPE) that does not contaminate the material being processed. The major parts of the polymer turbomill are shown in Figure 1, after 40 hours of milling α-SiC particulate. For larger scale versions, metal parts coated with polymer would be required for structural reasons. A detailed report by Wittmer[6] discusses the development of the polymer turbomill and the preparation of ultrafine high-purity α-SiC powders (BET surface areas of 30 to 35 m^2/g for 3 to 6 hours milling time).

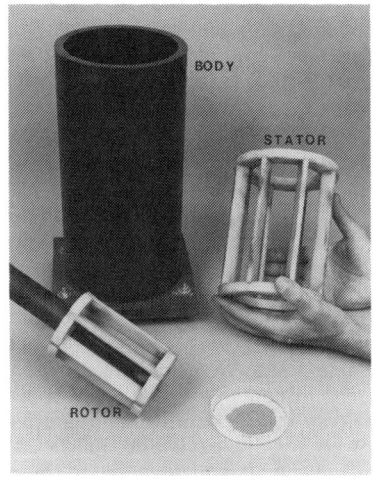

Fig. 1. Rotor, stator and body for all polymer turbomill.

Hoyer and Petty[7] used the same polymer turbomill to investigate turbomilling of several hard to grind ceramic materials and the influence of various dispersion agents. Recently Prochazka[8] reported the use of a polymer turbomill with a polymer, ceramic or metal rotor utilizing metal or ceramic coarse grinding media on the effectiveness of grinding boron carbide.

This paper will present the processing methods used for the production of ultrafine α-SiC powders (with and without boron and carbon additions), the sonic modulus and modulus of rupture (MOR) results for hot-pressed specimens, and some comments on characterization and identification of flaw origins in fracture specimens.

EXPERIMENTAL

A commercial purity green α-SiC[†] was used in this study. A standard 30-mesh grade was used for the coarse milling media and a 200-mesh grade was the material milled. Sieve analyses for these materials are given in table 1.

Turbomilling was conducted in a 12.7 cm diameter mill constructed of UHMWPE polymer as described in[6]. The specification of this unit are given in table 2. The lab scale turbomill is powered by a 2 H.P. D.C. motor with infinitely variable speed control that facilitates loading at low speeds.

A typical turbomilling run consists of adding 2 liters of distilled water to the mill and adjusting the speed to 500 r.p.m. The fine 200-mesh SiC (772.5 g) is then added and the speed increased to 1200 r.p.m., followed by the addition of the 30-mesh SiC (1930.5 g). Finally, 10% (200 ml) of a 1% solution of tetra-sodium pyrophosphate is added as a deflocculant and the speed increased to 1600 r.p.m.. Air bubbles trapped in the slurry can be released by lowering the speed to about 200

Table 1. Screen analyses for starting SiC materials.

SiC Mesh	Screen Size	% Retained On
30	+30	87.2
	−30 +40	12.1
	−40 +60	0.4
	−60 +100	0.1
	−100	0.2
200	+100	3.9
	−100 +200	83.6
	−200 +325	7.5
	−325	5.1

Table 2. All polymer turbomill specifications.

Turbomill Specifications
Container 12.7 cm ID
Stator 10.2 cm ID
Rotor 8.9 cm OD
Volume 3.6 liter max.
2.5 liter min.
Speed 0–1750 RPM
Transducers RPM, HP, Torque

Silicon Carbide

r.p.m. and quickly returning to the higher speed. It is important that the speed is always kept high enough to prevent the mill rotor from locking up due to the course material settling to the bottom of the mill body. After the milling cycle (4 hours for this study) has been completed, the mill is turned off and disassembled to recover the slurry. The slurry is passed through a screen stack containing 40, 100, 200 and 400 mesh screens and then poured into a sedimentation cell to separate the -5 micron SiC.

Sedimentation is accomplished by dilution with distilled water to less than 5 wt % total solids, followed by thorough agitation of the slurry. After the proper settling time (determined by Stoke's method as described in[6]), the fines are drawn off and the coarse settlings are again dispersed in distilled water and a second settling is made. The fines are then flocculated with a 3% HCl solution and allowed to settle, and the supernatant water is removed. The slurry at this stage is about 15 wt % or 5 vol % solids. Further dewatering can be accomplished by oven drying, vacuum filtering or centrifugation followed by drying, or spray drying. The cake formed by conventional drying is then ground or lightly milled to -200 mesh to form a flowable powder.

A process flow diagram for the powders prior to hot-pressing is given in figure 2. This diagram shows two paths for the addition of carbon and boron sintering aids. The boron and carbon can be added to the dried SiC powder produced by the methods just described or in situ during the turbomilling process.

The dry SiC powder is placed in a beaker with a volatile solvent (benzene, acetone, toluene, etc.) in the amount of 20 ml of solvent per gram of powder and magnetically stirred on a hot plate set at 40 to 50°C while the boron[†] and carbon[§] are added. The material is stirred until dry and then granulated to pass a 100-mesh screen. The powder is ready to press into a preform or go directly into the hot-press.

An alternative is to add the boron and carbon during the turbomilling process. Since the processing of this particular grade of SiC has been well characterized, it was possible to predict the yield of ultrafine (>5 micron fraction) SiC powder, and therefore, predetermine the measure of the boron and carbon additions. The boron and carbon were added after 2 hours of turbomilling and the run continued for an additional 2 hours. The assumption was made that the boron and carbon would remain in the fines during the dewatering phase. Dewatering consisted of passing the slurry through a 400-mesh screen, followed by settling in a sedimentation column for an appropriate time to allow for separation of the coarse fraction. The slurry containing the fines was flocculated with a 3% HCl solution, dried, and granulated to -100 mesh prior to hot-pressing.

A vacuum hot-press with a 5.08 cm graphite die and graffoil spacers and liner was used to press discs approximately 1 cm thick. The powder was loaded between graffoil spacers and a pressure of 1.2 MPa was applied from room temperature up to 1200°C. The temperature was then ramped to 1500°C in one hour, while the pressure was ramped to 21 MPa. Finally, the temperature was ramped to 1900°C in 1 hour and the pressure ramped to 34.5 MPa. After holding for a period of 30 to 45 min, the temperature was raised slightly. If no further densification was observed, the run was terminated and the furnace allowed to cool at the furnace rate.

Following hot-pressing, the discs were processed according to the flow diagram given in figure 3. The surfaces of the discs were cleaned with a surface grinder by rough grinding with a 180 grit diamond wheel, followed by finish grinding with a 600 grit diamond wheel. After cleaning in acetone, the sonic modulus values were

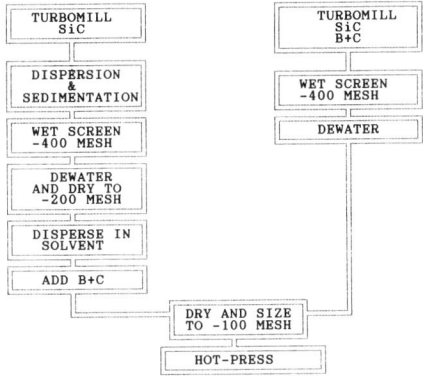

Fig. 2. Powder process flow diagram prior to hot-pressing SiC.

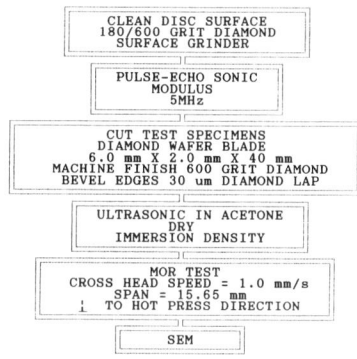

Fig. 3. Flow diagram for test specimen preparation and testing.

determined by pulse-echo overlap techniques from the shear and longitudinal wave speed at a frequency of 5 MHz.

Following sonic testing, the discs were cut into MOR test specimens with a diamond wafering saw, the surfaces were finished with a 600 grit diamond wheel, and the edges were beveled on a 30 micron diamond lap. After machining, the MOR bars were ultrasonically cleaned in acetone to remove any grease or grinding fluids, dried and checked for density by the standard immersion method.

MOR was determined by three-point bending perpendicular to the hot-pressing direction. Testing was conducted at room temperature on a bench top Instron mechanical test unit at a cross head speed of 1.0 mm/s with a span of 15.65 mm. Ten MOR bars for each hot-pressed disc were broken, and the fractured surfaces were examined by SEM to identify the fracture origins.

Results and Discussion

A turbomilling time of 6 hours was used initially to determine the optimum milling time for the production of high surface area SiC powder. SEM photomicrographs of the starting 200-mesh α-SiC and the submicron powders produced by turbomilling for 6 hours are given in figures 4 (*A*) & (*B*), respectively. The sample for (*B*) was prepared by heating the SEM stub to 150°C in a small lab oven prior to placing several drops on it of the flocculated and dewatered -5 micron slurry that was diluted 20:1 with 2-propanol. This caused rapid drying of each floc which enabled easier viewing in the SEM. From the photomicrograph the flocs appear to be about 2 microns and are made up of submicron powder with an esd (equivalent spherical diameter as calculated from the BET surface area of 31 meter squared per gram) of about 0.06 microns.

Figure 5 is a typical plot of surface area and the percent of coarse material lost to fines as a function of milling time for a six hour turbomilling run. As seen in the figure, the surface area increases to a maximum of 33.8 m^2/g at 3 to 4 hours milling time, and due to fracture of the larger milling media, the surface area is observed to decreases for longer milling times. This data was generated for unmilled coarse material. In practice, new material is added to the oversized material recovered during separation to make up for the coarse SiC lost to fines. Several runs are required to break in or preround the coarse SiC to reduce the coarse fraction lost to fines and the amount of intermediate coarse material. For this study, unmilled coarse SiC was used for each run to maintain consistency between runs and a time of 4 hours was used for each run to maximize the resulting surface area.

 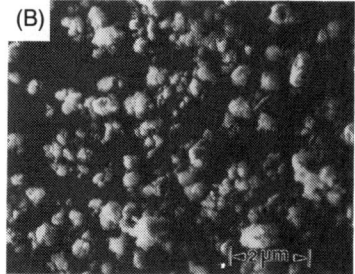

Fig. 4. SEM photomicrographs which show (*A*), the as-received 200-mesh α-SiC and (*B*), the ultrafine powder produced by turbomilling (*A*) for 6 hours.

Typical analyses for the starting α-SiC, and ultrafine powder produced in the metal turbomill and the polymer turbomill after 6 hours of turbomilling are presented in Table 3. After turbomilling, the powder prepared in the metal mill picked up almost 1% Fe and some Ni, while the powder turbomilled in the polymer mill appeared to have slightly higher purity than the starting material. It is thought that some of the fine particulate and surface impurities might be solubilized during

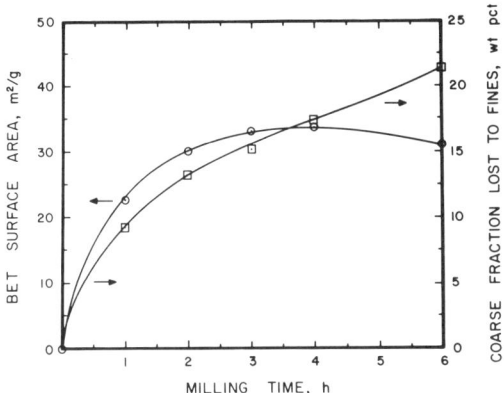

Fig. 5. BET surface area and coarse fraction lost to fines as a function of turbomilling time.

the turbomilling process. Oxygen analysis was about 0.7 to 0.8% for the starting powders and 1.3 to 1.6% for the polymer turbomilled powders milled in distilled water.

One lot of ultrafine SiC powder prepared by turbomilling for 4 hours in the stainless steel turbomill lined with polyurethane, after sedimentation, dewatering and drying, was acid leached for 2 hours at 50°C with a 0.5 normal HCl solution. Following acid leaching, the powders were washed and filtered 5 times with distilled water. The Fe was reduced by this process to about 0.1% prior to the addition of boron and carbon for hot-pressing.

The results for the hot-pressed α-SiC powders containing boron and carbon are given in Table 4, and appear to be in agreement with published values. Density, elastic modulus (E), shear modulus (G), and MOR were lower for the acid leached powder produced in the metal turbomill which contained 1.0% boron and 1.0% carbon (ID M3011).

As mentioned previously, boron and carbon were added to the SiC powder by stirring in benzene or acetone. Processing in acetone at the 1.0% addition level (ID P3013) appears to have produced slightly better results than benzene (ID P3012), especially in decreasing the standard deviation for MOR. When the boron and carbon are added by stirring in benzene or acetone it would appear that the boron and/or carbon may be agglomerated in an already agglomerated powder. Agglomeration is a recognized universal problem for all ultrafine dry powders, and agglomerates are known sources of critical flaws. Some but not all of the agglomerates could be crushed during hot-pressing, and the agglomerates remaining may act as flaws. In some cases the agglomerates also might initiate exaggerated grain growth, and the large grains in a fine grained matrix can act as critical flaws or stress concentrators.

The best results were obtained for the addition of 0.5% boron and 1.0% carbon added during the turbomilling process (ID P3015) compared to the same amounts added by stirring in acetone (ID P3014). The density, elastic modulus, shear modulus, and MOR were the highest and the standard deviation for MOR was the

Table 3. Typical analyses of SiC before and after turbomilling.

Constituent	Before Turbomilling	After 6 hr. Turbomilling Metal	Polymer
α SiC	98.60	{ Bal }	{ Bal }
Si, free	0.15	{ }	{ }
SiO_2, free	0.63	{ }	{ }
C. free	0.36	{ }	{ }
Fe	0.08	1.0	0.08
Al	0.08	0.04	0.03
Ca	0.05	0.01	<0.01
Mg	0.03	0.02	<0.01
Ni	ND	0.08	ND

Table 4. Properties of hot-pressed turbomilled SiC powders.

ID	%B	%C	DENS.(g/cc)	%TD	E(GPa)	G(GPa)	MOR	(MPa)
M3011*	1.0	1.0	3.185	99.2	431	186	337	+/−43
P3012*	1.0	1.0	3.194	99.4	439	189	358	+/−41
P3013‡	1.0	1.0	3.195	99.5	472	202	369	+/−33
P3014‡	0.5	1.0	3.190	99.4	469	201	362	+/−27
P3015†	0.5	1.0	3.195	99.5	474	203	378	+/−22

M = metal turbomill/acid leached   P = polymer turbomill
* B and C added by stirring in benzene
‡ B and C added by stirring in acetone
† B and C added by turbomilling in water

lowest for the turbomill addition. Since equivalent or slightly enhanced properties were obtained for lower boron content (0.5% for the turbomill addition compared to 1.0% for stirring additions), it is thought that the addition of boron and carbon during the turbomilling process provides for better dispersion and homogenization than addition by stirring. In addition, the boron should be ground to a much finer size during turbomilling which would also aid in promoting dispersion. During flocculation after turbomilling, it is believed that the boron and carbon are trapped within the flocs and kept from heterogeneous separation. This would be expected to produce more uniform microstructures with corresponding improvement or uniformity of properties as observed for sample P3015.

Since SiC has a fracture toughness of about 3.0 MPA square root meter, the calculated theoretical critical flaw size for bulk flaws would be about 40 microns for strengths of 350 to 400 MPa. In the weaker fractured MOR specimens where the boron and carbon were added by stirring, some carbon agglomerates and/or large grains (about 50 microns) were observed at or near the suspected fracture origins (which is in agreement with the calculated theoretical critical flaw size).

For stronger specimens where the boron and carbon were added in situ during turbomilling, fracture appears to follow a tortuous path and no obvious fracture origins could be identified. This latter behavior can be observed in figure 6 which is a SEM photomicrograph of the fractured surface of P3015, where (T) is the tensile face of the specimen. The fracture contours in figure 6 were followed back to a point of intersection and the area observed by SEM at higher magnification (figure 7). As seen in figure 7, there is some evidence of preferred orientation for transgranular fracture, microporosity, and some free carbon (the small dark spots), but no microstructural irregularities were observed. The average grain size is about 10 microns which is much lower than the calculated critical flaw size; however, lack of positive flaw identification is common for dense, brittle ceramics with no distinguishing features.

The results of this work indicate that additions of carbon and boron can be made in situ to α-SiC during the turbomilling process. Advantages over other methods are improved processing rates for production of ultrafine powders with well dispersed additions and the resultant improvement in properties and microstructural uniformity, reduced number of processing steps, and proven upscale potential. Although not fully optimized for this system, these results encourage the use of turbomilling techniques for grinding and/or homogenization of other hard to grind and/or disperse advanced ceramic powders, suspensions and composites.

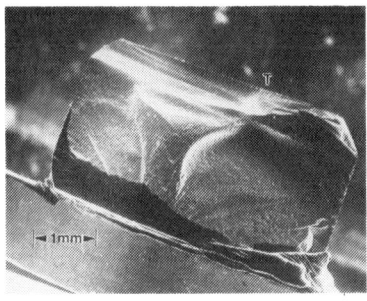

Fig. 6. SEM photomicrograph of P3015 showing fracture topography and macrostructure.

Conclusions

1. Ultrafine α-SiC powders with surface areas above 30 m^2/g were produced by turbomilling in water for milling times of 2 to 6 hours.
2. Boron and carbon hot-pressing aids were added both by stirring in benzene or acetone and in situ during the turbomilling process.
    a. Dispersions of B and C by stirring in acetone appeared to produce slightly improved properties over powders dispersed in benzene.
    b. Additions of 0.5% B and 1.0% C in situ during turbomilling produced the best properties due to more homogeneous dispersion of the hot-pressing aids.

3. Physical results for specimens prepared from the acid leached powders produced in the metal turbomill were poor compared to all of the powders produced in the polymer turbomill.

4. Intra- and inter-granular fractures were observed to occur in all specimens. Indications are that bulk flaws near the calculated critical flaw size in the form of agglomerates and large grains were responsible for initiation of fracture in the weaker specimens prepared by additions of B and C by stirring. No microstructural irregularities were observed for fractured specimens where the B and C were added during turbomilling.

5. Further work is required for developing optimization of the turbomilling process for grinding and/or homogenization of advanced ceramic powders, suspensions and composites.

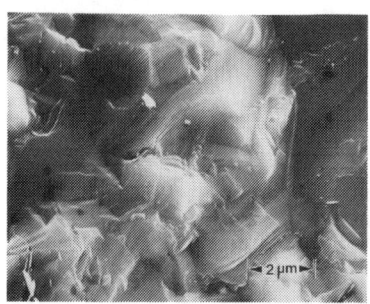

Fig. 7. SEM photomicrograph of P3015 showing evidence of intra- and inter-granular fracture and microporosity.

Acknowledgements

The author would like to thank the U.S. Bureau of Mines Tuscaloosa Research Center, Tuscaloosa, AL, especially the Ceramics Research Group, for their support in providing materials, technical assistance and time on their equipment, Norton Company, Worcester, MA, for the SiC raw material, and Naval Research Laboratory, Washington, D.C. for their assistance hot-pressing, physical property determination and fractography.

---

*Recently renamed turbomilling to avoid confusion with rotary arm ball mill type attritor mills.
‡Norton Company, Worcester, MA
†Boron 99.5% pure from United Minerals Corporation, New York, N.Y.
§HP Carbon black from Ultra-Carbon Corp., Bay City, MI

References

[1] I. L. Feld and B. H. Clemmons, Process for Wet Grinding Solids to Extreme Fineness. U.S. Pat. 3,075,710, Jan. 29, 1963.

[2] M. H. Stanczyk and I. L. Feld, "Investigation of Operating Variables in the Attrition Grinding Process", BuMines RI 7168, 1968, 28 pp.

[3] ——. "Comminution by the Attrition Grinding Process", BuMines Bulletin 670, 1980, 43pp.

[4] L. Y. Sadler, III, D. L. Stanley and D. R. Brooks, "Attrition Mill Operating Characteristics", *Powder Technology*, 12, 1975, pp. 19–28.

[5] D. L. Stanley, L. Y. Sadler, III, D. R. Brooks, and M. A. Schwartz, "Production of Submicron Silicon Carbide Powders by Attrition Milling", Proc. Second Intern. Conf. on Fine Particles. Electrochem. Soc., Boston, MA, Oct., 1973, pp. 331–36.

[6] D. E. Wittmer, Sr., "Use of Bureau of Mines Turbomill to Produce High-Purity Ultrafine Nonoxide Ceramic Powders", BuMines RI 8854, 1983, p. 12.

[7] J. L. Hoyer and A. V. Petty, "High Purity, Fine Ceramic Powders Produced in the Bureau of Mines Turbomill", Ceramic Proceedings, Vol. 6, 9-10, 1985, pp. 1342–1355.

[8] S. V. Prochazka, "Grinding of Hard Substances to Submicron Grain Size", Advances in Ceramics, Vol. 21, Am. Ceram. Soc. Publ., 1987. Proc. of Ceramic Powder Science and Technology, Synthesis, Processing and Characterization Conf., Boston, MA, Aug. 3-6, 1986, pp. 311–320.

Turbomilling Parameters Affecting the Ultrafine Grinding of Alpha-SiC

Jesse L. Hoyer

U.S. Department of Interior
Bureau of Mines
Tuscaloosa Research Center
Tuscaloosa, AL 35486

Abstract

The Bureau of Mines, U.S. Department of the Interior, turbomill is being evaluated for use in the production of ultrafine high-purity ceramic powders. In recent years interest has increased in the use of α-SiC as a construction material for use in applications where resistance to severe thermal, pressure, and environmental conditions is necessary. The effects of milling parameters, such as type of dispersant, temperature, pH, and gas environment on the milling of α-SiC were studied.

The grindability of α-SiC was improved by the addition of a dispersant. Of the six dispersants studied, the best results were obtained using Marasperse N-22. Increasing the temperature and making the slurry slightly basic also increased grinding efficiency; however, the gas environment of the mill had an adverse effect.

Optimum grinding was obtained at a slurry temperature of 50°C, pH 9.5 using Marasperse N-22 as the dispersant. A minus 100- plus 200-mesh α-SiC was ground to 80 pct less than 1 μm in 4 h.

Introduction

Interest in structural materials that can withstand severe conditions of temperature, pressure, and environment and can substitute for high-temperature alloys that require imported materials such as chromium, cobalt, and nickel has been increasing in recent years.

As a result, the utilization of ultrafine, high-purity ceramic powders in advanced and composite ceramics is increasing. Nonoxides such as SiC and Si_3N_4 have shown promise in meeting the requirements for use in heat engines, heat exchangers, and other areas where high temperatures and pressures are required.

The Bureau of Mines, U.S. Department of the Interior, turbomill (also known as an attrition grinder) has been used in the past to produce a wide variety of ultrafine materials[1-13]. This technique, originated and patented by the Bureau[3], consists of intense agitation of a milling medium, the material to be milled, and a suspending liquid. A previous report summarized research on turbomilling and discussed commercial applications of the process[14].

In a recent report[15], Wittmer discussed the development of an all-polymer mill, which would eliminate metal contamination of the mill product. Wittmer's report also describes the use of autogenous milling, in which the milling medium and the material to be milled are of the same or similar composition. In this study, the effects of dispersants, temperature, pH, presence of different gases, and milling time on particle size of alpha-silicon carbide (α-SiC) were investigated.

BUREAU OF MINES TURBOMILL

The turbomill (figs. 1 and 2), a 12.7-cm-diam unit, consists of four main parts: 1) a rotor (8.7 cm diam) composed of vertical bars fixed to upper and lower disks, the upper one attached to the drive shaft, 2) a cagelike stator (18.8 cm height) composed of vertical bars attached to rings at the top and bottom, 3) a cylindrical container (32.5 cm height) with a 3,400 mL working volume, and 4) a frame which holds the motor and the machine components. The rotor, stator, and container were constructed of ultra-high-molecular-weight polyethylene.

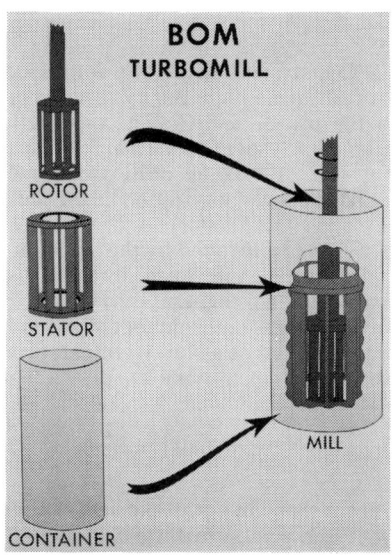

Fig. 1. Bureau of Mines turbomill (12.7 cm diam).

MATERIALS AND CONDITIONS

The starting material used in the tests was green α-SiC* of two different size fractions. A minus 20- plus 30-mesh fraction was used as the milling medium, and the material being milled was a minus 100- plus 200-mesh fraction (average diameter of 99 μm). The suspending medium was distilled water with 1 vol pct of a 10 wt pct solution of dispersant added. The six dispersants used were tetrasodium pyrophosphate, TSPP; Darvan #7, a sodium salt; Marasperse N-22, a sodium lignosulfonate; Nopcosperse 44, an ammonium salt; Aerosol OT, an anionic disodium sulfosuccinate; and Norlig NH, an ammonium lignosulfonate.† Other

Fig. 2.   All-polymer container, stator, and rotor components of the turbomill.

parameters for milling were total solids, 52 wt pct, with a coarse to fine ratio of 2.5:1; rotor speed, 1,500 rpm; bottom clearance of the rotor, 0.2 cm; temperature, 25°C; and a milling time of 4 h unless otherwise noted.

Experimental Work

During each milling test, a 250-mL sample was withdrawn each hour from the mill by syphon and passed over a 325-mesh screen. The oversize material was returned to the mill along with 250 mL of water. Particle size of the minus 325-mesh fraction was monitored with equipment (Microtrac Particle Size Analyzer, L&N Instruments, North Wales, PA) which utilizes a laser. This instrument does not distinguish soft agglomerates and particulates. Soft agglomerates are made up of particulates that do not successfully disperse. Prior to sizing, the samples were submerged in an ultrasonic cleaner to help break up these soft agglomerates. However, soft agglomerates can remain; therefore, the particle size is a measurement of the particulates and any remaining agglomerates.

Surface area was measured with a Brunauer-Emmett-Teller (BET) surface area analyzer. The equivalent spherical diameter (ESD), in μm, was determined from the specific surface area and the density of SiC (3.21 g/cm^3) by the equation

$$ESD = \frac{6}{S\rho}$$

where S = specific surface area, m^2/g,
and ρ = material density, g/cm^3.

The ESD is a measurement of an ultimate particle size which is the size of the particles present as actual particles or as part of an agglomerate. Since particle size measured using the laser technique does not differentiate between soft agglomerates and particulates, a comparison of particle size values determined by BET and laser techniques can indicate the presence of agglomeration. Scanning electron microscopy (SEM) was used to visually measure particle size and determine agglomeration characteristics of the powders.

After 4 h milling, the mill was washed out with distilled water. The discharge was wet screened through a 325-mesh screen. The plus 325-mesh material was dried and weighed and the percent minus 325-mesh material was determined by difference. This represents the amount of minus 100-plus 200-mesh starting material reduced to minus 325-mesh. The amount of minus 1 μm material was calculated using the Microtrac data. These values were used to compare milling efficiencies of different test runs.

The zeta potential was measured (using a ZM-80 Zeta Meter[†]) for selected samples that had been dried and screened through a 200-mesh screen.

Oxygen analysis was done by the analytical group at the Bureau's Albany (Oregon) Research Center using an inert gas fusion technique. The samples had been dried and screened to pass through a 200-mesh screen.

EVALUATION OF DISPERSANTS

Seven milling tests were run using six different dispersing agents and one control test was run with no additions. The average particle size values, based on laser measurements, of the α-SiC milled powders are shown in figure 3. The average size of the α-SiC ground using no dispersant, Marasperse N-22, Norlig NH, Aerosol OT, and Darvan #7 decreased gradually during the 4 h milling. The α-SiC ground with Marasperse N-22 showed the most rapid decrease. Particle size of the SiC prepared using TSPP and Nopcosperse 44 as the dispersant increased with time. Inadequate dispersion of the material resulted in the formation of soft agglomerates which could not be broken by ultrasonic treatment. Table 1 lists BET data for the six dispersant tests and shows that the ESD of the powders milled in TSPP and Nopcosperse 44 actually decreased. This indicates that the particle size actually decreased and the increase shown in figure 3 is actually a result of the formation of agglomerates during milling. SEM micrographs in figure 4 show the as-received SiC grain and the powder after turbomilling using TSPP as a dispersant. Agglomeration of the powder is evident.

Particle size distribution data, listed in table 2, for the Marasperse N-22 indicate that the amount of <1 μm material is about 55 pct after 1 h of milling; the amount does not increase significantly over the next 3 h. An SEM micrograph (fig. 4C) shows the powder prepared using Marasperse N-22 as the dispersant.

Contamination of the SiC powders prepared using Marasperse N-22 as the dispersing agent was negligible as determined by spectrographic analysis. Results are reported in table 3. Contamination of the SiC using the other dispersants was also insignificant, with the iron content decreasing with time. During milling, the surface of the particles is scrubbed resulting in removal of the iron, which is a surface contaminant. The iron cleaned from the surface of the particles during milling is dissolved in the milling medium and removed during washing. The decrease in iron content is advantageous because iron must often be leached from ceramic powders after grinding. Sodium contamination of some powders was a

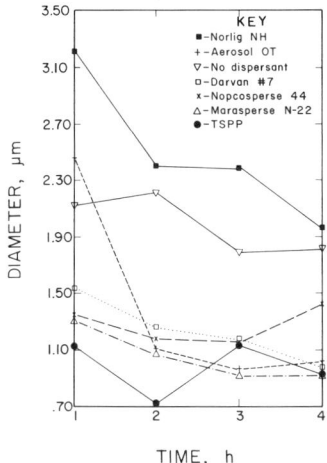

Fig. 3. Average particle size of α-SiC powders prepared by turbomilling.

Table 1. Particle size of α-SiC prepared using different dispersants.

Dispersant	ESD, μm 1 h	4 h
None	0.21	0.13
TSPP	.27	.16
Darvan #7	.22	.16
Marasperse N-22	.24	.17
Nopcosperse 44	.22	.18
Aerosol OT	.32	.20
Norlig NH	1.05	.28

result of the dispersant used. If Na^+ were found to be detrimental to the high-temperature properties of a ceramic body, milling with a nonsodium dispersant such as Nopcosperse 44 or Norlig NH would be necessary.

Effect of Temperature

Three tests were made with the grinding temperature of the slurry maintained at 25°, 50°, and 70°C. Marasperse N-22 was used as the dispersant. Table 4 shows the average particle size determined by laser and by BET methods. As the temperature of milling increases, the particle size decreases slightly.

However, the milling temperature does affect the grinding efficiency as shown by the percent of the material ground to minus 325-mesh and minus 1 μm after 4

Fig. 4. (A) Minus 200-mesh, α-SiC grain (X 150); (B) turbomilled α-SiC using TSPP (X 5,000); (C) turbomilled α-SiC using Marasperse N.22 (X 5,000).

Table 2. Particle size distribution data for milled α-SiC with Marasperse N-22 as dispersant.

Diam, μm	1 h[†]	pct[*] less than 2 h[†]	3 h[†]	4 h[†]
2.21	84	90	96	96
1.30	71	80	86	86
0.80	51	59	65	65
0.55	30	34	38	38
0.39	12	12	13	13
0.30	3	3	3	3
0.20	1	1	1	1

[*]Percent of minus 325-mesh fraction of milled material.
[†]Milling time.

Table 3. Spectrographic analysis of α-SiC with Marasperse N-22 as dispersant, percent.

Milling time, h	Fe	Al	Ca	Na	Mg
0	0.83	<0.10	<0.30	0.18	<0.01
1	.33	<.10	<.02	.23	.11
2	.21	<.10	<.02	.17	<.02
3	.23	<.10	<.02	.26	<.02
4	.26	<.10	<.02	<.23	<.02

Table 4. Particle size, grinding efficiency, and chemical analysis of α-SiC with Marasperse N-22 as dispersant after 4 h milling at different temperatures.

Milling temperature, °C	ESD, μm	Average diam, μm	pct* less than 325 mesh	1 μm	pct Fe	Na
25	0.18	0.87	60.6	43.8	0.18	0.13
50	.19	.76	78.0	57.5	.05	.14
70	.15	.73	82.1	63.1	.00	.19

*Percent of starting material.

h of milling listed in table 4. Table 4 also includes chemical analysis data for these tests. The iron content decreases with increasing milling temperature. Another advantage of milling at increased temperature is that the amount of cooling water required to adequately cool the system is reduced. Increasing the milling temperature from 50° to 70°C did not result in a significant increase in the grinding efficiency. Since mill wear increases as the temperature approaches the softening point of the polymer, running at 50°C would prolong the life of the polymer mill.

Effect of pH

Tests were run at pH values of 3.6, 6.5, and 9.5 to determine the effect of slurry pH on milling. An additional test was made at 9.5 and 50°C to evaluate the combined effects of pH and temperature. Table 5 includes particle size data for these runs. The data indicate that a change in pH does not affect the particle size of the material significantly.

The efficiency of grinding, table 5, is increased when the SiC is milled in a slightly basic environment. The dispersing efficiency of Marasperse is affected by pH, according to its manufacturers, with best results reported between pH 7 and 10[16]. This increased dispersion in a basic environment would permit greater particle/particle contact and increased grinding efficiency. The grinding efficiency of the silicon carbide increased slightly when the milling temperature was raised to 50°C. The combined effects of pH and temperature result in an increase in the grinding efficiency of the α-SiC.

Table 5. Particle size of α-SiC after 4 h milling at different pH values with Marasperse N-22 as dispersant.

Milling medium, pH	Milling temperature, °C	ESD, μm	Average diam, μ	pct* less than 325 mesh	1 μm
3.6	25	0.22	0.90	63.0	49.9
6.5	25	.18	.87	60.6	43.8
9.5	25	.20	.82	86.5	67.9
9.5	50	.21	1.14	92.6	79.7

*Percent of starting material.

EVALUATION OF DIFFERENT MILLING GAS ENVIRONMENTS

Four tests were performed in which argon was bubbled through the milling system, through a port on the side near the bottom of the container, at different flow rates. In these tests, a dispersant was not used, in order to determine the effect of the gas alone. The particle size data for these tests, listed in table 6, indicate no significant change with the addition of argon to the system. The percent of the 100- by 200-mesh material that passed through a 325-mesh screen after 4 h of milling and the amount of less than 1 μm are also listed in table 6. Argon did not appear to have a significant effect on the grinding efficiency of the SiC.

The addition of an emulsion stabilizer, EC-111§, to stabilize the bubbles in the system was also investigated. The SiC was milled using three levels of EC-111 addition at a flow rate of 1.0 scfh of argon. Table 6 includes particle size and grinding efficiency data for these three tests. The particle size of the SiC is not changed significantly by the addition of the emulsion stabilizer. However, the addition of the EC-111 increases the grinding efficiency slightly.

The effect of the presence of oxygen and nitrogen was also investigated. A flow rate of 1.0 scfh was used with no dispersant added to the system. Table 6 includes particle size data for the powders milled in the presence of these gases. The particle size data indicate that milling in oxygen or nitrogen has an adverse effect on milling efficiency of α-SiC.

Zeta potentials of the powders prepared using the different gases are listed in table 7. Zeta potential is a measurement of the effect of electrostatic charge in a colloidal system. A net repulsive force will cause the particles to repel each other as they approach each other resulting in less agglomeration and a more stable, better dispersed system. Due to less agglomeration in the system, the number of particle/particle contacts will increase resulting in an improved grinding action. With greater dispersion, (i.e. a more negative zeta potential measurement), the grinding efficiency should increase. As table 7 indicates, the grinding efficiency is greatest for the material with the lowest zeta potential.

Oxygen analysis data, table 7, for the powders indicate that milling in any atmosphere increases the oxygen content of the material. As a material is milled, new surfaces are formed that can adsorb oxygen; therefore, as the surface area increases, the oxygen content increases. The sample prepared in Marasperse N-22

Table 6. Particle size of α-SiC after 4 h milling in presence of different gas environments with Maraspere N-22 as dispersant.

Gas environment	Flow rate, scfh	ESD, μm	Average diam, μ	pct* less than 325 mesh	pct* less than 1 μm
No gas	NAp	0.13	1.81	58.6	16.8
Argon	0.5	.16	1.77	62.4	20.2
Do	1	.12	1.47	57	17.2
Do	1.5	.16	1.57	66.2	21.1
Do	2	.14	1.5	47.9	22
Argon + 3.25 mL of EC-111	1	.17	1.41	66.1	29.2
Argon + 6.5 mL of EC-111	1	.13	1.05	60	27.9
Argon + 13 mL of EC-111	1	.12	1.46	68.8	29
Oxygen	1	.13	1.66	39.5	19.3
Nitrogen	1	.13	1.64	40.1	23.4

NAp Not applicable.
*Percent of starting material.

Table 7. Zeta potential, oxygen content, and grinding efficiency of α-SiC milled under different conditions.

Milling conditions	Gas, 1 scfh	Zeta potential	O_2, pct	pct* less than 1 μm
No milling, as received	None	-19.5	0.31	0
No dispersant	None	-9.5	1.61	16.8
Do	Argon	-15.9	1.47	17.2
Do	Oxygen	-19.1	1.38	19.3
Do	Nitrogen	-20.7	1.37	23.4
Do	Argon + 6.5 mL of EC-111	-20.7	1.39	27.9
Maraspere N-22	None	-23.4	2.23	43.8

*Percent of starting material.

Silicon Carbide

showed the greatest increase in oxygen. This sample also has the largest amount of submicron powder after 4 h of milling.

CONCLUSIONS AND RECOMMENDATIONS

A study of the grinding of α-SiC in a turbomill using different dispersants, temperatures, pH values, and gas environments indicated the following results:

1. The addition of a dispersant increased grinding efficiency; the best results were obtained using Marasperse N-22.
2. The grinding efficiency of α-SiC is increased when the temperature of the slurry is increased from 25° to 50°C.
3. The grinding efficiency is increased when the pH is changed from acidic to basic with Marasperse N-22 as a dispersant.
4. In an argon environment the addition of an emulsion stabilizer, EC-111, increased the grinding efficiency.
5. Optimum milling efficiency was obtained using Marasperse N-22 as the dispersant at a pH of 9.5 and a temperature of 50°C. Under these conditions minus 100- plus 200-mesh SiC was ground to 80 pct less than 1 μm in 4 h.
6. Contamination of the powders produced in the turbomill was negligible, with iron content decreasing with time. The reduction of iron reduces the need for HCl leaching after milling.
7. Continuous removal of the submicron particles during milling and addition of unground material would lead to improved grinding efficiency. Studies of a continuous process for the production of ultrafine SiC should be evaluated.

---

*Norton Co., Worcester, MA. Reference to specific products does not imply endorsement by the Bureau of Mines.
†TSPP, Fisher Scientific, Fair Lawn, NJ; Darvan, R.T. Vanderbilt Co., Norwalk, CT; Marasperse and Norlig NH, Reed Lignin, Inc., Atlanta, GA; Nopcosperse, Diamond Shamrock, Inc., Morristown, NJ; and Aerosol OT, American Cyanamid, New York, NY.
‡Zeta Meter, Inc., Long Island City, NY.
§Sipex EC-111, Alcolac Inc., Baltimore, MD.

References

[1] E. G. Davis, Beneficiation of Olivine Foundry Sand by Differential Attrition Grinding. U.S. Pat. 4,039,625, Aug. 2, 1977.

[2] E. G. Davis, E. W. Collins, and I. L. Feld. Large-Scale Continuous Attrition Grinding of Coarse Kaolin. BuMines RI 7771, 1973, 22 pp.

[3] I. L. Feld and B. H. Clemmons. Process for Wet Grinding Solids to Extreme Fineness. U.S. Pat. 3,075,710, Jan. 29, 1963.

[4] I. L. Feld, T. N. McVay, H. L. Gilmore, and B. H. Clemmons. Paper-Coating Clay From Coarse Georgia Kaolins by a New Attrition Grinding Process. BuMines RI 5697, 1960, 20 pp.

[5] J. L. Hoyer and A. V. Petty, Jr. High-Purity, Fine Ceramic Powders Produced in the Bureau of Mines Turbomill. *Ceramic Engineering and Science Proceedings*, 6, [9-10] 1985, pp. 1342-1355.

[6] W. E. Lamont, G. V. Sullivan, E. G. Davis, and S. D. Sanders. Olivine Foundry Sand From North Carolina Dunite by Differential Grinding. Pres. at Soc. Min. Eng. Fall Meeting and Exhibit, St. Louis, MO, Oct. 19-21, 1977, Soc. Min. Eng. AIME preprint 77-H-369, 22 pp.

[7] L. Y. Sadler III, D. A. Stanley, and D. R. Brooks. Attrition Mill Operating Characteristics. Powder Technol., v. 12, 1975, pp. 19-28.

[8] M. H. Stanczyk, and I. L. Feld. Continuous Attrition Grinding of Coarse Kaolin (In Two Parts). 1. Open-Circuit Test. BuMines RI 6327, 1963, 14 pp.

[9] ———. Continuous Attrition Grinding of Coarse Kaolin (In Two Parts). 2. Closed-Circuit Tests. BuMines RI 6694, 1965, 13 pp.

[10] ———. Investigation of Operating Variables in the Attrition Grinding Process. BuMines RI 7168, 1968, 28 pp.

[11] ———. Ultrafine Grinding of Several Industrial Minerals by the Attrition Grinding Process. BuMines RI 7641, 1972, 25 pp.

[12] D. A. Stanley, L. Y. Sadler III, D. R. Brooks, and M. A. Schwartz. Attrition Milling of Ceramic Oxides. *Am. Ceram. Soc. Bull.*, 53, 1974, pp. 813-829.

[13] ———. Production of Submicron Silicon Carbide Powders by Attrition Milling. Paper in Proceedings of the Second International Conference on Fine Particles (Boston, MA, Oct. 7-11, 1973). Electrothermal and Metallurgical Division, The Electrochem. Soc., Inc., Boston, MA, 1974, pp. 331-336.

[14] M. H. Stanczyk and I. L. Feld. Comminution by the Attrition Grinding Process. BuMines B 670, 1980, 43 pp.

[15] D. E. Wittmer, Use of Bureau of Mines Turbomill To Produce High-Purity Ultrafine Nonoxide Ceramic Powders. BuMines RI 8854, 1984, 12 pp.

[16] Reed Lignin, Inc. The Chemistree Book, Handbook of Lignin Chemicals, p. 11.

ULTRASONIC IMPACT GRINDING

DAVID O. MOORE

Vice President, Bullen Ultrasonics, Inc.
4613 Camden Rd.
Eaton, OH 45320

ABSTRACT

Ultrasonic Impact Grinding is a non-traditional machining technique which can be applied in the machining of hard, brittle materials. Engineering ceramics, various kinds of glass, silicon, and fused silica are materials typically machined. The advantages of Ultrasonic Impact Grinding are: 1) low stress and/or damage caused in the material; 2) the ability to machine various and unique configurations; 3) the ability to machine multiple cavities or configurations at once and; 4) surface finishes lower than .4 microns.

This paper discusses the machines used, the tooling required, and the materials machined with this process and will look at some recent applications in the aerospace, automotive and electronic industries.

INTRODUCTION

Ultrasonic Impact Grinding is the use of ultrasonically induced vibrations delivered to a designed tool, combined with an abrasive slurry, to produce accurate cavities of regular and odd shapes in hard, brittle materials. This machining process is non-thermal, non-chemical, non-electrical and creates no change in the metallurgical, chemical, or physical properties of the material machined. Typical materials machined are fused quartz, glass, single crystal silicon, engineering ceramics, carbides and various metals.

Ultrasonic Impact Grinding is the conversion of a high frequency electrical signal into mechanical motion which is acoustically transmitted thru a metal toolholder and cutting tool. This linear oscillation is typically at a rate of 20,000 times per second and is used with an abrasive slurry flowing around the cutting tool to microscopically chip the material away. The machined area becomes an exact counterpart of the cutting tool used in the operation; therefore, the types and shapes of cuts to be made are almost limitless.

Conversion of Electrical Energy To Mechanical Motion

Ultrasonic vibrations are produced by coupling an electronic generator, a transducer package (either magnetostrictive or piezoelectric), a transmitting connecting body, and a toolholder-tool combination. The generator converts typical line voltage into the voltage and frequency required to energize the transducer coupled to it. The typical nominal frequency used is 20KHz.

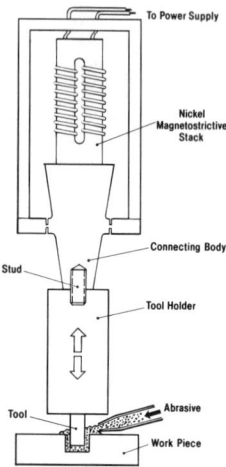

Fig. 1. Ultrasonic transducer and toolholder assembly.

In many Ultrasonic Impact Grinding applications a magnetostrictive transducer is used because of its wide bandwidth in frequency range. A transducer designed for 20KHz nominal operating frequency may have a bandwidth of 17KHz to 23KHz. This allows much more design flexibility in toolholders and allows tooling to wear and be dressed many times before the tooling is too short to tune in and resonate.

Most transducers are connected to a transmitting connecting body. This connecting body is designed to resonate at the same nominal frequency of the transducer and provide the means for which the toolholder can be connected. In most cases a threaded stud is used on the end of the connecting body to connect the toolholder.

Toolholder Design

The metal toolholder and tool combination must be designed to transmit the acoustic energy properly and resonate within the bandwidth of the transducer used. The resonant length is partially dependent on the material used for the toolholder-tool. For Ultrasonic Impact Grinding toolholders, 304 Stainless Steel or Monel are typically used because of their soldering and brazing characteristics.

The stroke (actual oscillating movement of the tool) of the toolholder must be capable of handling the particular abrasive size used in the grinding operation. Because of this, toolholders fall into two categories: non-amplifier and amplifier. A non-amplifier will reproduce the stroke of the connecting body. If the connecting body is set to have a stroke of .013 mm (.0005"), then the toolholder will have a stroke of .013 mm (.0005"). An amplifier will increase the stroke. The maximum stroke usually desired for grinding applications is .064 mm (.0025"). This can be achieved by designing the toolholder as an amplifier and multiplying the input stroke from the connecting body to the desired output of the tool. Toolholders made from high strength materials can have an output stroke as high as .203 mm (.008") without being over stressed.

The attachment of the cutting tool to the toolholder can be accomplished by several methods depending on the desired operation parameters. First of all, the cutting tool can be machined directly into the end of the toolholder. This is the strongest joint because the cutting tool is held by the strength of the parent material itself.

Soldering and silver brazing are commonly used to attach cutting tools. Silver brazing will provide a bond that should allow the tool to have a stroke of .06 mm (.0025"). Low temperature soldering will give a bond to allow a stroke of .04 mm (.0015"). The advantage of using a soldering bond is that the toolholder can be used again after the tool has worn out.

High strength adhesives can be used as long as the stroke is below .013 mm (.0005"). This would be used when a fine abrasive is required for a desired finish. Again, the tools are easily changed, thus saving the toolholder body.

Mechanical connection of the toolholder to the tool is another option. This allows the tool designer to use different materials for the tool and toolholder. An example would be using a titanium toolholder and a stainless steel tool. The maximum stroke recommended in this application is .064 mm (.0025"). Mechanically held tools are very useful for high volume applications. The two grooves shown in the quartz housing in Figure 2 are being machined with mechanically held carbide inserts. The inserts cost less than $1.00 and can machine 30 parts before being discarded.

The Abrasive Element

The abrasive is the most important item of Ultrasonic Impact Grinding. Commonly used abrasives are: boron carbide, silicon carbide, and aluminum oxide. Boron carbide is the hardest abrasive and will last the longest. Silicon carbide can be used for such materials as glass, quartz, and single crystal silicon. Boron carbide is recommended for machining the engineering ceramics (silicon carbide, silicon nitride, boron carbide, aluminum oxide, zirconium oxide, etc.) and metal parts.

The abrasive is suspended in water at approximately 50% by volume and is pumped to the part at a rate of 1-3 gallons per minute. This high volume of slurry acts to cool the tool and workpiece as well as supplying the abrasive and removing the cut particles. The fast motion of the tool impacts the abrasive into the workpiece, thus chipping or grinding away the material. For deep cuts, vacuum assistance can be used to help pull the abrasive into the cut, and then out thru the middle of the toolholder.

The abrasive size will determine the surface finish and speed of the cut. Table No. 1 compares the typical average finish obtained with various abrasive sizes. A comparison of cutting speed versus abrasive size is shown in Table No. 2. The only variable is the abrasive size. All other parameters (tool pressure, stroke, workpiece material, etc.) remain constant.

Applications

Today Ultrasonic Impact Grinding is being used heavily in the electronics, aerospace, and automotive industries. The electronic industry uses the process for

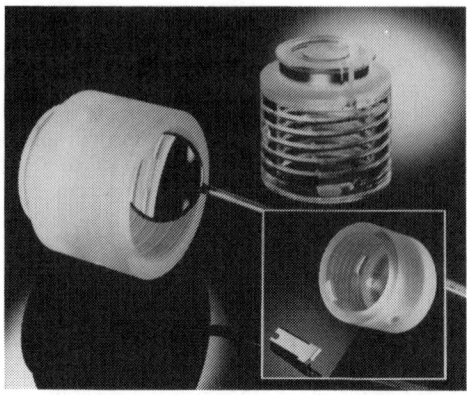

Fig. 2. Quartz housing with two internal grooves cut ultrasonically.

Table 1. Abrasive size vs finish.

50 micron (240 grit)	1 micron (44 micro inch)
38 micron (320 grit)	.9 micron (36 micro inch)
30 micron (400 grit)	.7 micron (26 micro inch)
20 micron (600 grit)	.4 micron (18 micro inch)
9 micron (800 grit)	.2 micron (8 micro inch)

Table 2. Abrasive size vs cutting speed.

30 micron	.026 cm^3/min.
20 micron	.011 cm^3/min.

machining small holes thru ceramic substrates. When plating thru these holes is required, the hole surface must be very good for the plating to adhere. Because Ultrasonic Impact Grinding is a non-chemical, non-thermal, non-electrical process, it can achieve stress-free, good quality holes (see Fig. 3).

Figure 4 shows a close up view of a .64 mm (.025) thick ceramic substrate with a very tight pattern of .038 mm (.015) diameter holes that were machined at once with one tool. Tooling has been made to machine over 1200 holes in a 76 mm square area.

Figure 5 shows another application where square cavities were machined to a certain depth and holes machined the rest of the way thru. In this application the square cavities and thru holes were machined with two different tools. The locational tolerance of the square cavities was .05 mm (.002") true position.

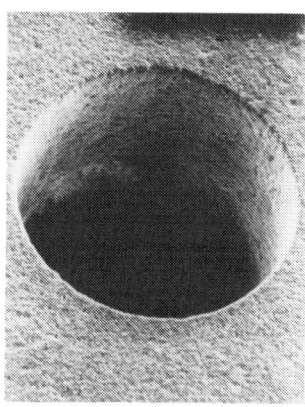

Fig. 3. 100X view of ultrasonically machined hole.

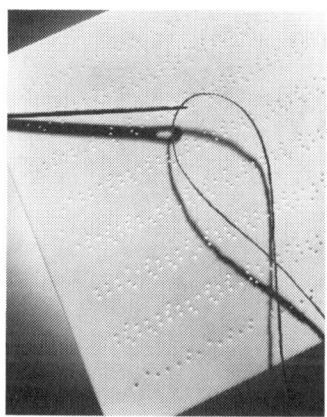

Fig. 4. Alumina substrate with .38 mm (.015") holes.

Large diameter holes can also be machined. In machining the wrist pin hole in a silicon nitride piston, the tool had to cut a 64 mm (2.5") diameter plug out of the middle. Tools for cutting up to 127 mm (5.0") diameter have been developed.

Because of the requirements for higher operating temperatures, Ultrasonic Impact Grinding is used extensively in machining components for gas turbines. These components are being tested for automotive and aerospace applications.

Figure 6 shows several silicon carbide parts and associated tooling. These parts are being tested in an automotive gas turbine program directed by NASA.

Fig. 5. Ceramic pallet with square cavities and thru holes.

Fig. 6. Silicon carbide parts for AGT Program.

Equipment

Ultrasonic Impact Grinders rated from 200 watt to 2400 watt are available. These machines can be bench mounted or complete floor standing models with full C.N.C. control of axis movement. Special models are built for unique applications to fit the part being machined or suit the machine for a high volume application.

Summary

Because of its stress-free machining, and the increasing use of ceramics, glass, and graphite compounds in the electronics, aerospace, and automotive fields, Ultrasonic Impact Grinding is experiencing continued growth. The ability to machine an almost limitless variety of sizes and shapes makes this process unique in machining and brittle materials.

Suspension Processing of Beta-SiC Powders

Bruce A. Bishop and H. Kent Bowen

Ceramics Processing Research Laboratory
Materials Processing Center
Massachusetts Institute of Technology
Cambridge, MA 02139

Abstract

Nonaqueous dispersions of as-received and centrifugally size-classified (0.2-μm) β-SiC powders were prepared separately in heptane or toluene using the commercial dispersant OLOA™ 1200. SiC suspensions were doped with B_4C and carbon black (in heptane) or B_4C and polyphenylene (in toluene) in the dispersed state. Composite dispersions were colloidally pressed to form green compacts. Parts were subsequently isopressed, heat-treated for binder removal, isopressed again, then sintered at either 2000°C or 2100°C. Densities exceeding 99% of theoretical were achieved. The effects of dispersion quality, particle size, carbon source, binder removal, isopressing, and sintering temperature on densities and microstructure development are discussed.

Introduction

Pressureless sintered silicon carbide is an important structural ceramic that combines in one material excellent high-temperature strength retention[1], good resistance to oxidation[2] and creep[3], and good tribological properties[4]. It is therefore an excellent candidate for high-temperature, load-bearing applications. Furthermore, as is the case with most ceramics, raw materials for the production of SiC (i.e., silica and carbon) are cheap and plentiful.

Processing-induced defects, however, limit the effective strength of polycrystalling SiC parts. Processing flaws of 20-80 μm in one case[5] and over 100 μm in another[2] limited the room-temperature flexural strengths of test bars to the 40-50 ksi range. Some of the flaws exceeding 100 μm were found to be caused by agglomerates in the green microstructure[2]. Because ceramics processing affects both micro- and macroscale structure and subsequent properties, research in this area is essential to increased performance and economic efficiency.

Most SiC ceramics processing is done with SiC powders mixed with a particulate boron source and a particulate or polymeric carbon source. The dispersion of powders through repulsive forces to improve particle packing and increase dopant homogeneity is not always considered. Recently, however, Freedman and Millard[6] used both aqueous and alcoholic dispersions of SiC to improve the uniformity of SiC suspensions, with subsequent improvements in densification and mechanical properties. Nonaqueous dispersions of SiC in low-dielectric-constant media can also be used to process SiC ceramics.

Processing can be improved by incorporating two plans of action. First, eliminate ceramic-powder-packing defects by using colloidal dispersions. Electrostatic or steric repulsion or a combination of both in colloidal dispersions serves to separate the SiC into single particles that pack uniformly to a high density. This packing of discrete particles minimizes the number and size of defects in the body and results in increased strength.

Second, disperse sintering aids more uniformly through the use of particulate suspensions and/or molecular chemical additives. Homogeneously doping SiC with the sintering aids necessary for densification prevents defects from forming and allows complete densification to occur more easily. Thus, improved structural properties are realized.

Dispersions

OLOA™ 1200, an effective dispersant for submicrometer powders in low-dielectric-constant media, was used exclusively in this research. Interaction of the OLOA™ 1200 molecule with surface acidic sites resulted in dispersion of a carbon black by both steric and electrostatic mechanisms[7,8]. This combination of repulsive mechanisms can also be produced in aqueous systems using water-soluble dispersants[9]. Based on stability measurements in the OLOA™ 1200/carbon black system, a concentration of approximately 4 wt% OLOA™ 1200 by weight of powder was needed to disperse a carbon black that had a surface area of 25 m^2/g. Similar observations about the repulsion mechanism were made for Betarundum Ultrafine β-SiC, which also has acidic sites in the form of silanol groups[10]. In fact, the amount of dispersant required for the two powders varied in proportion to the difference in surface area.

As shown in Figure 1, the stable, Newtonian suspensions became shear thinning at higher solids contents. There are two possible causes for this rheological change. First, the combination of 1) the reduction in the electrostatic component of the repulsive force imparted by OLOA™ 1200 in a low-dielectric-constant medium with increasing solids content[11] and 2) the increase in flocculation rate with the square of the number concentration of particles could produce a softly flocculated suspension from the stable dispersion. Second, shear thinning could result from the secondary electroviscous effect, i.e., from the repulsive interaction of electrical double layers that forces particles out of their flow paths by an amount that decreases with increasing shear rate[12]. The secondary electroviscous effect is prominent in aqueous, high-solids-content dispersions containing low concentrations of electrolyte. A high-solids suspension made with a low-dielectric-constant medium could mimic such a dispersion because the medium cannot support ionic species in high concentrations.

The change from Newtonian to shear thinning can be advantageously used in ceramics processing. The increase in viscosity as the shear rate approaches zero hinders differential settling of the powder in suspension, while the shear thinning of the dispersion allows easier processing at some finite shear rate. Also, less liquid needs to be removed from the dispersion to form a compact. More uniform processing could result from this system.

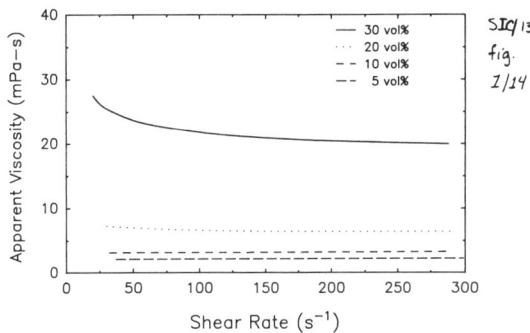

Fig. 1. Apparent viscosity as a function of shear rate for various volume fraction solids in the SiC/OLOA™ 1200/dodecane system.

Particle Size Classification by Centrifugation

In order to improve sintering kinetics, microstructural development, and ease of processing, the as-received Betarundum Ultrafine β-SiC powder was size classified into a narrow-size-distribution powder. The as-received powder ranged in size from less than 0.1 μm to ~2.0 μm in diameter with an average size of 0.35 μm as measured by the cross-sectional area, and exhibited morphologies from equiaxed to anisotropic. Particles less than 0.5 μm in diameter constituted >90% of the classified powder by cross-sectional area with an average size of 0.16 μm, and the morphology was equiaxed. Consequently, the classified powder is much closer to ideal than the as-received powder[13]. Details of the classification procedure are reported elsewhere[14].

The remainder of this report describes both the initial processing sequence used and the second, improved, sequence.

SUSPENSION PROCESSING: SEQUENCE I

Procedure

The suspension processing procedure for Sequence I is shown in Figure 2. Separate dispersions of as-received SiC, B_4C, and carbon black were prepared in heptane using OLOA™ 1200[*] as the dispersant (Table I). OLOA™ 1200 concentrations for the B_4C and carbon black were initially calculated relative to the SiC, based on the respective surface areas of the powders derived from the manufacturer's literature. The high solids content of the SiC suspension was used to take advantage of the shear thinning in the OLOA™ 1200 system. The dispersions were mixed at least one day, sonicated in an ultrasonic bath for 5 min, then used to prepare composite dispersions containing 0.5 wt% B_4C and 2.5 wt% carbon. The composite dispersions were mixed for several days, then sonicated for 1 h before colloid pressing (Fig. 3) at 69 MPa with a 20 min hold. The resulting pellets were isopressed, heat-treated to remove organics, and isopressed again before firing at

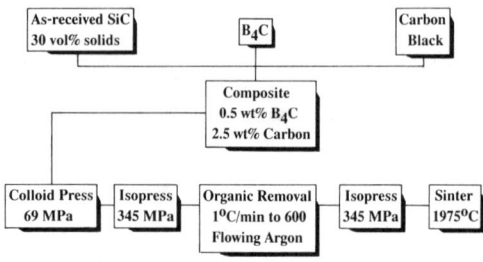

Fig. 2. Suspension Processing Sequence I.

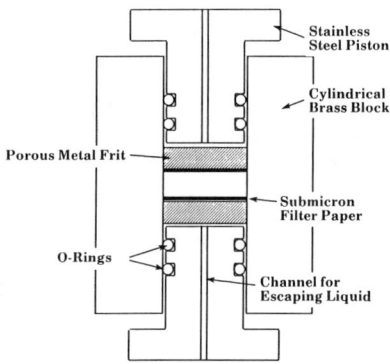

Fig. 3. Colloid press used to make 1.9-cm-diameter compacts.

Table I. Material concentrations in powder dispersions used for Sequence I.

Material	OLOA™ 1200 (wt%)*	Solids content (vol%)
SiC[†]	4	30
Boron carbide[‡]	18	1
Carbon black[§]	50	9

*By weight of powder.
[†]Betarundum Ultrafine β-SiC, Ibiden Co., Ltd., Ogaki, Japan.
[‡]Callery Chemical Co., Callery, PA 16024.
[§]Monarch 905 Carbon Black, Cabot Corp., Specialty Blacks Div., Boston, MA 02110.

1975°C for 20 min under flowing argon. The reasons for using the isopress-burnout-isopress route are discussed under SUSPENSION PROCESSING: SEQUENCE II Results and Discussion.

Results and Discussion

The green densities of these parts were 65% and the sintered densities were 83% of theoretical for SiC, based on weight and dimensional measurements. A representative fracture surface of the compacts is shown in Figure 4A. Finely textured agglomerates 10-30 μm in diameter were present throughout the bodies. The accompanying silicon x-ray dot map in Figure 4B shows that the agglomerates are silicon-deficient; therefore, they are either carbon black or boron carbide. The agglomeration effectively decreases the amount of dopant in the system, and any significant decrease in dopant level would result in lower part densities. Fired densities of these parts were indeed low.

The amount of dispersant used with the carbon black and boron carbide powders was determined by comparing each powder's surface area to that of SiC, for which the amount of dispersant required had already been determined, and adjusting the amount of dispersant accordingly. Since agglomerates in the compact microstructures were observed, the surface areas of the carbon black and boron carbide were directly measured by multipoint BET analysis and found to be 346 and 72 m^2/g, respectively, as compared to the 230 and 80 m^2/g values derived, incorrectly, from the manufacturer's literature. Therefore, not enough OLOA™ 1200 had been used to disperse the carbon black. This calculation assumes that the surface area of the carbon black was completely available to the dispersant. This may not be the case if the carbon contains pores smaller than the OLOA™ 1200 molecules and these pores make up a significant amount of the surface area. However, an insufficient amount of dispersant is a possible cause of agglomerate formation.

A second reason for agglomerate formation may have been imperfect powder dispersion. Very fine powders are difficult to disperse completely. More work on the dopant powder suspensions may be necessary in order to completely disperse the particles. Also, the composite dispersion was mixed and sonicated at about 30 vol% solids; more deagglomeration could be achieved if these processes were performed at a lower solids content.

SUSPENSION PROCESSING: SEQUENCE II

Procedure

An alternative processing sequence, with changes incorporated to address the difficulties encountered with Sequence I, is shown in Figure 5. Four different compositions were prepared by this sequence: A) as-received SiC with B_4C and carbon black, B) classified SiC with B_4C and carbon black, C) as-received SiC with B_4C and polyphenylene†, and D) classified SiC with B_4C and polyphenylene. Compositions A and B were prepared in heptane; compositions C and D were prepared in toluene. All systems used OLOA™ 1200 as the dispersant.

The B_4C and carbon black dispersions were improved over those used for Sequence I processing. Dispersion series were prepared based on the experimentally measured powder surface areas. Supernatants of the dispersions were observed by

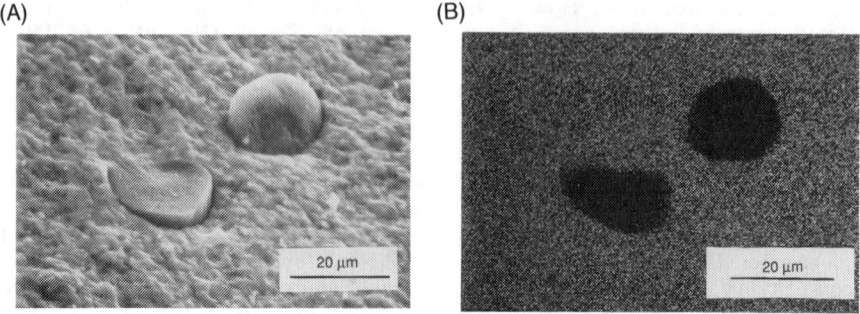

Fig. 4. (A) Fracture surface of compact produced using Sequence I, showing agglomerates resulting from processing; (B) silicon x-ray dot map of area shown in (A), indicating that the agglomerates are silicon-deficient.

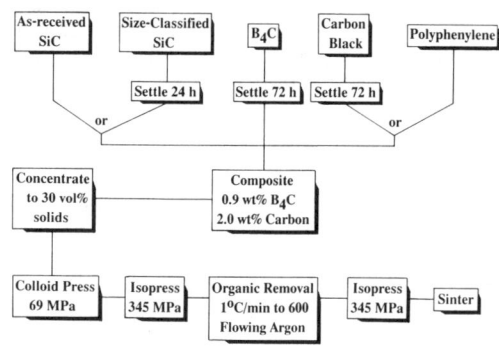

Fig. 5. Suspension Processing Sequence II.

SEM to check the degree of dispersion. The best dispersions (18 wt% and 72 wt% OLOA™ 1200 by weight of powder for B_4C and carbon black, respectively) were used in subsequent processing.

All particulate suspensions were made in the 5–10 vol% range in order to take advantage of the lower viscosity for better deagglomeration. Dispersions were mixed at least 64 h, then either sonicated for 3 min in the ultrasonic probe cup horn or for 30 min in the bath. Dispersions of the classified SiC, B_4C, and carbon black were allowed to settle in order to remove any remaining agglomerates or other large particles. This same procedure has been used to improve Al_2O_3-ZrO_2 composites[15].

Composite dispersions containing 0.9 wt% B_4C and 2 wt% carbon were prepared by mixing the dopants in either 5.7 wt% OLOA™ 1200/as-received SiC powder dispersions or ≈8 wt% OLOA™ 1200/classified SiC dispersions for at least 48 h, followed by sonication, as before. The amount of dispersant in the classified powder dispersions was a result of the classification procedure itself. Dispersions were then put under a nitrogen blanket and the solvent evaporated. The final solids content was approximately 30 vol%.

The consolidation route was that of Processing Sequence I. Densities were obtained after each operation, from colloidal pressing through the second isopressing step, by measuring weights and dimensions. Pore-size distributions were measured by mercury porosimetry† for B_4C- and carbon black-doped parts that had only been colloidally pressed (at 5 vol%) and burned out and for parts that had been formed by Sequence II. Values of 130° and 0.485 J/m² were used for the contact angle and the mercury surface tension, respectively.

Samples were randomly stacked in graphite crucibles and placed in a vertical carbon-tube furnace§ to be sintered at 2000°C or 2100°C for 40 min under 2.8 L/min of flowing argon. The heating rates used were 10°C/min to a temperature of 1500°C, then 25°C/min to the sintering temperature.

After sintering, densities were measured by liquid immersion using *1*-butanol. Fracture surfaces for each materials system sintered at 2000°C or 2100°C were observed by SEM.

Results and Discussion

Powder packing and volume and weight changes are shown in Table II. First, as-received powders packed 4-7% more densely than did classified powders. Second, volume shrinkage during the first isopressing procedure was proportional to the amount of organic in the system. Third, a volume expansion was observed for all samples after organic removal. Finally, the second isopressing procedure reversed the volume expansion which had occurred during binder removal.

The higher green density of the as-received powder is due to its wider particle-size distribution, since small particles can pack within the interstices of large particles. This can cause problems (e.g., bridging can occur) if the packing is not complete, resulting in a wide pore-size distribution with some large pores so highly coordinated that they remain stable until grain growth or rearrangement decreases the number of grains around the pores and makes them thermodynamically unstable. Also, the pore concentration throughout the part will not be uniform, since the large particles have no pores within them while the small particles have many pores about them. A nonuniform pore concentration can hinder sintering[16].

Powder consisting of small particles with a narrow size distribution can pack to densities well above 70% if the particles are allowed to rearrange upon settling[17]. If the classified powder is contained within a viscous, possibly softly flocculated suspension and is not allowed to rearrange, however, relatively low densities (55-60%) result. This lack of rearrangement is not surprising, too, because of the high levels of organic used: the classified-powder compacts contain about 25 vol% organics and the as-received powder compacts contain 15-17 vol% organics. Again, voids in the packing can result in stable pores, but the pore concentration should be uniform.

When isopressed, the particles in the green bodies rearrange under the applied pressure. Particle rearrangement in the system studied was enhanced by the lubricating nature of the dispersant. During isopressing, the largest pores are preferentially eliminated, thereby decreasing the pore coordination number[18].

OLOA™ 1200, however, causes a problem upon removal. When a solid organic is removed from a body as a vapor, the volume of the organic increases. This can

Table II. Results of Sequence II powder packing.

	Materials System*			
Processing Step	A	B	C	D
1. Colloid Pressing Density (%)	64	57	63	59
2. Isopressing Density (%)	67	62	66	62
Vol% Change	-5.3	-8.0	-4.2	-6.2
3. Density after Burnout (%)	63	55	62	57
Vol% Change	+0.4	+0.9	+1.0	+0.5
Wt. Loss (%)	6.3	10.1	4.9	7.9
4. Isopressing Density (%)	63	56	62	58
Vol% Change	-0.5	-0.7	-0.9	-.09

*A: As-received silicon carbide, boron carbide, and carbon black
B: Classified silicon carbide, boron carbide, and carbon black
C: As-received silicon carbide, boron carbide, and polyphenylene
D: Classified silicon carbide, boron carbide, and polyphenylene

result in pressure increases within the part that can form defects. The volume expansions recorded for these parts may seem small, but if the volume increase is assumed to be in the form of 50-μm-diameter voids, the number of voids in these parts would be on the order of 10^5. Organic removal from green ceramic ware is no trivial problem in either a structural or a chemical sense, especially in oxide- and noncarbide-based systems.

Because of this volume expansion upon organic removal, a second isopressing step was included in the processing sequence. As seen in Table II, this additional isopressing procedure resulted in volume shrinkage on the order of the binder-removal expansion. It was assumed that the additional isopressing step removed the packing defects that had formed during organic removal (a similar isopressing step was used to eliminate intentionally formed, controlled-size voids from Al_2O_3-ZrO_2 composites[19]).

The effects of powder classification and cold isostatic pressing (CIP) on packing are apparent in the pore-size distributions (Fig. 6). (It should be noted that when mercury porosimetry is used to measure pore sizes, the channel sizes between pores rather than the actual pores are measured; this method assumes that pore size scales with the channel size.) With or without isopressing, classified powders produce pore-size distributions that are narrower and pores that are, on the average, smaller than what can be obtained with unclassified powders. Cold isostatic pressing eliminates the largest pores, thereby reducing the average pore size and narrowing the pore-size distribution. The combination of powder classification and isopressing produces a pore-size range from 23 to 5 nm with a maximum at 17 nm; in

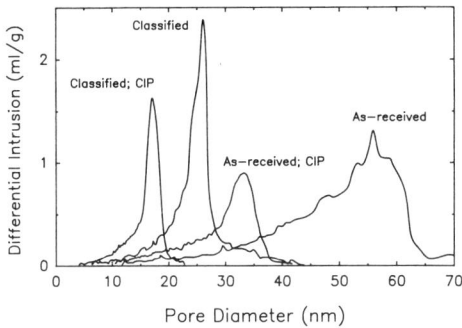

Fig. 6. Pore-size distributions of boron carbide- and carbon black-doped silicon carbide compacts made with as-received and classified powders, with and without the isopressing steps included in Sequence II.

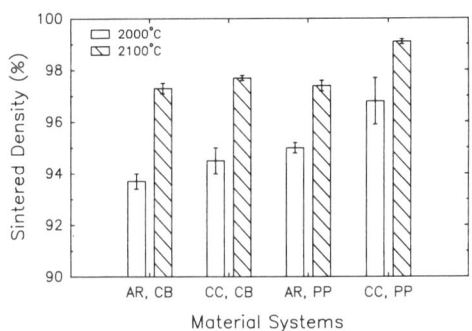

Fig. 7. Sintered densities of compacts made from the four material systems at 2000°C and 2100°C. (Error bar is the standard deviation obtained using three samples.)

comparison, as-received, nonisopressed compacts contain pores ranging from greater than 70 nm to less than 15 nm, with most pores having a size of 56 nm.

Isopressing and powder size classification should improve the densification of ceramic compacts. Isopressing either classified or as-received powder compacts reduces their pore sizes and size distributions and increases their densities. As a consequence, the number of grains surrounding a pore (i.e., the pore coordination number) should decrease and the number of particle-particle contact points should increase. Because the larger pores would be eliminated, fewer pores would be thermodynamically stable and less grain growth would be required to lower the pore coordination number. Therefore, isopressed compacts should densify more readily than compacts that have not been isopressed.

The use of classified powder to make compacts should increase the driving force for densification, decrease the driving force for grain growth, and produce smaller pores with narrower pore-size distributions and a more uniform pore concentration throughout the body. These effects combined with the isopressing procedure will promote sintering to higher densities.

The actual sintering results obtained in this study are shown in Figure 7. At 2000°C, densities ranged from about 94 to 97% of theoretical. Classified-powder compacts had higher densities than did compacts made from as-received powder and the same dopants. Also, samples doped with polyphenylene had higher densities than did those doped with carbon black for the same type of powder. These same trends held for specimens sintered at 2100°C.

Figures 8–11 reveal microstructures obtained in the various specimens by sintering at 2000°C. Figure 8 shows the fracture surface of a 93.6% dense compact made with as-received SiC, B_4C, and carbon black. This microstructure is much less dense than the one shown in Figure 9, from a compact which was 96.7% dense and consisted of classified SiC, B_4C, and polyphenylene. Figures 10 and 11 show defects observed in the SiC bodies. Figure 10 shows a large grain which had grown in the midst of fine matrix grains of a carbon black-doped, classified-powder compact; large grains made up a very small fraction of the total sintered volume. Figure 11 shows a crack that had formed in a carbon black-doped, classified-powder compact. A green part made within this material system was found to contain similar cracks.

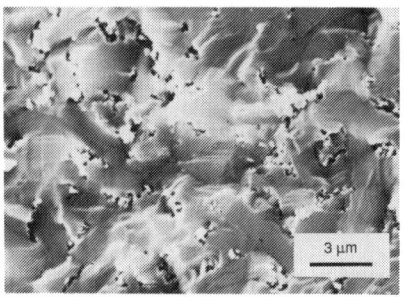

Fig. 8. Fracture surface of an as-received powder compact doped with 0.9 wt% B_4C and 2 wt% carbon black, sintered at 2000°C.

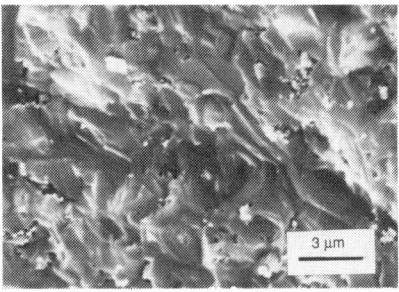

Fig. 9. Fracture surface of a classified-powder compact doped with 0.9 wt% B_4C and 2 wt% carbon in the form of polyphenylene, sintered at 2000°C.

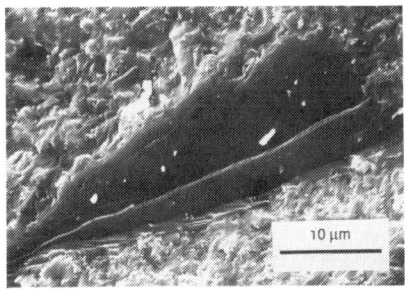

Fig. 10. Fracture surface of a classified-powder compact doped with 0.9 wt% B_4C and 2 wt% carbon black, sintered at 2000°C. Note the 35-μm platey grain in the midst of a fine-grain matrix.

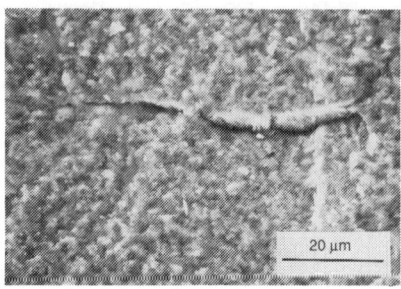

Fig. 11. Fracture surface of a classified-powder compact doped with 0.9 wt% B_4C and 2 wt% carbon black, sintered at 2000°C. Note the 50-μm crack.

The microstructures shown in Figures 12 through 14 were prepared at 2100°C. Figure 12 is a 97.3% dense body made with as-received SiC, B_4C, and carbon black. Small spherical pores are evident, but the grain size is not discernible. This is an indication that the compacts have been overfired and produced exaggerated grains through the β-α phase transformation. The microstructure in Figure 13 is found in a 99.1% dense compact made with classified SiC, B_4C, and polyphenylene, fired at 2100°C. It too has no observable grain size and is assumed to be overfired. Evidence of this phase transformation is shown by the feather-like structure in Figure 14[20].

With regard to densification, a finer particle size with a narrower particle-size distribution and a polymeric carbon dopant gave the best results. As expected, classified-powder compacts yielded better results than did as-received powder compacts containing the same dopants. The observation that a polymeric carbon source produces higher densities than does a carbon-black dopant has been made before[21]. This result was attributed by Prochaska to the greater dopant uniformity provided by the polymeric carbon source. In Process Sequence II, an attempt was made to disperse the carbon black as well as possible. Consequently, the densities were not as different as previously observed, but carbon-black doping still gave lower densities than did doping with a polymeric carbon source.

Three types of structural defects were found in the parts. At 2000°C, grains on the order of 10-35 μm were observed in the classified-powder parts. These grains were smooth and flat compared to the finer matrix. When approximately 10 μm in size, the grains appear elliptical. As they become larger, the aspect ratio increases, and ridges down the length of the grains become prominent. Based on their morphologies, the grains are thought to be alpha grains in the beta matrix. Three possible causes for the formation of these grains in the classified-powder compacts are: 1) the growth rate of the platey grains may have been higher than that in the unclassified-powder compacts due to the possibly finer matrix in the former compacts, 2) impurities such as aluminum may have located preferentially in the smaller particles, thereby promoting the transformation, and 3) the smaller size-classified particles may have contained a higher concentration of defects (e.g., stacking faults) than did the as-received powder, resulting in more nuclei at which the β-α transformation could occur. Density does not seem to play a role in the

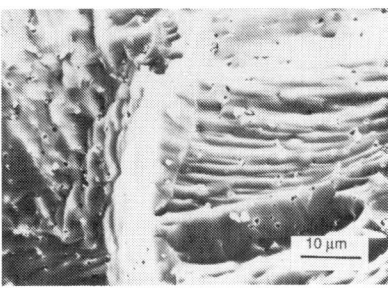

Fig. 12. Fracture surface of an as-received powder compact doped with 0.9 wt% B_4C and 2 wt% carbon black, sintered at 2100°C.

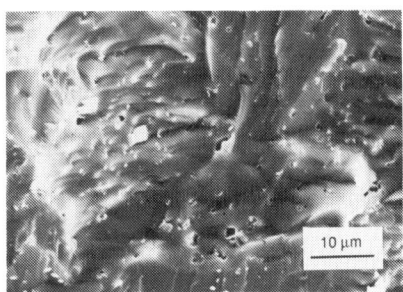

Fig. 13. Fracture surface of a classified-powder compact doped with 0.9 wt% B_4C and 2 wt% carbon in the form of polyphenylene sintered at 2100°C.

platey-grain formation because the parts made with as-received powder and polyphenylene had a higher density than did the classified-powder and carbon-black parts, but showed no such grains.

Cracks were observed that ranged in size from 50–100 μm to about 10 μm. Also, small radial drying cracks were observed. Both types of cracks were found predominately in the classified-powder compacts, which contained the most dispersant. The carbon-doped, classified-powder parts exhibited the most cracks. Since the OLOA™ 1200 remains solvated after pressing and the classified-powder compacts contain the smallest pores, the greater amount of retained solvent in these compacts could be the cause of their cracking. Any excess dispersant remaining from the classification process would only make the cracking worse. Consequently, the OLOA™ 1200 concentration should be held to a minimum.

The phase transformation at 2100°C can also be considered a defect, in that the feather structure has been reported to decrease the strength of SiC by about 30%[20]. The transformation from β to α SiC can be controlled by the choice of temperature,

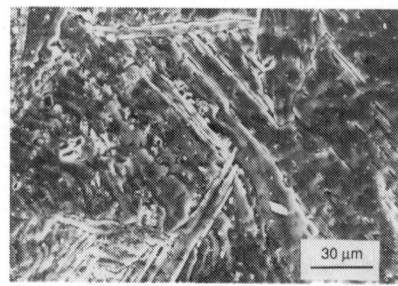

Fig. 14. Feather-like structure found in the fracture surface of a classified-powder compact doped with 0.9 wt% B_4C and 2 wt% carbon in the form of polyphenylene, showing evidence for the β-α phase transformation.

time, atmosphere, and sintering aid. In order to produce high-quality SiC parts reproducibly from β-SiC powders, this transformation will have to be suppressed.

SUMMARY

Stable dispersions that became shear thinning at high solids content were produced through the use of a commercial dispersant, OLOA™ 1200, and used to make nonagglomerated SiC green compacts. When poor suspensions of dopants were used, carbon-rich agglomerates resulted, which lowered the fired densities. Isopressing both before and after organic removal improved packing densities and removed defects caused by binder removal. However, still present were cracks believed to be associated with amounts of dispersant greater than the minimum amount required for dispersion, and large platey grains believed to have resulted from the β-α transformation. Classified powders, as compared to as-received powder, produced parts with smaller pores and narrower pore-size distributions. When isopressed, pore-size distributions from both powders were narrowed and shifted to smaller pore sizes. When sintered, classified-powder compacts produced higher densities. Also, polyphenylene-doped compacts were slightly more dense than compacts doped with carbon black. Consequently, compacts of classified powder doped with B_4C and polyphenylene produced the highest densities. Sintering at 2000°C resulted in densities as high as 97% and grain sizes on the order of a few micrometers. Compacts sintered at 2100°C exhibited densities as high as 99%, but underwent the beta-alpha phase transformation and probably have very large grains.

ACKNOWLEDGMENTS

I would like to thank Standard Oil for the opportunity to undertake this research at their Warrensville Heights facility. The assistance of all the scientists and technicians there was truly appreciated. I would especially like to thank Dr. Joseph

Fox for his invitation to visit Standard Oil and for his guidance throughout this research project.

This work was sponsored by the Air Force Office of Scientific Research, Basic Sciences Division, under Contract No. F49620 84 C 0097.

---

*Chevron Chemical Co., Oronite Additives Div., San Francisco, CA 94105.
†Polysciences Inc., Warrington, PA 18976.
‡Model 9220 Autopore II, Micromeritics Inc., Norcross, GA 30093.
§Model No. 1000-3060-P20 Graphite Element Furnace, Astro Industries, Inc., Santa Barbara, CA 93101.

References

[1] F. Thummler, "Sintering and High Temperature Properties of Si_3N_4 and SiC"; pp. 247-77 in Materials Science Research, Vol. 13. Edited by G. C. Kuczynski. Plenum Press, New York, NY, 1980.

[2] S. Dutta, "Sinterability, Strength, and Oxidation of Alpha Silicon Carbide Powders," *J. Mat. Sci.*, 19, 1307-13 (1984).

[3] S. Prochazka and P. C. Smith, "Investigation of Ceramics for High-Temperature Turbine Vanes," AD-779-053 (G.E. Corporate R&D, P.O. Box 8, Schenectady, NY 12301), April 1974, as cited in V. Krishnamachari and M. R. Notis, "Interpretation of High-Temperature Creep of SiC by Deformation Mapping Techniques," *Mat. Sci. Eng.*, 27, 83-88 (1977).

[4] K. Miyoshi, D. H. Buckley, and M. Srinivasan, "Tribological Properties of Sintered Polycrystalline and Single-Crystal Silicon Carbide," *Am. Ceram. Soc. Bull.*, 62 [4] 494-500 (1983).

[5] R. K. Govila, "Phenomenology of Fracture in Sintered Alpha Silicon Carbide," *J. Mat. Sci.*, 19, 2111-20 (1984).

[6] M. R. Freedman and M. L. Millard, "Improved Consolidation of Silicon Carbide," *Ceram. Eng. Sci. Proceed.*, 7 [1-2] 884-92 (1986).

[7] R. J. Pugh, T. Matsunaga, and F. M. Fowkes, "The Dispersibility and Stability of Carbon Black in Media of Low Dielectric Constant. 1. Electostatic and Steric Contributions to Colloidal Stability," *Colloids and Surfaces*, 7, 183-207 (1983).

[8] R. J. Pugh and F. M. Fowkes, "The Dispersibility and Stability of Carbon Black in Media of Low Dielectric Constant. 2. Sedimentation Volume of Concentrated Dispersions, Adsorption, and Surface Calorimetry Studies," *Colloids and Surfaces*, 9, 33-46 (1984).

[9] M. Persson, A. Forsgren, E. Carlstrom, L. Kall, B. Kronberg, R. Pompe, and R. Carlsson, "Steric Stabilization of Silicon Carbide Slips"; pp. 623-32 in High Tech Ceramics. Edited by P. Vincenzini. Elsevier Science Publishers B.V., Amsterdam, 1987.

[10] B. Bishop and H. K. Bowen, "Nonaqueous Dispersions of Beta-Silicon Carbide Powder," to be submitted to *Adv. Ceram. Matls*.

[11] W. Albers and J. Th. G. Overbeek, "Stability of Emulsions of Water in Oil II. Charge as a Factor of Stabilization Against Flocculation," *J. Colloid Sci.*, 14, 510-18 (1959).

[12] J. G. Brodnyan and E. L. Kelley, "The Effect of Electrolyte Content on Synthetic Latex Flow Behavior," *J. Colloid Sci.*, 20, 7-19 (1965).

[13] H. K. Bowen, "Basic Research Needs on High Temperature Ceramics for Energy Applications," *Mat. Sci. and Eng.*, 44, 1-56 (1980).

[14] P. Nahass, R. Pober, and H. K. Bowen, "Semicontinuous Classification of Ceramic Powders"; submitted to *Adv. Ceram. Matls.* (1987).

[15] W. C. Moffatt and H. K. Bowen, "Composite Ceramic Production by Precipitation of Polymer Solutions Containing Ceramic Powder," CPRL Report #66, MIT, pp. 1-6 (1986); accepted for publication in *J. Mat. Sci. Lett.*

[16] W. D. Kingery and B. François, "Sintering of Crystalline Oxides I. Interactions Between Grain Boundaries and Pores"; pp. 471-98 in Sintering and Related Phenomena. Edited by G. C. Kuczynski, N. A. Hooten, and C. F. Gibbon. Gorden and Breach Science Publishers, New York, NY, 1967.

[17] T. Vasilos and W. Rhodes, "Fine Particles to Ultrafine-Grain Ceramics"; pp. 137-72 in Ultrafine-Grain Ceramics. Edited by J. J. Burke, N. L. Reed, and V. Weiss. Syracuse University Press, Syracuse, NY, 1970.

[18]F. F. Lange, "Sinterability of Agglomerated Powders," *J. Am. Ceram. Soc.*, **67** [2] 83–89 (1983).

[19]F. F. Lange, B. J. Davis, and E. Wright, "Processing-Related Fracture Origins: IV, Elimination of Voids Produced by Organic Inclusions," *J. Am. Ceram. Soc.*, **69** [1] 66–69 (1986).

[20]C. A. Johnson and S. Prochazka, "Microstructures of Sintered SiC"; pp. 366–77 in Ceramic Microstructures '76. Edited by R. M. Fulrath and J. A. Pask. Westview Press, Boulder, CO, 1977.

[21]S. Prochazka, "Sintering of Silicon Carbide"; pp. 421–31 in Mass Transport Phenomena in Ceramics, Materials Science Research Vol. 9. Edited by A. R. Cooper and A. H. Heuer. Plenum Press, New York, NY, 1975.

Slip Casting and Sintering of Silicon Carbide

Elis Carlström, Michael Persson, Eva Bostedt, Annika Kristoffersson and Roger Carlsson

Swedish Institute for Silicate Research
Box 5403
S-402 29 Göteborg, Sweden

Abstract

An α-SiC powder coated with polyphenylene was dispersed in water and slip cast. A stable dispersion was achieved by the use of a wetting agent and pH adjustment. Particles or agglomerates, in the polyphenylene coated powder, with a size greater than 1 μm were removed by using sedimentation.
　　Slip cast samples with the agglomerates removed showed a higher sintered density. By removal of the agglomerates and by the use of slip casting the sintering temperature could be reduced by approximately 100°C to 2025°C. In this way can exaggerated grain growth be avoided.

Introduction

Both α and β-silicon carbide can be pressureless sintered to form high performance ceramic materials[1,2]. The most common route is additions of boron and excess carbon as sintering aids. Sintered silicon carbide materials manufactured in this way usually have little or no grain boundary phase and a low fracture toughness[3]. The low fracture toughness makes removal of defects an important objective in the fabrication of silicon carbide materials.
　　Typical defects that are found on fracture origins in silicon carbide are agglomerates, large grains, iron inclusion, boron and carbon inclusions. The boron and carbon inclusions originate from excess or inhomogeneously distributed sintering aids. Carbon has a low diffusivity along the grain boundaries and therefore is often added as a polymer. The polymer is dissolved in a solvent and mixed with the powder. The solvent is evaporated leaving a coating of polymer on the powder. Upon heating during sintering the polymer pyrolyses to carbon. Boron has a high surface diffusivity on SiC[4] and is readily distributed along the powder surfaces. Boron is often added as a powder.
　　Large grains in the microstructure originate from exaggerated grain growth or from large grains in the starting powder. α-SiC exhibits a thermal anisotropy and this will degrade the strength of a material that contains large grains by causing residual stresses in the material[5]. Also a crack that nucleates in a large grain will encounter a lower resistance to growth (single crystal K_{Ic} for SiC is lower than that for the polycrystalline material) and will reach significant dimensions before being affected by the general microstructure[6]. The strength dependence on grain size due to the change from single grain behaviour to polycrystalline behaviour will

disappear above a certain grain size. Above this grain size the size dependence might still be in effect due to for example a difference in thermal expansion. Such a difference exists in sintered SiC between the SiC matrix and inclusions of excess sintering aids[7].

Agglomerates tend to cause crack like voids in the sintered material, thus causing mechanical failures[8]. Agglomerates are usually present in all commercial powders but can also be created during the processing before firing. Mechanical failures from agglomerates, originating from the granulation of the powder for pressing, have been observed in SiC as well as in other ceramics. Agglomerates or large grains also inhibit the final densification of a ceramic material[9,10].

In this work large grains and agglomerates in the starting powder have been removed. The following processing has been done in a liquid media in order to avoid the reformation of agglomerates. Green microstructures with favourable sintering behaviour can be formed by using slip casting. A more narrow pore size distribution with smaller pores can be obtained by slip casting compared with dry pressing. Voids created by remanent granule boundaries, that have not been restructured during pressing, can often be found in pressed bodies[11].

Slip casting of pure silicon carbide can be done by using pH adjustment that is a purely electrostatic mechanism for stabilization[12]. SiC has a relatively low iso-electric point around pH 4. The iso-electric point moves to lower pHs when the oxygen content of the powder increases approaching the iso-electric point for silica at about pH 2.5. The nature of the α-SiC surface at lower oxygen contents is not well known. The present knowledge indicates that the surface consists of a thin amorphous SiO layer covered (at least in water) with hydroxyl groups[13,14]. The SiO layer has somewhat different surface characteristics from a SiO layer on SiO_2 as can be seen by the shift of the iso-electric point.

Electrostatic stabilization of dilute as well as concentrated slips can easily be achieved in the high pH range where the zeta potentials have a high absolute value. At too high pHs (pH > 13) the addition of base alone increases the ionic strength of the suspension so that the resulting compression of the electrical double layer results in a reduced stability. Corrosion of the plaster moulds also prevents practical use of extreme pHs. Electrosteric mechanisms for stabilizing SiC by using polyethyleneimines have also been found[12]. In this work pH adjustment and addition of a wetting agent were used to stabilize an α-SiC with added sintering aids.

MATERIALS

An α-silicon carbide[*] with a specific surface are of $15 m^2/g$ and an oxygen content of 0.8% was used. Amorphous boron[†] was used as a boron additive and polyphenylene[‡] was used as a carbon source.

A naphthalene sulphonic acid condensate[§] was used as a wetting agent.

ANALYTICAL

The zeta potential of the polymer pure SiC as well as coated SiC particles with addition of a wetting agent was measured as a function of pH. NaCl was used as an electrolyte to keep a constant ionic strength of 0.01 M and the pH was adjusted with HCl and NaOH. The electrophoresis equipment was of the microscope type with a planar glass cell and blackened platinum electrodes[¶].

Viscosity of the slip was measured by using a rotational viscosimeter[**]. A cylindrical measuring system with a narrow shear gap and a volume of 100 cm^3 was used. Particle size distribution was measured on samples of the diluted slip with an x-ray sedimentometer[††].

The carbon yield of polyphenylene during sintering was estimated by using the weight loss during pyrolysis of pure polyphenylene in argon at 1000°C for 1 hour. The densities of the sintered compacts were measured by using the water immersion method. The theoretical density was taken to be 3.21g/cm^3. Polished and etched surfaces (Murakami solution) were observed with a scanning electron microscope (SEM).

PROCESSING

Polyphenylene was dissolved in toluene and mixed with silicon carbide. The toluene was evaporated in a rotary evaporator leaving a coating of polyphenylene on the SiC particles. 5 weight% of polyphenylene (corresponding to 4% excess carbon) was added to the SiC powder in this way. Polyphenylene was chosen because of its high-carbon yield and its low reactivity with water, regardless of pH. This way of coating the particles results in an even distribution of carbon but has the disadvantage of making the particles hydrophobic. Agglomerates are also formed during the coating process, and these need to be redispersed.

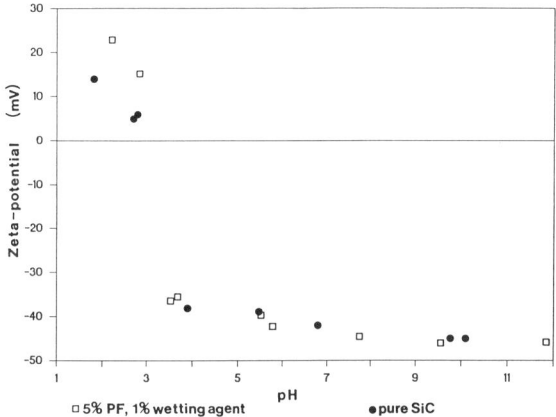

Fig. 1. Zeta potential of pure α-SiC compared with zeta potential of α-SiC coated with polyphenylene with addition of 1% wetting agent.

The particles were dispersed in water by adding 1 weight% of a commercial wetting agent (naphthalene sulphonic acid condensate) to the powder and adjusting pH to 10 by additions of NH_4OH.

The coarse particles were left to sediment from a dilute dispersion (3 vol%). All particles with an equivalent spherical diameter above 1μm, as calculated by Stokes law, were sedimented out. The fines were kept, siphoned off and concentrated by flocculation. Flocculation was achieved by adjusting the pH to a pH near the isoelectric point by additions of HCl. The flocced particles were washed with distilled water to reduce the electrolyte concentration of the suspension.

From this suspension of particles smaller than $1\mu m$, slips were prepared. Stable dispersions were obtained by pH adjustment to a pH in the range of 10 to 11. One weight% boron was added to the SiC in the slip by stirring. Sedimentation tests in the pH range used for casting were done with boron showed no indication of flocculation. The solid contents of the slips were increased by centrifugation.

Bars (80×11×8mm) were cast in plaster moulds. The mould surfaces were treated with an ammoniumalginate solution to enhance mould release. After drying, the slip cast bodies were sintered in a carbon resistance furnace. The atmosphere was 0.1 MPa argon and the heating rate was approximately 20°C/min. The samples were kept at the top temperature for 30 minutes. The temperature was monitored by using a two-colour optical pyrometer. The polyphenylene polymer was pyrolysed during this heat treatment.

Results and Discussion

The zeta potential curve of the coated powder with addition of wetting agent shows the same behaviour as that of the pure SiC. This gives the possibility to disperse and flocculate the powder by adjustment of the pH (figure 1).

The densification during sintering shows that the distribution of carbon by coating the particles with polyphenylene is acceptable. There is however reason to suspect that the coating is not perfectly even. The mixing in the rotary evaporator is good in the initial stage but as soon as a powder cake is formed migration of the polyphenylene during evaporation is possible. If an adsorption exists, the adsorbed part of the polyphenylene will not migrate. This has, however, not been investigated.

Viscosity of slips with different amounts of wetting agent (naphthalene sulphonic acid condensate) was measured. A maximum of the viscosity associated with foaming of the slip was observed for 2-3 weight% additions of wetting agent at pH 10 (figure 2). The foaming can be explained by a stabilizing effect on foam lamellas by partially wetted particles.

Incomplete wetting of hydrophobic particles in water can make the particles work as foam stabilizers. Completely hydrophobic particles tend to stay in the air. Completely wetted particles prefer the liquid. Partly wetted particles however tend to collect at the interface. Thin liquid lamella (that is the foam bubbles) are stabilized by these particles (figure 3). Foam increases the viscosity and creates large pores in slip cast bodies.

When the particles were wetted completely by sufficient wetting agent addition they could be dispersed at pH 10 to 11. A suspension of 67 weight% of SiC (as received powder) with 3% wetting agent added had a viscosity below approximately 10 mPas (the lower measuring limit of the viscosimeter).

5 weight% wetting agent was added to the SiC in the slips used for slip casting and prepared from the as-received powder. The total amount of wetting agent for the slips made from classified powder was approximately 1 weight%. This figure is somewhat uncertain as wetting agent was added during the sedimentation procedure and may not have been removed fully by the washing of the powder with distilled water. Some of the removed agglomerates may contain high

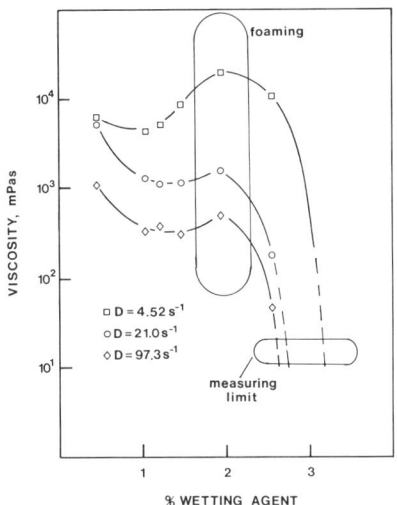

Fig. 2. Viscosity for suspension of SiC-particles coated with polyphenylene (single point measurements) as a function of wetting agent addition at three different shear rates. Solid content of slip was 67 weight% and pH 10.

concentrations of polyphenylene. This might also be a reason for the lower demand for wetting agent for the classified powder. The free carbon content during sintering could then also be lower than the assumed. The classification only marginally increases the specific surface of the powder even if it decreases the average particle size. The reason for this is that the removed agglomerates have high specific surface areas.

Initial foaming was observed during slip preparation with the as-received powder. Slow stirring removed the foaming. Stable dispersions with a solid contents in the range of 62 to 63 weight% were prepared with the polyphenylene coated as received and classified SiC powder. No new agglomerates are formed during the slip preparation. This was shown by measurements of the particle size distribution of diluted slips prepared from the classified powder. The particle size curve shows that the cut off size of 1 μm from the sedimentation is retained during slip preparation (figure 4).

It cannot be excluded that the naphthalene sulphonic acid condensate could contribute to the stability of the slips. However, the drop in viscosity with increasing pH and the zeta-potential curve is very similar to the behaviour exhibited by pH adjustment to a pure SiC powder. For this reason it is referred to as wetting agent in this context. The coating of the powder does not seem to influence the electrostatic stability of the slip. The reason for this could be that the polyphenylene does not cover the complete particles. At the edges and corners of the particles the layer might wear off during mixing. These parts of a particle play an important role in the charging of particles.

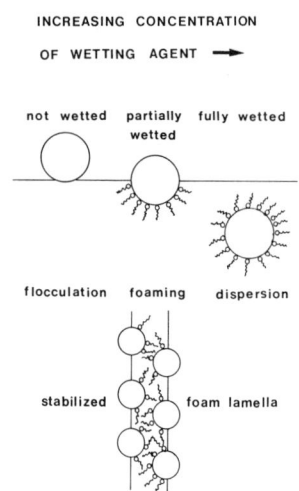

Fig. 3. Particle stabilization of foam.

Sintered densities were measured after sintering in argon at 2125°C for 30 minutes. Slip cast samples prepared from powders with the agglomerates removed sinter to a density of 98.4% of theoretical. Samples prepared with the as-received powder sinter to a density of 94.1% under the same conditions.

Dense materials can be reached at lower temperatures when classified powders and slip casting are used. Figure 5 shows the sintered density as a function of sintering temperature for the classified powder. A high sintered density can be obtained at sintering temperatures so low as 2025°C.

The SEM micrographs show micrographs polished and etched (Murakami solution) samples. The samples were slip cast from classified powders and sintered in argon (figure 6-9).

The maximum size of carbon and boron rich inclusions that are found in the microstructure is below 10 μm.

Exaggerated grain growth is found in α-silicon carbide at 2125°C. The large grains have tabular shapes. The feather shaped clusters of the tabular grains that are seen in β-SiC samples are not observed in α-SiC and are probably associated with the β to α transformation.

Conclusions

Classified SiC powders reach higher sintered densities than powders in which the agglomerates have not been removed. The sintering temperature could be lowered by approximately 100°C by classifying the powder and forming by slip casting. Polyphenylene coated SiC particles can be dispersed in much the same way as pure SiC if a wetting agent is added.

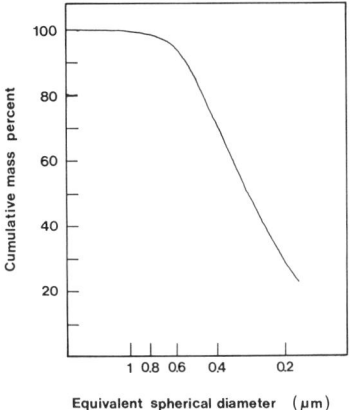

Fig. 4. Particle size of a diluted SiC slip as shown by the continuous tracing from one measurement with the x-ray sedimentometer.

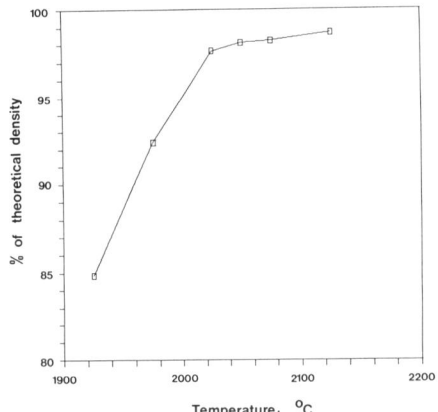

Fig. 5. Sintered density vs temperature for individual samples.

Silicon Carbide

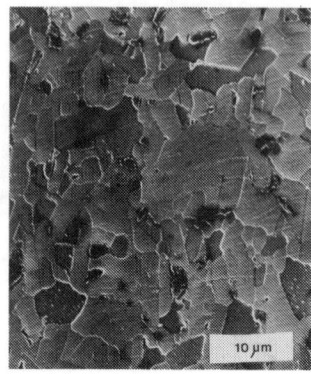

Fig. 6. SEM-micrograph of sample sintered at 2125°C for 30 minutes.

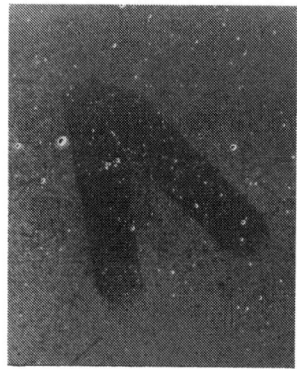

Fig. 7. SEM-micrograph of sample sintered at 2125°C for 30 minutes. Notice the exaggerated grain growth with mm sized grains.

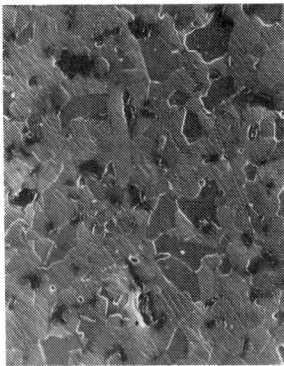

Fig. 8. SEM-micrograph of sample sintered at 2075°C for 30 minutes. At this temperature exaggerated grain growth is not observed on polished surfaces.

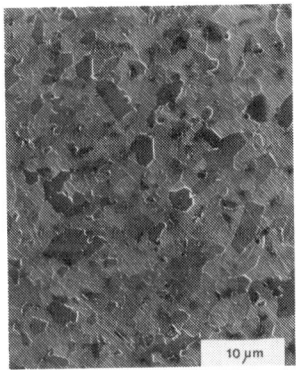

Fig. 9. SEM-micrograph of sample sintered at 2025°C for 30 minutes. Most of the porosity is already eliminated at this temperature.

Acknowledgements

The authors wish to thank Martin Sjöstedt for the help with the sintering experiments and Lars Eklund for the SEM work. This work was sponsored by the Swedish Board for Technical Development.

---

*Carbogran UF-15, Lonza-Werke, Fed. Rep. of Germany
†HCST 1549, HC Starck, Fed. Rep. of Germany
‡Poly(p-phenylene) Polysciences Inc., Pennsylvania, USA
§Lomar D, Diamond Shamrock Corp., Ohio, USA
¶REPAB AB, Sweden
**Rheomat-30, Contraves AG, Switzerland
††Sedigraph 5000ET, Micromeritics Inc., USA

References

[1] S. Prochazka, "Investigation of Ceramics for High Temperature Turbine Components", U.S. N.T.I.S. AD/A Report No. 005 830, 1975.

[2] J. A. Coppola and C. H. McMurtry, "Substitution of Ceramics for Ductile Materials in Design"; National Symposium on Ceramics in the Service of Man. Washington D.C., Carnegie Institution, 1976.

[3] M. Srinivasan and S. G. Seshadri, "Application of Single Edge Notched Beam and Indentation Techniques to Determine Fracture Toughness of Alpha Silicon Carbide"; pp. 46-68 in Fracture Mechanics Methods for Ceramics, Rocks and Concrete. ASTM STP 745. Edited by S. W. Freiman and E. R. Fuller Jr. American Society for Testing and Materials, 1981.

[4] H. Suzuki and T. Hase, "Some Experimental Considerations on the Mechanism of Pressureless Sintering of Silicon Carbide"; pp. 345-65 in Proceedings of the International Symposium of Factors in Densification and Sintering of Oxide and Non-oxide Ceramics. Edited by S. Somiya and S. Saito. Gakujutsu Bunken Fukyu-Kai, Tokyo, 1978. Japan (1978) 345-365.

[5] S. Prochazka and R. J. Charles, "Strength and Microstructure of Dense Hot-Pressed Silicon Carbide", pp. 579-88 in Fracture Mechanics of Ceramics. Edited by R. C. Bradt, A. G. Evans, D. P. H. Hasselman and F. F. Lange. Plenum Press, New York 1974.

[6] J. P. Singh, V. Virkar, D. K. Shetty and R. S. Gordon, "Strength-Grain Size Relations in Polycrystalline Ceramics", *J. Am. Ceram. Soc.*, **62** [3-4] 179-183 (1979).

[7] H. Suzuki and K. Kijima, "Mechanical Strength of Sintered Silicon Carbide"; pp. 490-504 in Proceedings of the International Symposium on Ceramic Components for Engines I. Edited by S. Somiya, E. Kanai and K. Ando. KTK Scientific Publishers, Tokyo, 1983.

[8] F. F. Lange and M. Metcalf, "Processing-Related Fracture Origins: II, Agglomerate Motion and Cracklike Internal Surfaces Caused by Differential Sintering", *J. Am. Ceram. Soc.*, **66** [6] 398-406 (1983).

[9] W. H. Rhodes, "Agglomerate and Particle Size Effects on Sintering Yttria-Stabilized Zirconia", *J. Am. Ceram. Soc.*, **64** [1] 19-22 (1981).

[10] L. Hermansson, E. Carlström, M. Persson and R. Carlsson, "Sintering of Silicon Carbide with a Narrow Particle Size Range"; pp. 538-47 in Proceedings of the International Symposium on Ceramic Components for Engines I. Edited by S. Somiya, E. Kanai and K. Ando. KTK Scientific Publishers, Tokyo, 1983.

[11] I. A. Aksay, "Microstructure Control Through Colloidal Filtration"; pp. 94-104 in Forming of Ceramics. Edited by J. A. Mangels and G. L. Messing. Advances in Ceramics, 9 94-104 (1984).

[12] M. Persson, A. Forsgren, E. Carlström, L. Käll, B. Kronberg, R. Pompe and R. Carlsson, "Steric Stabilization of Silicon Carbide Slips", pp. 623-32 in High Tech Ceramics. Edited by P. Vincenzini. Elsevier Science Publishers B. V., Amsterdam, 1987.

[13] P. K. Whitman and D. L. Feke, "Colloidal Characterization of Ultrafine Silicon Carbide an Silicon Nitride Powders," *Advanced Ceramic Materials*, **1** [4] 366-70 (1986).

[14] M. N. Rahaman, Y. Boiteux and L. C. De Jonghe, "Surface Characterization of Silicon Nitride and Silicon Carbide Powders", *Am. Ceram. Soc. Bull.*, **65** [8] 1171-76 (1986).

Nuclear Magnetic Resonance Imaging for Detecting Binder/Plasticizers in Green-State Structural Ceramics

W. A. Ellingson

Materials and Components Technology Division
Argonne National Laboratory
Argonne, IL 60439

J. L. Ackerman and L. Garrido

Massachusetts General Hospital/Harvard Medical School
Boston, MA

S. Gronemeyer

Siemens Medical Systems
St. Louis, MO

Abstract

The distribution of binder/plasticizers (B/Ps) in green-state ceramics is important because locally high concentrations can cause porous regions after densification. Proton nuclear magnetic resonance imaging is a chemically sensitive nondestructive, noninvasive imaging technique which is becoming well-established in the medical community. To date, however, little work has been done to apply this technique to solid state imaging such as imaging the distributions of organic B/Ps in green-state ceramics. Polyethylene glycol (PEG, also called Carbowax) is a typical organic binder for ceramics. In the work reported in this paper, we have explored both a small-bore (<10 cm) experimental imaging system and a state-of-the-art medical imaging system relative to detecting the soft-solid (wax-like) B/Ps used in ceramics. The ability to detect B/Ps systems was evaluated on a 1.5-T medical imager (Siemens Magnetom™) using T_1-weighted imaging techniques and a 10-cm eye coil standard. The ability to detect B/Ps was also studied with a modified small-bore coil Technicare Facility using special RF and gradient coils. The initial results show that a medical system may not be able to detect B/Ps unless elevated temperatures are used, whereas the experimental small-bore system shows B/P distribution quite well. In addition, higher magnetic field strength should be more effective for ceramics, since proton signal strength increases rapidly with the magnetic field strength.

Introduction

Nuclear magnetic resonance (NMR) imaging (also called MRI) is a chemically sensitive, noncontacting, and nondestructive imaging technique which has been routinely used in chemistry, physics, and biology for over 35 years. In these areas, its main applications are in determining chemical composition, molecular or crystal structure, and molecular dynamics.[1,2] The use of NMR to produce tomographic images of objects was reported in 1973.[3] Efforts to apply NMR imaging in the materials sciences have begun only recently.[4-6] NMR imaging clearly holds potential for aiding in the development of ceramics processing. It is a particularly promising technique for investigating the spatial distribution of organic binders and plasticizers (B/Ps), which are of great interest in green ceramics because concentration distributions affect local densification rates during densification processes. Since NMR is non-contacting, delicate green-state and partially densified ceramics can be easily studied without damage to the material.

Application of NMR imaging for B/P distribution is perhaps most attractive for injection-molded ceramics because of the very long B/P burnout time required and thus the related costs. Several injection-molding problems have been identified[7] which can be related to B/P distribution and which cause defects in final parts. Table I lists typical defects and some of the potential causes of these defects, many of which can be traced to a poor distribution of organic B/P; for example: (1) incomplete parts, due to an improper feed material which is the result of poor mixing of the ceramic powder and the B/P; (2) large failure-causing pores, created either by large pockets of the B/P itself (left by poor mixing) or by agglomerates (which may tend to be high in B/P); (3) so-called knit lines (regions within the part where injected material folds on itself but does not join completely), possibly caused by improper feed material; and (4) cracks in the final product, perhaps caused by a high local concentration of B/P (low ceramic green packing density) with resultant locally large differences in densification rate and thermal expansion. Clearly, the ability to map the distribution of organic B/Ps would improve our understanding of the injection molding process as well as other ceramic processes.

Considerations in NMR Imaging of Organics in Ceramics

Because the polymeric organic B/Ps used in green state compacts contain a high concentration of protons, they are potential candidates for direct NMR imaging; thus we have explored the direct imaging of these components. However, the B/Ps used for ceramics are often "soft solids" (wax-like); and, for a number of reasons, NMR measurement in solids generally requires more specialized experimental techniques and places greater demands upon instrumentation than NMR measurement in fluids.[8,9] Several factors exist that make the absence of large-scale and isotropic molecular motion the key element that affects NMR measurement in solids (including "soft solids"). These factors are:

(1) The lack of substantial molecular motion in solids tends to lengthen spin lattice relaxation times (T_1) considerably, especially for spins that are not part of an abundant reservoir (a collection of spins at relatively high concentration and strongly coupled to each other). Protons in typical organic solids form an abundant spin reservoir and can therefore relax by means of spin diffusion to rapidly relaxing T_1 "sinks" (e.g., paramagnetic centers or rapidly

Table I. Injection Molding Defects and Causes[7].

Type of Defect	Causes
Incomplete part	Improper feed material Poor tool design Improper material and/or tool temperature Inadequate tool lubrication
Large pores	Entrapped air Improper material flow and consolidation during injection Agglomerates Large pockets of organic binder/plasticizer due to incomplete mixing
Knit lines	Improper tool design or feed material Incorrect temperatures
Cracks	Sticking during removal from tool Improper tool design Improper extraction of binder/plasticizer

rotating methyl groups). Rare (chemically or isotopically dilute) spins whose quantum number, $I = 1/2$ do not possess any such mechanism and can have exceedingly long T_1s, although spins whose quantum number $I > 1/2$ may relax by means of the quadrupolar interaction. Since protons are the most abundant nuclei in most B/Ps used in ceramics, they are the best candidates for NMR imaging, and so spin lattice relaxation is not expected to be a major problem. For NMR imaging of ^{13}C or ^{29}Si (^{29}Si would be useful for silicon-containing ceramics), the exceedingly long relaxation times[8] would have to be dealt with by special techniques.[10]

(2) Static direct dipole-dipole coupling produces linewidths ranging from a few kHz to tens of kHz. Since the size of a coupling is proportional to the magnetogyric ratio of each spin of a coupled pair and inversely proportional to the cube of their separation, couplings involving abundant protons tend to produce line widths at the upper end of the scale and shortened relaxation times (T_2s) on the order of tens of microseconds. This extreme shortening of spin-spin T_2 is most severe for completely rigid proton-containing solids. In the case of polyethylene glycol (Carbowax), polyvinyl alcohol, and similar waxy polymeric B/Ps, the small amount of molecular motion present creates T_2s on the order of hundreds of microseconds in length, which makes imaging possible with special apparatus.

(3) The anisotropy of the chemical shift produces broad "powder pattern" lineshapes. Fortunately, this broadening interaction can be refocused with a 180° RF pulse that is normally used in the production of a spin echo.

These three factors conspire to produce a low signal-to-noise ratio per unit of data acquisition time, and large intrinsic spectral bandwidths, for solids. Large spectral bandwidths make imaging-plane selection difficult and degrade the spatial resolution of frequency-encoded dimensions.

With standard imaging equipment, such as that used for medical NMR imaging, the shortest echo times, TE, that can be obtained are in the neighborhood of 12-17 msec. Since the T_2s of polymeric B/Ps that we have measured tend to be on the order of 1 msec or less, we expect that it will be essentially impossible to obtain images of such B/Ps with conventional room temperature NMR imaging techniques. This has been confirmed by our experiments.[11] We will demonstrate, however, that B/Ps with a high plasticizer (e.g., water) content can be imaged.

Test Results

Using both a medical imager and an experimental small-bore imager, we recently conducted a series of tests on a number of specimens which contained various amounts of B/P. In the first series of tests, we used a medical 1.5-T superconducting-magnet NMR imaging system (Siemens Magnetom™) to evaluate the feasibility of imaging the B/P directly.

For these tests, a green-state specimen was made by mixing 25 wt % B/P with SiC powder. Stark SiC powder was used as the starting powder. The B/P was PEG with mw = 8000. The SiC/PEG mixture was cold-pressed to 517 MPa (7500 psi) at room temperature to obtain a right circular cylinder 25 mm in diam and 25 mm high. As shown in Fig. 1, five 5-mm-diam glass beads (large defects) were incorporated at the midsection of the specimen to simulate defects. Standard (medical imaging) spin echo and inversion recovery pulse sequences were used to obtain T_1-weighted, proton density weighted, and T_2-weighted images. The slice thickness was 5 mm; the in-plane resolution was 0.59 mm.

Figures 2 and 3 are axial and transverse images obtained with a T_1-weighted spin echo sequence [repetition time = 0.5 sec, echo time (TE) = 17 msec, pixel size = 600 μm]. The imaging time for Figs. 2 and 3 was 4.3 min [4 excitations × 128 phase-encoding steps × 0.5 sec (TR repetition time)]. Not only are the five glass beads well-visualized, but the image inhomogeneity in the surrounding B/P suggests nonuniformity in its distribution. The T_2-weighted (second echo) images had very low S/N ratios, indicating a relatively short T_2 relaxation time on the order of one TE for the B/P. The S/N ratios were also low for the proton density weighted images which were made with a longer TE of 28 msec. The in-plane gradients used for phase and frequency encoding were 0.52 gauss/cm. The field homogeneity of the magnet itself in the imaging volume was ~1 ppm. In this first series of tests, we also examined a similar specimen on the small-bore experimental imager. Figure 4 shows the resulting image. It must be noted that this concentration of binder and plasticizer (water, in this case) is unrealistically high.

To improve the imaging of conventional polymeric B/Ps in green ceramics, we have employed a special NMR probe for the experimental small-bore machine. This probe has several important features. First, it has an RF coil which is substantially smaller than the 10-cm-diam coil normally used; i.e., it has an active

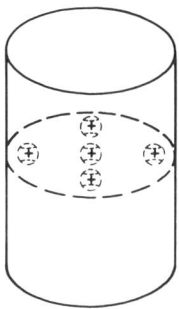

Fig. 1. Schematic Diagram of SiC/25 wt % PEG Green Ceramic Used in Initial Test of Direct NMR Imaging of B/P.

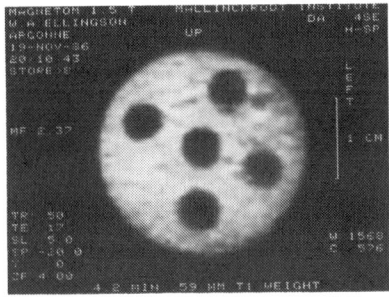

Fig. 2. Tomographic NMR Image of B/P Distribution and Defects at Midsection of Specimen Shown Schematically in Fig. 1. The dark-colored circles are the 5-mm-diam glass beads intentionally introduced at the midsection to simulate known defects.

volume which is cylindrical in shape, with a diameter and length each equal to 3 cm. The coil axis is oriented perpendicular to the direction of the main magnetic field (solenoidal configuration) to enhance the sensitivity of the measurement. With 100 W of RF power (our usual maximum level), we obtain a 90° pulse length of 6 μsec; this compares with an RF pulse length of 50 μsec for a standard whole-body coil. This should represent nearly an order-of-magnitude improvement in RF performance. The detected S/N ratio is expected to behave similarly. The short RF pulse length is necessary for uniform excitation of broad spectral lines associated with solid materials.

The second feature of this probe is that it has its own set of magnetic field gradient coils which are used in place of the standard coils. Because the coil volume is smaller, identical currents will produce much more intense field gradients. Thus, while the normal field gradients in the small-bore machine are 1 G/cm maximum,

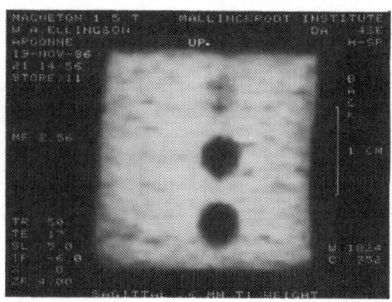

Fig. 3. Direct NMR Image of B/P Distribution and Intentional Defects in the Specimen Shown in Fig. 2. In this case, the image is taken sagittally (see Fig. 2 for sagittal plane location). The axial location of the 5-mm glass beads is clearly detected.

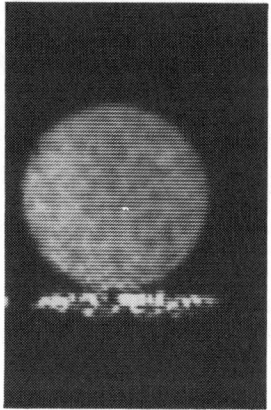

Fig. 4. Conventional Spin Echo Tomographic Image of a SiC Compact Prepared with a 25 wt % PEG (8000 mw) Binder and Water Plasticizer. NMR image slice thickness was ~15 mm; TE = 15 msec.

the maximum field gradients with the smaller coil are on the order of 10 G/cm. In reality, the probe is capable of producing field gradients on the order of 60 G/cm at low duty cycles.

This new set of gradient coils reduces the minimum echo time for the 10-cm-diam gradient coil set from about 15 msec to below 2 msec (the 2-msec limitation is due not to the new gradient coils, but rather to the minimum time interval allowed in the pulse programming in this system).

Fig. 5. Spin Echo Image of Al_2O_3 Compact with 2.5 wt % B/P (Very Low B/P Content). The image was obtained with the special RF/gradient coil probe described in the text (high-sensitivity RF coil; intense and rapidly switched gradients). TE = 2 msec.

Figure 5 shows a room temperature spin echo (conventional) image of a green Al_2O_3 compact containing 2.5 wt% B/P. A TE of 3.16 msec was used to obtain this image, which is not obtainable with any standard commercial clinical or research NMR imaging system. The S/N ratio, however, is somewhat less than satisfactory, and the resolution is presently not very high.

In a second series of experiments, we again wanted to compare the imaging capability of a state-of-the-art commercial medical imager with that of the small-bore experimental unit with the special RF/gradient coil probe. For this series of tests, Si_3N_4 was used as the powder together with a proprietary B/P. Two test specimens, each 25 mm in diameter and 25 mm high, were made from pelletized material.*

Prior to the imaging tests, we conducted a series of linewidth tests in a variable-temperature NMR spectrometer to establish the T_2 values. We suspected that elevated temperature would be necessary to increase T_2 to values close to those necessary for medical instruments, i.e., ~10-15 msec. The T_2 tests were conducted in a Brucker 7.1-T, 300-MHz variable-temperature NMR spectrometer. We first conducted tests on the B/P itself with the results shown in Table II. To get approximate T_2 values, we used the relationship

$$\Delta^{1/2} \approx \frac{1}{\pi T_2},$$

where Δ = full width at half maximum of the amplitude vs frequency trace and T_2 = spin-spin relaxation time desired. This assumes a Lorentzian lineshape, which usually has reasonable validity for liquids and semi-solids.

We also conducted a series of tests on the combined Si_3N_4-B/P system. The results of these tests are shown in Table III. The T_2 values are at best <1 msec even at elevated temperatures. These very low T_2 values are essentially nearly 2 orders

Table II. T_2 Measurements on Binder/Plasticizer.

Temperature (°C)	FWHM (Hz)	T_2 (msec)
23	1000	0.3
42	1000	0.3
47	800	0.4
49	700	0.45
51	640	0.5
53	580	0.55
55	550	0.58

Table III. T_2 Measurements on Si_3N_4 + Binder.

Temperature (°C)	FWHM (Hz)	T_2 (msec)
30	3700	0.9
45	2840	0.11
60	1490	0.2
70	1420	0.22
80	1520	0.21
90	1510	0.21
100	1550	0.21
	1350	0.24*

*A different pulse width was used.

of magnitude less than those necessary for reasonable S/N ratios on a commercial medical NMR imager. It is not totally clear why there is such a large difference between the T_2 values for the B/P alone as opposed to the B/P-ceramic system, but it may come from a surface reaction. We will pursue this in further work. We experimentally verified the imaging at low T_2 by placing the specimen in a glass vial and heating it by immersion in hot water. The vial and hot water were placed in a Styrofoam container which fit snugly inside the circular eye coil (Fig. 6).

We conducted two imaging studies, one at 50°C and one at 80°C. In the first test, routine medical imaging techniques were used with T_1-weighting,[†] TR = 0.5 sec, TE = 17 msec, 5-mm slice thickness, and 2 acquisitions in a spin echo mode.

Figure 7 shows the resulting NMR image. This image shows the B/P itself (very bright, 100–200 times background), the ceramic-B/P system (no detectable signal), and a test specimen of 8000-mw PEG. Clearly, at the 15 wt % B/P loading at 50°C, the protons are not sufficiently mobile to produce a useful image.

Figure 8 shows an NMR image of a set of samples similar to the one shown in Fig. 7 but at 80°C. The signal strength again is 100–200 times the background for B/P alone and 10 times the background for B/P in the ceramic, but still not large enough to be useful. It is conjectured that the light region around the ceramic-B/P specimen might be water vapor, since the specimen was heated in water.

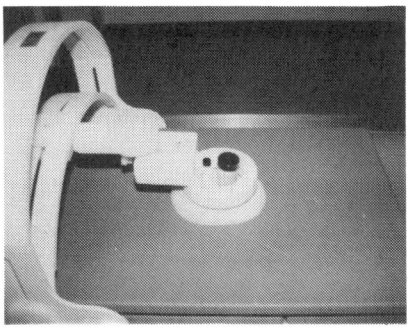

Fig. 6. Photograph Showing Glass Vial Containing Ceramic-B/P System in Orbit Coil Prior to Placement in Primary Magnet Gantry.

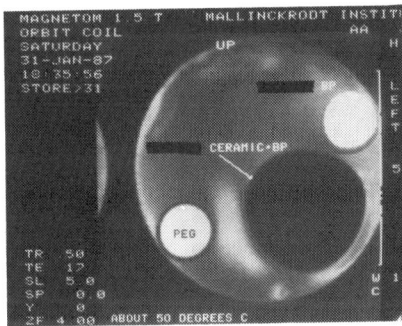

Fig. 7. NMR Image of Proprietary Ceramic-B/P System, B/P Itself, and a Test Specimen of 8000-mw PEG. Image obtained with test samples at ~50°C.

Using the small-bore experimental machine, we compared the signal acquisition from a specimen like the one used to obtain Fig. 8, but under different conditions. We placed the same 15-wt % proprietary B/P specimen in the special RF coil at room temperature. A pulse sequence was selected which yielded a TR of 4.5 msec. Figure 9 shows the resulting image.

SUMMARY AND CONCLUSIONS

Direct NMR imaging of organic B/Ps in ceramic materials can be accomplished. However, for medical systems to be useful, techniques for increasing T_2 to the 10-msec range will be necessary. One potential method is heating. On the other hand, special RF and gradient coil configurations, which obviate the necessity to heat the specimens, can be developed. We have shown that it is possible to image

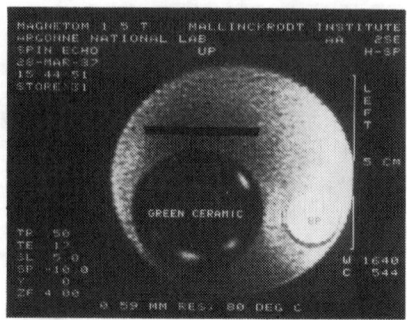

Fig. 8. NMR Image of Proprietary Ceramic-B/P System and B/P Itself. Image obtained with test specimen at 80°C.

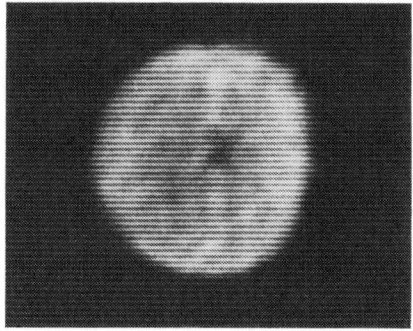

Fig. 9. NMR Image of Proprietary Ceramic-B/P System Taken in Experimental Small-Bore Machine. No slice selection employed.

with as little as 2.5 wt % B/P at 2-T field strength but presently with a sacrifice of tomographic image quality. In addition, we have noted a change in NMR parameters of B/P when it is studied alone as opposed to after mixing with ceramic powder. This difference may suggest reactions at the surface between the B/P and the powder.

Acknowledgements

Work jointly supported by the U.S. Department of Energy, Office of Fossil Energy, Advanced Research and Technology Development Materials Program, and the Office of Energy Conversion and Utilization Technologies Division (ECUT) Materials Project, under Contract W-31-109-ENG-38.

---

*Pelletized material is the ceramic-B/P previously mixed, then ground to small (1-3 mm) irregularly shaped pieces.
†T_1-weighting in medical terminology, i.e., reduced T_1 effect. T_2 values in human tissues are of order 50-25 msec, so in the medical imaging context, the combination of short (<~1 sec) TR and short (<~30 msec) TE produces a T_1-weighted image. In the context of B/Ps with T_2 of the order of 1 msec, such parameters produce an image which is heavily T_2-weighted.

References

[1] E. D. Becker, High Resolution NMR, Academic Press, New York, 1980.

[2] C. P. Slichter, Principles of Magnetic Resonance, 2nd ed., Springer-Verlag, New York, 1978.

[3] P. C. Lauterbur, *Nature* **242**, 190 (1973).

[4] W. A. Ellingson, J. L. Ackerman, J. D. Weyand, R. A. DiMilia, and L. Garrido, "Characterization of Porosity in Green-State and Partially Densified Al_2O_3 by Nuclear Magnetic Resonance Imaging," in *Ceram. Eng. and Sci. Proc.*, Vol. **8**, No. 7-8 (1987), pp. 503-512.

[5] J. L. Ackerman, W. A. Ellingson, J. A. Koutcher, and B. R. Rosen, "Development of Nuclear Magnetic Resonance Imaging Techniques for Characterizing Green-State Ceramic Materials," in Proc. 2nd Intl. Symp. on the Nondestructive Characterization of Materials, Montreal, Canada, Plenun Press, NY, 1987, pp. 129-137.

[6] L. B. Welch, S. T. Gronczy, M. T. Mitsche, L. J. Bauer, J. Dworkin, and A. Giambalvo, "Proton NMR Imaging of Green-State Ceramics," in Proc. Quantitative NDE Conference, La Jolla, CA, D. O. Thompson and D. E. Chimenti, eds., Plenum Press, New York, 1987, pp. 441-456.

[7] H. C. Yeh, J. M. Wimmer, M. E. Huang, M. E. Rorabaugh, J. Schienle, and K. H. Styhr, Improved Silicon Nitride for Advanced Heat Engines, Ann. Tech. Report submitted to NASA by AiResearch Casting Co., A Div. of the Garrett Corp., NASA-CR-175006 (October 1985).

[8] M. Mehring, High Resolution NMR Spectroscopy in Solids, Springer-Verlag, Berlin, 1976.

[9] D. E. Axelson, Solid State Nuclear Magnetic Resonance of Fossil Fuels, Multiscience Limited, Canada, 1985.

[10] J. L. Ackerman, D. P. Rayleigh, R. G. Griffin, and M. J. Glimcher, "A Phosphorous-31 Magnetic Resonance Imaging of Solid Calcium Phosphates: Potential for Chemical Imaging of Bone," in Proc. 6th Ann. Meeting, Society of Magnetic Resonance in Medicine, New York, August 12-21, 1987 (in press).

[11] W. A. Ellingson, J L. Ackerman, L. Garrido, P. S. Wong, and S. Gronemeyer, Development of Nuclear Magnetic Resonance (NMR) Imaging Technology for Advanced Ceramics, Argonne National Laboratory report ANL-87-53 (in press).

Section III
Mechanical Properties and Environmental Stability

Effects of Various Consolidation Techniques on Microstructure, Strength, and Reliability of Alpha-SiC

Sunil Dutta

National Aeronautics and Space Administration
Lewis Research Center
Cleveland, OH 44135

Abstract

Silicon carbide has strong potential for heat engine hardware and other high temperature structural applications. It has low density, good strength retention at high temperatures and excellent oxidation resistance. However, SiC like other ceramics, shows strength variability due to process related flaws such as shrinkage cracks/voids, isolated large pores, inclusions, etc. In order to minimize such defects, improved processing such as slurry pressing, hot isostatic pressing (HIPing), and sinter-HIPing were investigated. This paper examines the effect of these various consolidation techniques on strength, microstructure, and process related defects. Also, the feasibility of glass encapsulation was determined. Densification by sintering was carried out at temperatures ranging from 1900 to 2200°C for a period of 10 to 240 min under 0.1 MPa in argon. By contrast, a much lower temperature (1850 to 1900°C) was required to achieve a final density >97 percent of theoretical by HIPing under 138 MPa argon. The HIPed silicon carbide exhibited an extremely fine-grained microstructure with grain size varying between 0.2 to 5.0 μm compared to 1 to 30 μm in pressureless sintered material. By contrast, no significant microstructural changes were observed between sinter-HIPed and sintered silicon carbide. A duplex microstructure consisting of equiaxed and elongated grains was observed in both sintered and sinter-HIPed material, while in HIPed silicon carbide, only ultra-fine equiaxed grains were predominant. The fine-grained HIPed silicon carbide exhibited significantly higher (60 percent) average flexure strength (655 MPa) as compared to (415 MPa) for the sintered material. Process-related defects such as large voids, shrinkage cracks etc., were not observed in HIPed silicon carbide.

Introduction

Silicon carbide's unique properties, such as excellent high-temperature fast fracture strength, and creep and oxidation resistance, make it a promising material that is being considered and tested for application in heat engine systems. High-density silicon carbide bodies can be produced by pressureless sintering[1,2]. However, the sintered material often contains process-related flaws such as shrinkage cracks, isolated large voids, and pore clusters that result in poor reliability. These process-related flaws are caused by the presence of agglomerates in the as-received powder[3-6]. Earlier work[5,7] has shown that improved cold forming, such as slurry pressing, facilitates dispersion and uniform particle packing and thereby minimizes the number of process-related flaws as well as critical flaw size. Hot isostatic pressing also further eliminates process-related flaws and thereby results in improved strength[8]. This study examined the effects of various consolidation techniques on strength, microstructure, and processing flaws in α-SiC material.

EXPERIMENT

Commercial high-purity α-SiC powder* designated as "Type 2" was used. The powder contained premixed sintering aids (boron and carbon) as received from the manufacturer. Chemical analysis of the powder is shown in Table I. As-received powder was sieved through a 100-mesh screen, and 3 g of powder was dry pressed to bars (3.81 by 0.79 by 0.45 cm) at 41.3 MPa in a double-acting, tungsten carbide-lined die. Pressed bars were then vacuum sealed in a thin-walled latex tube and isostatically pressed at 413 MPa.

Table I. Analysis of starck as-received α-SiC powder.

Element	Impurity Analysis (PPM) Type 2 α-SiC (B, C)
Al	140
Ca	40
Fe	10
Ti	30
V	20
B	*0.60
Free C	*7.31
Surface Area (BET) m^2/gm	22

*Wt Percent

For slurry pressing, 200 g SiC powder was milled with 225-ml solution of water/ammonium hydroxide (pH = 11) in a polyethylene jar with 200 g SiC grinding media. After the slurry had been mixed for 48 hr, it was pressed at 14 MPa. The disk-shaped specimens (4.7 cm in dia. by 0.6 cm thick) were slowly dried and then isostatically pressed at 413 MPa. Earlier work extensively discusses the slurry pressing process[5,7].

Pressureless sintering was carried out on both dry-pressed and slurry-pressed specimens at 1900 to 2200°C for 30 min, under 0.1 MPa flowing argon pressure. For comparison, both dry-pressed and slurry-pressed specimens were placed together in a crucible and sintered under identical conditions.

Hot isostatic pressing was carried out with slurry-pressed disks (4.7 cm dia. by 0.6 cm thick). The disks were encapsulated with tantalum cans. After outgassing for 6 to 8 hr at 1100°C, the cans were vacuum sealed. After a thorough leak check, the cans were placed in a conventional HIP furnace. After an initial pressure of 2000 psi was applied, both pressure and temperature were raised simultaneously until the desired pressure and temperature were reached. The cans were HIPed at different temperatures varying from 1850 to 2000°C for 30 to 60 min at 138-MPa argon gas pressure.

Sinter-HIPing was carried out on pre-sintered disks having densities greater than 90 percent of theoretical, where almost all open pores were eliminated. The pre-sintered disks were placed in the HIP furnace without any encapsulation. After an initial pressure of 2000 psi was applied, both pressure and temperature were raised simultaneously, and the pre-sintered disks were further HIPed at 2100 to 2200°C for 30 to 60 min at 138-MPa argon gas pressure.

Sintered, HIPed, and sinter-HIPed specimens were machined into test bars (2.54 by 0.64 by 0.32 cm), and the surfaces were ground with a 400-grit fine diamond wheel to a final surface finish of 8 rms. Density was measured by the water immersion method. Microstructural characterization was made by optical and electron microscopy. Flexure strength tests were conducted by four-point loading using a 0.95 cm loading span and a 1.87-cm support span test fixture. Testing was conducted at room temperature with a crosshead speed of 0.05 cm/min at room temperature. Fracture surfaces of selected test specimens were examined by scanning electron microscopy to identify critical flaws.

RESULTS AND DISCUSSION

Sintered Silicon Carbide

The densification behavior of both dry-pressed and slurry-pressed specimens sintered together at 1900 to 2200°C for 30 min is shown in Fig. 1. The dry-pressed specimens had relatively lower final densities than the slurry-pressed specimens after sintering at 1900 to 2000°C. However, at higher temperature (~2150°C) the final densities are equivalent for both dry-pressed and slurry-pressed specimens. This suggests that a sintering temperature of approximately 2150°C is adequate to achieve a final density equal to 96 percent of theoretical.

Figure 2 shows typical microstructures of dry-pressed and slurry-pressed specimens sintered at 2200°C for 30 min. The microstructures consisted of predominantly equiaxed grains with an average grain size of 5 to 6 μm. Both types of specimens had more or less identical grain morphology.

The room temperature flexure strengths of dry-pressed and slurry-pressed specimens sintered at 2150 or 2200°C for 30 min, are shown in Fig. 3. Up to 23 percent improvement in strength was obtained in slurry-pressed specimens over dry-pressed specimens. Earlier work[7] reported that the higher strength was due to more homogeneous pore distribution in the slurry-pressed specimens resulting from improved dispersion of agglomerates in the powder. The examination of fracture surfaces (Fig. 4) indicated that slurry pressing, in general, reduced the frequency of large (100 μm) flaws in the sintered material. However, the reduction of ~100 μm flaws by slurry pressing had apparently no effect on the Weibull modulus (m = 7.9) between dry-pressed and slurry pressed materials (Fig. 5).

The specimens sintered at 2200°C clearly demonstrated a definite improvement in strength as well as a reduction in the average size of critical flaws, using slurry pressing as compared to dry compaction.

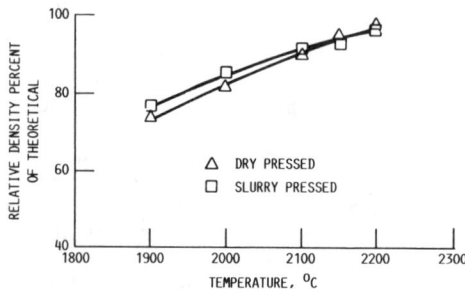

Fig. 1. Relative density of dry-pressed/vs slurry-pressed α-SiC powders sintered for 30 min at different temperatures.

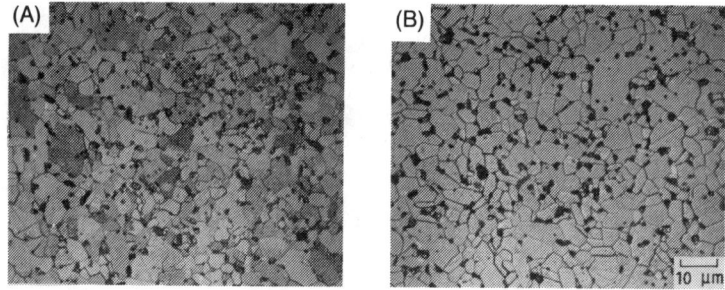

Fig. 2. Microstructure development in (*A*) dry-pressed vs (*B*) slurry-pressed α-SiC (sintered for 30 min at 2200°C).

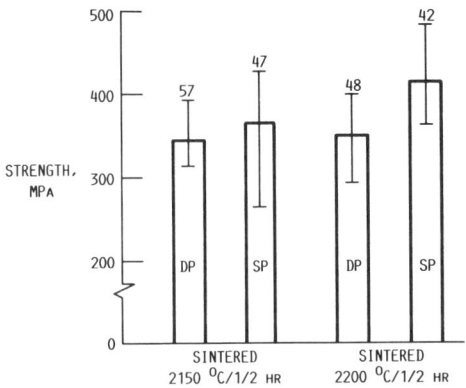

Fig. 3. Room temperature flexure strength of dry-pressed/sintered vs slurry-pressed/sintered α-SiC.

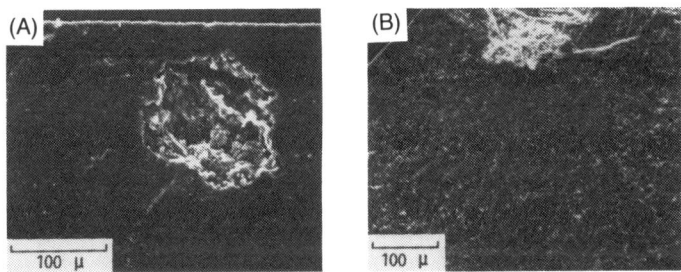

Fig. 4. Room temperature fracture of (A) dry-pressed/sintered (σ_f-275 MPa) vs (B) slurry-pressed/sintered (σ_f-340 MPa) α-SiC.

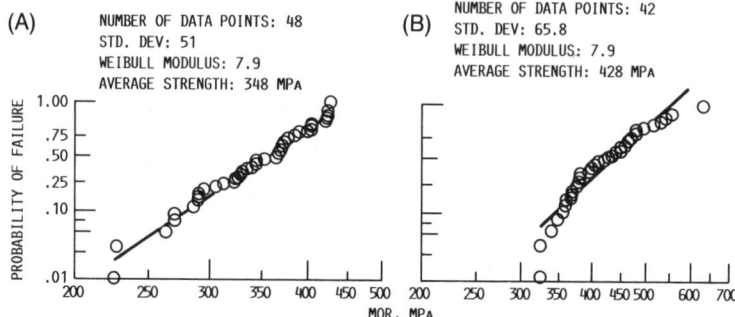

Fig. 5. Room temperature Weibull probability chart for (A) dry-pressed vs (B) slurry-pressed α-SiC sintered at 2200°C for 30 min.

Hot-Isostatic-Pressed Silicon Carbide

The densification behavior of hot isostatic pressed (HIPed) α-SiC (Ta encapsulation) as a function of temperature is shown in Fig. 6. The relative density is compared with that of the dry-pressed/sintered and slurry-pressed/sintered specimens. HIPed α-SiC achieved a final density equal to 98 percent of theoretical at 1900°C, as compared with 2200°C required for pressureless sintering. No further increase in final density was obtained by increasing the HIPing temperature to 2100°C.

The final density of HIPed α-SiC with glass† encapsulation was determined to be 98 percent of theoretical.

Figure 7 shows typical microstructures of α-SiC HIPed (Ta-encapsulation) at 1900 and 2000°C for 30 to 120 min. The microstructures consist of ultrafine, equiaxed grains varying between 0.3 and 5 μm as determined by electron microscopy. Grain size analyses were performed† for HIPed α-SiC (Figs. 8 and 9). The mean grain sizes were estimated to be 2.8 μm for Ta-encapsulated and HIPed α-SiC and 3.6 μm for glass-encapsulated and HIPed SiC. Some exaggerated grains ~10 μm were observed in glass-encapsulated HIPed α-SiC (Fig. 9). This was attributed to interaction of the molten glassy phase with the α-SiC matrix at the high HIPing temperature. By contrast, little grain growth was observed in Ta-encapsulated α-SiC HIPed between 1900°C for 30 min and 2000°C for 120 min (Fig. 7).

The room-temperature flexure strengths of α-SiC HIPed with both Ta and glass encapsulation are shown in Fig. 10. The data are compared with the strengths of dry-pressed/sintered and slurry-pressed/sintered specimens. A significantly higher average strength of 576 MPa (84 ksi) was achieved in Ta-encapsulated and HIPed α-SiC. A further increase in strength to 655 MPa (95 ksi) was obtained by annealing the HIPed specimens at 1200°C for 2 hr in air.

The increase in strength for annealing was attributed to surface oxidation of machining flaws/damage which enables the material to exhibit higher strength. This strength was 90 percent higher than the dry-pressed/sintered strength and 60 percent higher than the slurry-pressed/sintered strength.

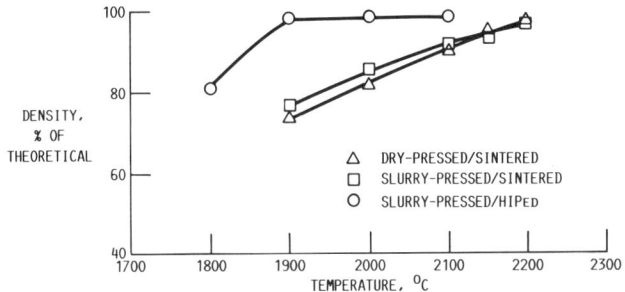

Fig. 6. Relative density of α-SiC powders sintered and HIPed for 30 min vs sintering temperature.

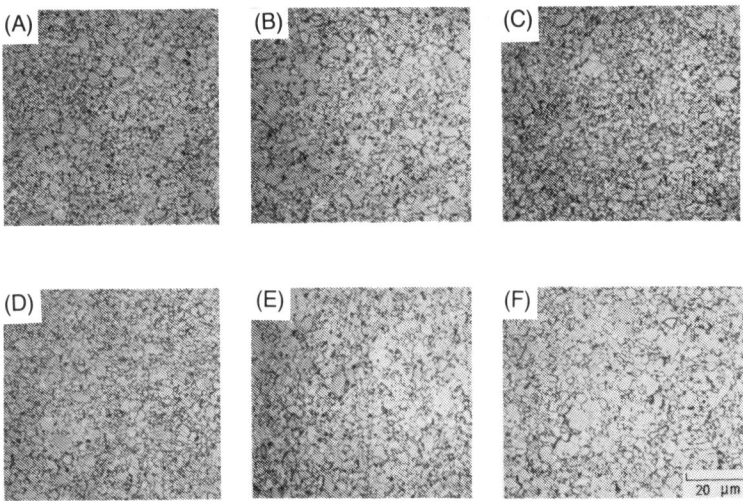

Fig. 7. Microstructures of hot isostatic pressed α-SiC (Ta-encapsulation) at 1900°C for (A) 30 min (B) 60 min (C) 120 min and at 2000°C for (D) 30 min (E) 60 min and (F) 120 min.

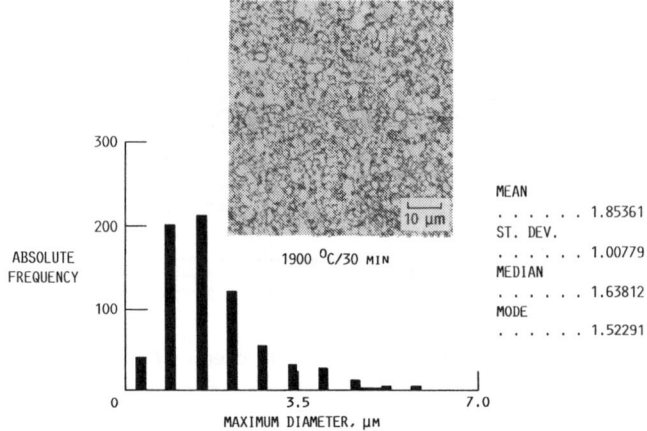

Fig. 8. Grain size analysis of (Ta-encapsulated) HIPed α-SiC.

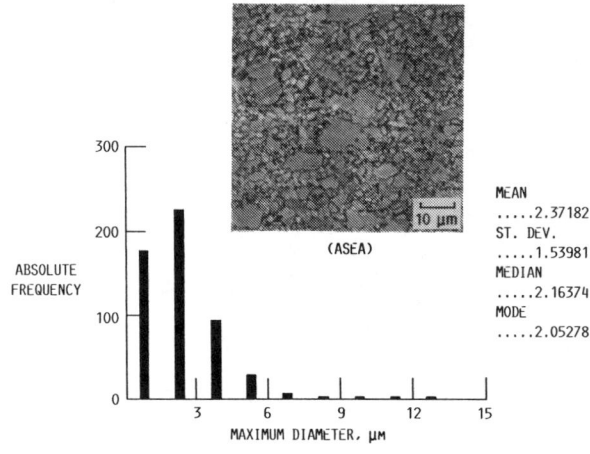

Fig. 9. Grain size analysis of (glass-encapsulated) HIPed α-SiC.

By contrast, the average flexure strength of α-SiC HIPed with glass encapsulation was found to be 458 MPa (66 ksi) (Fig. 10). Thus, there was no improvement over slurry-pressed/sintered strength. Fracture surface examination (Fig. 11) indicated a considerable amount of glassy-phase inclusions, suggesting interaction of molten glass with the SiC matrix along with exaggerated grain growth. No such inclusions were found in Ta-encapsulated and HIPed α-SiC (Fig. 12). Overall, processing flaws typical of sintered material, such as shrinkage cracks and large voids, were not observed in HIPed specimens. Instead, the critical flaws were surface related (Fig. 12).

The Weibull moduli of Ta-encapsulated and glass-encapsulated and HIPed α-SiC specimens were estimated to be 9.8 and 8.8, respectively. Weibull probability charts (Fig. 13) show little difference in Weibull moduli in spite of differences in flaw population and strength between the two sets of specimens. The Weibull numbers were equivalent to that of sintered α-SiC (Fig. 5). Also, similar Weibull moduli have been reported for commercial sintered α-SiC[9,10]. Therefore, it is evident that although various consolidation techniques significantly improve flexure strength and result in different flaw populations, they have no effect on Weibull modulus.

Sinter-HIPed Silicon Carbide

In general, density improvements were observed in sintered α-SiC specimens sinter-HIPed at 2100 to 2200°C. For example, a final density equal to 97.4 percent of theoretical was achieved by HIPing 95-percent dense presintered specimens at 2150°C for 30 min at 138 MPa argon gas pressure. Sinter-HIPing 96.5-percent dense slurry-pressed/sintered specimens at 2200°C for 30 min produced a final density equal to 98.5 percent of theoretical. Density improvement by sinter-HIPing was also reported by Watson et al.[11], who attributed it to further reduction of residual porosity in sintered α-SiC.

Average room-temperature flexure strengths of sinter-HIPed α-SiC are shown in Fig. 14. An average strength of (436±75) MPa was achieved in specimens sinter-HIPed at 2150°C for 30 min, but sinter-HIPing at 2200°C for 30 min produced a lower average strength (336±41) MPa. The lower strength was attributed to large grain growth at the higher temperature. At 2150°C the grain structure (Fig. 15) was primarily equiaxed, with grain size ranging from 3 to 15 μm; at 2200°C the grain structure was completely elongated, with aspect ratio varying from 1:2 to 1:9. Grain size was 5 to 130 μm.

Sinter-HIPed α-SiC had considerably lower strength than HIPed α-SiC specimens (Fig. 14). Furthermore the sinter-HIPed strengths were more or less equivalent to the pressureless-sintered strengths (Fig. 3). This suggests that although sinter-HIPing improved the density of sintered α-SiC, it made no noticeable improvement in its strength over that of pressureless-sintered material. Also, there was no apparent improvement in Weibull modulus. Figure 16 shows room-temperature Weibull probability charts for α-SiC sinter-HIPed at 2150 and 2200°C respectively, for 30 min. The Weibull values of 6.4 and 9.3 are more or less equivalent to those for sintered α-SiC (Fig. 5) and HIPed α-SiC (Fig. 13). These results further confirm that various consolidation techniques influence strength, microstructure, and residual flaws, but have little if any influence on Weibull modulus.

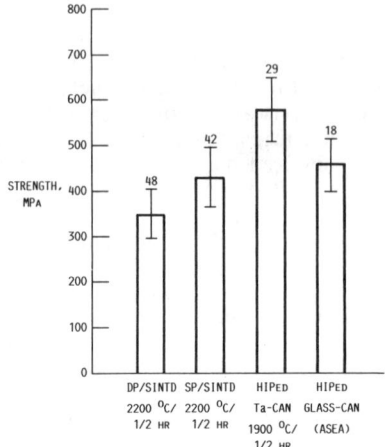

Fig. 10. Room temperature flexure strength of sintered vs HIPed Silicon Carbide.

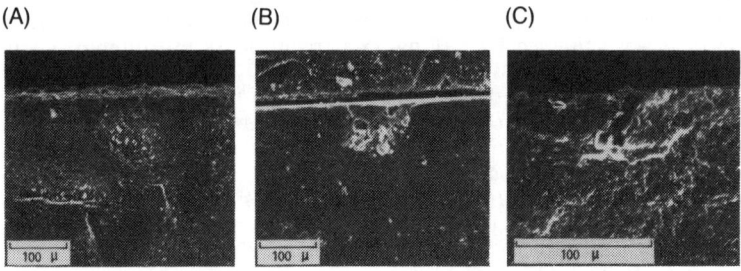

Fig. 11. Room temperature fracture of hot isostatic pressed (glass encapsulation) α-SiC, (A) σ_f-358 MPa (B) σ_f-390 MPa and (C) σ_f-357 MPa.

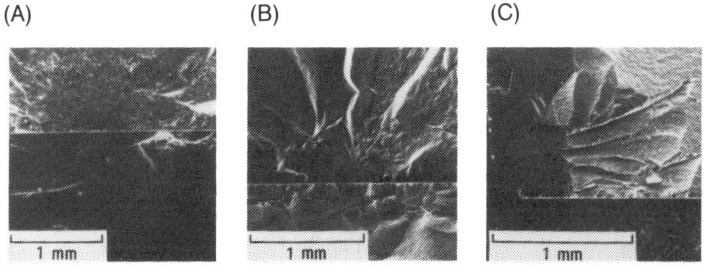

Fig. 12. Room temperature fracture of hot isostatic pressed (Ta-encapsulation) α-SiC, (A) σ_f-445 MPa (B) σ_f-480 MPa and (C) σ_f-530 MPa.

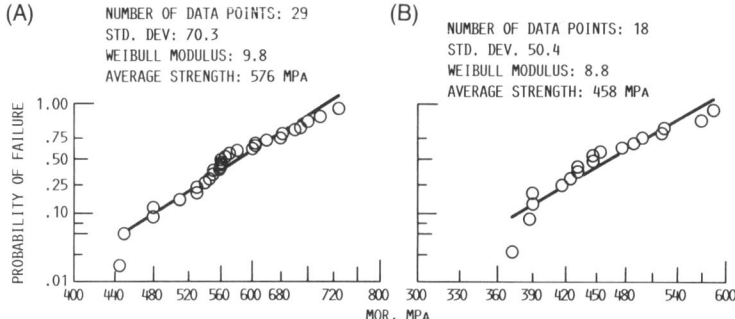

Fig. 13. Room temperature Weibull probability chart for hot isostatic pressed α-SiC, (A) HIPed with Ta-encapsulation (1900°C/30 min) and (B) HIPed at ASEA with glass-encapsulation.

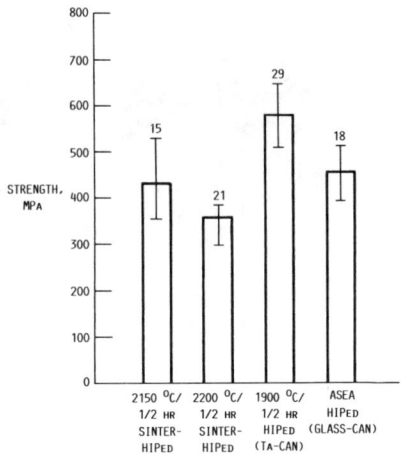

Fig. 14. Room temperature flexure strength of sinter-HIPed vs HIPed Silicon Carbide.

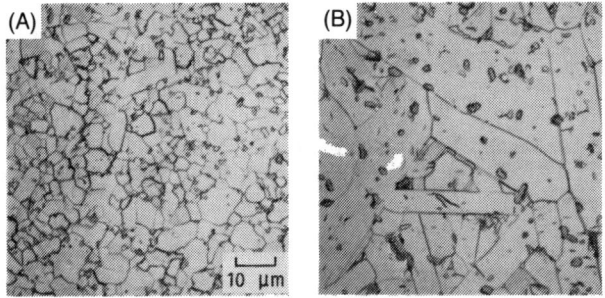

Fig. 15. Microstructure development in sinter-HIPed α-SiC (A) at 2150°C for 30 min, D=97% and (B) at 2200°C for 30 min, D=98.5%.

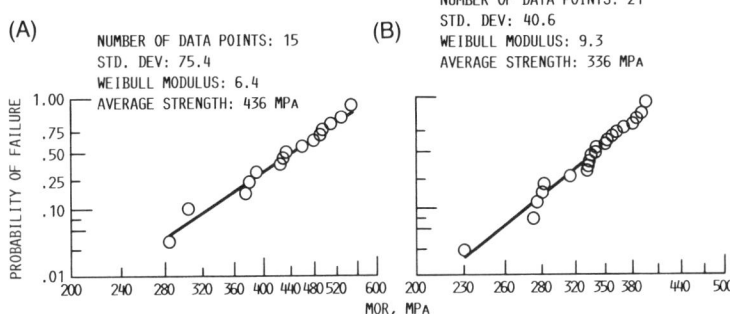

Fig. 16. Room temperature Weibull probability chart for sinter-HIPed α-SiC (A) at 2150°C for 30 min and (B) at 2200°C for 30 min.

Conclusion

The studies reported herein show a significant effect of various consolidation techniques on strength, microstructure, and critical processing flaws in α-SiC. By using slurry pressing, instead of dry pressing, an improvement of approximately 25 percent in baseline strength was achieved in pressureless sintered α-SiC. Further significant improvement in average strength to as high as 655 MPa (95 ksi) was achieved by using hot isostatic pressing (HIPing). This strength value is 60 percent higher than the slurry-pressed/sintered strength and 90 percent higher than the dry-pressed/sintered strength. By contrast, sinter-HIPing did not noticeably improve average strength. Process-related flaws typical of sintered material, such as shrinkage cracks and large voids, were not observed in HIPed α-SiC. The critical flaws were generally surface related. However, the differences in strength as well as in the type of flaw population between sintered and HIPed specimens were not reflected in the Weibull moduli. It is anticipated that surface finishing such as lapping, polishing, and heat treating might reduce the effect of surface-related flaws, thereby improving the Weibull modulus in HIPed α-SiC material.

---

*Herman C. Starck, Berlin, Goslar, West Germany.
†HIPed at ASEA, Robertsfors, Sweden.
‡Zeiss Videoplan Analyzer.

References

[1] S. Prochazka, "Sintering of Silicon Carbide," Ceramics for High Temperature Applications, Eds: J. J. Burke, A. E. Gorum, and R. N. Katz (Brook Hill Publishing Co., Chestnut Hill, MA, 1974) p. 239-252.

[2] J. A. Coppola, H. A. Lawler, and C. H. McMurty, U.S. Patent 4123286, October (1978).

[3] F. F. Lange, B. I. Davis, and I. A. Aksay, "Processing-Related Fracture Origins: III, Differential sintering of ZrO_2 Agglomerates in Al_2O_3/ZrO_2 Composite," *J. Amer. Cer. Soc.*, 66, 6, 407-408 (1983).

[4] S. Dutta, "Densification and Properties of α-SiC," *J. Amer. Cer. Soc.*, 68, 10, C 269-270 (1985).

[5] M. R. Freedman and M. L. Millard, "Improved Consolidation of Silicon Carbide," *Ceram. Eng. Sci. Proc.*, 7, 4, 884-892 (1986).

[6] S. Dutta, "Sinterability, Strength, and Oxidation of Alpha Silicon Carbide Powders," *J. Mater. Sci.*, 19, 4, 1307-13 (1984).

[7] J. B. Hurst and S. Dutta, "Simple Processing Method for High Strength Silicon Carbide," *J. Amer. Cer. Soc.*, 70, 11 (1987).

[8] S. Dutta, "Strength Optimization of α-SiC by Improved Processing," NASA Conference Publication 2427, May 20-21, pp. 89-98 (1986).

[9] R. K. Govila, "High Temperature Strength Characterization of Sintered Silicon Carbide," AMMRC TR-82-51, pp. 1-78 (1982).

[10] S. Dutta, "Reliability of Commercial Sintered Silicon Carbides," (submitted to the *J. Amer. Cer. Soc.*).

[11] G. K. Watson, T. J. Moore, and M. L. Millard, "Effect of Hot Isostatic Pressing on the Properties of Sintered Alpha Silicon Carbide," *Am. Cer. Soc. Bull.*, 64, 9, 253-56 (1985).

Microstructure and Mechanical Properties of Pressureless Sintered Alpha-SiC

S. G. Seshadri, M. Srinivasan, and K. Y. Chia

Standard Oil Engineered Materials Company
Niagara Falls, NY 14302

Abstract

The influence of microstructural parameters such as the grain size and shape distributions and porosity on the mechanical behavior of silicon carbide materials is discussed. The flexural strength is generally not sensitive to bulk porosity up to 7% and the presence of large isolated grains; The elastic constants and the fracture toughness, however, are influenced significantly by the porosity and the distribution of pores.

Introduction

The silicon carbide ceramics, especially single phase sintered α-SiC, are finding increasing use in many wear, corrosion and some structural-related components in the automotive, chemical pumping and heat exchanger applications.[1] The increased success of this material in these applications depends to a considerable extent on the optimization of the microstructure and properties to suit performance demands in these applications.

The common factors that control the microstructural development of ceramics are: (a) the physical, chemical and crystallographic characteristics of starting powder and sintering aids, (b) the extent of homogeneous mixing and packing during green forming and (c) sintering parameters such as temperature, time and furnace atmosphere including heating and cooling cycle. All these factors significantly influence the mechanical, thermal and physical properties of the resulting material. Correlation of the resulting microstructure to the properties of the material should take into account contributions from many of these variables.

This study is primarily concerned with the effects of microstructural variables like pore size, and grain size and shape distributions on the flexural strength, fracture toughness and elastic modulus of the sintered alpha silicon carbide.

Experimental Conditions

Hexoloy SA (Standard Oil Engineered Materials Company) is a single phase silicon carbide ceramic with greater than 99% SiC content. In this study, the densities and the microstructural variables (pore and grain size distributions) were varied by altering the sintering parameters and carbon additions.[2] By using selective sintering conditions (primarily temperature), the porosity was varied between 1 to 7.5% in the sintered α-SiC. The effect of different fabrication techniques such as unidirectional cold die pressing, isostatic pressing and injection molding and subsequent sintering have also been evaluated. The α-SiC plates made by the former two processes were subsequently machined to provide flexural specimens with surface roughness of about 0.2 to 0.4 μm. Additional test bars were also made by direct injection molding and sintering to yield specimens with as-fired surfaces (surface roughness—0.5 to 0.8 μm) with no machining.

The bulk density of the specimens were measured by water immersion technique. A quantitative image analysis system is used to determine the microstructural variables such as the size and shape distributions of grains and porosity. The flexural strength measurements were made in four point flexure. Both resonant frequency and ultrasonic velocity techniques were used to determine elastic moduli. The fracture toughness values were measured by single edge notched beam and vicker's indentation techniques.

Results and Discussion

Effects of Background Porosity

The variation of the flexural strength with porosity is shown in figure 1. The influence of background porosity on flexural strength is generally not clearly understood in sintered α-SiC with densities greater than 90% of theoretical value. There is no significant trend in flexural strength both at room temperature and at 1371°C, regardless of the fabrication technique used. This finding is supported well by extensive fractography results which show that the strength limiting flaws in these materials are primarily large complex voids (related to processing and fabrication) rather than any feature related to the general porosity (fig. 2-3). In the range of densities considered, these strength limiting flaws seem to be unchanged and the failure mode still related to the critical flaw defined by the Griffith's criterion for brittle fracture, but the failure mechanism is expected to change with further increase of porosity resulting in reduction in flexural strength.

The elastic modulus determined by resonant beam technique changes as predicted by micromechanical theories with a linear decrease with porosity (fig. 4). The variation of the Young's modulus with porosity for sintered α-SiC can be described in an in a linear fashion for up to 8% porosity as shown below;

$$E \text{ (GPa)} = 419 - 838\ P$$

where P is the volume fraction of porosity. It should be noted that this relationship may not be valid for non-uniformly distributed porosity in the material.

The fracture toughness (nominal) of sintered α-SiC as measured by single edge notched beam (with diamond wheel machined wide notches) is plotted as a function of porosity in fig. 5. The scatter in the data overwhelms the minor decreasing trend in the K_{Ic} with increased porosity. This drop is very consistent with the reduction in actual fracture area due to increased volume fraction of porosity. The fracture toughness is expected to be generally invariant if the correction for porosity is taken into account.

The fracture toughness as measured by vicker's indentation fracture technique shows a surprising trend. The K_{Ic} increases dramatically with porosity (fig. 6). In addition the hardness of the material decreases as expected with porosity (fig. 7). As a consequence the brittleness index parameter as proposed by Lawn and coworkers[3] decreases with porosity. The effect of surface residual stresses may be

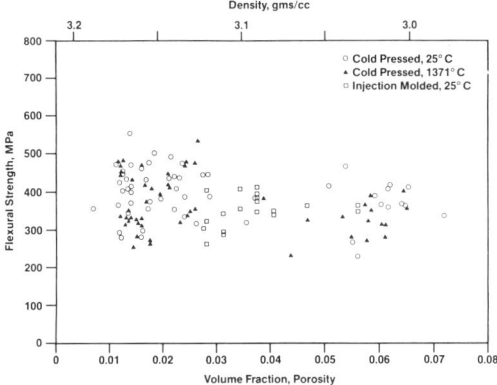

Fig. 1. Variation of flexural strength with porosity/density in sintered α-SiC at room temperature and at 1371°C.

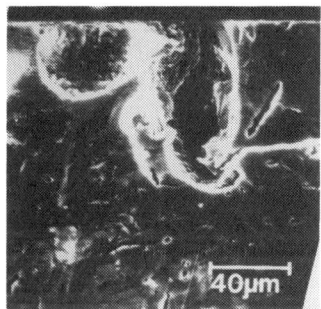

Fig. 2. Fracture origins for uniaxial pressed sintered α-SiC, average grain size = 7.5 μm, fracture strength = 291 MPa.

Silicon Carbide 217

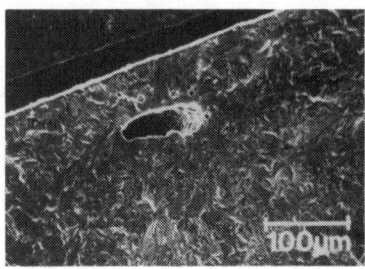

Fig. 3. Fracture origins for injection molded α-SiC, average grain size = 5.5 μm, fracture strength = 484 MPa.

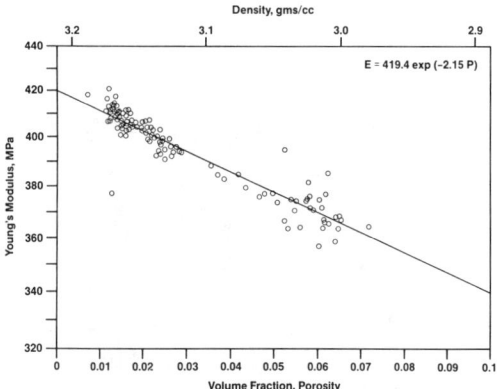

Fig. 4. The dependence of Young's modulus on porosity for sintered α-SiC.

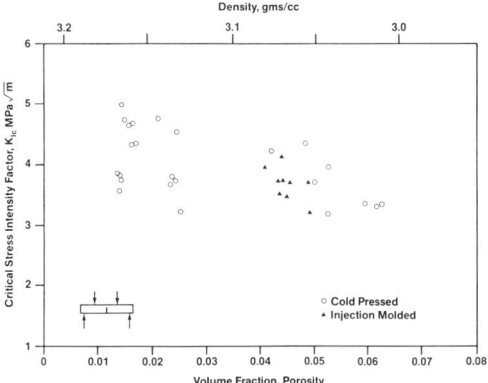

Fig. 5. Effect of porosity on single edge notched beam fracture toughness of sintered α-SiC.

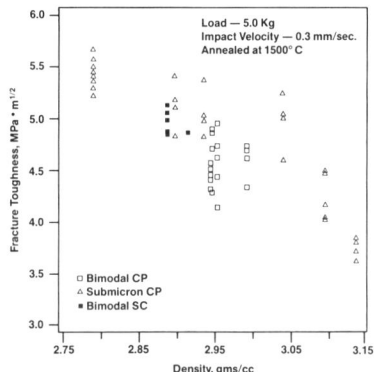

Fig. 6. The dependence of vicker's microindentation fracture toughness on density for sintered α-SiC.

Silicon Carbide

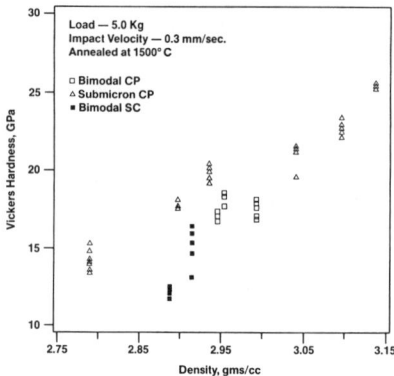

Fig. 7. Variation of vicker's hardness with density sintered α-SiC.

ruled out as many subsequent annealing treatments and specimens from different batches and of different geometries and thickness have followed the same master plot with respect to hardness, toughness and brittleness index. These findings are also supported by many qualitative experiments to assess the chipping and crater formation on the surface by indentation loads. These damage mechanisms were greatly minimized for the material with *controlled* porosity of 8%.

It is not the intent of this work to explain the discrepancies between the toughness measurements by using single edge notched beam and indentation methods and many past works concerning fracture mechanics have dealt with the aspects of macro- and micro- methods of toughness determinations. However, the Vicker's indentation measurements seem to be very significant in qualitative assessment of the surface response such as resistance to chipping. The increased porosity with uniform spatial distribution has resulted in improved chipping resistance in practice.

Effects of Grain Size

By selective sintering and post-sinter heat treatment procedures, a limited, but exaggerated grain growth was achieved in injection molded, cold pressed and isopressed sintered α-SiC. The procedure did not result in the growth of all the grains simultaneously, but resulted in isolated grains with large aspect ratios (the length 50-60 times the width) surrounded by randomly oriented more or less equiaxed grains (fig. 8).

The flexural strength and fracture toughness, measured via SENB technique, of sintered α-SiC are plotted as an inverse function of square root of the average grain size in fig. 9. Each datum point represents an average of 6 to 10 specimens. There is no apparent effect of grain size on the flexural strength of the material. This is consistent with the nature of critical flaw types revealed by extensive fractographic evaluation. The critical flaws are generally large voids and are unrelated to the large isolated grains present (figs. 2-3). A thorough survey

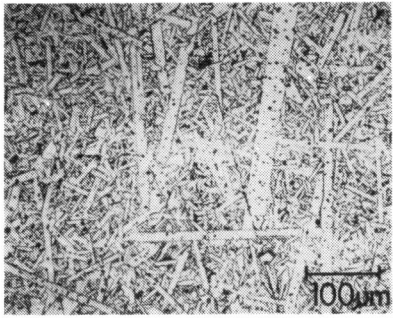

Fig. 8. Microstructure showing the exaggerated grain growth in sintered α-SiC.

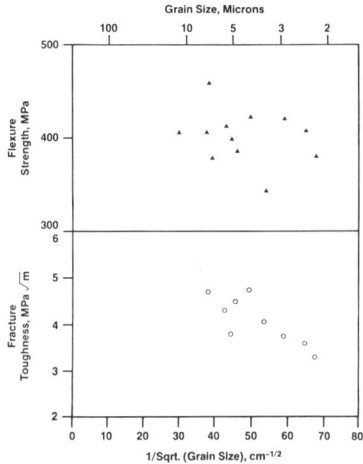

Fig. 9. Flexural Strength and SENB fracture toughness versus the average grain size of sintered α-SiC.

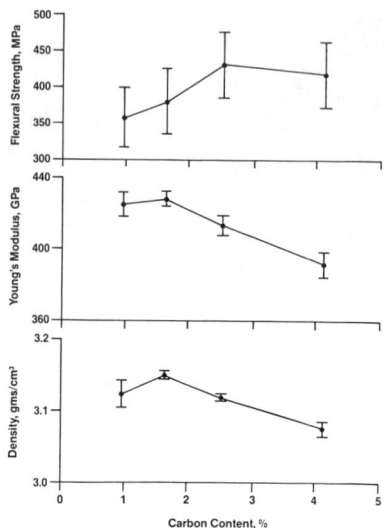

Fig. 10. Effect of carbon addition on the properties of sintered α-SiC.

regarding the effects of the grain size on the strength of ceramics was recently made by Rice[4]. The expected increase in strength with decreasing grain size is not observed in this material and the strength is more or less constant in the grain size range considered. This result is consistent with the observation that the critical flaws are seldom related to the large grains.

The fracture toughness, however, shows a much stronger dependence on the grain size as shown in figure 9, where average values of K_{Ic} of 4 or 5 specimens are plotted as a function of grain size. The fracture toughness shows a clear increase with increased grain size. This trend is partially due to general density increase which accompanies the grain growth in this material.

Effect of Carbon Addition

Additional carbon in the form of an organic resin may be added to the starting powder to control grain growth and densification in the sintered α-SiC, resulting in independent control of the density and grain size. The initial addition of carbon enhances densification and further additions result in a steady decrease in density (fig. 10). The Young's modulus of these materials varies in the same fashion as density indicating the strong dependence of the modulus on density. The flexural strength, however, shows a significant increase at carbon additions over 3 wt %. This increase is clearly related to the uniform reduction in the grain size. However, generally no increase in apparent porosity is seen as seen in the microstructures (fig. 11). There is also significant decrease in the average grain size and the spread in the

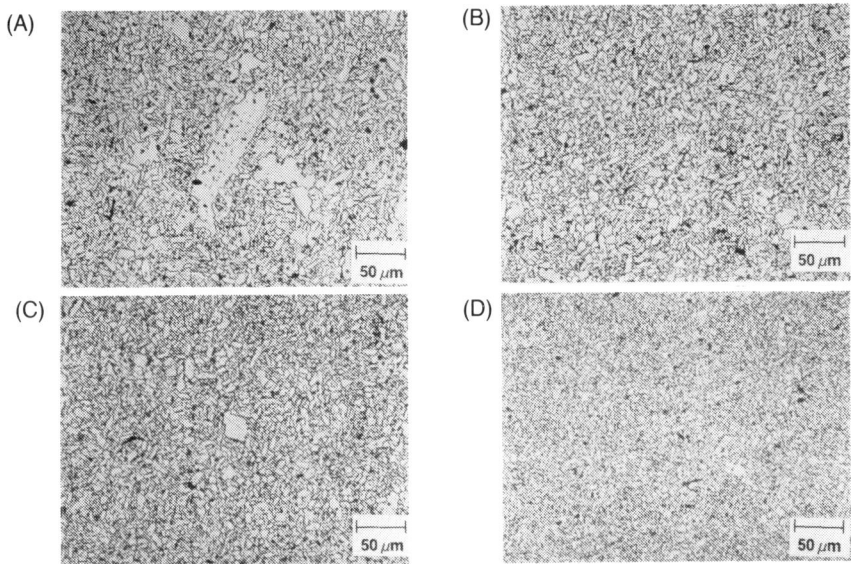

Fig. 11. The changes in grain size with increase in carbon content in sintered α-SiC, (A) 0.95%C (B) 1.62% C(C) 2.51%C and (D) 4.10%C. No changes in porosity were noticed.

size distribution (fig. 12), consistent with earlier reported observations by Murata and Smoak.[5]

Similar observations were also made by Prochazka et al[6]. They indicate that the large exaggerated grains in the microstructure is due to β to α SiC transformation and the carbon additions suppress this anomalous grain growth, probably by the removal of oxygen. However, our results show that the average shape of the grains is generally unchanged for a starting powder consisting of entirely the alpha polymorph (with minor 4H and 15R modifications) and the size distribution itself has shifted towards the lower size, indicating that the suppression of grain growth is not just limited to the large exaggerated grains. Similar changes in grain size distributions were also noted by Schwetz et al[7]. The figures 13 and 14 clearly show the decrease of the dimensions of the grain with little change in the aspect ratio or roundness of the grains. This is a significant new finding with implications in expected strength levels once processing related flaws are reduced in size.

CONCLUSION

The flexural strengths of sintered α-SiC is generally invariant with density (for materials with controlled porosity) above 3.0 gm/cm^3 as the strength limiting flaws tend to be large processing related voids (agglomerates were not observed in this study and in general presence of these lead to lower strengths). Similarly, no significant changes in strength were also seen with varying grain size. However,

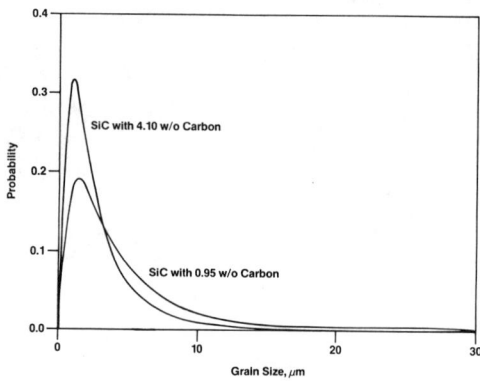

Fig. 12. Typical grain size distribution of sintered α-SiC with 0.95 and 4.10 wt % carbon.

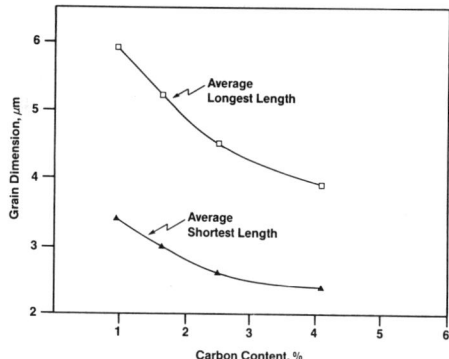

Fig. 13. The effect of carbon addition on the grain size of sintered α-SiC.

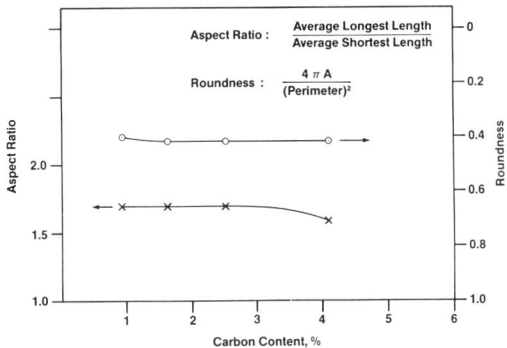

Fig. 14. The influence of carbon, as an additive, on the shape parameters of the grains in sintered α-SiC.

significant increase in strength is achieved by microstructural modification by the addition of excess carbon to the material, which causes uniform reduction in grain size and *insignificant change in grain shape.*

The fracture toughness as measured by single edge notched beam shows a slight decrease with porosity. This is contrasted by the increase in K_{Ic} seen by the indentation fracture method. The K_{Ic} may also be increased by uniform reduction in grain size.

The Young's modulus is strongly dependent on the density and varies linearly with density in the range above 3.0 gm/cm^3, consistent with the typical behavior of ceramics.

Acknowledgments

The authors thank Dr. Boecker and Mr. Hailey in preparation of some of the specimens and Mr. K. Selkregg and Mr. L. D. King for their assistance in the image analysis and mechanical testing.

References

[1] R. Ohnsorg, M. Teneyck, J. Zanghi, and T. B. Sweeting, "Ceramic Component Fabrication," Proc. 23rd Automotive Technology Development—Contractors Coordination Meeting, Dearborn, MI, 1985.

[2] W. D. G. Boecker, and L. N. Hailey, "Sintered Silicon Carbide/Graphite/Carbon Composite Ceramic Body Having Ultrafine Grain Microstructure," U.S. Patent No. 4525461, June 25, 1985.

[3] B. R. Lawn, and D. B. Marshall, "Hardness, Toughness, and Brittleness: An Indentation Analysis," *J. Amer. Ceram. Soc.*, **62**, 7-8 (1979) 347-350.

[4] R. W. Rice, "Machining Flaws and the Strength Behavior of Ceramics," NRL Memorandum Report 4076, Naval Research Laboratory, Washington DC, (1979).

[5] Y. Murata, and R. H. Smoak, "Densification of Silicon Carbide by the Addition of BN, BP, and B_4C and Correlation to Their Solid Solubilities," Intl. Symp. in Densification and Sintering, Hakone, Japan. Oct. (1978).

[6] S. Prochazka, and W. J. Dondalski, "Improvements in Silicon Carbide," U.S. Patent No. 1473911, May 18, 1977.

[7] K. A. Schwetz, and A. Lipp, "The Effect of Boron and Aluminum Sintering Additives on the Properties of Dense Sintered Alpha Silicon Carbide," *Science of Ceramics* **10**.

SiC-Based Ceramics with Improved Strength

T. Barrett Jackson, Andrew C. Hurford, Susan L. Bruner,
and Raymond A. Cutler

Ceramatec, Inc.
2425 South 900 West
Salt Lake City, UT 84119

Abstract

Silicon carbide was sintered using a transient liquid phase produced by the carbothermal reduction of alumina. Sintering occurred rapidly between 1850 and 1950°C due to the presence of the liquid phase. The resulting ceramic was fine grained (average grain size less than 5 microns) and contained SiC (starting polymorphs) as well as Al_2OC and AlN (when sintering in N_2) or solid solutions composed of SiC, Al_2OC, and AlN. Room temperature strengths ranged from 500 to 700 MPa, with strength retention to 1200°C. Bulk fracture toughness values for materials sintered in Ar were higher than conventional SiC, while the same composition sintered in N_2 gave toughness values comparable to SiC. Sintering in N_2 was more difficult than in Ar due to aluminum oxynitride formation. Mechanical properties of liquid phase sintered SiC are discussed relative to conventionally sintered silicon carbide.

Introduction

Silicon carbide has been a leading candidate for applications requiring high temperature creep and oxidation resistance but has lower strength and fracture toughness than silicon nitride. The higher thermal conductivity of SiC, as compared to Si_3N_4, is attractive in many high temperature applications. The most significant achievement in processing SiC was Prochaska's success in pressureless sintering β-SiC powder by adding small amounts of boron and carbon[1]. Coppola, et al.[2] showed that α-SiC powder could be densified using the same additives. Boecker, et al.[3] added small amounts of Al and C to densify α SiC and obtained fine-grained microstructures. Tanaka, et al.[4] showed that strength retention to 1600°C was possible by doping β silicon carbide with boron, aluminum and carbon. All of these compositions sintered to densities near 97% of theoretical after heating to temperatures of 2050 to 2100°C for 20 minutes to 1 hour.

Cutler, et al.[5] showed that wurtzite (2H) SiC can incorporate substantial amounts of AlN and Al_2OC in solid solution and coined the acronym "SiCAlON" to describe these materials in analogy to "SiAlON" ceramics. Rafaniello, et al.[6] reported on properties of SiC·AlN solid solutions made by hot pressing submicron powders. Huang, et al.[7] showed that pressureless sintering of SiC was possible when working with SiC-Al_2OC ceramics. The liquid phase was formed by the reaction between Al_2O_3 and C during the reactive sintering process. They showed that both α and β SiC could be densified without boron additions. Strengths in excess of 300 MPa and 600 MPa were achieved by pressureless sintering and hot pressing, respectively[7]. The purpose of this paper is to give further details on processing and report on strength improvement in SiC-Al_2OC ceramics.

EXPERIMENTAL PROCEDURES

Raw materials used (α SiC[*], β SiC[†], Al_2O_3[‡], Al_4C_3[§], Al[¶], C[**], SiO_2[††], and Si[‡‡]) were all less than 325 mesh in size with similar purities and surface areas as those reported previously[7]. The powders were vibratory milled in n-hexane (20 volume % solids) using 1 wt. % dispersant[§§] for 48 hours using high purity alumina media. The powders were wet screened −325 mesh and dry screened −80 mesh after drying in a rotary evaporator. Alternatively, powders were milled inside a stainless steel mill using WC-6 wt. % Co media at a ball to charge ratio of 10:1 (by weight) for 24 hours. Specimens approximately 4 mm × 5 mm × 50 mm were uniaxially pressed at 35 MPa followed by isostatic pressing at 200 MPa. Sintering was performed by placing the bars inside a closed crucible (threaded graphite cap and body were sealed using graphite foil) to limit volatilization, and heating under N_2 or Ar in a resistance heated graphite furnace. Alternatively, hot pressing was performed at temperatures between 1750 and 2150°C in a N_2 or Ar environment using graphite dies loaded uniaxially under a pressure of 35 MPa. Young's modulus was determined by strain gaging strength bars and measuring the stress-strain response of the bars in four point bending. Linear thermal expansion measurements were made between 25 and 1125°C using a dilatometer. Additional characterization and physical property testing were performed as described previously[7].

RESULTS AND DISCUSSION

Liquid Phase Sintering

In order to show that the eutectic between Al_2O_3 and Al_4O_4C [8] (see pseudobinary phase diagram between Al_2O_3 and Al_4C_3 in Figure 1) was responsible for liquid phase sintering, a wide variety of compositions (calculated weight percent Al_2OC in the final composition varied between 5 and 75) based upon the reaction

$$x(SiC) + z(Al_2O_3 + Al_4C_3) \rightarrow x(SiC) \cdot 3z(Al_2OC) \qquad (1)$$

were sintered at temperatures between 1800 and 2100°C. Powder processing was carried out in stainless steel ball mills with WC-Co media instead of Al_2O_3 media to ensure that the Al_2O_3/Al_4C_3 ratio remained constant. Beta SiC powder was used for all compositions and specimens were sintered within a closed graphite crucible (the crucible was still permeable and allowed some exchange with the atmosphere) in either 1 l/min flowing Ar or N_2. All specimens were held at temperature for 5 minutes.

Specimens heated in Ar sintered rapidly at temperatures between 1850 and 1950°C (see Figure 2) nearly irrespective of starting composition. Beta SiC (with 1 wt. % B and 1 wt. C added as sintering aids) showed no densification under these conditions and had a density of 1.8 g/cc. Densification was clearly aided by the liquid phase formed between the reaction of Al_2O_3 and C. When identical specimens were sintered in N_2 instead of Ar, densification occurred more gradually

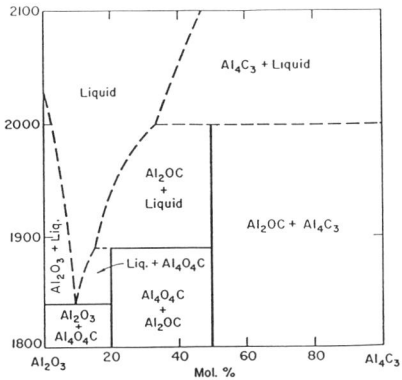

Fig. 1. Pseudobinary Al_2O_3-Al_4C_3 phase diagram after Foster, et al.[8].

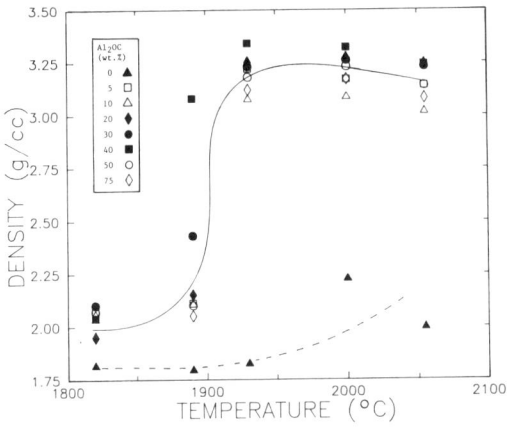

Fig. 2. Densification of SiC-Al_2OC ceramics in Ar as a function of temperature (5 minutes at temperature). Note rapid densification of specimens containing Al_2OC (solid line) between 1850 and 1950°C as compared to SiC with B and C additions (dotted line).

(see Figure 3). Specimens sintered in nitrogen showed AlN formation at temperatures above 2000°C due to the reactions

$$Al_2OC + N_2 \rightarrow 2AlN + CO \qquad (2)$$
$$Al_4C_3 + 2N_2 \rightarrow 4AlN + 3C \qquad (3)$$

while aluminum oxynitride cubic spinel, referred to as AlON[9], was detected by x-ray diffraction (XRD) in specimens sintered in N_2 between 1800 and 2000°C. These results suggest that AlON formation, which is an intermediate in the nitridation of Al_2O_3 to form AlN, competes with sintering and is the reason for the sluggish densification in N_2 as compared to Ar. XRD showed that the short sintering time did not allow for substantial solid solution formation. Sintered specimens contained primarily β SiC, with Al_2OC formation when sintered in Ar, or AlN when sintered in N_2. Specimens sintered in argon retained some Al_2O_3 whereas the alumina was fully converted to AlN for specimens sintered at temperatures above 2000°C in nitrogen.

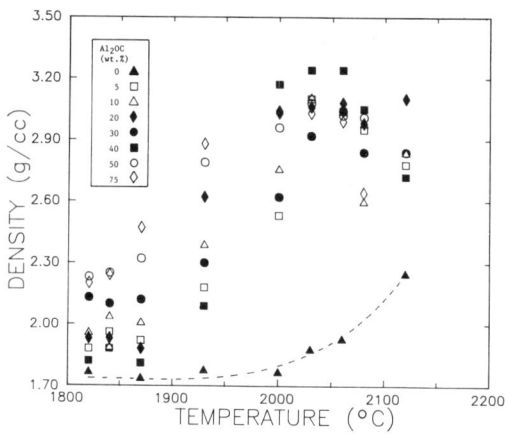

Fig. 3. Densification of SiC-Al_2OC ceramics in N_2 as a function of temperature (5 minutes at temperature). Sluggish densification (compare with Figure 2) is due to AlON formation. Dotted line shows SiC with B and C additions.

Reaction Path

Since it was desired to take advantage of the liquid phase to promote particle rearrangement during liquid phase sintering no efforts were directed towards prereacting SiC, Al_2O_3, and Al_4C_3 to form SiC·Al_2OC solid solutions prior to sintering. Instead, commercial powders were reacted in-situ during sintering. In order to determine if mechanical properties (i.e., strength, hardness and toughness) were sensitive to the reaction path taken, four methods for making SiC-10 wt. % Al_2OC were investigated as shown below:

$$SiC + 0.018Al_2O_3 + 0.018Al_4C_3 \rightarrow SiC \cdot 0.054Al_2OC \quad (4)$$

$$SiC + 0.018Al_2O_3 + 0.072Al + 0.054C \rightarrow SiC \cdot 0.054Al_2OC \quad (5)$$

$$0.973SiC + 0.027SiO_2 + 0.027Al_4C_3 \rightarrow SiC \cdot 0.054Al_2OC \quad (6)$$

$$0.973Si + 0.027SiO_2 + 0.108Al + 1.054C \rightarrow SiC \cdot 0.054Al_2OC \quad (7)$$

All powders were prepared in a vibratory mill using alumina media. Alpha SiC was used as the silicon carbide source for Reactions (4)-(6). Hot pressing at 1800°C for 10 minutes in N_2 was chosen as the method for densification since Reaction (7) had low green density due to the selection of acetylene black as the carbon source. X-ray diffraction patterns of hot pressed specimens made by Reactions (4)-(6) were identical and the same α SiC polytypes present in the starting powder were identified along with AlN or SiC·AlN formation. It is not surprising that the XRD patterns were similar for Reactions (4)-(6), as Reaction (5) only differs from Reaction (4) in that aluminum and carbon are substituted for aluminum carbide, and Reaction (6) will form alumina as an intermediate reaction species since the reduction of SiO_2 by Al_4C_3 to form Al_2O_3 and SiC is strongly exothermic[10]. Reaction (7) showed a much different XRD pattern indicating that the specimens contained either β SiC and AlN formation or β SiC and SiC·AlN formation. This is consistent with β SiC formation occurring when Si and C are exothermically reacted at temperatures above 1400°C[11].

The strength of specimens made by the four reaction paths are compared in Table 1, with strengths ranging between 525 and 725 MPa. Due to the small number of specimens tested, it is only possible to conclude that all four of the reaction paths result in similar strengths. Grains were typically equiaxed with average grain size less than 2 microns in size. Reaction (7) produced substantially finer grains with an average grain size of 0.2 microns. Fractography showed that the failure sites were primarily internal voids or areas of tabular grains up to 10 microns in size, with typical flaw sizes of 15 to 30 microns. The fractured surfaces showed a mixture of transgranular and intergranular fracture for each of the reaction paths investigated. Reaction (7) had fine porosity uniformly dispersed throughout the microstructure indicating the need for a higher hot pressing temperature.

Table 1. Density and strength of SiC-10 wt. % Al_2OC specimens made by reactions (4)-(7).

Reaction	Density* (g/cc)	Strength (MPa)†		
		No.	Mean	Std. Dev.
4	3.21	4	731	81
5	3.17	5	555	181
6	3.18	5	624	69
7	3.08	4	522	17

*All compositions hot pressed at 1800°C in N_2 for 10 minutes.
†Four-point bend strength on 6 × 12 × 43 mm bars.

The hardness for specimens made by Reactions (4)-(6) were similar with typical values ranging between 22 and 25 GPa for Vicker's indents made with a 75 N load. The hardness of specimens made by Reaction (7) was only 15 GPa due to the finely dispersed porosity. Indentation fracture toughness measurements[12] varied between 3 and 5 MPa·m$^{1/2}$ but were generally greater than 4 MPa·m$^{1/2}$. Based on the variability in indentation fracture toughness measurements[13], it was not possible to differentiate between specimens made by different reaction paths.

Further investigation of these four reaction paths is merited to optimize processing and understand differences in microstructures (i.e., extent of solid solution formation, if any, in these rapidly densified compositions and distribution of phases). It can be concluded, however, that acceptable mechanical properties (i.e., bend strengths exceeding 500 MPa and fracture toughness greater than 3.0 MPa·m$^{1/2}$) are achievable by any of the four reaction routes. It is widely known that conventional pressureless sintering of SiC powders is sensitive to both free silicon and surface oxygen (i.e., SiO_2 on the surface of SiC particles) and that sintering is difficult if the C/O molar ratio is not near unity and/or if there is free silicon present. Reactions (6) and (7) show that the SiO_2 can be the oxygen source instead of Al_2O_3, and Reaction (7) indicates that free Si will not hinder densification when using this liquid phase sintering approach.

The Al_2O_3/Al_4C_3 ratio was varied in Reaction (4) to determine the sensitivity of mechanical properties to carbon content. The Al_2O_3/Al_4C_3 ratios investigated were based on the Al_2O_3-Al_4C_3 pseudobinary phase diagram by Foster, et al.[8] and were designed to give SiC-10 vol. % (Al_2O_3-Al_4C_3) compositions where the Al_2O_3/Al_4C_3 ratio was 9.0 (eutectic composition between Al_2O_3 and Al_4O_4C), 4.0 (Al_4O_4C), 1.5 (Al_4O_4C-Al_2OC phase field), 1.0 (Al_2OC) and 0.67 (Al_2OC-Al_4C_3 phase field). Alumina media wear made the actual ratios approximately 9.0, 4.4, 2.0, 1.4, and 1.0. Sintering in nitrogen was insensitive to the Al_2O_3/Al_4C_3 ratio with all compositions being sintered to closed porosity by 2050°C. Hot pressing in nitrogen at 1800°C showed that strength, hardness, and fracture toughness were insensitive to Al_2O_3/Al_4C_3 ratio. XRD showed that all compositions were similar with AlN forming at the expense of all aluminum species.

Effect of Al₂OC Content

In order to determine if there is any beneficial effect of Al_2OC other than the transient liquid phase which forms during the reaction between Al_2O_3 and Al_4C_3, the effect of Al_2OC content on mechanical properties was investigated. Beta SiC was ball milled (WC-Co media) with Al_2O_3 and Al_4C_3 to make compositions containing 0, 5, 10, 15, 20, 30, 40, 50, and 75 wt. % Al_2OC based on Reaction (1). These compositions were densified by hot pressing in an Ar atmosphere at temperatures between 1850 and 1900°C for compositions forming Al_2OC with hold times at temperature of approximately 5 minutes. For densifying the specimens containing no Al_2OC, 2.5 wt. % C and 0.5 % B were added as densification aids. This composition was hot pressed at 2160°C for 10 minutes. Al_2OC formation was clearly shown by XRD for compositions containing more than 40 wt. % Al_2OC. It should also be emphasized that the Al_2OC content is not exact, as Al_2O_3 and C were detected in XRD patterns of compositions containing more than 40 wt. % Al_2OC.

Density as a function of Al_2OC content is shown in Figure 4. With the exception of the baseline SiC composition, densities were higher than theoretically predicted (Al_2OC has a density of 3.0 g/cc[8]) due to WC-Co contamination during

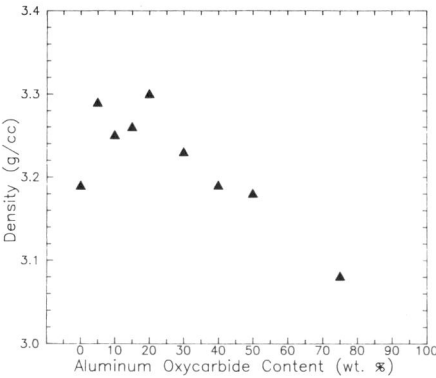

Fig. 4. Density as a function of Al_2OC content for hot pressed SiC-Al_2OC specimens. Density increases with 5 wt. % Al_2OC due to enhanced densification aided by liquid phase sintering.

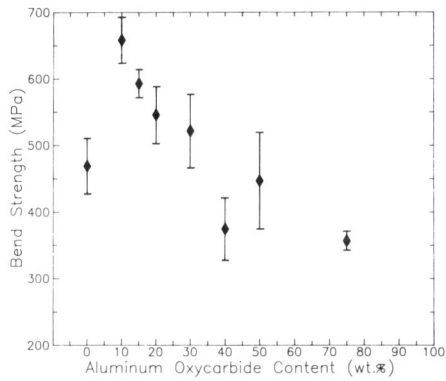

Fig. 5. Strength as a function of Al_2OC content for SiC-Al_2OC specimens hot pressed in Ar. Each data point is the mean of six specimens (error bars represent two standard deviations).

milling. Prior studies on Si_3N_4-based ceramics showed that WC-Co inclusions did not affect room temperature strength[14,15]. Extensive microscopy is required before theoretical density can be calculated for these materials, due to the presence of secondary phases. The important point is that the density sharply increases with as little as 5 wt. % Al_2OC, indicating that the liquid phase promotes densification. While it is well known that high density can be achieved by obtaining the right C/O ratio in SiC, two different carbon levels (0.5 and 2.0 wt. %) did not improve the

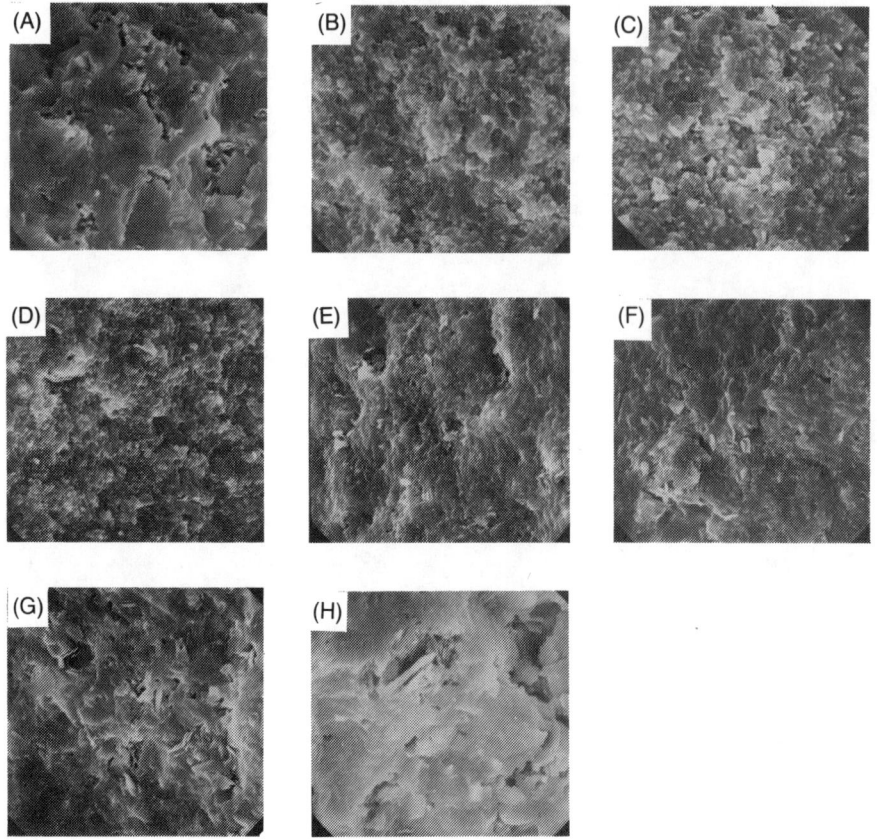

Fig. 6. SEM micrographs of fracture surfaces of SiC-Al$_2$OC strength bars. (A) SiC-2.5 wt. % C-0.5 wt. % B, (B) SiC-5 wt. % Al$_2$OC, (C) SiC-10 wt. % Al$_2$OC, (D) SiC-20 wt. % Al$_2$OC, (E) SiC-30 wt. % Al$_2$OC, (F) SiC-40 wt. % Al$_2$OC, (G) SiC-50 wt. % Al$_2$OC, and (H) SiC-75 wt. % Al$_2$OC.

density for the 0 % Al$_2$OC (i.e., SiC with 0.5 % B and 2.5 % C). No boron or carbon were added to the compositions containing Al$_2$OC.

Strength was significantly improved for 10 wt. % Al$_2$OC as compared to β SiC, but there was a significant decrease in strength with further increases in Al$_2$OC content (see Figure 5). Fracture surfaces of strength bars (see Figure 6) indicated that the improved strength was due to both an improvement in density, as well as a much finer grain size for specimens containing Al$_2$OC, as compared to conventionally sintered SiC. The fine grain size is the result of lower sintering temperatures limiting grain growth. (It is interesting to note that Lange[16] obtained fine grained SiC-10 vol. % Al$_2$O$_3$ when using Al$_2$O$_3$ as a liquid phase forming additive for densifying SiC via hot pressing). There is also a distinct switch from mainly transgranular fracture in SiC to a mixed mode of intergranular and transgranular fracture for specimens containing Al_2OC. $SiC-Al_2OC$ specimens containing higher amounts of Al_2OC (i.e., specimens with larger volumes of liquid phase during sintering) showed tabular SiC grains up to 10 microns in size (see Figure 6). The strength also drops with increasing Al_2OC due to the lower Young's modulus of Al_2OC as compared to SiC, as shown in Figure 7. The Young's modulus measured for SiC (corrected for density) is lower than the value of approximately 450 GPa reported by other investigators[6,17]. It is evident, however, that the modulus of Al_2OC is substantially lower than SiC, and that high Al_2OC contents are not desireable for structural applications.

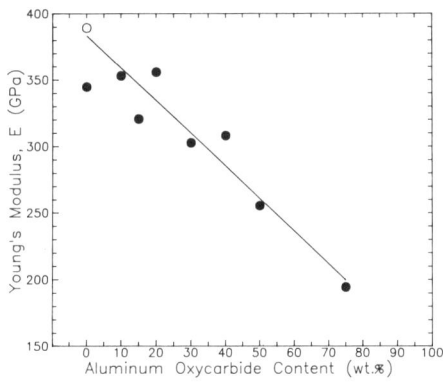

Fig. 7. Young's modulus as a function of Al_2OC content as measured by strain gaging $SiC-Al_2OC$ strength bars. Line shows least squares fit of data after correcting for porosity (open circle) of SiC composition.

Figure 8 gives hardness data as a function of Al_2OC content showing the sharp increase in hardness for 5 wt. % Al_2OC as compared to SiC, due to the lower porosity and finer grain size of the liquid phase sintered ceramic. The hardness decreases with increasing Al_2OC content due to the decreasing modulus and increasing grain size. The advantage of the liquid phase is it promotes sintering and allows sintering to occur at low temperatures and short times so that grain growth is minimized.

Figure 9 shows thermal expansion data as a function of Al_2OC content. The coefficient of thermal expansion of $SiC-Al_2OC$ ceramics increases linearly with increasing Al_2OC content.

Indentation fracture toughness data are shown as a function of Al_2OC content in Figure 10. While the scatter in the data preclude firm conclusions, fracture toughness appears to be independent of Al_2OC content. Short-rod measurements were made to obtain a bulk fracture toughness comparison between conventionally sintered SiC and SiC containing 10 wt. % Al_2OC. The SiC (0% Al_2OC) specimens gave a toughness of 2.39 ± 0.06 $MPa \cdot m^{1/2}$. SiC-10 wt. % Al_2OC specimens hot

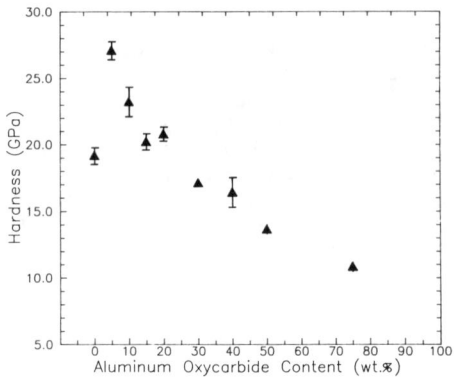

Fig. 8. Hardness as a function of Al_2OC content for SiC-Al_2OC specimens. Data points are mean values of 5 measurements (error bars represent two standard deviations and are less than symbol width when missing).

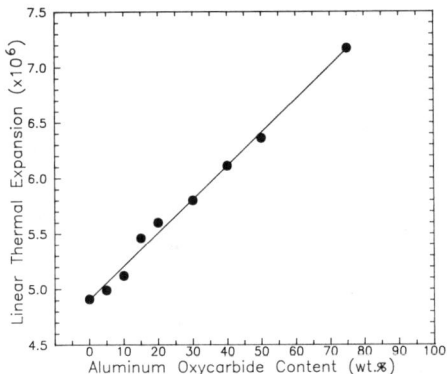

Fig. 9. Linear thermal expansion between 25 and 1125°C as a function of Al_2OC content for SiC-Al_2OC ceramics. Line shows least squares fit of data.

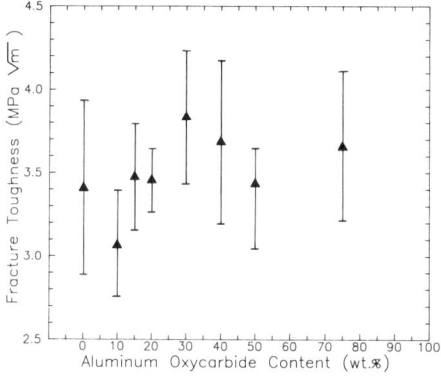

Fig. 10. Indentation fracture toughness data for SiC-Al_2OC specimens hot pressed in an Ar environment (see Fig. 8 for explanation of error bars).

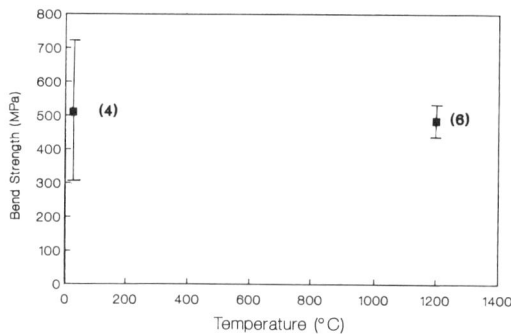

Fig. 11. Strength data for SiC-10 wt. % Al_2OC hot pressed in N_2 and tested in air.

pressed in Ar had a toughness of 3.97 ± 0.26 MPa·m$^{1/2}$ but the same composition hot pressed in N_2 had a toughness of 2.78 ± 0.21 MPa·m$^{1/2}$. While these data should be verified by double cantilever beam tests, they indicate that bulk fracture toughness of SiC-10 wt. % Al_2OC sintered in Ar is substantially improved over conventional SiC, whereas SiC-10 wt. % Al_2OC sintered in N_2 is comparable in toughness to sintered SiC. Visually, the fracture surfaces of the SiC-10 wt. % Al_2OC specimens sintered in Ar showed more surface roughness than specimens of the same composition sintered in N_2. When viewed under the electron microscope, however, there was no apparent difference in fracture mode except that the specimens sintered in Ar showed more transgranular fracture.

Strength data for SiC-10 wt. % Al_2OC specimens hot pressed in N_2 are shown both at 25 and 1200°C in Figure 11. The room temperature strength is retained to 1200°C. Examination of these specimens using high resolution TEM has shown grain boundaries devoid of amorphous phases and triple points containing crystalline phases[18] consistent with strength retention. Creep measurements on specimens of the same composition hot pressed in Ar showed creep rates of 10^{-9} to 10^{-6}/s at 1575°C in Ar at applied stresses between 38 and 200 MPa, respectively, resulting in a stress exponent of 1.7[19]. These data suggest that the liquid phase which promotes sintering is transient.

Pressureless Sintered SiC-Al₂OC

The main difficulty in obtaining high strength in the SiC-Al_2OC system is avoiding decomposition. Decomposition is believed to be affected by the high reactivity between SiC and Al_2O_3 making reactions such as

$$SiC + Al_2O_3 \rightarrow SiO + Al_2O + CO \tag{8}$$

thermodynamically favorable when the partial pressures of the gaseous specimens (i.e., SiO, Al_2O, and CO) are low. The reaction between Al_2O_3 and Al_4C_3, which produces the liquid phase, is very sensitive to the partial pressure of CO[7]. The closed graphite crucible, therefore, limits decomposition, which competes with sintering and helps to establish a P_{CO} suitable for sintering. The fact that the crucible is permeable, as demonstrated by N_2 diffusion to form AlN, limits conventional processing (i.e., sintering at high temperatures for extended periods of time) for SiC-Al_2OC compositions. Typical sintering cycles involve heating at rates above 30°C/minute (at temperatures in excess of 1000°C). This rapid heating limits decomposition, but also makes sintering in large production furnaces difficult.

To demonstrate the advantage of liquid phase sintered SiC, eight bars of SiC-10 wt. % Al_2OC were rapidly heated to 2200°C inside a closed crucible and the furnace was immediately shut off. Four of the bars were tested in the "as-sintered" condition and four were tested after subsequent HIPing at 1820°C for 5 minutes in 200 MPa Ar. The density of the "as sintered" bars was 3.15 g/cc and the density of the "HIPed" bars increased slightly to 3.18 g/cc. The bars were ground to a 30 micron finish and tested in four point bending. The four "as-sintered" bars had a mean strength of 639 MPa (std. dev. of 71 MPa) and the "HIPped" bars increased in strength to 733 MPa (std. dev. of 26 MPa). While the number of tested bars is small, the strengths are comparable to that obtained by hot pressing. The reason for the high strength is fine grain size and low porosity.

The sensitivity to sintering parameters is shown by the strength of 10 bars processed in an identical manner to those above, except densified at 2170°C (furnace shut off immediately upon reaching temperature). They had a density of 3.08 g/cc and strength of 541 MPa (std. dev. of 70 MPa). As expected, strength is dependent on density and high strengths can only be obtained when high densities are reached. In spite of the difficulties associated with volatility in this system, the strengths of SiC-10 wt. % Al_2OC have increased dramatically (from 320 MPa[7] to over 700 MPa) due to improved processing and further understanding of the competing reactions.

While the volatility of the SiC-Al$_2$OC system at high temperature limits its attractiveness for commercialization, the excellent mechanical properties obtained suggest that liquid phase sintering is a viable approach for obtaining stronger SiC by controlling grain size and achieving high densities. Other liquid phase forming additives for SiC are presently under investigation which do not rely upon contained sintering, but still result in fine grain size and high density.

CONCLUSIONS

Both α and β SiC were densified without boron additions by taking advantage of a liquid phase in the Al$_2$O$_3$-Al$_4$C$_3$ system. Rapid densification occurred at temperatures between 1850 and 1950°C in Ar. Sintering in nitrogen was hindered by aluminum oxynitride formation. Volatilization of the liquid phase and reactions between SiC and Al$_2$O$_3$ compete with sintering. Powder embedding or encapsulation techniques to limit volatilization and short sintering times were necessary to limit weight loss. Any of four reaction paths could be used to obtain bend strengths greater than 500 MPa. Strength was retained to 1200°C.

Engineering of properties is possible with SiC-Al$_2$OC ceramics as with SiCAlON ceramics[6]. Density, strength, hardness, Young's modulus and thermal expansion data showed expected trends as a function of Al$_2$OC content. SiC-Al$_2$OC compositions near 10% Al$_2$OC gave the best mechanical properties. Bulk fracture toughness measurements for specimens sintered in Ar suggested that SiC-10 wt. % Al$_2$OC specimens have higher toughness (approximately 4 MPa·m$^{1/2}$) than conventionally sintered SiC.

Strengths in excess of 700 MPa were achieved by pressureless sintering followed by uncontained HIPing of SiC-10 wt. % Al$_2$OC. Strengths in excess of 600 MPa were measured on sintered bars (no post sintering HIPing). The high strength was the direct result of liquid phase sintering limiting porosity and grain size. Flaw sizes were generally below 25 microns, average grain size was less than one micron, and high densities were achieved.

ACKNOWLEDGEMENTS

This work was supported by DOE under the Small Business Innovation Research (SBIR) Program (contract DE-AC03-84ER80191). Helpful discussions with Professor Anil V. Virkar of the University of Utah and Professor D. Lynn Johnson of Northwestern University are gratefully acknowledged.

---

*Grade A-10, Hermann C. Starck Inc. (New York, NY).
†Betarundum ultra-fine (Ibigawa Electric Industry Co. (Ogaki, Japan).
‡HP-DBM, Reynolds Metals Co. (Bauxite, AK).
§Hermann C. Starck Inc. (New York, NY).
¶Grade 123 Alcoa Metals Co. (Pittsburgh, PA).
**Acetylene black, Chevron Chemical Co. (Cedar Bayou, TX).
††OX-50, Degussa Corp. (Teterboro, NJ).
‡‡Si-244, Atlantic Equipment Engineers (Bergenfield, NJ).
§§PA-78B, Witco Chemical Co. (New York, NY).

References

[1] S. Prochazka, Proc. of the Conference on Ceramics for High Performance Applications, Hyannis, MA, 1973, edited by J. J. Burke, A. E. Gorum and R. M. Katz (Brook Hill Publ. Co., 1975).

[2] J. A. Coppola, L. N. Hailey and C. N. McMurtry, "Process for Producing Sintering Silicon Carbide Ceramic Body," U.S. Patent No. 4,124,667, November 7, 1978.

[3] W. Boecker, H. Landfermann and H. Hausner, "Sintering of Alpha Silicon Carbide with Additions of Aluminum," Pow. Met. Int., 11 [2] 83-85 (1979).

[4] H. Tanaka and Y. Inomata, "Normal Sintering of Al-doped Beta SiC," *J. Mater. Sci. Letters*, 4 315-317 (1985).

[5] I. B. Cutler, P. D. Miller, W. Rafaniello, H. K. Park, D. P. Thompson and K.H. Jack, "New Materials in the Si-C-Al-O-N and Related Systems," *Nature*, 275 (5679) 434-435 (1978).

[6] W. Rafaniello, K. Cho and A. Virkar, "Fabrication and Characterization of SiC-AlN Alloys," *J. Mater. Sci.*, 16 3479-3488 (1981).

[7] J. L. Huang, A. C. Hurford, R. A. Cutler and A. V. Virkar, "Sintering Behavior and Properties of SiCAlON Ceramics," *J. Mater. Sci.*, 21 1448-1456 (1986).

[8] L. M. Foster, G. Long and M. S. Hunter, "Reactions Between Aluminum Oxide and Carbon," *J. Am. Ceram. Soc.*, 39 [1] 1-11 (1956).

[9] J. W. McCauley and N. D. Corbin, "Phase Relations and Reaction Sintering of Transparent Cubic Aluminum Oxynitride Spinel (ALON)," *J. Am. Ceram. Soc.*, 62 [9-10] 476-479 (1979).

[10] R. A. Cutler, A. V. Virkar and J. B. Holt, "Synthesis and Densification of Oxide-Carbide Composites," Ceram. Eng. and Sci. Proc., 6 [7-8] 715-728 (1985).

[11] K. R. Rigtrup and R. A. Cutler, "Synthesis of Submicron Silicon Carbide Powder," this volume.

[12] G. R. Anstis, P. Chantikul, B. R. Lawn, and D. B. Marshall, A Critical Evaluation of Indentation Techniques for Measuring Fracture Toughness," *J. Am. Ceram. Soc.*, 64 [9] 533-543 (1981).

[13] R. F. Cook and B. R. Lawn, "A Modified Indentation Toughness Technique," *J. Am. Ceram. Soc.*, 66 [11] C200-C201 (1983).

[14] H. R. Baumgartner and D. W. Richerson, "Inclusion Effects on the Strength of Hot-pressed Si_3N_4," Fracture Mechanics of Ceramics, Vol. 1, ed. by R. C. Bradt, D. P. H. Hasselman and F. F. Lange, Plenum Press, New York and London (1974) pp. 367-386.

[15] A. G. Evans, "Structural Reliability: A Processing-Dependent Phenomenon," *J. Am. Ceram. Soc.*, 65 [3] 127-137 (1982).

[16] F. F. Lange, "Hot-pressing Behavior of Silicon Carbide Powders with Additions of Aluminum Oxide," *J. Mater. Sci.*, 10 314-320 (1975).

[17] R. A. Giddings, C. A. Johnson, S. Prochazka and R. J. Charles, "Fabrication and Properties of Sintered Silicon Carbide," General Electric report No. 75CRD060 (April 1975).

[18] R. W. Carpenter and R. A. Cutler, to be published.

[19] Z. C. Jou, A. V. Virkar, and R. A. Cutler, to be published.

The Fracture Resistance of a Sintered Silicon Carbide Using the Chevron-Notch Bend Specimen

Michael G. Jenkins, Asish Ghosh, Ken W. White, Albert S. Kobayashi, and Richard C. Bradt

College of Engineering
University of Washington
Seattle, WA 98195

Abstract

The fracture resistance of a dense sintered alpha silicon carbide was determined for temperatures from 20°C to 1400°C using the chevron-notch three-point bend specimen. Crack mouth opening displacement (CMOD) was continuously monitored throughout the elevated temperature fracture tests using a laser interferometric strain gage (LISG). The fracture toughnesses and the work-of-fracture values were observed to be independent of temperature from 20°C to 1400°C. Crack growth resistance curves in the form of G_R-curves were also determined from established relationships between the CMOD, the specimen compliance, the effective crack length and the load point displacement (LPD). In all instances the G_R-curves were observed to be flat, that is G_R was independent of the increasing crack length, Δa. The G_R-curves were also independent of temperature between 20°C and 1400°C. These fracture results are compared with those of other published studies on similar SiC materials.

Introduction

Ceramic materials, because of their high temperature strengths, low densities and low thermal conductivities are receiving increased attention for application at elevated temperatures, such as for utilization for the components of advanced heat engines[1-3]. The resistance of structural ceramics to crack growth is an important consideration for engineering design of long term survivability. If the resistance of a material to fracture can be completely characterized, then the damage tolerance of the structure in the presence of flaws can be accurately assessed and lifetimes can be satisfactorily predicted. At room temperature, most ceramics are traditionally considered to be brittle materials. However, at elevated temperatures, mechanical properties may often be expected to change. In some instances the fracture characteristics, expressed as either K_{IC} or in the form of crack growth resistance curves, or R-curves may be expected to vary with temperature. In the interest of design for elevated temperature applications, the resistance to fracture of potential structural ceramic materials must be determined over the temperature range of technological interest.

In this study, a commercially available, sintered alpha silicon carbide* was investigated for its fracture resistance at elevated temperatures. The density of this silicon carbide is 3.16 g/cm^3. The elastic modulus is 430 GPa at room temperature and the Poisson's ratio is 0.22. The coefficient of thermal expansion is reported to be $4.02 \times 10^{-6} K^{-1}$. The microstructure is shown as fracture surfaces in Figure 1. Previous studies[4-8] have not always been conclusive as to the character of the fracture parameters of this sintered silicon carbide from room temperature to elevated temperatures. The modest level of the fracture toughness and the high elastic modulus have precluded more extensive investigations of the crack growth resistance, i.e. the R-curve behavior. However, the recent combination of finite element methods, fracture mechanics techniques and a laser interferometric strain gage (LISG) system has facilitated the direct measurement of the fracture parameters of this material. The advantage is that the LISG technique is not subject to experimental errors from the test system loading compliance changes at elevated temperatures[9,10], thus the crack growth resistance or G_R-curves can be accurately determined.

This paper summarizes the results of an investigation utilizing the chevron-notch, three-point bend test specimen to determine the fracture resistance of a sintered alpha silicon carbide for the temperature range from 20°C to 1400°C. The results clearly illustrate that the fracture characteristics of this particular sintered silicon carbide are independent of temperature over that temperature range.

FRACTURE TESTING

Specimen Preparation

Chevron-notch three-point bend specimens were prepared by diamond sawing and then tested over a loading span of 40 mm at temperature intervals of 20°C from room temperature to 1400°C. These macroflaw test specimens had the nominal dimensions of 6.48 × 6.48 × 76 mm. The specimen and the chevron dimensional nomenclatures are illustrated in Figure 2. The initial notch depth ratios, $\alpha = a_o/W$, were approximately 0.44 for the chevron notches, which were machined with a rotating diamond-tipped saw blade to yield a notch radius of about 0.15mm.

Testing Procedures

Fracture testing was conducted on a screw-type, displacement-controlled load frame[1], at a cross head speed of 0.01 mm/min. A stiff test fixture of dense, silicon carbide was utilized. Once the loading sequence was initiated, it was maintained continuously until the complete fracture of the test specimen was achieved. Elevated temperature testing was completed in a refractory lined resistance furnace with silicon carbide heating elements. The tests were conducted at the specified test temperatures in air and ambient relative humidity conditions. Ports from the exterior of the furnace to the interior specimen test chamber were created to allow for the passage of the laser beams of the LISG system. The laser beam ports were covered with fused silica windows to minimize the convective air currents within and through the test chamber.

Application of the LISG technique to the elevated temperature fracture testing of ceramic specimens has been previously described by Jenkins, et al[10]. Interference fringes are created by the reflection of a incident laser beam from platinum

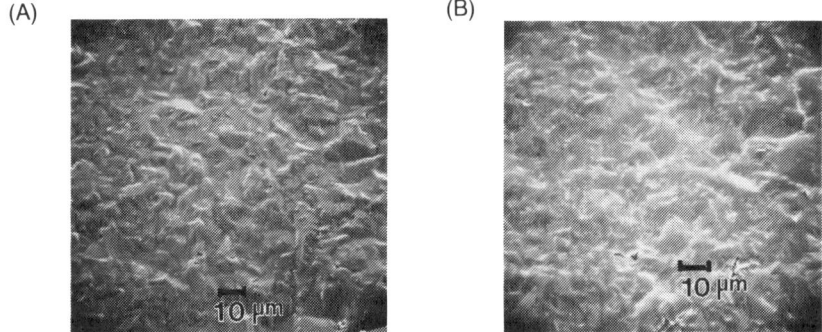

Fig. 1. Fracture surfaces of Silicon Carbide at (A) 20°C and (B) 1400°C.

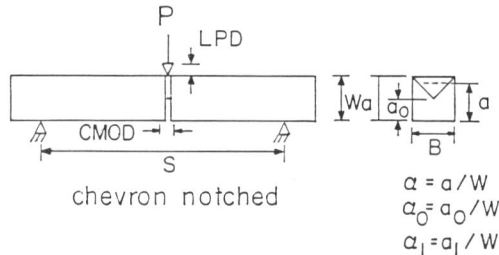

Fig. 2. The chevron-notch three-point bend specimen.

diffraction targets located on the opposite sides of the specimen notch. As the crack mouth opening displacement (CMOD) changes, the interference fringe patterns translate and are monitored by photodiodes. The passage of each fringe is equal to approximately 1.0 micrometer of crack mouth opening displacement, although a resolution of less than 0.25 micrometers is readily achieved. The fringe motion and the applied test load are simultaneously recorded during each individual fracture test. These are then utilized along with a 3-D finite element analysis of the specimen to produce plots of the load versus the load point displacement.

Results and Discussion

The average results for the individual test temperatures are summarized in Table I. Sustained stable crack growth was consistently observed for all of these chevron-notch test specimens. Figure 3 depicts a representative load versus CMOD history similar to that observed at each of the test temperatures for these chevron-notch test specimens.

Table I. Summary of fracture parameter results.

Temperature (°C)	Average K_{Ic} (MPa m$^{1/2}$)	Average WOF (J/m^2)	K_{Ic} from WOF (Eqn. 3) (MPa m$^{1/2}$)
20	2.91 ± 0.31	8.81 ± 0.64	2.82 ± 0.76
200	2.64 ± 0.33	6.39 ± 1.39	2.40 ± 1.12
400	2.88 ± 0.41	5.76 ± 0.80	2.26 ± 0.84
600	3.19 ± 0.49	7.98 ± 0.53	2.58 ± 0.68
800	2.71 ± 0.68	5.31 ± 1.03	2.10 ± 0.95
1000	2.76 ± 0.68	7.11 ± 2.22	2.42 ± 1.38
1200	3.37 ± 0.05	6.33 ± 1.69	2.26 ± 1.2
1400	3.03 ± 0.96	4.64 ± 0.54	1.93 ± 0.68
Overall Averages	2.94 ± 0.26	6.54 ± 1.38	2.35 ± 0.28

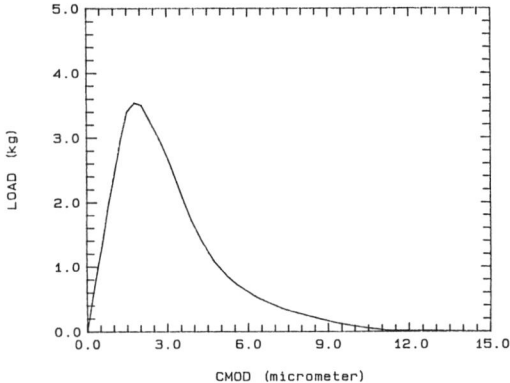

Fig. 3. Plot of typical load versus CMOD.

Fracture toughness values for the chevron-notch test specimens were calculated from the load maximum, P_{max}, and the minimum geometry correction factor for the stress intensity factor, Y_{min}, as has been extensively discussed by Pook[11] and by Munz, et al.[12]:

$$K_{Ic} = \frac{Y_{min} \, P_{max}}{B \, W^{1/2}} \qquad (1)$$

where K_{Ic} is the fracture toughness, B is the specimen thickness and W is the specimen width. Relationships for the stress intensity geometric correction factor,

Y, were determined from the 3-D finite element analysis previously reported by Jenkins, et al.[13]. The measured fracture toughnesses versus the testing temperatures for the chevron-notch test specimens are shown in Figure 4. From these results, it is evident that the fracture toughnesses are essentially independent of the test temperature through 1400°. The chevron-notch geometry allows for the initiation and stable growth of a real, "atomistically-sharp" macrocrack in the specimen. The fracture toughness values which are determined from the chevron-notch specimens may be considered representative of the fracture behavior of this material in the presence of a large sharp crack. The average chevron-notch fracture toughness for this SiC from room temperature through 1400°C is 2.94 ± 0.26 MPa m$^{1/2}$.

Once a crack has been initiated at the apex of the chevron-notch bend specimens, stable crack growth can be readily sustained due to the geometry of the chevron notch[13-17]. As a result, reliable values of the work-of-fracture and also the crack growth resistance curves can be readily obtained using these chevron-notch test specimens[13]. The work-of-fracture was determined from the area under the load versus the load-point-displacement curve divided by twice the projected area of one chevron of the fracture face. Schematic representation of the calculation of the work-of-fracture, after Nakayama[14] and Tattersall and Tappin[15], is shown in Figure 5. Relationships between the CMOD and the LPD from the finite element analysis of Jenkins et al.[13] were applied to generate the load versus LPD curves for the determination of the work-of-fracture from the chevron-notch specimens. The equation used for the work-of-fracture calculation was

$$\gamma_{WOF} = \frac{\int Pdu}{2A} \qquad (2)$$

The average work-of-fracture values versus temperature are shown in Figure 6 and are also summarized in Table I at each of the individual test temperatures. With the two apparent exceptions of a slightly higher work-of-fracture value at room temperature and a slightly lower value at 1400°C, the work-of-fracture remains essentially constant over the temperature range of measurement with an average value of 6.54 ± 1.38 J/m^2.

Fracture toughness values can also be calculated from these work-of-fracture results. Although the work-of-fracture measurement is not usually considered in specific terms of either plane stress or plane strain conditions, it appears that the plane strain condition prevails in the chevron-notch specimen geometry in this case. Therefore, fracture toughness values were calculated using the standard formula.

$$K_{Ic} = \sqrt{\frac{2E\,\gamma_{WOF}}{(1-\nu^2)}} \qquad (3)$$

A value for ν of 0.22 was used for all of the test temperatures. The calculated fracture toughness values from the work-of-fracture results are shown in Figure 4 and are comparable to the chevron-notch test specimen values at all test temperatures, although slightly lower. The average fracture toughness calculated from the work-of-fracture measurements is 2.35 ± 0.28 MPa m$^{1/2}$.

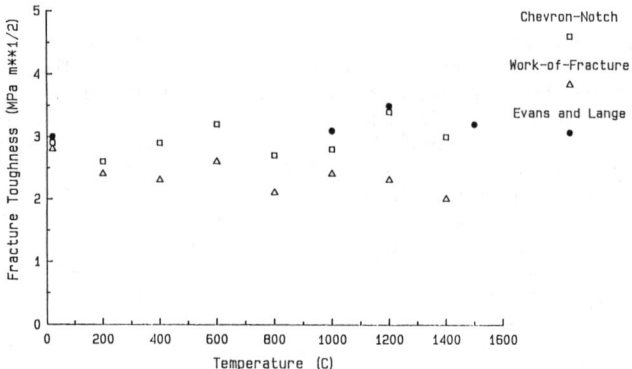

Fig. 4. Plot of fracture toughess versus temperature.

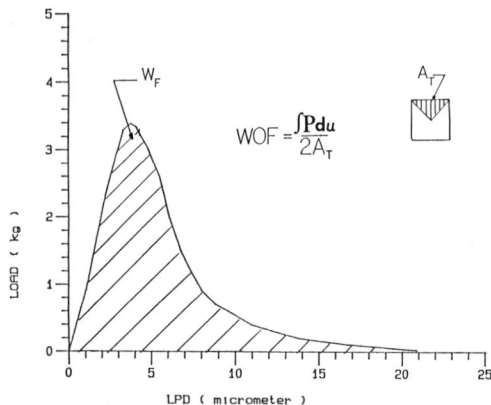

Fig. 5. Schematic diagram showing the calculation of the G_R curve.

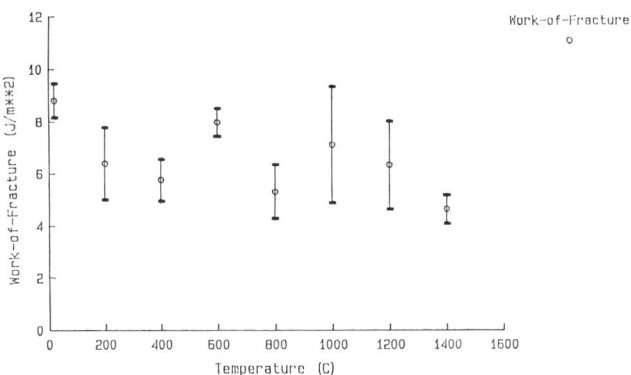

Fig. 6. Plot of work-of-fracture versus temperature.

Crack growth resistance or G_R-curves were developed from the stable crack growth load versus load point displacement curves for these chevron-notch bend specimens using the compliance versus crack length relations determined by the 3-D finite element analysis of the chevron-notch test specimen[13]. The compliance relation was applied to calculate the effective crack length during the stable crack growth process. The effective crack lengths and the load displacement history were then utilized to determine the change in the total strain energy for an incremental amount of crack extension, using the equation:

$$G = \frac{\Delta U}{\Delta A} \qquad (4)$$

where G is the strain energy release rate, ΔU is the change in total strain energy, and ΔA is the incremental increase in the fracture surface area. This procedure is shown schematically in Figure 7.

A typical G_R-curve with individual data points as a function of Δa is shown in Figure 8. As is evident from Figure 8, G_R remains constant with increasing crack length, indicating the existence of the flat R-curve behaviour for this material. G_R-curves for all of the test temperatures to 1400°C are depicted in Figure 9. However, the G_R-curves on the summary plot do not contain the individual data points, in order to maintain the clarity of the plots, as each G_R-curve was produced from a least squares analysis of all of the individual G_R-curves for that particular test temperature. It is significant to note that irrespective of the test temperature, G_R was found to be about the same, 14.03 ± 2.04 J/m^2. It also remained constant with increasing crack length.

The G_R-curves and the work-of-fracture are related, as plotting G_R versus Δa is essentially plotting a number of individual slices from the work-of-fracture curve. Conversely, integrating the G_R values over the entire fracture surface should yield the work-of-fracture. Instantaneously, the value of G_R should be $2\gamma_{WOF}$ for a flat G_R-curve material such as this SiC. The results in Figures 6 and 9 and the summary in Table 1 generally confirm this. The average work-of-fracture at room

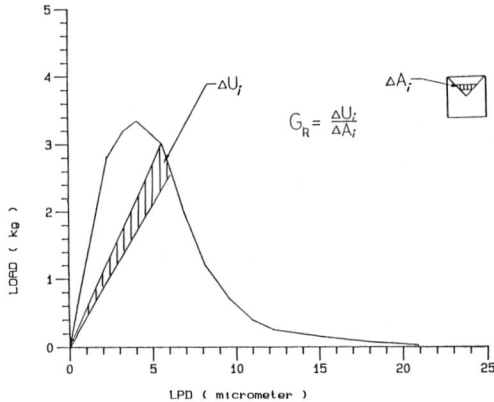

Fig. 7. Schematic diagram showing the calculation of the G_R-curve.

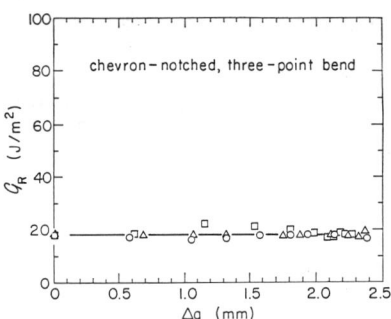

Fig. 8. Plot of room temperature G_R-curve.

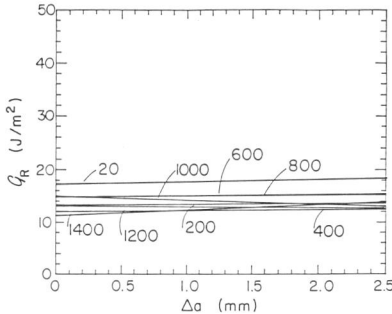

Fig. 9. Plot of G_R-curves at elevated temperatures.

temperature is 8.81 ± 0.64 J/m² while the flat G_R-curve at room temperature in Figure 8 yields a G_R value of 17.80 ± 1.52 J/m², as expected. The overall summary values in Table I and in Figure 9 further confirm the relationship of these quantities.

The fracture toughness results of this study are summarized in Table 1 and are illustrated in Figure 4 for the range of temperatures from 20°C to 1400°C. These fracture toughnesses values appear to be essentially constant. Table II lists a comparison of the fracture toughness results of this study with results of previously published elevated temperature investigations of the same or similar sintered silicon carbides. Previously reported values of the fracture toughness range from about two to six MPa m$^{1/2}$. From this summary it is evident that the reported fracture toughness values vary considerably for this material at the different test temperatures and also with the testing technique that is utilized by the different investigators.

Shih and Opoku[1] are the only researchers to have extensively compiled the elevated temperature fracture toughnesses of similar sintered silicon carbides. They confirm that the toughness remains essentially constant from 20°C to 1400°C in quite good agreement with the values measured in this study. Although for a sintered beta-SiC, Evans and Lange[8] have reported a value of 3.08 MPa m$^{1/2}$ at room temperature and an average of 3.0 through 1500°C using the double torsion method. For a similar, sharp crack growth measurement, again only at room temperature, a fracture toughness value of 3.08 MPa m$^{1/2}$ has been reported by Zhang and Chau[18] also using the chevron-notch test specimen. These results clearly illustrate that virtually identical values of the fracture toughness of silicon carbide can be consistently obtained by different investigators using different techniques, provided that the techniques utilize sharp cracks.

Summary and Conclusions

The fracture resistance of a dense, sintered alpha silicon carbide was measured by the chevron-notch three point bend specimen and found to be independent of temperature from 20°C to 1400°C. The fracture toughness, K_{Ic}, was constant, and flat G_R-curves existed for the complete temperature range, indicating a constant

Table II. Summary of elevated temperature studies.

Reference	Comment	Fracture Toughness (MPa m$^{1/2}$) Temperature (°C)							
		20	200	400	600	800	1000	1200	1400
This Study	CHV	2.9	2.6	2.9	3.2	2.7	2.8	3.4	3.0
	WCV	2.8	2.4	2.3	2.6	2.1	2.4	2.3	2.0
McHenry, et al[4]	DT	4.6				4.9 (900°C)	5.0	5.4 (1100°C)	
	CFKI	3.5				2.6 (900°C)	2.6	2.6 (1100°C)	
Coppola, et al[5]	CFKI	2.7						2.7	3.4
Henshall, et al[19]	SNBB	3.5			3.3		5.5 (1100°C)		5.8 (1500°C)

DT: Double Torsion
WCV: Work of Fracture Calculated
CHV: Chevron Notch Bend Bar
SNBB: Straight Notch Bend Bar
CFKI: Controlled Microflaw Knoop Indentation

resistance to fracture for this material. The fracture toughness as measured by the chevron-notch technique was 2.94 ± 0.26 MPa m$^{1/2}$.

The experimental fracture toughness values measured in this study compare favorably with appropriate previous literature values. The chevron-notch test method is a reliable technique for measuring the fracture toughness at elevated temperatures. Comparable results suggest that the double torsion method may also be viable specimen geometry, in spite of the concern over the nature of the crack front shape and character in that specimen. The existence of flat G_R-curves that do not vary significantly from room temperature to 1400°C suggests that any toughness contributions from the following wake region mechanisms are minimal in this material and that plastic flow mechanisms of toughening are not active below 1400°C either.

Acknowledgement

The research reported in this paper was supported by the Department of Energy, ORNL, Contract No. 86A-00209C. The authors are grateful to M. Srinivasan of Standard Oil Engineered Materials, Niagara Falls, N.Y. for assistance with providing the test samples.

---

*Hexoloy SA, Standard Oil Engineered Materials Company, Niagara Falls, NY 14302.
†Instron Corporation, Canton, MA 02021. Model TTDML

References

[1] T. T. Shih and J. Opoku, "Application of Fracture Mechanics to Ceramic Materials—A State-of the-Art Review," *Eng. Fract. Mech.*, 12, 479-498 (1979).

[2] R. N. Katz, "Application of Nitrogen Ceramics Gas Turbines: U.S. National Program," pp. 643-646, Nitrogen Ceramics, Ed. by F. L. Riley, Nordoff, Leyden, (1977).

[3] D. J. Godfrey, "Ceramics for High Temperature Engineering," *Proc. Brit. Ceram. Soc.*, No. 22, 1-25, (1973).

[4] K. D. McHenry and R. E. Tressler, "Fracture Toughness and High Temperature Slow Crack Growth in SiC," *J. Amer. Ceram. Soc.*, 63, (3-4) 152-156 (1980).

[5] J. A. Coppola, M. Srinivasan, K. T. Faber and R. H. Smoak, "High Temperature Properties of Sintered Silicon Carbide," The Carborundum Company, Niagara Falls, NY (1979).

[6] K. T. Faber and A. G. Evans, "Intergranular Crack Deflection Toughening in Silicon Carbide," *J. Amer. Ceram. Soc.*, 66, (6) C94-C96, (1983).

[7] J. L. Henshall and C. A. Brookes, "The Measurement of K_{Ic} in Single Crystal SiC Using the Indentation Method," *J. of Mat. Sci. Lttr.*, 4, 783-786, (1985).

[8] A. G. Evans and F. F. Lange, "Crack Propagation and Fracture in Silicon Carbide," *J. Mat. Sci.*, 10, 1659-1664, (1975).

[9] W. N. Sharpe Jr., "High Temperature Strain/Displacement Measurement," pp. 587-609 in Handbook on Experimental Mechanics, Ed. by A. S. Kobayashi, (1986).

[10] M. G. Jenkins, A. S. Kobayashi, M. Sakai, K. W. White and R. C. Bradt, "Fracture Toughness Testing of Ceramics Using a Laser Interferometric Strain Gage," *Bull. Amer. Ceram. Soc.*, 66 [12] 1734-1738 (1987).

[11] L. P. Pook, "An Approach to a Quality Control K_{Ic} Testpiece," *Int. J. Fract.*, 8 103-108, (1972).

[12] D. G. Munz, J. L. Shannon Jr. and R. T. Bubsey, "Fracture Toughness Calculation from Maximum Load in Four Point Bend Tests of Chevron Notch Specimens," *Int. J. Fract.*, 16 R137-R141, (1980).

[13] M. G. Jenkins, A. S. Kobayashi, K. W. White and R. C. Bradt, "A 3-D Finite Element Analysis of a Chevron-Notched, Three-Point Bend Fracture Specimen for Ceramic Materials," *Int. J. Fract.*, 34, 281-295, (1987).

[14] J. Nakayama, "Bending Method for Direct Measurement of Fracture Energy of Brittle Materials," *Jap. J. App. Phys.*, 3, 422-423, (1964).

[15] H. G. Tattersall and G. Tappin, "The Work of Fracture and its Measurement in Metals, Ceramics and Other Materials," *J. Mat. Sci.*, 1 296-301, (1966).

[16] J. Nakayama, H. Abe, and R. C. Bradt, "Crack Stability in the Work-of-Fracture Test: Refractory Applications," *J. Amer. Ceram. Soc.*, 64, (11) 671-675, (1981).

[17] J. C. Newman Jr. "A Review of Chevron-Notched Fracture Specimens," pp. 5-31 in Chevron-Notched Specimens: Testing and Stress Analysis, ASTM STP 855, Eds. J. H. Underwood, S. W. Freiman, F. I. Baratta, Amer. Soc. for Test. and Mat., Philadelphia, PA. (1984).

[18] Q. Zhang and S. Chau, "Indentation Fracture Behaviour and Measurement of Fracture Toughness of Si_3N_4 and SiC Ceramics," *J. of Inorg. Mat.*, 1, (3), 251-261 (1986).

[19] J. L. Henshall, D. J. Rowcliffe and J. W. Edington, "Fracture Toughness of Single Crystal Silicon Carbide," *J. Amer. Ceram. Soc.*, 60, [7-8] 373-375 (1977).

DEFORMATION BEHAVIOR OF REACTION-BONDED, CHEMICALLY VAPOR DEPOSITED AND SINTERED SILICON CARBIDES AT ELEVATED TEMPERATURES

ROBERT F. DAVIS, C. H. CARTER, JR., AND J. E. LANE*

Department of Materials Science and Engineering
North Carolina State University - Box 7907
Raleigh, NC 27695-7907

*Current Address: Westinghouse Electric Corporation
Research and Development Center, 1310 Beulah Road
Pittsburgh, PA 15235

ABSTRACT

Silicon Carbide is the generic name for a host of materials fabricated by several processing routes which result in a variety of microstructures and therefore differences in their respective creep properties. Specifically, reaction-bonded SiC possesses a microstructure of large and small hexagonal α-SiC grains and a separate phase of free Si. Creep occurs by glide and the controlling process of climb of dislocations. Polygonized subboundaries occur between the host SiC and that formed during reaction; additional subboundaries form as a result of climb. Deformation in the predominantly cubic SiC produced by CVD occurs solely by dislocation glide controlled via Peierls stress. Sintered α-SiC contains the additional phases of B_4C and C. The onset of creep occurs by dislocation glide; however, subsequent formation of Si-containing B_4C precipitates and their interaction with moving dislocations causes a change to occur in the controlling mechanism. The kinetic data shows creep in this material to be controlled by lattice and grain boundary self-diffusion above and below 1920K ± 20K, respectively. However, the parallel mechanism of dislocation glide contributes increasingly to the total strain as the number/volume of precipitates declines with increasing temperature.

INTRODUCTION

In certain close-packed structures such as SiC, there exists a special one-dimensional type of polymorphism called polytypism. Polytypes are alike in the two dimensions of the close-packed planes but differ in the stacking sequence in the dimension perpendicular to these planes. In SiC, the stacking sequence of the close-packed planes of covalently bonded primary coordination tetrahedra (either SiC_4 or CSi_4) can be described by the ABC notation. If the pure ABC stacking is repetitive, one obtains the zincblende structure. This is the only cubic SiC polytype and is referred to as 3C or β-SiC, where the three refers to the number of planes in the periodic sequence. The hexagonal (..ABAB..) sequence is also found in SiC. Furthermore, both can also occur in more complex, intermixed, forms yielding a wider range of ordered, large period, stacked hexagonal or rhombohedral structures of which 6H is the most common. All of these noncubic structures are known collectively as α-SiC.

Although SiC is produced by several processing routes, the characteristics of the final products are normally high strength and hardness as well as an excellent resistance to corrosion, erosion and thermal shock. For these reasons, several forms of SiC are candidate materials for use in structural applications at elevated temperatures. However, it is in these relatively severe environments that creep processes can play an important role in many temperature dependent changes in a materials physical and even chemical properties. Thus, it is important to have both constitutive equations which relate the kinetics of deformation to the applied stress and temperature and an understanding of the mechanisms which are active and controlling within the matrix of these conditions. As such, it is the objective of this paper to review the previous and ongoing creep research in the three most common forms of SiC: reaction-bonded, chemically vapor deposited (CVD) and sintered. Although the microstructures of these materials differ markedly, all three processing schemes produce materials which are essentially or completely dense. Quantitative characteristics of this microstructure and values of density are given in Table I.

Reaction-bonded SiC is produced by slip casting shapes from a slurry of colloidal graphite and Acheson-derived SiC, with the latter having a bimodal grain-size distribution ($\approx$ 100 μm and < 10 μm). The plate is then heated in a Si-containing N_2 atmosphere at T < 2273K to obtain reaction with the free C and resultant bonding. Free Si results as a continuous non-grain boundary second phase which varies in amount throughout each sintered piece. Almost all previous research concerned with deformation or creep in reaction-bonded SiC has been conducted in four-point bending at temperatures below the melting point of Si (1696K). Rumsey and Roberts[1] reported the observation of both primary and steady-state creep; they attributed the cause of this deformation to the flow of free Si on the SiC grain boundaries. Marshall and Jones[2] found only transient creep behavior; thus no conclusions concerning mechanisms were proposed. Larsen and co-workers[3-4] studied the deformation of several reaction-bonded silicon carbides of which all had a stress exponent of $\simeq$ 1.0. Although no activation energies were determined, diffusion was suggested as the rate-controlling process. Krishnamachari and Notis[5] reported an activation energy of 146 kJ/mol and concluded the controlling mechanism to be Coble grain boundary diffusion with Si being the rate controlling species. Schnürer et al.[6,7] observed the stress exponent to increase from one to four with an increase in stress. This was originally attributed to the increased mobility of dislocations with applied stress but has since been attributed to free Si at the grain boundaries.

Although several of the previously mentioned researchers have suggested that creep in reaction-bonded SiC is caused by the Si phase at the grain boundaries, TEM by Carter et al.[8] revealed no grain boundary second phase. Furthermore, as a result of the research of McHenry and Tressler[9] on hot pressed and sintered SiC stressed in the bending mode below 1673K, it is now believed that some, if not all, of the deformation reported above was the result of slow crack growth. However, the available data are insufficient to confirm this point.

In the CVD process, several combinations of source gases and procedures may be used; however, the most common commercial scenario and the one employed for the material discussed in this paper is the pyrolysis of methyltrichlorosilane ($CH_3SiCl_3 \rightarrow$ SiC + 3HCl) on graphite at $\simeq$ 1673K. The only deformation research

Table I. Selected properties of the three SiC materials.

Phases Present (wt. %)	Avg. Grain Size (SiC) (μm)	Pref. Orient.	Density ($10^3 kg/m^3$)
Reaction Bonded ~ 91 SiC, 9 Si	~ 100 and < 10	None	3.103 – 3.114
CVD SiC	Extremely Variable Throughout	⟨111⟩ 45° to stress axis	3.207
Sintered 99.08 SiC, 0.42 B_4C, 0.50 C	3.7	None	3.158

on CVD SiC has been conducted[10] by the authors of this paper and is described below.

One of the earliest deformation studies on high density (> 99%) polycrystalline α-SiC produced by hot pressing at 2873K was conducted in four-point bending by Farnsworth and Coble[11]. No sintering aids were intentionally introduced into this material during processing. It was tentatively concluded that creep in this material in the temperature and stress ranges of 2173K-2473K and 20-200 MPa, respectively, was controlled by C diffusion in the grain boundaries. Subsequent work by Francis and Coble[12] on similar hot pressed material but containing varying amounts of porosity and grain sizes provided qualitative support for the aforementioned mechanism. In additional research[12], single crystals of α-SiC were stressed in bending to 276 MPa at 2400K for 36.0 ks. Transmission electron microscopy confirmed the generation and reluctant movement of a few dislocations on the nonbasal {4̄401} and {1̄101} planes; similar dislocation effects on the {0001} planes were not observed. Thus, it was concluded that dislocations movement was not important for deformation in polycrystalline α-SiC. This is the same reasoning reported earlier by Hasselman and Batha[13] who found no plastic flow in α-SiC single crystals subjected to four-point bending up to 2473K. By contrast, Frantsevich and coworkers[14-16], using bending tests analogous to those employed in similar studies by Shaffer and Jun[17], reported considerable plastic deformation via an unnamed slip mechanism in semi-conductor grade single crystals at 2273K-2373K.

One of the more recently developed process routes for producing sintered SiC employs the additions of small amounts of C and B- or Al- containing substances (usually the carbides or nitrides) to very fine SiC powder which is molded into the desired shape and sintered in an inert atmosphere at 2400K-2500K. The material investigated by us contained 0.5 wt. % B (in the form of B_4C) and 2.0 wt. % free C in the unsintered state; the mixture was densified at 2373K in 0.1 MPa Ar for 7.2 ks.

Constant load, compressive stress creep research on high density sintered or hot pressed alpha silicon carbides containing 1% B and 1% C (sintered material) or 1% B_4C (hot pressed material) and having a porosity and average grain size of ~ 4% and 3.5 μm, respectively, has been reported by Djemel et al.[18,19]. The applied stress and temperature ranges were 500-1,700 MPa and 1573K-1773K, respectively. These researchers found that below a critical stress, σ_0, the values of Q and n were

292 kJ/mol and ≃ 1.0, respectively. The value of Q is similar to that noted by Coble and coworkers[11,12]. The controlling mechanism was not determined; however, it was postulated to be Coble creep.

Above so, the strain rate became very high, leading ultimately to fracture as a result of vacancy coalescence along the grain boundaries in tension (which are parallel to the axis of applied stress in compressive loading) and subsequent cavitation which promoted crack initiation and growth. The stress exponent values were reported to be in the range of 10-20. Diffusion in the grain boundaries was also postulated to control the deformation in this regime. No TEM research was conducted. Grathwohl et al.[20] have provided additional information on the creep in vacuum within the temperature and stress ranges of 1743K-1993K and 100-190 MPa, respectively, of SiC material doped with 0.25 wt. % or 0.34 wt. % B. Both samples exhibited primary and steady-state creep regimes. The stress exponent and activation energy calculated for the latter regime were ≃ 1.0 and 796 kJ/mol, respectively. The governing mechanism was concluded to be lattice diffusion.

The grain boundary chemistry of sintered α-SiC containing B and C has been investigated by Tagima and Kingery[21], Hamminger et al.[22,23] and Davis and coworkers[24,25] using TEM[21,24,25], Auger spectroscopy[22-25] wavelength dispersive analyses[22,23] and microautoradiography[22,23]. No grain boundary film or impurity has been discerned by these investigators. This is in contrast to the results of similar studies on Al-containing SiC materials[21-23]. The presence of a boundary phase having a thickness of < 1.0 nm is always found in these latter materials and has a significant effect on high temperature deformation, as demonstrated in several studies[20,26-28].

The following sections (1) describe the experimental procedures used by us to determine more definitively the kinetics and mechanisms of steady-state creep in the three materials noted above, and (2) provide results, a detailed discussion and conclusions regarding these studies.

EXPERIMENTAL PROCEDURES

The sample size and shape are critical for the operation of the creep equipment used in this research and, therefore, accurate measurement of the deformation. Right circular cylinders having a final height and diameter of 7.62 mm and 5.08 mm or 3.8 mm, respectively, were produced by the sequential process of (1) diamond-grinding the as-received plate, (2) ultrasonically trepanning ≃ 5.10 mm diameter samples perpendicular to the ground surface of the plate, (3) diamond-honing the cylindrical surface of each sample to remove most of the taper produced by ultrasonic cutting, (4) simultaneously diamond-lapping both ends of the sample flat and parallel, and (5) cylindrical lapping of the round surface. The combined final out-of-roundness and taper of a sample was typically ± 0.75 μm, based on nine readings at separate points on the sample. The ends of the sample were parallel to ± 1 μm and flat to ± 0.03 μm.

The CVD and sintered alpha materials were annealed before deformation at 2373K for 28.8Ks in an envelope of SiC powder and 0.1 MPa purified Ar to remove residual stress (in the CVD SiC) and to establish a fixed grain size. Exaggerated grain growth occurred in the CVD material; however, no change in grain size was noted in sintered α-SiC. Figure 1 shows the initial microstructures of all three materials prior to creep.

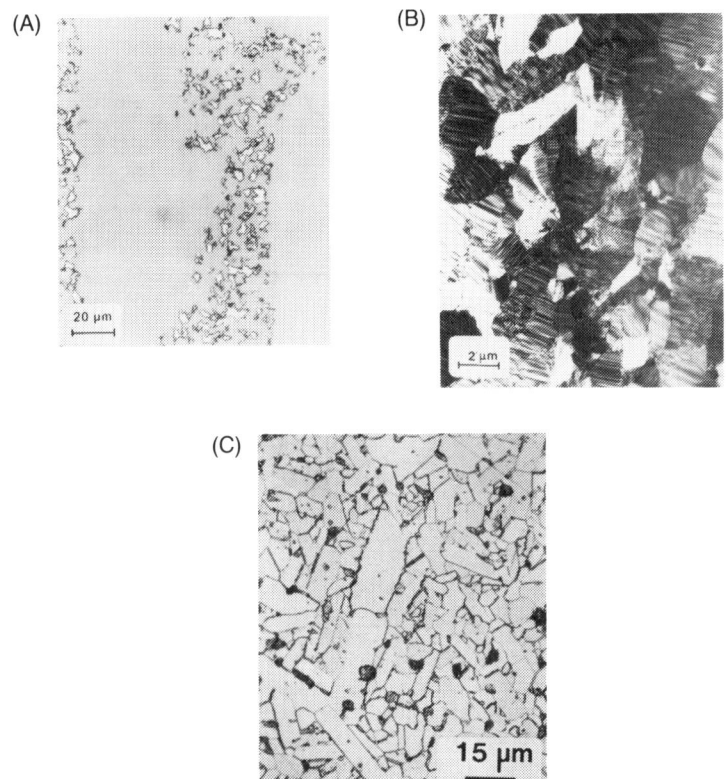

Fig. 1. Micrographs of the microstructures of uncrept (A) reaction-bonded NC-430 showing the free Si (small white areas) in SiC (gray areas); (B) CVD SiC showing heavy faulting within all the grains; and (C) sintered and annealed (2273K, 86.4 ks, and 0.1 MPa Ar) α-SiC (Hexoloy) showing equiaxed grains, moderately elongated crystals and lath- like particles.

The creep apparatus was specially designed and constructed for this research. It is described elsewhere[29]; however, a few of its cogent features are noted below. In this apparatus, uniaxial compressive stresses which are constant to within 1% for strains up to 10% can be applied to the sample, and strains can be read with an accuracy of 0.5 μm. Furthermore, loads as great as 440 kg can be applied to the samples, and the furnace can be operated in vacuum or inert gas to 2573K. Data acquisition, manipulation, and plotting are computer-controlled. The primary platens were polycrystalline CVD SiC, which had a preferred orientation of the (111) plane perpendicular to the applied stress.

Experiments to determine the temperatures and stresses required to cause measurable creep were conducted in 0.1 MPa Ar (all samples) at 1573K-1923K and

69-220 MPa (reaction-bonded[†]), 1573K-2023K and 69-220 MPa (CVD[‡]) and 1520K-2073K and 69-414 MPa (sintered alpha[§]). To determine the stress and temperature dependence of the material, each of these variables was changed in turn while the other was maintained constant. Both parameters were maintained at the desired values into the steady-state creep region until the data were sufficient to accurately measure the creep rate. To ensure that steady-state was achieved, the data was plotted in the form of $\dot{\varepsilon}$ vs. ε. Steady-state is indicated by the fact that $\dot{\varepsilon}$ becomes constant after a given amount of strain. All values of stress and temperature dependence were computed using steady-state creep rates. The TEM thin sections were prepared by (1) cutting an $\simeq$ 175 μm slice from a creep specimen using a low-speed diamond saw, (2) ultrasonically trepanning a 3 mm diameter disk from the slice, (3) sequential lapping and polishing of each face of the disk to a 1 μm diamond finish and the disk to a thickness of ~ 75 μm, and (4) ion-milling to electron transparency.

Results and Discussion

Reaction-Bonded SiC

Density values of the individual creep samples of the as-received material ranged from $(3.103-3.114) \times 10^3$ kg/m^3. Since the samples were essentially porosity free ($\simeq$ 0.03 vol %, (see Fig. 1A)) this corresponds to a variation in the free Si content of 10.9 to 12.2 vol %. Both the SiC and Si phases are continuous; thus, the free Si supports no load above its melting point. The material also contained 0.4 wt. % Fe, 0.1 wt. % Al, and < 50 ppm B as its major impurities.

In the determination of the stress exponent, the plot of the data for $\dot{\varepsilon}$ as a function of σ/G showed considerable scatter. However, a plot of the raw data as a function of the density of the uncrept sample showed that much of the scatter was caused by the aforenoted sample-to-sample density variations in the as-received material. To compensate for these density differences among samples, two mathematical procedures were used. Initially, a linear regression was performed on the plot of the values of the log of the steady-state strain rate vs. the measured density values. The slope and the intercept of the resulting linear curve were found to be -0.073 and 218.8, respectively. Second, the compensated strain rate for a given sample was calculated by (1) computing the strain rate for the mid-range value of density (3.108×10^3 kg/m^3) using the equation (log $\dot{\varepsilon}$ = 218.8 + 0.073(ρ)) from the above regression, (2) taking the ratio of this value to the strain-rate value calculated from this last equation using the measured bulk density of the sample, and (3) multiplying this ratio by the sample strain rate taken from the raw data. The log of this density-compensated steady-state strain rate was then plotted as a function of log σ/G, as shown in Fig. 2. The stress exponent was calculated from this compensated data to be 5.7.

The average value of the activation energy for creep of this material, determined from the standard plots of $\dot{\varepsilon}$ vs. 1/T shown in Fig. 3, was calculated to be 711 ± 21 kJ/mol from data collected at 147 MPa and 182 MPa.

The as-received NC-430 SiC contained a large number of dislocations as well as low angle boundaries, as can be seen in Fig. 4. These boundaries are the interfaces between the Acheson-derived SiC and the SiC that was formed during reaction-bonding. This was determined by observing the boundaries at low magnification and noting that they are approximately parallel to the edge of the grains and are at a distance from the edge that is approximately the thickness of the

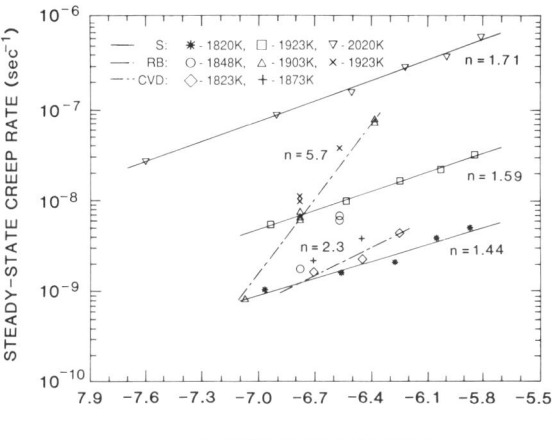

Fig. 2. Curves of steady-state strain rate vs σ/G for reaction-bonded SiC (RB), CVD SiC and sintered α-SiC (s).

newly formed SiC. This thickness was determined from impurity-sensitive, secondary-electron imaged SEM micrographs of polished, unetched material. The materials on both sides of these boundaries was determined to be α-SiC. In addition, these boundaries were distinguishable form some other low-angle boundaries (to be shown later) by the cavities that lie along them.

The slip traces seen in Figs. 5A-B indicate that considerable dislocation glide is occurring. Other interesting features in Fig. 5A are the kinks (or jogs?) and the pile-up of the slip bands at some type of low angle boundary. Burgers vector analysis of the dislocations in the slip bands of Fig. 5B revealed that these dislocations are Shockley partials in the hexagonal system having the vector a/3 $\langle 110 \rangle$. A comparison of the stress exponent value of 5.7 with that of theoretical models indicates that dislocation climb by pipe diffusion, which has a stress exponent of 5, and dislocation glide/climb controlled by climb, with a stress exponent of 4.5, are the mechanisms having values closest to the experimental number.

The activation energies for self-diffusion of C and Si in α-SiC have been determined[30,31], with the values being as follows: pure SiC, 696 ± 157 kJ/mol for Si lattice diffusion; n-type (N-doped), 789 ± 10 kJ/mol for Si lattice diffusion; pure SiC, 714 ± 5 kJ/mol for C lattice diffusion; and n-type (N-doped), 790 ± 8 kJ/mol for C lattice diffusion. Silicon is the slower-diffusing and thus the rate-controlling species for lattice diffusion in both forms.

Correlation of the value of the activation energy of creep to that of self-diffusion of Si in N-doped α-SiC reveals that the former value is close to but slightly lower than that for lattice diffusion. (The activation energy of the N-doped SiC was chosen because the Acheson-derived SiC in NC-430 is produced in an N_2 atmosphere.) However, in dislocation climb by pipe diffusion the controlling factor is the pipe-diffusion coefficent[32]. Activation energies for pipe diffusion of Si or

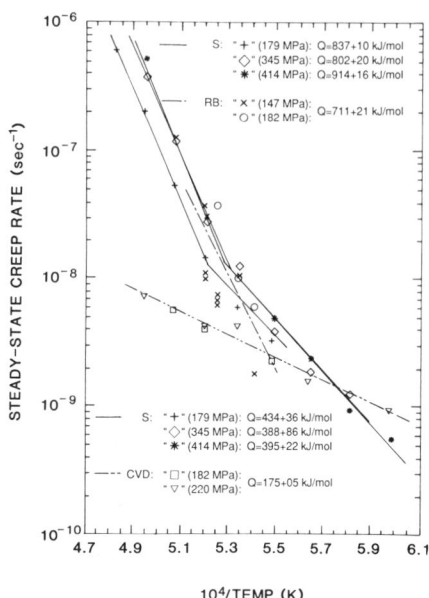

Fig. 3. Curves of steady-state strain rate vs 1/T for reaction-bonded SiC (RB), CVD SiC and sintered α-SiC (σ).

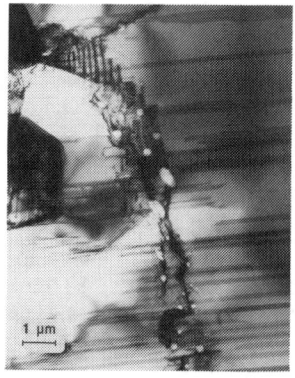

Fig. 4. High voltage TEM micrograph of as-received NC-430 SiC showing a low angle boundary between an original grain and material formed by reaction, cavities on the boundary and numerous stacking faults crossing the boundary.

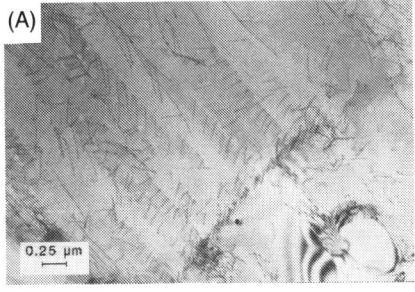 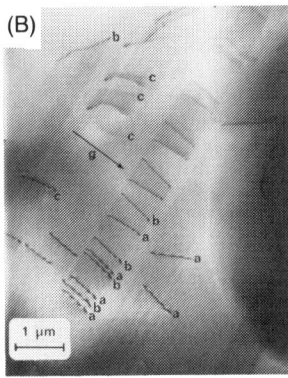

Fig. 5. TEM micrographs of NC-430 after creep showing (A) slip bands of which the majority end at a low angle boundary and (B) a slip band containing several pairs of Shockley partials having the Burgers vectors (a) a/3 [0$\bar{1}$10], (b) a/3 [10$\bar{1}$0] and (c) a/3 [1$\bar{1}$00]. All dislocations lie in the basal plane. g = [22$\bar{4}$3], z $\simeq$ [0$\bar{1}$14].

C in SiC are not available; however, some research has been performed on metals which invariably resulted in pipe-diffusion activation energies which are smaller than those for lattice diffusion. Gjostein[33] reported values of 0.3 to 0.7 for the ratio of the activation energy for dislocation pipe-diffusion (edge and screw) to that of lattice diffusion in metals. Volin et al.[34] found that, for Al which has a Burgers vector of a/2 $\langle 110 \rangle$ (the same as that for β-SiC), this ratio is 0.67. Although these figures are for pure metals and SiC is a covalent compound, this gives some idea of the range of possible ratios of the activation energy for pipe diffusion to that for lattice diffusion in SiC. The ratio of the activation energy for creep of NC-430 SiC to the activation energy for lattice diffusion of Si in N-doped α-SiC is 7.37/8.18 or 0.90.

Thus, although the value of the stress exponent points to dislocation climb by pipe diffusion as the controlling mechanism, the similarity in values of the activation energies for lattice diffusion of Si and for creep in α-SiC (the only form determined to be present in the creep samples) indicates that dislocation glide/climb controlled by climb is the controlling process. Thus, it is imperative that the evidence provided by electron microscopy is examined closely to resolve this question.

The TEM studies definitely indicate a dislocation mechanism. One of the most notable differences in the as-received and crept material in the TEM is the appearance of slip bands such as those seen in Figs. 5A-B. These slip bands are proof that there is considerable dislocation movement by glide occurring as a result of creep.

The primary indicator that climb also occurred in this material was the observation of the formation of low-angle boundaries (or cell walls) during the creep process, as shown by the horizontal boundary in Fig. 6. These boundaries can be formed only if either (1) climb is occurring or (2) there are enough slip systems and the resolved shear stresses are proper to achieve cross slip. There are two major reasons which make the former process much more likely. The first is that all of the dislocation bands were observed to lie in the basal plane. Therefore, not enough operative slip systems to allow cross slip were observed in this material at the temperatures and stresses used. Second, the cell wall contained many dislocation loops. These loops would occur in a low-angle boundary such as this only if climb were active.

Thus, the combination of the closeness in activation energy values to those of lattice self-diffusion and TEM observations in the SiC prove that steady-state creep in this reaction-bonded SiC occurs by dislocation glide/climb and is controlled by climb.

Chemically Vapor Deposited SiC

Density measurements for the CVD material supported the microstructural results which showed it to be theoretically dense ($\rho = 3.207 \pm 0001) \times 10^3$ kg/m^3 and without variation in C among samples. Thus, the creep data showed very little scatter relative to that observed in the NC-430. Spectrochemical and neutron activation analyses revealed only Cu (60ppm), Fe (140ppm) and Zn (303ppm) as major impurities. A major portion of the last two elements was derived from the hardened steel die and plunger used to pulverize that sample.

The stress exponent was found to vary from 2.3 for $T \leq 1923K$, as shown in Fig. 2; at $T = 1923K$, $n = 3.7$ (not shown). The average value of the activation energy calculated form the curves presented in Fig. 3 was 174.6 ± 4.8 kJ/mol.

Complete faulting existed in the as-deposited material, as shown in Fig. 1B and by the very high density of lattice fringes in Fig. 7A. The stacking sequence of the basal plane in the as-deposited material was almost totally random as was evident by the very considerable streaking in the diffraction pattern. Annealing allowed exaggerated grain growth caused by the movement of high angle boundaries in grains oriented off the preferred orientation. This step also produced a more ordered stacking sequence of the basal planes, particularly in grains experiencing exaggerated grain growth.

It was determined that the primary α polytype in this CVD material was 6H. Furthermore, there was roughly a 60:40 ratio of β - to - α which existed in the as-deposited material and which was maintained in the annealed and crept samples. There was, however, a redistribution of these polytypes during annealing, as shown in Fig. 7B.

The dislocations in the as-received material were determined to be primarily of the type $b = a/6 \langle 114 \rangle$ or $b = a/2 \langle \bar{1}10 \rangle$, both of which are produced by the combination of a Shockley partial and a Frank partial and are sessile. All dislocations were found to lie on the (111) plane[1].

Slip bands were the predominant dislocation structure observed in the as-crept material at all stresses and temperatures (see Fig. 8); however, they were never seen in the as-received material. As in the NC-430 SiC, these dislocations are Shockley partials which move in alternating pairs[1]. The specific Burgers vectors seen in the slip bands shown here are $a/6 [2\bar{1}\bar{1}]$ and $a/6[1\bar{2}1]$. Other dislocations seen in the deformed samples had Burgers vectors of $b = a/3 [1\bar{1}1]$ and $b = a/6 [114]$. The $a/6 [114]$ dislocations were observed only at the highest temperatures of deformation. Although these last dislocations were also observed in the as-received

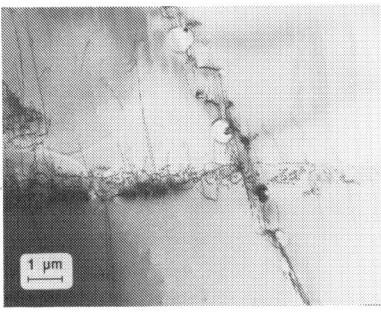

Fig. 6. TEM micrograph of NC-430 deformed at 1903K and 147 MPa showing a vertically oriented low angle boundary containing pores and formed during the original processing and a horizontal boundary formed during creep as a result of climb processes.

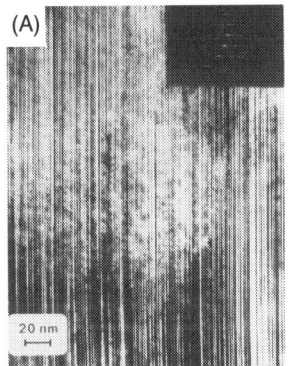 

Fig. 7. Transmission electron micrographs and associated electron diffraction patterns for CVD SiC. (A) Lattice image of as-deposited material containing an almost random stacking sequence. (B) Lattice image of material deformed at 1973K and 182 MPa; dark areas are α-SiC and light areas are β-SiC.

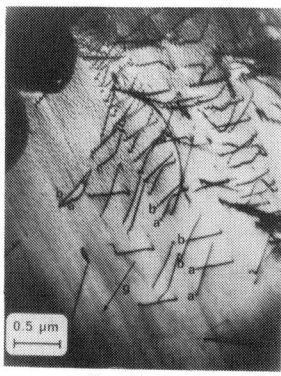

Fig. 8. TEM micrograph of CVD SiC deformed at 2023K and 220 MPa. The Burgers vectors of the labeled dislocations are as follows: (a) $b = a/6\,[2\bar{1}\bar{1}]$, (b) $b = a/6\,[1\bar{2}1]$. ($g = [\bar{2}20]$, $z \simeq [112]$.)

material, they appeared only in the glide bands in the crept samples and were of a different shape than those in the uncrept samples.

The calculated value of the activation energy for steady-state creep in this material of 174.6 kJ/mol is much lower than the activation energy values for lattice self-diffusion of Si or C in α- or β-SiC noted above. Even if the lowest value of 563.5 kJ/mol which is that for boundary diffusion of C in β-SiC35 is taken for comparison, this latter value is still 3.2 times higher than the value of the activation energy for creep in this material. On this basis it is deemed unlikely that the controlling creep mechanism is diffusion-controlled.

The range of values for the stress exponent indicates that the controlling creep mechanism is either dislocation motion or grain boundary sliding. The results of the TEM study proved that the creep mechanism definitely involves dislocation glide. Furthermore, no indication of grain-grain separation at triple points or movement of grains relative to one another was evident in any of the large number of micrographs examined. It was therefore concluded that below 1923K, the principal creep mechanism is dislocation glide controlled by the energy of the Peierls stress hills.

Employing an equation developed by Seeger[36] which estimates the activation energy for creep via a double kink mechanism (i.e., a double kink is formed in a dislocation acting against the Peierls stress), it was found that the experimental value (111 kJ/mol) roughly approximates the calculated value determined for the CVD SiC.

At high temperatures (> 1923K), the stress exponent increased to 3.7 indicating that the controlling creep mechanism may be changing at high-temperatures to one involving dislocation glide/climb controlled by climb.

Sintered α-SiC

Characterization of the Crept and Uncrept Materials: Density measurements on the initial α-SiC also revealed no variation greater than ± 0.001 Kg/m^3 among samples; quantitative microstructural evaluation showed ~ 0.2 % porosity in the material. The sintered material also contained residual free C (0.5 wt %), B (0.42 wt %) and the impurities (in ppm) of Fe (100), V (40), Se (20), Mg (9) and Cu (6), as determined by mass spectrographic analysis. Secondary ion mass spectrometry also identified N ($\simeq$ 100 ppm) as the only additional impurity. The optical micrograph of Fig. 1C illustrates the variety of grain sizes and shapes as well as the small amount of residual porosity which existed in the as-received sintered material. The average grain size was determined to be 3.70 μm.

Transient or primary creep was observed in all the α-SiC samples under all the conditions where deformation was observed. This phenomenon was most obvious during the first condition of each run; however, the magnitude of this regime steadily diminished with an increase in stress or temperature, although it did occur under all conditions employed in this study.

The stress exponent increased from 1.44 to 1.71 with temperature (see Fig. 2); it was not a function of stress at a given temperature. The curves of ln $\dot{\varepsilon}$ vs. 1/T showed a change in slope between $\simeq$ 1880K and $\simeq$ 1920K, as shown in Fig. 3. The respective activation energies below and above this temperature interval were 388-434 kJ/mol and 802-914 kJ/mol. These latter two sets of values are comparable to those noted above for the self-diffusion of C and Si through the lattice and along the grain boundaries of a polycrystalline sample of N-doped α-SiC, respectively.

Precipitation of Si-containing B$_4$C occurred throughout all the α-SiC grains (see Fig. 9) during creep as a result of supersaturation of B within this lower temperature range (relative to that used for sintering). This phenomenon was not stress-induced, as precipitation also occurred during regular annealing within the same temperature range. The global mean precipitate diameters, determined from all the data for each of the creep runs at 1820K and 2020K were 13.0nm and 24.7nm, respectively. The number of precipitates per unit volume, the surface-to-surface interparticle distance, the volume fraction and the surface-to-volume ratios of the precipitate phase for the sample crept at 1820K were (1060 ± 170)/μm^3, 41.3nm, 0.00122 and 0.5nm^{-1}, respectively; analogous values for the material crept at 2020K were (150 ± 29)/μm^3, 79.6nm, 0.00119 and 0.25nm^{-1}, respectively. Thus, while the mean diameter of the precipitates increased (with a corresponding decrease in the number of particles and an increase in the interparticle distance), the volume fraction was almost unchanged. This shows that considerable diffusion of the B, leading to coalescence of the precipitates and the resultant doubling of the mean precipitate diameters noted above, occurred at the higher temperature; however, the solubility of B in the α-SiC remained approximately the same.

Scanning Auger microprobe (SAM) analyses with in situ sample fracture capability and TEM were used to investigate the character of the grain boundaries in the uncrept and crept samples. The primary objective of this research was to confirm or refute the presence of the often suggested presence of an amorphous phase which has been proposed[36-41] as the vehicle by which both densification and plasticity are achieved (values of n of 1-2 may indicate viscous flow of a boundary phase as one possible mechanism of creep). No B$_4$C precipitates or any other amorphous or crystalline fugitive boundary phase was discerned by either technique (B detection limit in SAM analyses = 1.5 at %). In contrast, TEM revealed only

 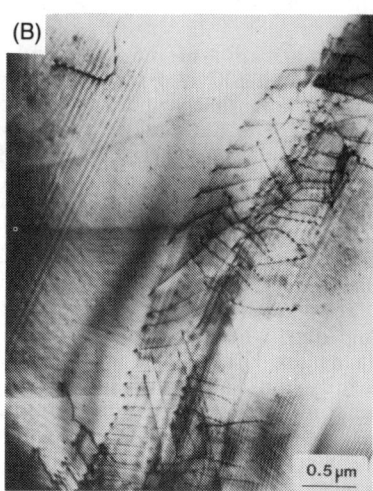

Fig. 9. Representative TEM micrographs of the crept (2020K - 414MPa - 6% strain) α-SiC. Figure (A) shows the high density of dislocation slip bands; Figure (B) shows a single slip band in this material as well as the occurrence of the Si-containing B_4C precipitates and their interaction with particular dislocations in the band.

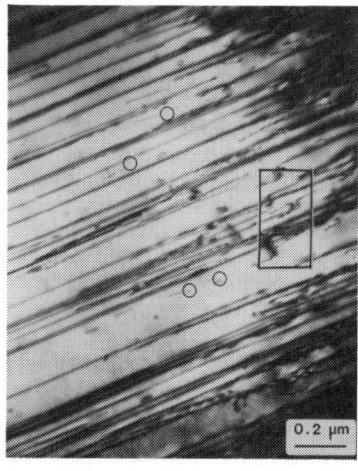

Fig. 10. TEM image showing considerable faulting, evidence for climb (rectangular areas) and "hairpin" dislocations (circled areas) in an α-SiC sample crept under the final conditions of 2020K-414 MPa to 8.2% strain.

boundary dislocations. Similar results were observed in a sample reheated to the sintering temperature of 2373K and quenched in flowing He.

A TEM examination of the as-received and annealed α-SiC material showed them to be virtually identical. They were essentially free of line defects; however, occasional stacking faults occurred in selected grains, and a few grains contained a substantial number of faults. By contrast, the crept samples contained a very high density of stacking faults caused by the movement of partial dislocations within slip bands in the material, as shown in Fig. 9. Burgers vector analysis assisted by the development of a Kikuchi map[42] for 6H-SiC revealed the dislocations to be Shockley partials having the vector $a/3\,[01\bar{1}0]$. Microtwins were also observed and served as companion mechanisms of deformation with the partial dislocations in the glide band.

Electron microscopy research also revealed two additional phenomena onging during creep, namely climb on the $\langle\bar{1}012\rangle$ second order pyramidal plane and precipitation. Climb was most frequently realized in samples having the highest strain. (It was, however, occasionally observed even in the micrographs of samples subjected only to the initial stage of primary creep). Microstructurally, it was initially discerned as dislocations which appeared to move orthogonal to their initial slip directions. Several examples of this feature are shown within the rectangular area in Fig. 10.

Burgers vector analysis was used to ascertain conclusively that climb was actually occurring. For this analysis, a set of micrographs were obtained of a dislocation structure, observed in a sample subjected only to primary creep at 1970K for 339.6 ks at an applied stress of 276 MPa. A partial series of this set is presented in Fig. 11. In this figure, the dislocations labeled "a" and "b" are Shockley partials with $b = a/3\,[\bar{1}010]$ and $a/3\,[01\bar{1}0]$, respectively. The dislocation segments labeled "c" - "e" are perfect dislocations with $b = a/3\,[\bar{1}\bar{1}20]$. In Fig. 11(A) each set of "a" and "b" partials and the segment denoted "d" lie in different (0001) planes. Climb occurred between the basal planes via segments "c", determined by trace analysis to lie in the $(\bar{1}012)$ plane, and "e", which lies in an undetermined plane. This phenomenon was a direct result of the interaction of an initial dislocation and the later segments with the precipitates, as shown most clearly in Fig. 11(E). The dislocation "c" occurred as a result of climb, as shown by the fact that its Burgers vector does not lie in the same plane as the dislocation. (If these features were cross slip, the dislocation and its associated Burgers vector would lie in the same plane). The examination of several hundred micrographs has led to the conclusion that this climb structure occurs infrequently and thus plays only a minor role in the deformation of α-SiC.

Direct evidence for the interaction of the precipitates with the gliding dislocations is suggested in Fig. 9(B). However, the micrograph of Fig. 12 confirms this interaction; it was observed in both the initial primary regime and in samples crept for longer times.

No evidence was observed in any sample of unaccommodated grain boundary sliding (Rachinger sliding). Indeed intentional observations of numerous grain boundaries and triple points revealed no evidence of grain boundary corrugation or pore formation or the occurrence of triple point voids. Moreover, there were no major microstructural differences among the samples crept above and below the transition range in activation energy or in those crept for very long time.

Analyses of the Results: The prevailing microstructural features throughout all the crept samples are the glide bands and tangles of dislocations, the Si-containing B_4C

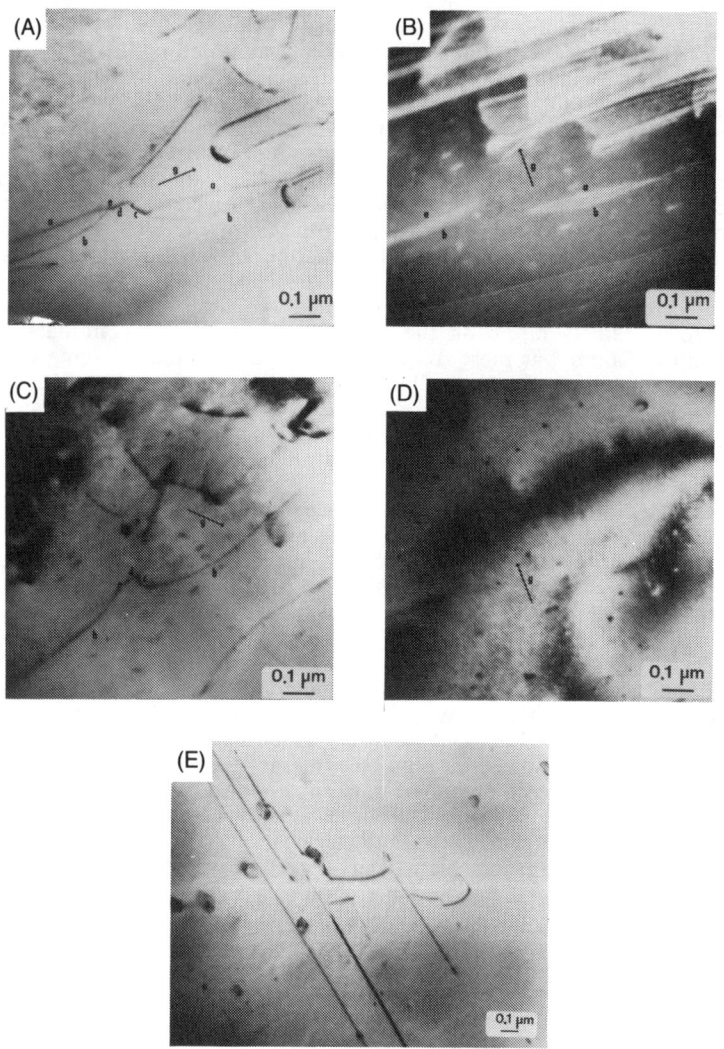

Fig. 11. Partial series of the set of TEM micrographs taken for Burgers vector analysis of dislocation segments labeled "a" - "e". Dislocations "a" and "b" have Shockley partials with b = a/3 [$\bar{1}010$] and b = a/3 [0$\bar{1}$10], respectively. Segments "c" - "e" are perfect dislocations with the same b = a/3 [$\bar{1}\bar{1}$20]. Note in Fig. (A) that the left and the right-hand sets of partials (a set is an "a" and a "b") as well as the segment labeled "d" all lie on different (0001) planes. Climb occurs between these basal planes via segments "c", determine by trace analysis to be in the ($\bar{1}$012) second order pyramidal plane, and "e" which lies in an

undetermined plane. The "a" partials of each set are out of contrast in (B). The dark field image in (C) shown the perfect dislocations of "c" and "a" to be out of contrast; (D) is edge-on and all dislocations are out of contrast. This analysis was conducted on sample which was annealed for 86.4 ks at an applied stress of 276 MPa. (A) g = [11$\bar{2}$0], Z ≈ [5$\bar{5}$02]; (B) g = [1$\bar{2}$16], Z ≈ [4221]; (C) g = [$\bar{1}$108], Z ≈ [4401]; (D) g = [006], Z ≈ [1100]. Figure (E) shows the climb process to result directly from the interaction of an initial dislocation with a B_4C precipitate.

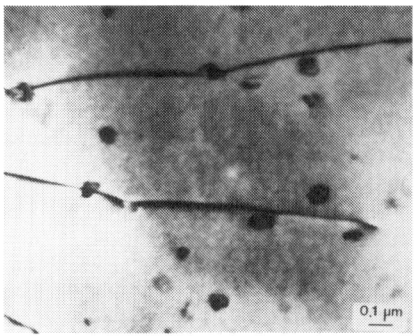

Fig. 12. TEM image which confirms both the occurrence of precipitates and the interaction of these precipitates with selected leading partial dislocations in the crept α-SiC.

precipitates and the interaction between these two individual entities. However, a comparison of the experimental values of the stress exponents (1.44 - 1.71) with the theoretical values of this parameter for various steady-state mechanisms of creep indicates that one or more grain boundary sliding mechanisms.(n = 1.0 - 2.0) are actually controlling deformation in this regime over the entire temperature range investigated. However, no mode of Rachinger sliding was observed in the α-SiC material studied in this research within the limits of temperature and stress employed. The absence of an amorphous grain boundary phase in our material has been proven, as described above; moreover triple point voids or folds or grain boundary porosity were not observed. Finally, the activation energy values are relatively close to those determined for C and Si self-diffusion through the lattice and along the grain boundaries. All of the above evidence strongly indicates that a Nabarro-Herring mechanism involving grain-boundary sliding accommodated simultaneously by lattice self-diffusion is controlling creep at the higher temperatures. Similarly, at the lower temperatures, the evidence indicates that Coble creep, i.e., grain boundary sliding accommodated simultaneously by grain boundary diffusion is controlling creep at the lower temperatures.

Both the continuum[43-46] and microscopic[47-53] theories of diffusional creep predict a stress exponent of 1.0 for both boundary and lattice mechanisms.

However, if the dislocation density in the boundaries and/or their mobility is low (i.e., the boundaries are poor sources and sinks for matter), creep is said to be "mobility limited" or "interface-reaction controlled" because its rate is determined by local processes occurring at the grain boundary rather than by the kinetics of long-range diffusion. Moreover, the value of the stress exponent will increase up to a maximum value of 2 (assuming no dispersed particles in the boundary which may severely limit nonconservative motion). This latter characteristic is common in pure materials at low stresses and/or low temperatures. However, the value of n normally decreases to $\simeq 1.0$ with an increase in one or both of these variables.

In contrast with the aforenoted theory and experimental results, the value of the stress exponent for the α-SiC increased with an increase in temperature but was unaffected by changes in stress at a given temperature. This variation in n values strongly suggests that at least two deformation mechanisms are operating in parallel. These results also indicate that mechanisms(σ) having an n value > 1.0 becomes increasingly important (but not controlling) at the higher temperatures. In addition, the constancy in slope at each temperature indicates that neither the dominant or the controlling mechanism is changing as a function of stress up to 414 MPa. This supposition is also supported by the activation energy data discussed above.

Dislocation processes are the obvious choice as secondary mechanisms, as the theoretical stress exponents are > 3.0 and the TEM research has revealed their considerable activity in the α-SiC during deformation. In fact the dislocation density in these samples appears to exceed that noted in the other silicon carbides described above and crept with approximately the same amount of strain. This results from the considerable interaction of the moving dislocations with the Si-containing B_4C precipitates. This type of phenomenon has received attention from the metallurgical community because of the importance of oxide dispersion strengthened (ODS) alloys for use at elevated temperatures (see Ref. 55-57 for reviews of this work).

A compilation and synthesis of the present data strongly indicate that deformation in sintered α-SiC in the ranges of temperature and stress used in this research occurs at the outset by the glide of Shockley partials on {0001} planes. The subsequent formation of the B_4C precipitates and their interaction with a majority of the moving dislocations within the initial segment of primary creep hinders this mode of deformation to the extent that grain boundary sliding accommodated at low and high temperatures by grain boundary and lattice diffusion, respectively, become the controlling mechanisms of steady-state creep. The increase in the value of both the stress exponent and the activation energy with temperature are indicative of parallel mechanisms of deformation. Moreover, the former data indicate the increasing contribution of dislocation glide to the total strain as a direct function of temperature. This relationship is to be expected as a result of the substantial decrease in the number of precipitates per unit volume which allows an increasing number of dislocations to pass without encounter through the grains.

Conclusions

1. The kinetics and mechanisms of steady-state creep have been investigated in three types of silicon carbides: reaction-bonded, chemically vapor deposited and sintered alpha. In the reaction-bonded material, an activation energy for creep of 711 ± 20 kJ/mol and a stress exponent of 5.7 were calculated from the density-compensated data. The integration of this data with that obtained via TEM showed the controlling creep mechanism within the ranges

of temperature and stress of 1848K-1923K and 110-220 MPa, respectively, to be glide/climb controlled by climb.

2. An activation energy of 174 ± 5 kJ/mol was calculated from the creep data for the CVD material. A stress exponent of 2.3 was calculated for 1873K; a value of 3.7 from the data at 1923K. The stress range was 110-147 MPa. The controlling creep mechanism at $T \leq 1873K$ was determined to be dislocation glide controlled by the Peierls stress; above this temperature, the evidence suggests that dislocation glide/climb controlled by climb may become an increasingly important mechanism.

3. For the sintered α-SiC within the respective temperature and stress ranges of 1670K-2073K and 138-414 MPa; the values of the stress exponent increased from 1.44 to 1.71. The activation energy exhibited a knee between $\simeq$ 1880K and $\simeq$ 1920K, depending on the stress. Precipitation of Si-containing B_4C occurred throughout all the α-SiC grains during creep. However, neither this phase nor any other amorphous or crystalline fugitive boundary phase was discerned via TEM or SAM. The most prominent microstructural features of the crept samples were the extensive number of dislocation slip bands and resultant stacking faults and the interaction of these dislocations with the B_4C precipitates. As a result of this dislocation/precipitate interaction, the controlling creep mechanisms in this material were grain boundary sliding accommodated by grain boundary diffusion at $T \leq 1800K$ and lattice diffusion at $T \geq 1920K$. These mechanisms occurred in parallel with dislocation glide. The contribution of this latter mechanism increased with temperature due to substantial coalescence of the precipitate phase and a resultant 6-fold decrease in their number per unit volume.

Acknowledgements

The authors acknowledge support of the National Science Foundation (DMR 802-2197 and 812-0804), the Army Research Office (DAAG29-79-G-006 and DAAL03-86-K-0013) and the Basic Energy Sciences Division, U.S. Department of Energy through the SHARE program (EY-76-C-05-0033 and DE-76-C-050033) with Oak Ridge Associated Universities. Appreciation is also expressed to Deposits and Composites for the CVD SiC, to SOHIO Engineered Materials Corporation for the α-SiC samples, to Perkin Elmer and L. LaVanier for the use of and assistance with the SAM 600, to Drs. J. Bentley and E. Kenik of Oak Ridge National Laboratory, K. More of the Oak Ridge Associated Universities Program and Dr. S. Chevacharoenkul of NCSU for assistance with the TEM and to Drs. J. Russ and T. Hare of NCSU for guidance and the use of their equipment in the microstructural analyses of the precipitates in the α-SiC.

[†]NC-430; Norton Company, Worcester, MA 01606
[‡]Deposits and Composites, Inc. Reston, VA 22070
[§]Hexoloy, SOHIO Engineered Materials Company, Niagara Falls, NY 14302
[¶]The Burgers vectors and diffraction vectors are given as cubic for this material; since, the cubic β-SiC is dominant.

References

[1] J. C. Rumsey and A. L. Roberts, "Delayed Fracture and Creep in Silicon Carbide," Proc. Br. Ceram. Soc., 2, 233 (1967).

[2] P. Marshall and R. B. Jones, "Creep of Silicon Carbide," *Powder Metall.*, 12, 193 (1969).

[3] D. C. Larsen and G. C. Walther, "Property Screening and Evaluation of Ceramic Vane Materials," AFML Rept. No. IITRI-D6114-ITR-24, October 1977.

[4] D. C. Larsen and J. W. Adams, "Property Screening and Evaluation of Ceramic Turbine Materials," Rept. No. 11, Contract F33615-79-C-5100, November 1981.

[5] V. Krishnamachari and M. R. Notis, "Interpretation of High Temperature Creep of SiC by Deformation Mapping Techniques," *Mater. Sci. Eng.*, 27, 83 (1977).

[6] K. Schnürer, F. Thümmler, and G. Grathwol, Universtat Karlsruhe; personal communication, June 1979.

[7] K. Schnürer, G. Grathwol and F. Thümmler, "Kriechverhalten Verschnedier SiC-Werstaff, (in Ger.); pp. 645-652 in Science of Ceramics 10. Deutsche Keramische Gesellschaft, Berlin, 1980.

[8] C. H. Carter, Jr., R. F. Davis and J. Bentley, "Kinetics and Mechanisms of High Temperature Creep in Silicon Carbide: I, Reaction Bonded," *J. Am. Ceramic. Soc.*, 67 [6] 409-417 (1984).

[9] K. D. McHenry and R. E. Tressler, "Fracture Toughness and High-Temperature Slow Crack Growth in SiC," *J. Am. Ceram. Soc.*, 64 [3-4] 152-156 (1980).

[10] C. H. Carter, Jr., R. F. Davis and J. Bentley, "Kinetics and Mechanisms of High Temperature Creep in Silicon Carbide: II, Chemically Vapor Deposited," *J. Am. Ceram. Soc.*, 67 [11] 732-740 (1984).

[11] P. L. Farnsworth and R. L. Coble, "Deformation Behavior of Dense Polycrystalline SiC," *J. Am. Ceram. Soc.*, 49 [5] 264-268 (1966).

[12] T. L. Francis and R. L. Coble, "Creep of Polycrystalline Silicon Carbide," *J. Am. Ceram. Soc.*, 51 [2] 115-116 (1968).

[13] D. P. Hasselman and H. D. Batha, "Strength of Single Crystal Silicon Carbide," *Appl. Phys. Lett.*, 2 [6] 111-113 (1963).

[14] I. N. Frantsevich, V. A. Kravets, L. O. Egoroy, K. V. Nazarenko and V. Z. Sushkevich, "High Temperature Deformability of α-SiC," *Sov. Powd. Met. Metal Ceram.*, 10 [8] 229-231 (1979).

[15] I. N. Frantsevich, V. A. Kravets, K. V. Nazarenko and V. Z. Sushkevich, "Deformation Structure of α-SiC Single Crystals," *Sov. Powd. Met. Metal Ceram.*, 12 [8] 654-663 (1973).

[16] I. N. Frantsevich, V. A. Kravets and K. V. Nazarenko, "Plastic Deformation of α-SiC," *Sov. Powd. Met. Metal Ceram.*, 14 [8] 679-682 (1975).

[17] P. T. B. Shaffer and C. K. Jun, "The Elastic Modulus of Dense Polycrystalline Silicon Carbide," *Mat. Res. Bull.*, 7 [1] 63-70 (1970).

[18] A. Djemel, J. Cadoz and J. Philibert, "Deformation of Polycrystalline α-SiC," pp. 381-394 in Creep and Fracture of Engineering Materials and Structures, Edited by B. Wilshire and D. R. J. Owen. Pineridge Press, Swansea, U. K., 1981.

[19] A. Djemel, B. Pellissier, J. Castaing and J. Cadoz, Unpublished research presented at the 83rd annual meeting of the American Ceramic Society, Washington, DC. See Abstract #111-B-81, *Bull. Am. Ceramic. Soc.*, 60 [3] 381 (1981).

[20] G. Grathwohl, T. H. Reets and F. Thummler, "Creep of Hot-Pressed and Sintered SiC with Different Sintering Additives," *Science of Ceramics* 11, 425-431 (1981).

[21] Y. Tajima and W. D. Kingery, "Grain Boundary Segregation in Aluminum-Doped Silicon Carbide," *J. Mat. Sci.*, **17** [8] 2289-2297 (1982).

[22] R. Hamminger, G. Grathwohl and F. Thummler, "Microanalytical Investigation of Sintered SiC Part I: Bulk Material and Inclusions," *J. Mat. Sci.*, **18** [2] 353-364 (1983).

[23] R. Hamminger, G. Grathwohl and F. Thummler, "Microanalytical Investigation of Sintered SiC, Part II: Study of the Grain Boundaries of Sintered SiC by High Resolution Auger Electron Spectroscopy," *J. Mat. Sci.*, **18** [10] 3154-3160, (1983).

[24] R. F. Davis, J. E. Lane, C. H. Carter, Jr., J. Bentley, W. H. Waldin, D. P. Griffis, R. W. Linton and K. L. More, "Microanalytical and Microstructural Analyses of Boron and Aluminum Regions in Sintered Alpha Silicon Carbide," Scann. Electron. Micros., 1984/III, 1161-1167 (1984).

[25] K. L. More, C. H. Carter, Jr., J. Bentley, W. H. Wadlin. L. LaVanier and R. F. Davis, "Occurrence and Distribution of Boron-Containing Phases in Sintered α-SiC," *J. Am. Ceram. Soc.*, **69** [9] 695-698 (1986).

[26] C. Carry and A. Mocellin, "High Temperature Creep of Dense Fine Grained Silicon Carbides," pp. 391-404 in Deformation of Ceramic Materials II, Materials Science Research Vol., 18, Edited by R. E. Tresler and R. C. Bradt. Plenum Press, New York, 1984.

[27] R. Moussa, J. L. Chermant and F. Osterstock, "Creep and Creep Rupture of HP-SiC Containing an Amorphous Intergranular Phase," pp. 617-630 in Deformation of Ceramic Materials II, Materials Science Research Vol. 18, Edited by R. E. Tressler and R. C. Bradt. Plenum Press, New York, 1984.

[28] J. L. Chermant and F. Osterstock, "Creep Behavior of SiC-Al Materials," *Mater. Sci. and Eng.*, **71** [complete] 147-158 (1985).

[29] C. H. Carter, Jr., C. A. Stone, R. F. Davis and D. R. Schaub, "High Temperature, Multi-Atmosphere, Constant Stress Compression Creep Apparatus," *Rev. Sci. Instrum.*, **51** [10] 1352-1357 (1980).

[30] J. D. Hong and R. F. Davis, "Self-Diffusion of Carbon-14 in High Purity and N-Doped α-SiC Single Crystals," *J. Am. Ceram. Soc.*, **63** [9-10] 546-542 (1980).

[31] J. D. Hong, D. E. Newberry and R. F. Davis, "Self-Diffusion of 30Si in α-SiC Single Crystals," *J. Mater. Sci.*, **16** [12] 2485-2494 (1981).

[32] F. R. N. Nabarro, "Steady-State Diffusional Creep," *Philos. Mag.*, **16**, 231-237 (1967).

[33] N. A. Gjostein, "Short Circuit Diffusion," pp. 241-274 in Diffusion, American Society of Metals, Metals Park, Ohio, 1973.

[34] T. Volin, K. Lie, and R. Balluffi, "Measurement of Rapid Mass Transport Along Individual Dislocations in Aluminum," *Acta Metall.*, **19**, 263-274 (1971).

[35] M. Hon and R. F. Davis, "Self-Diffusion of 14C in Polycrystalline β-SiC," *J. Mater. Sci.*, **14** [12] 2411-2421 (1979).

[36] S. Prochazka, pp. 171-182 in Special Ceramics 6, Edited by P. Popper. British Ceramic Research Association, Stoke-on-Trent, 1975.

[37] H. Suzuki and T. Hase, "Some Experimental Considerations on the Mechanism of Pressureless Sintering of Silicon Carbide," pp. 345-365 in Proceedings of the International Symposium of Factors in Densification and Sintering of Oxide and Non-Oxide Ceramics, edited by S. Somiya and S. Saito. Gakujuts Bunken Fukyukai, β-SiCOokayama, Japan, 1979.

[38] T. Hase and H. Suzuki, "Initial-State Sintering of β-SiC with Concurrent Boron and Carbon Additions," *Yogyo Kyokai Shi*, **88** [5] 225-230 (1980).

[39] H. Suzuki and T. Hase, "Boron Transport and Change of Lattice Parameter During Sintering of β-SiC," *J. Am. Ceram. Soc.*, **63** [5-6] 349-350 (1980).

[40]F. F. Lange and T. K. Gupta, "Sintering of SiC with Boron Compounds," *J. Am. Ceram. Soc.*, **59** [11-12] 537-538 (1976).

[41]A. H. deA. Bressiani, p. 92 in "Das Verdichtungsverhalten von β-SiC beim Heisspressen mit verschiedenen Sinterhilfen," Doktors der Naturwissen-schaften Thesis, University of Stuttgart, 1984.

[42]J. E. Lane, C. H. Carter, Jr., K. L More and R. F. Davis, "A Kukuchi Map for 6H Alpha Silicon Carbide," *J. Mater. Res.*, **1** [6] 737-739 (1986).

[43]F. R. N. Nabarro, "Deformation of Crystals by the Motion of Single Ions," Rep. Conf. Strength Solids (Bristol) The Physical Society, London, 1948 pp. 75-90.

[44]F. R. N. Nabarro, "Steady-State Diffusional Creep," *Philos. Mag.*, **16** [2] 231-237 (1967).

[45]C. Herring, "Diffusional Viscosity of a Polycrystalline Solid," *J. Appl. Phys.*, **21** [5] 437-445 (1950).

[46]R. L. Coble, "A Model for Boundary Diffusion-Controlled Creep in Polycrystalline Materials," *J. Appl. Phys.*, **34** [6] 1679-1682 (1963).

[47]W. Bollmann, Crystal Defects and Crystalline Interfaces, Springer, Berlin, 1970.

[48]M. F. Ashby, F. Spaepen and S. Williams, "The Structure of Grain Boundaries Described as a Packing of Polyhedra," *Acta Metall.*, **26** [11] 1647-1663 (1978).

[49]Grain Boundary Structure and Kinetics, Edited by R. W. Balluffi, Am. Soc. Metals, Metals Park, OH, 1980.

[50]M. F. Ashby, "On Interface-Reaction Control of Nabarro-Herring Creep and Sintering," *Scripta Metal.*, **3** [11] 837-842 (1969).

[51]M. F. Ashby, "Boundary Defects and Atomistic Aspects of Boundary Sliding and Diffusional creep," *Surf. Sci.*, **31** [6] 498-542 (1972).

[52]B. Burton, "The Influence of Alumina Dispersions on the Diffusion-Creep Behavior of Polycrystalline Copper," *Metal Sci. Jour.*, **5** [1] 11-15 (1971).

[53]E. Arzt, M. F. Ashby and R. A. Verrall, "Interface Controlled Diffusional Creep," *Acta Metall.*, **31** [12] 1977-1989 (1983).

[54]R. M. Cannon, W. H. Rhodes and A. H. Heuer, "Plastic Deformation of Fine Grained Alumina (Al_2O_3) I, Interface-Controlled Diffusional Creep," *J. Am. Ceram. Soc.*, **63** [1-2] 46-53 (1980).

[55]L. M. Brown and R. K. Ham, "Dislocation-Particle Interactions," pp. 9-135 in Strengthening Methods in Crystals, edited by A. Kelly and R. B. Nicholson. Elsevier, Amsterdam (1971).

[56]J. W. Marton, Micromechanisms in Particle-Hardened Alloys, Cambridge University Press, Cambridge, UK, 1980.

[57]E. Arzt and D. S. Wilkinson, "Threshold Stresses for Dislocation Climb over Hard Particles: The Effect of an Attractive Interaction," *Acta Metall.*, **34** [10] 1893-1898 (1986).

The Behavior of SiC and Si_3N_4 Ceramics in Mixed Oxidation/Chlorination Environments

John E. Marra* and Eric R. Kreidler

The Ohio State University
Columbus, OH 43210

Nathan S. Jacobson and Dennis S. Fox

National Aeronautics and Space Administration
Lewis Research Center
Cleveland, OH 44135

Abstract

The behavior of silicon-based ceramics in mixed oxidation/chlorination environments was studied. High pressure mass spectrometry was used to quantitatively identify the reaction products. The quantitative identification of the corrosion products was coupled with thermogravimetric analysis and thermodynamic equilibrium calculations run under similar conditions in order to deduce the mechanism of corrosion. Variations in the behavior of the different silicon-based materials are discussed. Direct evidence of the existence of silicon oxychloride compounds is presented.

Introduction

The recent developments in advanced ceramic materials have made these materials candidates for applications previously restricted to metals and superalloys. In particular, Si-based ceramics are prime candidates for use in advanced heat engines and heat exchangers due to their excellent high temperature oxidation resistance.

During the operation of these devices the materials are often exposed to severe oxidative and/or corrosive gases. Marine based turbine engines ingest chlorine from sea salt spray making the mixed oxidation/chloride reactions of these materials may be important. Similarly, Si-based ceramics used in heat exchanger technology may be exposed to a variety of halide gases (typically fluorine and chlorine) when used in the refining of Ti and Zr and the remelting of aluminum[1]. Silicon-based ceramics are inherently unstable in oxidizing atmospheres, but obtain their oxidation resistance by the formation of a protective silica layer at the gas-solid interface. The mixed oxidation/chlorination behavior of these materials is further complicated by the volatility of the silicon chloride compounds.

The mixed oxidation/chlorination behavior of metals and superalloys at high temperatures has been studied by various researchers.[2-4] Jacobson et al.[5-7] have made use of a high pressure mass spectrometer sampling system in conjunction with conventional thermogravimetric analysis to study these reactions. The coupling of these two analytical techniques allows the corrosion products to be unambiguously identified. This identification, coupled with kinetic information, leads to a further understanding of the mechanisms of the corrosion process.

Although the behavior of metals in high temperature environments containing both chlorine and oxygen has been examined, little work has been done to investigate the behavior of silicon-based materials in these environments. Chlorine, however, is widely used in the thermal oxidation of high-purity Si for the production of gate oxide devices. The chlorine acts to limit the amount of sodium impurities present in the growing oxide and thereby increases the performance characteristics of the device. As a result of this practice, the role of chlorine in the thermal oxidation of silicon has been widely studied.[8-11] However the majority of the previous work has focused on the characteristics of the oxide film and more specifically on chlorine incorporation into the oxide. The previous research has not dealt with the mixed oxidation/chlorination reaction from an environmental stability standpoint.

Recently McNallan and his coworkers[12] have studied the behavior of silicon carbide materials in mixed oxidation/chlorination environments because of their recent application in advance heat exchanger technology. Using thermogravimetric analysis they analyzed various types of low-cost silicon carbide in various mixtures of chlorine and oxygen at 900°C. Their work showed that the observed corrosion rates were complex functions of the oxygen content of the gas stream and the processing route and additives of the silicon carbide material.

Previous work suggests several areas which require further examination. There is very little information on the behavior of Si-based ceramics in the presence of small amounts of chlorine. Also the volatile reaction products need to be better characterized. In this paper the reaction of SiC and other silicon-containing ceramics with mixed oxygen/chlorine gases is discussed. Kinetic and volatile product identification data are presented and possible reaction mechanisms are postulated.

THERMODYNAMIC CALCULATIONS

A series of thermodynamic equilibrium calculations was performed using the SOLGASMIX-PV[13] computer program. Based upon available thermodynamic data and the relative amount of each phase present, the program predicts the phase assemblage resulting in the minimum total system free energy. Conditions for performing the analyses were chosen to simulate the conditions of the actual corrosion experiments.

The program predicts that large amounts of $SiCl_4$ are produced when SiC or Si_3N_4 are exposed to mixtures of chlorine and argon. Similar analysis for silica shows little generation of volatile silicon tetrachloride. This implies that the silica layer present on silicon carbide or silicon nitride ceramics is not susceptible to attack by chlorine and will act to protect the substrate material from chlorination.

The protective nature of the silica surface layer was also shown when reactions between SiC and Si_3N_4 and mixed oxygen/chlorine gases were examined. These calculations showed decreasing amounts of volatile $SiCl_4$ with increasing amounts of oxygen in the gas stream. This decrease in the amount of chlorination is accompanied by a larger amount of silica present in the minimum free energy phase assemblage.

EXPERIMENTAL

A variety of silicon-based materials were analyzed in this research, ranging from high-purity silicon[†] and silica[‡] to less pure SiC and Si_3N_4 materials. Four types of SiC were used in this study. The materials tested included:

1. Norton NC203[§]—hot-pressed material to which Al_2O_3 has been added as a densification aid.

2. Sohio Hexoloy[¶]—sintered sample to which boron has been added as a sintering aid.

3. Norton NC430[§]—reaction sintered material densified with free silicon. Contains approximately 15% free silicon.

4. Single crystal material obtained as a by-product of the Acheson process.

Various silicon nitride materials were also analyzed, however for the purpose of this paper only the behavior of a hot-pressed material containing 6 weight percent yttria (GTE AY-6)[**] will be discussed.

Thin plate samples of approximate dimensions 13 × 6 × 1 mm were examined. Immediately before testing, the plates were lightly ground on a 15 micron diamond wheel in order to eliminate any surface roughness resulting from the cutting process. The samples were sequentially ultrasonically washed in a soap solution, acetone, and alcohol.

Thermogravimetric experiments were performed using a gold-plated Cahn RH electrobalance. The gas mixtures listed in Table I were made by electronically metering appropriate amounts of premixed 2% Cl_2/Ar, pure O_2, and/or pure Ar into a mixer filled with glass beads. A total flow velocity of 7.7 cm/sec past the sample at 950°C was maintained in both the mass spectrometer and thermogravimetric analysis experiments. In order to protect the balance from corrosive gases, a counter stream of argon was passed through the balance mechanism from the top. The objective of these experiments was to analyze the oxidation/chlorination behavior of the unreacted silicon-based materials. In order to avoid oxide film formation, the samples were brought to temperature in a 5% H_2/Ar mixture before the reaction gases were admitted. Two or three runs were performed for each condition.

The volatile corrosion products were identified with the mass spectrometer sampling system shown in fig. 1. This system consisted of two parts—a small tube furnace surrounding a quartz reaction tube, in which the reaction occurred, and the actual mass spectrometer. Conditions in the tube were as close to the balance apparatus as possible. The only difference was in the method of initiating an experiment. The sample was suspended from a hook at the end of an alumina rod which could be moved into the hot zone of the furnace. Experiments were initiated by flowing the appropriate gas mixture through the tube. When a stable gas signal was obtained on the mass spectrometer trace, the sample was raised from the cold zone into the hot zone of the furnace. This procedure prevented the formation of a substantial oxide film on the sample before the start of an experiment.

The mass spectrometer shown in fig. 1 has been described in detail elsewhere[14] and will only be briefly discussed here. During the course of an experiment the

Table I. Gas Mixtures.

2% Cl_2/Ar
1% Cl_2/1% O_2/Ar
1% Cl_2/2% O_2/Ar
1% Cl_2/4% O_2/Ar
1% Cl_2/10% O_2/Ar
1% Cl_2/20% O_2/Ar

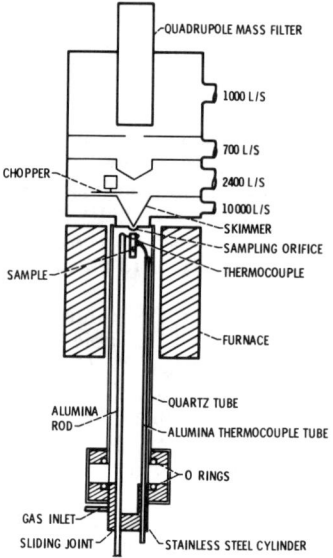

Fig. 1. Schematic of high pressure mass spectrometer sampling system.

specimen was held in close proximity to a sampling cone which contains a 0.022 cm orifice. Volatile species generated by the sample/gas interaction entered the sampling cone and underwent free jet expansion and proceeded as a molecular beam through a series of differentially pumped vacuum chambers to a mass filter. A quadrupole mass filter, operating at 10^{-10} atm, was used to directly identify the volatile species generated by an atmospheric pressure chemical process. Source and background signals were distinguished by chopping the molecular beam and using a lock-in amplifier for detection. The peaks obtained in the mass spectrometer trace were identified both by their mass-to-charge ratio (m/e) and their "isotopic fingerprint" which is quite distinctive for simple chlorides based upon the natural abundances of the chlorine isotopes (75% at ^{35}Cl and 25% at ^{37}Cl). Ion intensities were calculated from the following expression[15]:

$$I = V/CRg$$

Where:

V = the sum of the voltage peaks for each of the major isotopes

C = ionization cross-section of the molecule determined from additivity rules[16]

R = Input resistor

g = instrument parameter

It is important to note that these intensities do not represent partial pressures, but rather only a semiquantitative measure of the primary vapor species. Quantitative pressure values are difficult to obtain using the sampling configuration employed due to the problems associated with accurate determination of the ionization cross-sections of the various species, the differing degrees of fragmentation, and different mass filter transmissions and Mach focusing numbers for the various ions.

Condensed phase products were examined using scanning electron microscopy (SEM). In order to eliminate the problems associated with the hydration of the reaction products, specimens for SEM analysis were immediately placed in a desiccator after a run.

RESULTS

SiC

The predominant gaseous corrosion product was determined to be $SiCl_4$ based upon the analysis of spectra obtained for actual corrosion experiments and that obtained when an argon gas stream saturated with $SiCl_4$ (*l*) was examined. Through analysis of the fragmentation patterns, it was determined that both $SiCl_3$ and $SiCl_4$ are formed by the reaction of silicon-based materials with chlorine at 950°C. This analysis also indicated that the amount of silicon trichloride produced was small when compared to the amount of silicon tetrachloride formed. As a result of this determination, the amount of corrosion occurring was measured in the mass spectrometer studies by examining the ratio of the intensity of the $SiCl_4^+$ product signal to the intensity of the Cl_2^+ reactant signal.

Table II shows the intensities of the mass spectrometer signals for the various types of SiC materials as a function of gas stream composition at 950°C. As expected, increasing the oxygen content of the gas stream results in a decrease in the amount of volatile product observed presumably as a result of the formation of a protective oxide layer on the sample surface. This oxide layer limits the contact of the chlorine with the silicon carbide substrate and thereby limits the amount of $SiCl_4$ gas produced. Although this rule generally holds true, the amount of volatile product observed for any one gas stream composition varies significantly depending on the sample. Table III shows reaction rates determined by thermogravimetric

Table II. Relative intensities of $SiCl_4^+$ produced for various SiC materials as a function of gas stream composition at 950°C.

Gas Mixture	I_{SiCl_4}/I_{Cl_2}			
	NC203	Hexoloy	NC430	Single Xtal
2% Cl_2	9.4×10^{-3}	2.4×10^{-3}	4.8×10^{-3}	1.1×10^{-3}
1% Cl_2/1% O_2	2.0×10^{-3}	2.5×10^{-3}	9.7×10^{-4}	4.7×10^{-4}
1% Cl_2/2% O_2	7.5×10^{-4}	1.8×10^{-3}	7.2×10^{-4}	5.0×10^{-4}
1% Cl_2/4% O_2	1.3×10^{-4}	7.2×10^{-4}	1.4×10^{-4}	1.5×10^{-4}
1% Cl_2/10% O_2		2.5×10^{-5}	1.3×10^{-5}	
1% Cl_2/20% O_2		9.4×10^{-6}		

Table III. Effect of gas composition on rate of weight loss for various types of Silicon Carbide at 950°C.

Gas Stream	-k $\frac{1}{4}$mg/cm^2/hr		
	NC203	Hexoloy	NC430
2% Cl_2	6.927 + 0.496	3.591 + 0.8874	11.07 + 0.5
1% Cl_2/2% O_2	1.316 + 0.045	3.461 + 0.344	0.5275 + 0.1674
1% Cl_2/4% O_2	0.067 + 0.019	0.156 + 0.0407	0.0380 + 0.0146
1% Cl_2/10% O_2	0.028 + 0.013	0.024 + 0.129	

analysis for conditions identical to those used in the mass spectrometer experiments. These results show the same trends as those illustrated in Table II.

The effect of oxygen content on the observed corrosion rate is also shown in figure 2. This figure shows the weight change versus time behavior for the hot-pressed SiC (NC203). The figure clearly shows that the addition of oxygen to the gas stream slows the rate of weight change.

Another interesting feature of figure 2 is the lack of a parabolic component to the curves, when oxygen is present. This linear dependency shows that, under the conditions studied, the oxide scales which form on the SiC samples remain pervious to gases throughout the duration of the experiments. Although the protection afforded by the scales increases as the oxygen content of the gas in which they were grown increases, none of the oxide scales completely suppresses corrosion. This behavior, combined with the fact that the corrosion rate changed with varying gas flow rates, indicates that gas transport of $SiCl_4$ away from the sample and/or Cl_2 to the sample is the rate controlling step in the corrosion reaction.

In a 2% Cl_2/Ar gas stream the siliconized material (NC430) produced the greatest amount of volatile corrosion product followed by the hot-pressed material

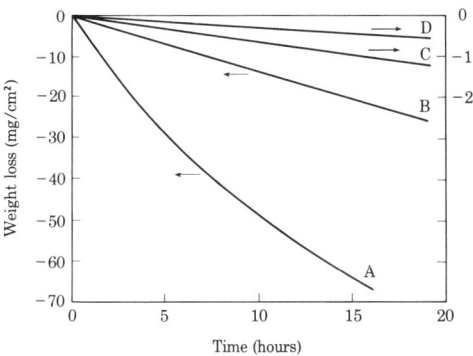

Fig. 2. Weight loss of hot-pressed SiC (NC 203) at 950°C when heated in gas streams containing argon and: (A) 2% Cl_2; (B) 1% Cl_2, 2% O_2; (C) 1% Cl_2, 4% O_2; and (D) 1% Cl_2, 10% O_2. Note enlarged scale for curves C and D.

(NC203), the sintered specimen (Hexoloy), and the single-crystal material. However, this sequence changes when oxygen is introduced into the system. When a gas stream of 1% Cl_2/1% O_2/Ar is considered, the sintered sample shows the larger amount of corrosion product and the higher rate of weight loss followed by the hot-pressed, siliconized, and single-crystal materials. In fact, the intensity of the $SiCl_4^+$ signal observed for the sintered specimen is slightly higher in the 1% oxygen-containing gas than for the gas stream consisting of 2% Cl_2 in argon. The linear rate constants determined by thermogravimetric analysis show no significant variation for the sintered sample when the gas stream is changed from 2% Cl_2/Ar to 1% Cl_2/2% O_2/Ar.

For all of the SiC samples, increasing the oxygen content of the gas stream above 4 percent results in the observance of very small amounts of volatile product and negligible rate of weight loss at 950°C.

The lack of reaction on the siliconized sample after increasing the oxygen content of the gas stream is illustrated in fig. 3. Figure 3 shows the surface morphology (by SEM) of the siliconized SiC sample as a function of gas stream composition at 950°C. In gas streams of 2% Cl_2/Ar and 1% Cl_2/2% O_2/Ar a significant amount of surface relief is evident. This relief, due to the preferential attack of the free silicon phase present in these materials, becomes less when the oxygen content of the gas stream is increased to four percent. In fact, gas streams containing 10 and 20 percent oxygen show surface characteristics nearly identical to the as-prepared sample. This lack of attack indicates that in gas streams containing 4% or more oxygen, sufficient oxidation of the sample surface occurs which limits the attack by chlorine. Similar results were obtained at 800 and 1100°C for the various SiC materials.

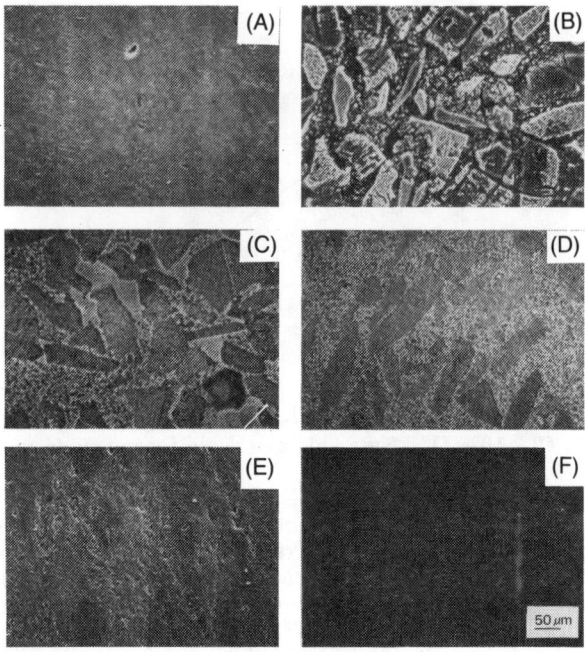

Fig. 3. Surface morphology of siliconized SiC (NC 430) after corrosion at 950°C in gas streams containing argon and: (B) 2% Cl_2; (C) 1% Cl_2, 2% O_2; (D) 1% Cl_2, 4% O_2; (E) 1% Cl_2, 10% O_2; and (F) 1% Cl_2, 20% O_2. Micrograph "A" is an "as prepared" specimen surface which has not been subjected to corrosion.

Si, SiO_2, and Si_3N_4

The mass spectrometric and thermogravimetric results for the silicon, silica, and silicon nitride samples are shown in Tables IV and V respectively. The results for silicon show a severe rate of attack in a 2% Cl_2/Ar gas stream but very little reaction after oxygen is introduced into the system. This indicates that a very protective oxide layer forms on the silicon material even in the presence of small amounts of oxygen.

The lack of a chlorination reaction in the case of the silica sample indicates that any SiO_2 formed on the surface of silicon-containing samples should act to protect these materials from attack by chlorine.

The absence of a reaction in the case of silicon nitride (AY-6) was somewhat unexpected. However, similar experiments on chemically vapor deposited Si_3N_4 show this same behavior suggesting that it is intrinsic to the material. It may be that Si_3N_4 forms an SiO_2 layer with less pores and microcracks for chlorine penetration. It may also be that the thin Si_2N_2O layer between the oxide and nitride provides further protection and contributes to the lower reactivity of Si_3N_4.

Table IV. Relative intensities of $SiCl_4^+$ generated during the reaction of Si, SiO_2, and Si_3N_4 with various gas streams at 950°C.

	I_{SiCl_4}/I_{Cl_2}		
Gas Stream	Silicon	Silica	Silicon Nitride
2% Cl_2	4.3×10^{-2}	1.1×10^{-5}	4.5×10^{-5}
1% Cl_2/1% O_2	3.0×10^{-5}		

Table V. Effect of gas stream composition on rate of weight loss for Silicon and Silicon Nitride at 950°C.

	$-k$ [mg/cm^2/hr]	
Gas Stream	Silicon	Silicon Nitride
2% Cl_2	25.91 + 3.37	0.03911
1% Cl_2/2% O_2	0.01855	

Silicon Oxychlorides

Silicon oxychloride compounds were directly identified during the mass spectrometric analysis of the Si-C-Cl-O-Ar system at 950°C. A detailed description of the analysis of these compounds is presented elsewhere[17] and will only be briefly discussed here. Mass spectrometric evidence of the existence of these compounds was first obtained when a sample of Norton NC203 was reacted in a 1% Cl_2/1% O_2/Ar gas stream at 950°C. Parent molecules with the formulae Si_2OCl_6 and Si_3OCl_8 were identified with the mass spectrometer. Fragments associated with these molecules were also observed. A typical spectrum showing the existence of the high mass silicon oxychloride compounds is schematically illustrated in fig. 4. The silicon oxychloride compounds appear to be formed by a gas phase reaction between the reactant oxygen gas and the silicon tetrachloride produced by chlorination of the silicon-based material.

DISCUSSION

As discussed by McNallan,[12] the large differences observed in the behavior of the various silicon carbide materials when exposed to mixed oxidation/chlorination conditions were somewhat unexpected. However, these variations may be explained through the analysis of the chemical composition of these materials and how this composition affects the microstructure. The fact that oxidation of the SiC samples results in an inhibition of the chlorination reaction indicates that the differences observed in the mixed oxidation/chlorination behavior are most likely a manifestation of the differences in the reaction of oxygen with the various substrate SiC materials. Since the gases react directly with the SiC and not the SiO_2, the

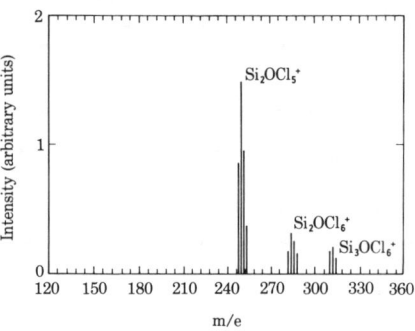

Fig. 4. Mass spectrum of silicon oxychloride molecules produced when hot-pressed SiC (NC 203) was heated at 950°C in a stream containing 98% Ar, 1% Cl_2 and 1% O_2.

differences in the behavior of the various materials must be associated with the physical and chemical nature of the growing oxide scale and the ability of the chlorine to penetrate the scale.

The hot-pressed sample (NC203) and the single-crystal material show very similar behavior in high temperature environments containing both oxygen and chlorine. These materials showed a moderate rate of attack in both the two percent chlorine and oxygen containing gases. The siliconized sample (NC430), on the other hand, showed a very high degree of attack in the 2% Cl_2/Ar gas stream but a very small degree of attack in environments containing even small amounts of oxygen. This occurrence may be explained by analyzing the role of the free silicon that is present in this material. As discussed previously, the silicon phase is more susceptible to attack by chlorine, yet also more readily oxidized than silicon carbide. This sample, therefore, shows a high amount of attack and weight loss in 2% Cl_2/Ar due to the preferential attack of the silicon phase by the chlorine gas, but in the presence of oxygen, a small amount of attack is observed due to the preferential oxidation of the silicon phase. This preferential oxidation of the silicon phase effectively decreases the surface area available for chlorination.

The sintered sample (Hexoloy) contains between 0.5 and 3 percent free carbon in addition to the boron that is added as a sintering aid. As a result of this fact the sample exhibits the lowest rate of weight loss in the 2% Cl_2/Ar gas stream and the highest rate of attack in gas streams containing a small amount of oxygen. Because of the positive free energy of formation of CCl_4 at 950°C there is a very low affinity of chlorine for the carbon-rich areas of the sample. This lower affinity effectively lowers the area available for the chlorination reaction and slightly lowers the amount of volatilization occurring. In gas streams containing small amounts of oxygen the free carbon phase reacts with the oxygen to form carbon monoxide or dioxide. This consumption of the reactant oxygen locally lowers the partial pressure of oxygen at the surface of the sample. The lower partial pressure prevents the protective oxide layer from forming and allows the chlorination reaction to continue even under normally oxidizing conditions.

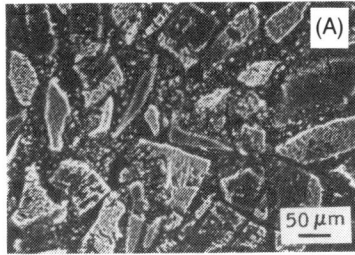
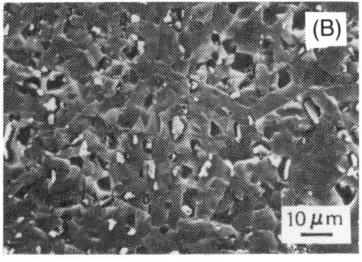

Fig. 5. Scanning electron micrographs of the corroded surfaces of (A) siliconized (NC 430) and (B) sintered (Hexoloy) silicon carbides. Corrosion conditions were 30 minutes at 950°C in 2% Cl_2, 98% Ar. The white particles in "B" are carbon.

The difference in the chlorination behavior of the siliconized and sintered SiC materials is illustrated in fig. 5. This figure shows the surface morphology of the samples after a 30 minute exposure to a 2% Cl_2/Ar gas stream at 950°C. The large amount of surface relief in the SEM photomicrograph clearly indicates that the silicon phase in the siliconized material is severely attacked by chlorine. The sintered material, on the other hand, does not exhibit this large amount of surface attack. In fact, the bright spots in this photomicrograph represent the free carbon particles that are resistant to chlorine attack.

CONCLUSIONS

Reaction of silicon-containing ceramics with chlorine gas produces volatile chloride corrosion products. The predominant product formed at temperatures in the vicinity of 950°C is silicon tetrachloride. The effect of oxygen on the silicon carbide-chlorine system depends greatly on the minor phases that are present in the silicon carbide material. The processing additives (Si, C, or B) are especially important when they affect the overall carbon or silicon activity of the silicon carbide base material. These additives are also particularly important when the oxygen content of the gas stream is relatively low (1-2%).

It has also been observed that a vigorous reaction occurs in the silicon-chlorine system at 950°C. This reaction ceases upon introduction of 1% oxygen to the gas stream due to the rapid formation of a coherent, protective oxide surface layer.

Very little reaction occurs when SiO_2 and Si_3N_4 are exposed to chlorine. The lack of reaction in the case of silicon nitride is most likely due to the presence of a less permeable SiO_2 layer or possibly a Si_2N_2O formed on silicon nitride.

The mass spectrometer studies have also shown that silicon oxychloride compounds are formed during the reaction of Si and SiC with mixtures of oxygen and chlorine at 950°C. Compounds with the formulae Si_2OCl_6 and Si_3OCl_8 have been directly analyzed for the first time. Preliminary results indicate that these compounds are formed primarily by a gas phase reaction between the reactant oxygen gas and the silicon tetrachloride corrosion product.

Acknowledgments

The initial financial support for work done at the Ohio State University was provided by the Army Research Office under contract number DAAG29-82-K-0149.

Support was also provided by The Edward Orton Junior Ceramic Foundation during the later stages of the study.

---

*Now with: E.I. du Pont de Nemours and Company, Savannah River Laboratory, Aiken, SC 29808
‡Crysteco Company, Wilmington, OH.
†Thermal American Fused Quartz Company, Montville, NJ.
§Norton Company, Worcester, MA.
¶Sohio Engineered Ceramics, Niagara Falls, NY.
**GTE Products Corporation, Towanda, PA.

References

[1] P. Elliot, C. J. Tyreman, and R. Prescott, "High Temperature Alloy Corrosion by Halogens," *J. Metals*, **37** [7] 20 (1985).

[2] J. M. Oh, M. J. McNallan, G. Y. Lai, and M. F. Rothman, "High Temperature Corrosion of Superalloys in an Environment Containing both Oxygen and Chlorine," *Met. Trans.*, **17A** 1087 (1986).

[3] R. P Viswanath, D. Rein, and K. Hauffe, "High Temperature Corrosion of NiCrAl Alloy in Oxygen-Chlorine Mixtures," Werkstoffe Und Korrosion, **31** 778 (1980).

[4] Y. Ihara, H. Ohgame, K. Sakiyama, and K. Hashimoto, "The Corrosion Behavior of Iron in Hydrogen Chloride Gas and Gas Mixtures of Hydrogen Chloride and Oxygen at High Temperatures," *Corros. Sci.*, **23** 805 (1982).

[5] N. S. Jacobson, "Reaction of Iron with Hydrogen Chloride-Oxygen Mixtures at 550°C," *Oxidation of Metals*, **26** [3-4] 157 (1986).

[6] N. S. Jacobson, M. J. McNallan, and Y. Y. Lee, "The Formation of Volatile Corrosion Products During the Mixed Oxidation-Chlorination of Cobalt at 650°C," *Met. Trans.*, **17A** 1223 (1986).

[7] N. S. Jacobson, "Application of an Atmospheric Pressure Mass Spectrometer to Chlorination Reactions, NASA TM87270, NASA-Lewis Research Center, Cleveland, OH 44135 (1986).

[8] Y. J. van der Meulen, C. M. Osburn, and J. F. Ziegler, "Properties of SiO_2 Grown in the Presence of HCl or Cl_2," *J. Electrochem. Soc.*, **122** [2] 284 (1975).

[9] A. Rhotagi, S. R. Butler, F. J. Feigl, H. W. Kraner, and K. W. Jones, "Chlorine Incorporation in HCl Oxides," *J. Electrochem. Soc.*, **126** [1] 143 (1979).

[10] J. Monkowski, R. E. Tressler, and J. Stach, "The Structure and Composition of Silicon Oxides Grown in HCl/O_2 Ambients," *J. Electrochem. Soc.*, **125** [11] 1867 (1978).

[11] K. Hirabayashi and J. Iwamura, "Kinetics of Thermal Growth of $HCl-O_2$ Oxides on Silicon," *J. Electrochem. Soc.*, **120** [11] 1595 (1973).

[12] M. J. McNallan, S. Y. Ip, S. Saam, and W. W. Liang, "High Temperature Corrosion of SiC Based Ceramics in Chlorine Containing Environments," p. 328 in High Temperature Materials Chemistry—III, Z.A. Munir and D. Cubicciotti, eds., The Electrochemical Society Press, Pennington, NJ (1986).

[13] T. M. Besmann, "SOLGASMIX-PV, A Computer Program to Calculate Equilibrium Relationships in Complex Chemical Systems," Oak Ridge National Laboratory Report, ORNL/TM-5775, Oak Ridge, TN 37830 (1977).

[14] C. A. Stearns, F. J. Kohl, G. C. Fryburg, and R. A. Miller, "A High Pressure Modulated Molecular Beam Mass Spectrometric Sampling System," NASA TM73726, NASA-Lewis Research Center, Cleveland, OH 44135 (1977).

[15] R. T Grimley, "Chapter 8—Mass Spectrometry," p. 195 in The Characterization of High-Temperature Vapors, J.L. Margrave, ed., John Wiley and Sons, NY (1967).

[16] J. B. Mann, "Ionization Cross Sections of the Elements Calculated from Mean-Square Radii of Atomic Orbitals," *J. Chem. Phys.*, **46** 1646 (1967).

[17] J. E. Marra, E. R. Kreidler, N. S. Jacobson, and D. S. Fox, "Direct Mass Spectrometric Identification of Silicon Oxychloride Compounds," submitted for publication to the Journal of The Electrochemical Society.

Oxidation of SiC Ceramic Heat Exchanger Materials in the Presence of Chlorine at 1300°C

S. Y. Ip and M. J. McNallan

Department of Civil Engineering, Mechanics, and Metallurgy
University of Illinois at Chicago
P.O. Box 4348
Chicago, IL 60680

M.E. Schreiner

Gas Research Institute
8600 W. Bryn Mawr Ave.
Chicago, IL 60631

Abstract

SiC based ceramic tubes can be used in high temperature heat exchangers to increase the efficiency of gas fired furnaces. In some industrial environments including aluminum remelting furnaces, the presence of chlorine contamination in the flue gases can cause the rate of oxidation of the tube to be accelerated. The effects of small amounts of chlorine contamination on the rate of high temperature oxidation of several low cost SiC based heat exchanger tube materials has been investigated by thermogravimetric analysis at 1300°C. The results are interpreted by considering the formation of volatile corrosion products at the interface between the SiC and the protective SiO_2 film and the disruption of the oxide film by these volatile species.

Introduction

Natural gas is a clean fuel which can be conveniently used to fire furnaces in a number of industrial operations. Waste heat recuperators can be used to significantly increase the efficiency of such furnaces[1]. In waste heat recuperation technology, heat from the combustion gases leaving the furnace is transferred through high temperature heat exchangers into the combustion air. This can increase the maximum operating temperature of the furnace and reduce the fuel requirement to maintain the furnace at a given temperature.

The improvement in the energy efficiency of the furnace is maximized when the heat exchanger is operated at the highest possible temperature. Many furnaces used in the metallurgical and glass industries would benefit from the use of recuperators operating at temperatures above the maximum use temperatures of high temperature metals[2]. Structural ceramic materials can be used at higher temperatures than metals and can be fabricated into tubing for use in heat exchangers[3-5]. SiC is particularly suitable for this application because it is oxidation resistant, has a high thermal conductivity and good high temperature strength.

SiC ceramic heat exchangers have a history of successful application in furnaces in a number of metallurgical operations, but preliminary tests have shown that they may be subject to accelerated corrosion when used on aluminum remelting furnaces[6-8]. Flue gases from aluminum remelting furnaces are particularly corrosive because they are contaminated by halogen compounds including chlorine which are used in the fluxing and degassing of the aluminum. This paper describes research performed on the corrosion of a series of low cost SiC materials in purified $Ar-O_2-Cl_2$ environments. The goal of this work is to help identify the mechanisms of high temperature corrosion of SiC based heat exchanger materials in the environments typically encountered in aluminum remelting systems.

MATERIALS AND PROCEDURES

In order to reduce processing costs, SiC based heat exhanger tubes are usually not pure SiC, but contain additives to increase the rate of sintering and to reduce the temperatures required in the sintering operation. The presence of the sintering aids may affect the subsequent corrosion behavior of the materials. Three types of SiC based ceramic materials have been included in this research. All three are commercially available in the form of approximately one inch diameter tubing suitable for use in high temperature heat exchangers. A letter designation system was created for the test materials in this study. Type A material is a sintered alpha SiC material produced by SOHIO which contains boron and excess carbon to accelerate the sintering process. Type D material is Coors SCRB 205, which is a reaction sintered SiC material containing approximately 15% excess Si by volume. Type F material is Asahi Roiceram which is a fine grained alpha SiC which contains Al_2O_3 as a sintering aid.

The experiments performed in this study consisted of thermogravimetric measurements in corrosive environments supplemented by examination of the corrosion products using optical and scanning electron microscopy. The apparatus used for the thermogravimetric experiments is shown schematically in Figure 1. In this apparatus, mass measurements are made using a fused quartz spiral housed in a water cooled tube to prevent temperature fluctuations from affecting the spring constant of the spiral. The weighing apparatus is coupled to a mullite reaction tube which is heated by a high temperature furnace with molybdenum disilicide elements. The reagent gases are Ar, O_2 and Cl_2 in the form of purchased premixed $Ar-Cl_2$ gas mixtures. The O_2 and Ar are purified by passage through anhydrous $CaSO_4$ and Ascarite for removal of H_2O and CO_2 respectively. The $Ar-Cl_2$ mixtures were purified by passage through concentrated H_2SO_4 to remove H_2O. After purification, the gases were mixed in the desired proportions in a packed column and supplied to the reaction tube from the bottom.

In each experiment, a ring segment of the SiC based ceramic was sliced from the tubing using a low speed diamond saw. The specimen was heated to the reaction temperature in flowing argon and the experiment was initiated by introducing the corrosive gas mixture. The change in mass of the specimen was monitored at frequent intervals during the experiment by reading the extension of the quartz spiral. All of the experiments were performed with gases flowing through the reaction tube at a superficial velocity of 1.5 cms^{-1}. At the conclusion

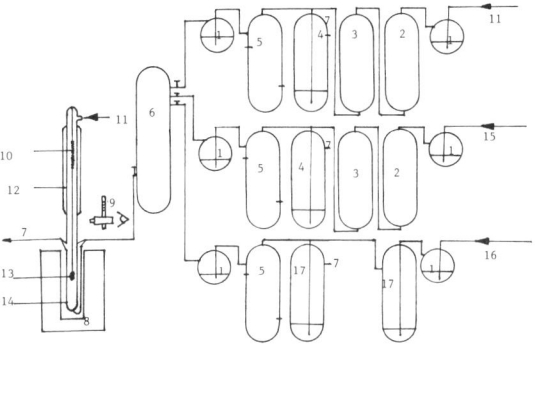

Fig. 1. Schematic diagram of apparatus used for thermogravimetric experiments in high temperature chlorine containing environments.

1. BUBBLER
2. DESSICANT COLUMN
3. ASCARITE COLUMN
4. DBP COLUMN
5. FLOWMETER
6. MIXING COLUMN
7. HOOD
8. FURNACE
9. CATHETOMETER
10. QUARTZ SPIRAL
11. ARGON CYLINDER
12. WATER JACKET
13. SAMPLE
14. REACTION TUBE
15. HYDROGEN OR OXYGEN CYLINDER
16. HCl OR CHLORINE CYLINDER
17. SULFURIC ACID COLUMN

of each experiment, the specimens were examined by optical and electron microscopy to determine the morphology of the corrosion products.

RESULTS AND DISCUSSION

A critical feature of high temperature corrosion in Cl_2 containing environments is the volatility of many chloride and oxychloride reaction products[9]. The complex kinetics observed in many mixed environments which contain both oxygen and chlorine is the result of the interactions between the formation of volatile species (mostly chlorides and oxychlorides), and condensed species (mostly oxides)[10]. Volatile oxide species such as SiO, CO, and CO_2 can also be formed during corrosion of SiC.

At temperatures below 1100°C in environments which are high in Cl_2 and low in O_2, SiC can be corroded rapidly by the formation of volatile reaction products[11,12]. Under these conditions no protective SiO_2 film can be formed. The kinetics of the corrosion of SiC under these conditions has been discussed elsewhere[13]. At higher temperatures and in environments where the partial pressure of oxygen is higher, SiO_2 grows rapidly enough to produce a protective oxide film on the corroding specimens. Chlorine is able to penetrate this film, however, and the kinetics of oxidation of SiC are affected by the presence of chlorine leading to higher corrosion rates in contaminated environments. This paper describes the results of a series of experiments performed under these conditions.

Figure 2 shows the results of thermogravimetric tests performed on the three materials at 1300°C in an environment consisting of Ar-0.1%Cl_2-20%O_2 by volume. The oxidation kinetics of these three materials in air at this temperature have been determined at the Center for Advanced Materials at the Pennsylvania State

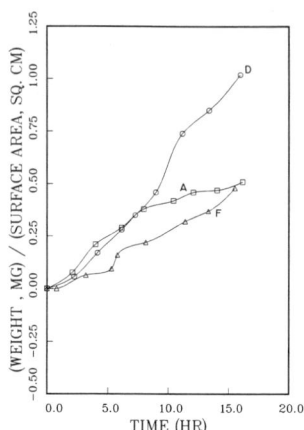

Fig. 2. Thermogravimetric results for SiC based materials exposed to Ar-0.1%Cl_2-20%O_2 at 1300°C.

University[14]. All three materials oxidize according to parabolic rate equations with rate constants which would lead to mass increases of 0.0475 mg cm^{-2} for material A, 0.16 mg cm^{-2} for material D, and 0.27 mg cm^{-2} for material F respectively. The presence of chlorine contamination in the environment has resulted in deviations from parabolic kinetics for materials D and F and substantial increases in the rates of oxidation for all three materials.

The morphologies of the corrosion products on the specimens were examined in order to identify the mechanism for the accelerated oxidation. The morphology of the type A specimen corroded in Ar-0.1%Cl_2-20%O_2 was not significantly different from that of the uncorroded specimen. The surface of the type F specimen after corrosion was covered by widely scattered aluminum rich particles which appeared to have formed during the corrosion. The surface of the type D specimen, shown in Figure 3, exhibited a number of bubbles in the oxide film. These ranged from 10 to 20 microns up to several millimeters in diameter and were located primarily on the exposed surfaces of the free silicon phase in this material as shown in Figure 3. Because the material D specimen showed the largest increase in corrosion in the presence of chlorine contamination, the formation of these bubbles is believed to be associated with the accelerated corrosion of the SiC based materials.

At higher levels of contamination by chlorine, the corrosion occurs much more rapidly. Figure 4 shows the thermogravimetric results obtained in Ar-2%Cl_2-20%O_2 by volume at 1300°C. All three of the materials are attacked more rapidly in this environment with the average increase in mass increased by approximately a factor of four. The difference is particularly large for material A, which also exhibits very irregular kinetics with alternating periods of increasing and decreasing mass.

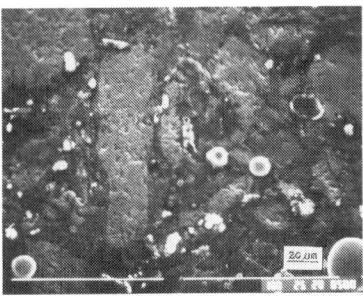

Fig. 3. Scanning electron micrograph of surface of type D material after exposure to Ar-0.1%Cl_2-20%O_2 at 1300°C showing bubble formation in SiO_2 film on free silicon phase.

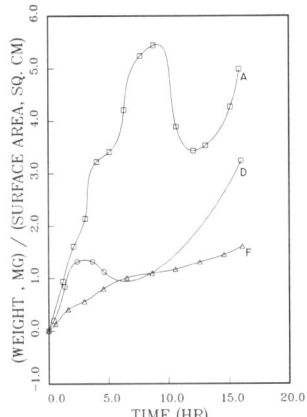

Fig. 4. Thermogravimetric results for SiC based materials exposed to Ar-2%Cl_2-20%O_2 at 1300°C.

The reason for the irregular kinetics can be identified from the morphology of the corrosion products found on material A after corrosion and shown in Figure 5. The condensed corrosion products consist of a film of glassy SiO_2 containing a number of bubbles with diameters up to 1 cm. The irregular kinetics observed for this material results from the formation and bursting of the bubbles and from the intermittent contact between the bubbles and the walls of the reaction tube. Similar

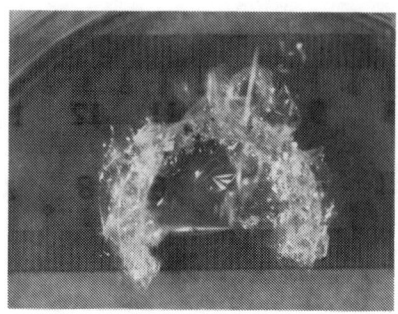

Fig. 5. Photograph of type A material specimen after exposure to Ar-2%Cl_2-20%O_2 at 1300°C showing large bubbles in SiO_2 corrosion products.

but slightly smaller bubbles were observed on the surfaces of the material D and F specimens corroded in this environment.

When the oxygen content of the corrosive environment is reduced, the magnitude of the net increase in mass is also reduced as shown in Figure 6 which shows the thermogravimetric results obtained in Ar-2%Cl_2-2%O_2 at 1300°C. This change may be related to the increase in the relative amount of volatile corrosion products which would be expected to be formed when the ratio of chlorine to oxygen in the environment is increased. The kinetics still generally obey linear rate equations, and the magnitude of the fluctuations in the mass are smaller than in the higher oxygen environment.

Figure 7 shows the surface of the type A specimen after corrosion in the Ar-2%Cl_2-2%O_2 environment at 1300°C. The size and number of bubbles in the oxide film is significantly smaller than that observed in the higher oxygen environment. The oxide layers formed on all three materials appear to be much thinner after corrosion in the Ar-2%Cl_2-2%O_2 environment than after corrosion in the Ar-2%Cl_2-20%O_2 environment.

When the materials are corroded in Ar-0.1%Cl_2-20%O_2, the quantity of volatile corrosion products formed appears to be smaller. Figure 8 shows the thermogravimetric results obtained in this environment. Both materials A and D exhibit larger increases in mass in this environment than in the higher chlorine environment at the same oxygen content. The mass change for material F is irregular. The size and number of bubbles formed in the oxide films on the specimens are much smaller in this environment than in the 2%Cl_2 environments. Figure 9 shows the irregular bubbles observed on material D after exposure to this environment, while neither material A nor material F showed any bubble formation after this exposure.

PROPOSED MECHANISM OF ACCELERATED CORROSION

In most mixed oxidation environments containing chlorine, the primary mechanism of accelerated corrosion is the formation of volatile corrosion products rather than condensed corrosion products which can contribute to the formation of a protective film[15]. During the corrosion of SiC based ceramics in gas mixtures

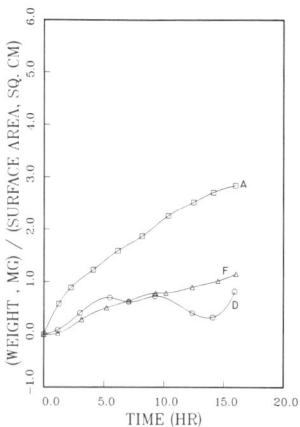

Fig. 6. Thermogravimetric results for SiC based materials exposed to Ar-2%Cl$_2$-2%O$_2$ at 1300°C.

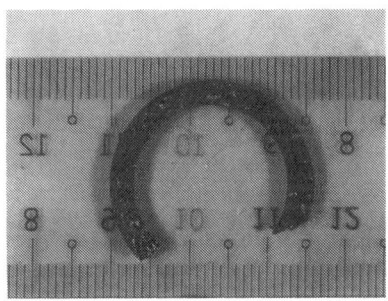

Fig. 7. Photograph of type A material specimen after exposure to Ar-2%Cl$_2$-2%O$_2$ at 1300°C showing smaller bubbles with thinner walls than on material exposed to Ar-2%Cl$_2$-20%O$_2$ at the same temperature.

Silicon Carbide

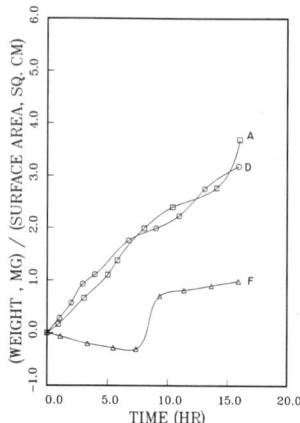

Fig. 8. Thermogravimetric results for SiC based materials exposed to Ar-0.1%Cl_2-2%O_2 at 1300°C.

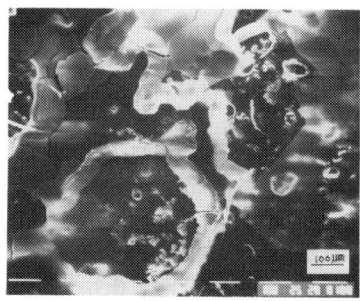

Fig. 9. Scanning electron micrograph of surface of type D material after exposure to Ar-0.1%Cl_2-2%O_2 at 1300°C showing bubble formation.

containing O_2 and Cl_2 at temperatures above 1100°C, volatile corrosion products are formed, but this does not result in active oxidation of the type observed at lower temperatures because a film of SiO_2 forms quickly enough on the material to separate it from the gaseous environment. The separation is not perfect, however, because chlorine is able to dissolve into the SiO_2 film and modify its properties[16].

On the basis of the apparent plasticity of the SiO_2 film formed on the SiC materials in the presence of chlorine at temperatures well below the melting point of SiO_2, chlorine appears to act as a network modifier in the SiO_2 and reduces its viscosity. From the Stokes-Einstein equation, this would also be expected to increase the diffusion coefficients of mobile species in the SiO_2[17], and this would result in a high parabolic rate constant for oxidation under conditions where the rate of oxidation is controlled by transport through a protective film. This phenomenon may be sufficient to explain the effects of chlorine on oxidation rates at low levels of contamination by chlorine (<0.1% by volume), but other mechanisms are also active when the level of contamination is higher.

Chlorine is able to diffuse through the oxide film and form volatile compounds (either chlorides or oxychlorides) at the interface between the SiO_2 and the SiC. These compounds have a lower solubility in the SiO_2 than the chlorine does, and rather than diffuse back out through the SiO_2, they form bubbles which disrupt the film and separate the oxide from the substrate. Similar phenomena have been observed during the oxidation of Si in the presence of chlorine[18,19]. This gas evolution causes a departure from parabolic oxidation kinetics for two reasons: 1) The formation of volatile reaction products leads to a decrease in the mass of the specimen rather than an increase in the mass. 2) The formation of bubbles results in a mechanical disruption of the SiO_2 film which exposes the underlying SiC and also affects the mass transport through the film so that the length of the diffusion path through the SiO_2 is no longer directly proportional to the mass of oxide formed on the surface of the SiC.

For all of the materials, the rate of reaction is highest in the environment where both the partial pressure of chlorine and that of oxygen is maximum (2%Cl_2-20%O_2). When the concentration of oxygen is reduced to 2%, the amount of condensed reaction products on the specimen is reduced so that the net increase in mass of the specimen is smaller. This may be because a larger fraction of the silicon is forming volatile corrosion products under these conditions. When the concentration of chlorine in the corrosive gas mixture is reduced to 0.1%, the quantity of volatile corrosion products is reduced and bubble formation is not observed for the A or F materials. When no bubbles are formed, the rate of oxidation may still be accelerated by the effect of chlorine on the diffusion coefficients in the SiO_2. More increase in mass is observed on material D after exposure to Ar-0.1%Cl_2-2%O_2 than after exposure to Ar-0.1%Cl_2-20%O_2 indicating that the ratio of Cl_2/O_2 in the gas mixture affects the ability of the chlorine to penetrate the SiO_2 and form bubbles.

Bubble formation in the SiO_2 film on material D occurred first on the free silicon phase indicating that a silicon containing species rather than CO or CO_2 is responsible for the bubble formation. Sintering aids in the SiC materials may affect the kinetics of mixed oxidation in O_2/Cl_2 environments by affecting the properties of the SiO_2 film. On the basis of these experiments, it appears that Al_2O_3 sintering aids in the SiC may reduce the rate of attack in O_2/Cl_2 environments by making the film more protective.

Acknowledgement

This research was supported by the Gas Research Institute through subcontract GRI-TPSU-UI-1302-390 from the Center for Advanced Materials at The Pennsylvania State University under prime contract No. 5084-238-1302.

References

[1] W. W. Liang, E. S. Tabb, "GRI's Advanced Heat Transfer Systems Program," Industrial Heat Exchangers, Edited by A. J. Hayes, W. W. Liang, S. L. Richlen, E. S. Tabb, ASM Internations, Metals Park, OH, 1985, pp. 29-36.

[2] C. J. Dobos, W. W. Liang, "High Temperature Materials for Gas Fired Industrial Applications," High Temperature Materials Chemistry—II, Edited by Z. A. Munir, D. Cubiciotti, The Electrochemical Society, Inc., Pennington, NJ 1983, pp. 401-405.

[3] M. C. Kerr, "Advanced Ceramic Heat Exchangers Utilizing Hexoloy SA Single Phase Silicon Carbide Tubes," Industrial Heat Exchangers, Edited by A. J. Hayes, W. W. Liang, S. L. Richlen, E. S. Tabb, ASM International, Metals Park, OH 1985, pp. 391-395.

[4] B. D. Foster, "Silicon Carbide Components for Heat Exchangers Design and Performance Testing," Industrial Heat Exchangers, Edited by A. J. Hayes, W. W. Liang, S. L. Richlen, E. S. Tabb, ASM International, Metals Park, OH, 1985, pp. 397-402.

[5] C. J. Dobos, K. Green, "Reaction Bonded SiC Components for High Temperature Energy Systems," Industrial Heat Exchangers, Edited by A. J. Hayes, W. W. Liang, S. L. Richlen, E. S. Tabb, ASM International, Metals Park, OH, 1985, pp. 403-409.

[6] C. E. Smeltzer, T. A. Argabright, M. E. Ward, W. W. Liang, "An Investigation of the Hot Corrosion of Silicon Carbide (SiC) Recuperator Tubes in Aluminum Remelt Furnace Stacks," Industrial Heat Exchangers, Edited by A. J. Hayes, W. W. Liang, S. L. Richlen, E. S. Tabb, ASM International, Metals Park, OH 1985, pp. 299-305.

[7] J. I. Federer, T. N. Tiegs, D. M. Kotchik, D. Petrak, "Analysis of Candidate Silicon Carbide Recuperator Materials Exposed to Industrial Furnace Environments," ORNL/TM-9677, 1985.

[8] J. I. Federer, P. J. Jones, "Oxidation/Corrosion of Metallic and Ceramic Materials in Aluminum Remelt Furnaces," ORNL/TM-9741, 1985.

[9] P. L. Daniel, R. A. Rapp, "Halogen Corrosion of Metals," Advances in Corrosion Science and Technology, Vol 8, Edited by M. G. Fontana, R. W. Steahle, Plenum Press, NY, 1976, pp. 55-172.

[10] N. S. Jacobson, M. J. McNallan, Y. Y. Lee, "The Formation of Volatile Corrosion Products During Mixed Oxidation-Chlorination of Cobalt at 650°C," *Metallurgical Transactions*, **Vol 17A**, 1986, pp. 1223-1228.

[11] S. Y. Ip, S. Saam, M. J. McNallan, W. W. Liang, "High Temperature Corrosion of SiC Based Ceramics in Chlorine Containing Environments," High Temperature Materials Chemistry—III, Edited by Z.A. Munir, D. Cubiciotti, The Electrochemical Society, Inc., 1986, pp. 328-338.

[12] J. E. Marra, E. R. Kreidler, N. S. Jacobson, D. S. Fox, "Direct Mass Spectrometric Identification of Silicon Oxychloride Compounds," submitted to *Journal of the Electrochemical Society*.

[13] S. Y. Ip, D. S. Park, M. J. McNallan, manuscript in preparation.

[14] Private communication from Center for Advanced Materials, Pennsylvania State University, State College, PA.

[15] Y. Y. Lee, M. J. McNallan, "Ignition of Nickel in Environments Containing Oxygen and Chlorine," *Metallurgical Transactions*, **Vol 18A**, 1987, pp. 1099-1107.

[16]D. W. Hess, B. E. Deal, "Kinetics of the Thermal Oxidation of Silicon in O_2/HCl Mixtures," *Journal of the Electrochemical Society*, **Vol 124**, 1977, pp. 735-739.

[17]G. H. Geiger, D. R. Poirier, Transport Phenomena in Metallurgy, Addison Wesley Publishing Co., Reading, MA, 1973, pp. 455-456.

[18]J. Monkowski, R. E. Tressler, J. Stach, "The Structure and Composition of Silicon Oxides Grown in HCl/O_2 Ambients," *Journal of the Electrochemical Society*, **Vol 125**, 1978, pp. 1867-1873.

[19]M. D. Monkowski, J. R. Monkowski, I. S. T. Tsong, J. Stach, R. E. Tressler, "Microstructure Development During the Thermal Oxidation of Silicon in Chlorine Containing Ambients," *J. Non-Crystalline Solids*, **Vol 49**, 1982, pp. 201-209.

Active Oxidation of SiC in Low Dew-Point Hydrogen Above 1400°C

Hyoun-Ee Kim[*] and D. W. Readey

Department of Ceramic Engineering
The Ohio State University
Columbus, OH 43210

Abstract

The corrosion of polycrystalline SiC was studied at temperatures between 1400°C and 1527°C in one atmosphere of hydrogen containing water vapor. Active oxidation occurred for H_2O/H_2 ratios between 2×10^{-5} and 5×10^{-2}. At low water vapor pressures, the rate of corrosion increased with water vapor content consistent with a model of gaseous diffusion of water vapor to the solid surface as the rate-controlling step. At high water vapor pressures, the rate of corrosion decreased with water vapor content consistent with the outward diffusion of gaseous diffusion products, SiO and CO.

Introduction

Silicon carbide is a leading candidate as a structural material in heat engines, heat exchangers, and similar applications[1,2,3]. An important property of SiC is its oxidation resistance imparted by a protective SiO_2 scale[4]. However, in atmospheres of low oxygen content, SiO_2 is unstable and active oxidation occurs through the formation of gaseous SiO[5] leading to rapid rates of material loss. The purpose of this investigation was to investigate active oxidation of SiC in nominally reducing hydrogen-water vapor mixtures.

Experimental

Thermodynamic equilibria for SiC in H_2 at 0.1 MPa with water vapor pressures from 10^{-7} to 10^{-2} MPa at temperatures from 1300 to 2000 K, were calculated with the SOLGASMIX-PV[6] program. Free energy data for all product gases and condensed species possible in the Si-C-O-H system for which thermodynamic data could be found[7] were used in the calculations.

Polycrystalline silicon carbide samples[1] near theoretical density (>97.5%) were used. Major impurities were less than 0.5 w/o carbon and 0.2 w/o boron. X-ray diffraction indicated that the samples were mainly the 6H polytype with small amounts of 4H and 33R polytypes. Disk-shaped samples were cut with a diamond wafering blade and then diamond polished. Typical final sample dimensions were 1.27 cm diameter and 0.03 ± 0.01 cm thick. A hole 0.05 cm diameter was diamond drilled near the edge of a sample to attach it to a microbalance.

A microbalance[†] above of a vertical tube furnace[§] was used to monitor the weight change of the samples during corrosion. The samples were suspended from the microbalance with a thin molybdenum wire chain. Hydrogen was purified by passing it through a catalyst unit[¶] and subsequently eliminating water vapor with calcium sulfate and a liquid nitrogen cold trap. After purification, controlled amounts of water vapor were added by gas mixing. The water vapor pressure of the gas both entering and exiting the furnace was monitored with an electronic hygrometer[**]. Before a corrosion run was begun, the furnace was heated and the gas allowed to flow through the furnace. When entering and exiting water vapor contents were the same, the sample was lowered into the furnace. This procedure was particularly important in attaining low water vapor pressures. The total pressure was always 0.1 MPa. Several corrosion experiments were periodically interrupted and the sample removed from the furnace for microscopic examination[††]. Efforts were made to observe the same region on the sample as a function of reaction time.

RESULTS

Thermodynamic Equilibria

The calculated equilibrium partial pressures of the product gases for the reaction of alpha SiC in H_2 as a function of the water vapor pressure at 1700 K are shown in Figure 1. These calculated thermodynamic equilibria show that the partial pressures of the oxygen-containing gases, such as SiO and CO, increase almost linearly with water vapor pressure, while those of the other species remain constant or decrease slightly. As a result, SiC will dissociate at low water vapor pressures into silicon and CH_4. As the water vapor pressure increases, the SiO pressure becomes large enough so that SiC will react stoichiometrically and SiO and CO become the main product gaseous species. This behavior continues until the partial pressure of H_2O becomes sufficiently high for solid SiO_2 to be in equilibrium with SiC. The data in Figure 1 suggest that the following reactions are probable, listed in order of increasing water vapor pressure:

$$SiC_{(s)} + 2 H_{2(g)} = Si_{(s,1)} + CH_{4(g)} \qquad (1)$$

$$SiC_{(s)} + H_{2(g)} + H_2O_{(g)} = SiO_{(g)} + CH_{4(g)} \qquad (2)$$

$$SiC_{(s)} + 2 H_2O_{(g)} = SiO_{(g)} + CO_{(g)} + 2 H_{2(g)} \qquad (3)$$

$$SiC_{(s)} + 3 H_2O_{(g)} = SiO_{2(s)} + CO_{(g)} + 3H_{2(g)} \qquad (4)$$

The experimental H_2O/H_2 range in this research covers Eq. (2) to Eq. (4). A final point worth noting in these calculated equilibria is that the H_2O pressure of the product gas is about three orders of magnitude smaller than that of the reactant gas. This implies that virtually all of the ambient water vapor is consumed by reaction with SiC.

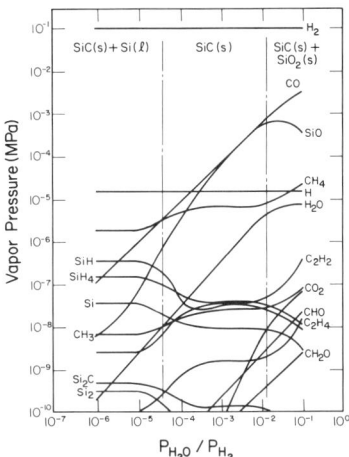

Fig. 1. Calculated equilibrium partial pressures of the product gases from the reaction of SiC in H_2 and H_2O at 1700 K.

Reaction Kinetics

For all temperatures and water vapor pressures investigated, the weight-loss of SiC as a function of time was linear up to several hours as shown in Figure 2. In all cases, the total weight loss was small and sample dimensions could be assumed to remain constant. From the weight loss rate and sample dimensions, the corrosion flux density as a function of temperature and water vapor pressure was calculated, Figure 3. These data were obtained at a constant gas velocity of 0.59 cm/s. The effect of gas velocity at two different water vapor pressures is shown in Figures 4 and 5.

DISCUSSION

Water Vapor Pressure and Temperature Dependence

There are several features in Figure 3 worth noting. First, the rate of corrosion increases almost linearly with water vapor pressure, reaches a maximum, and then decreases. Second, the maximum in the corrosion rate shifts to lower water vapor pressures as the temperature decreases. Third, at low water vapor pressures, the corrosion rate is weakly temperature dependent. Conversely, the corrosion rate is strongly temperature dependent at high water vapor pressures.

Corrosion at Low Water Vapor Pressures

The almost linear dependence on the water vapor pressure, weak temperature dependence, and the dependence on gas velocity all indicate that gaseous diffusion

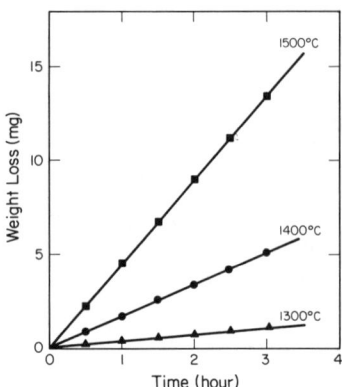

Fig. 2. Weight loss as a function of reaction time at $P(H_2O) = 4.0 \times 10^{-6}$ MPa.

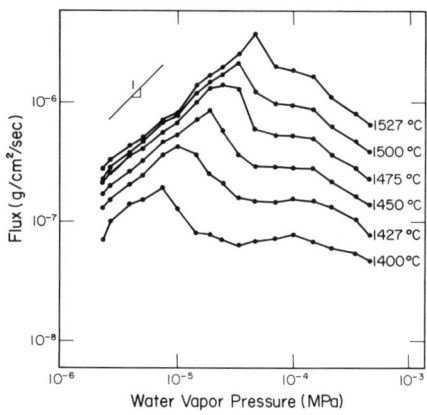

Fig. 3. Corrosion flux density as a function of water pressure with temperature as a parameter; gas velocity = 0.59 cm/s.

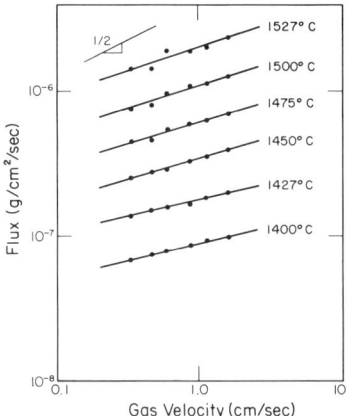

Fig. 4. Corrosion flux density as a function of gas velocity; $P(H_2O) = 1.0 \times 10^{-5}$ MPa.

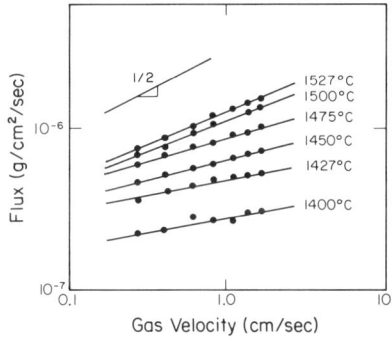

Fig. 5. Corrosion flux density as a function of gas velocity; $P(H_2O) = 1.0 \times 10^{-4}$ MPa.

is the rate-controlling mechanism in this temperature-pressure regime. At low partial pressures, the thermodynamic calculations suggest that the reaction controlling the rate of corrosion is that in either Eq. 2 or Eq. 3. Assuming the reaction in Eq. (3) dominates, the corrosion flux is related to the gas fluxes by:

$$J = J(SiC) = -J(SiO) = -J(CO) = 1/2\ J(H_2O).$$

Therefore, the rate of corrosion, J, is determined by the diffusion flux of H_2O:

$$J = -\frac{D}{2RT}\frac{dP}{dx}(H_2O) \quad (5A)$$

and can be expressed as[10]

$$J = \frac{h}{2RT} P°(H_2O). \quad (5B)$$

where h is the mass transfer coefficient, D the gaseous diffusion coefficient, R the gas constant, T the temperature, and $P°(H_2O)$ is the water vapor pressure in the ambient gas stream. The water vapor pressure at the solid surface can be ignored since it is about three orders of magnitude lower than that in the gas stream as seen in Figure 1. The mass transfer coefficient for the thin disk geometry used in this study is given by[8,9,10]:

$$h = D/d\ (2.55 + 1.478\ Re^{1/2}\ Sc^{1/3}) \quad (6)$$

where
d = disk diameter parallel to flow
Re = $(\Delta v)/v$, the Reynold's number
Sc = v/D, the Schmidt coefficient
v = kinematic viscosity
v = velocity of bulk gas stream

and D = $D_{H_2O-H_2}$ = interdiffusion coefficient for H_2O and H_2.

The first term in Eq. 6 is for diffusion from a finite thin disk into an infinite stagnant medium and was calculated[10] from an appropriate solution to Fick's second law[11]. As seen in Figure 3, at low water vapor pressures, the rate of corrosion is almost directly proportional to the ambient water vapor pressure as predicted by Eq. (5B).

The Reynold's number is proportional to the gas velocity, Eq. (6). Therefore, at constant water vapor pressure and temperature, the mass transfer coefficient depends on the gas velocity as:

$$h = A + B\ v^{1/2} \quad (7)$$

where A, B are constants. A plot of the corrosion rate versus $v^{1/2}$ at different temperatures is shown in Figure 4. The linear relationship between the corrosion rate and the square root of the gas velocity is apparent.

The temperature dependence of the corrosion rate provides additional insight into the corrosion mechanism. At constant water vapor pressure and gas velocity, the corrosion rate is a function of temperature only. The only temperature dependent term in Eq. (6) is the gaseous interdiffusion coefficient for two gases, A and B, which can be calculated from the Chapman-Enskog equation[9]:

$$D_{AB} = 1.85 \times 10^{-3} (1/M_A + 1/M_B)^{1/2} T^{3/2}/(P \, r_{AB}^2 \, X_{D,AB}) \quad (8)$$

where M_A, M_B = molecular weights of A and B
r_{AB} = collision diameter
$X_{D,AB}$ = collision integral
P = pressure.

The values of r_{AB} and $X_{D,AB}$ are material properties and are independent of temperature. Therefore, when the water vapor pressure and the gas velocity are fixed, the diffusion coefficient is proportional to $T^{3/2}$. Consequently, from Eqs. (5) and (8), the corrosion rate is proportional to $T^{1/2}$. The experimental corrosion rate versus $T^{1/2}$ at different water vapor pressures is indeed linear as shown in Figure 6.

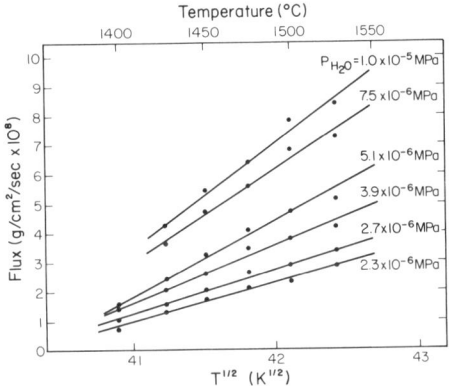

Fig. 6. Corrosion flux density as a function of $T^{1/2}$ at low water vapor partial pressures.

Finally, the experimental corrosion rates at 1527°C are compared with rates calculated from Eqs. (5), (6), and (8) in Figure 7. The agreement between the experimental data and theoretical predictions is excellent.

The thermodynamic calculations, the experimental corrosion rates as a function of water vapor pressure, gas velocity, and temperature, and the agreement between experimental and theoretical rates, strongly suggest that the corrosion of SiC in H_2 with a low H_2O content is controlled by transport of water vapor to the surface.

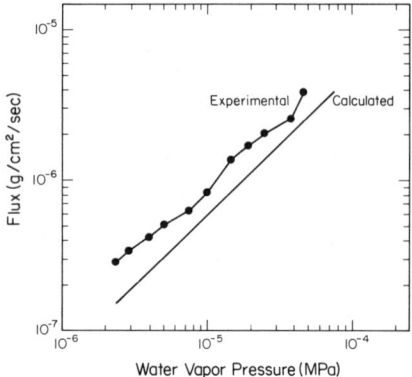

Fig. 7. Experimental versus calculated corrosion rates at 1527°C.

Figure 8 shows the evolution of the surface morphology of SiC reacted at low water vapor pressure at 1400°C. These micrographs show pits formed in the surface, some preferential grain boundary attack, and hillocks.

Corrosion at High Water Vapor Pressures

According to the model[12,13,14] developed by Wagner[12], active oxidation continues until the partial pressures of water vapor at the SiC surface becomes high enough for SiO_2 to be formed:

$$SiC_{(s)} + 3 H_2O_{(g)} = SiO_{2(s)} + CO_{(g)} + 3 H_{2(g)} \quad (9)$$

If SiO_2 forms a discontinuous layer on the SiC surface, active oxidation will continue. However, the rate of reaction will now decrease with water vapor pressure since continued corrosion depends on the reaction:

$$SiO_{2(s)} + H_{2(g)} = SiO_{(g)} + H_2O_{(g)} \quad (10)$$

For this reaction:

$$P(SiO)\ P(H_2O) = K_{eq} = \exp(-\Delta G°/RT) \quad (11)$$

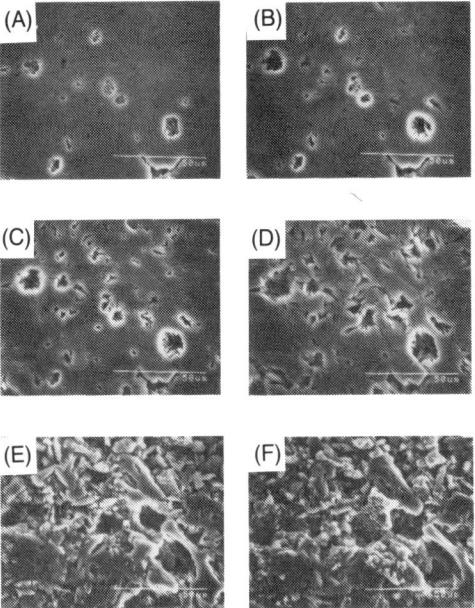

Fig. 8. Change is surface morphology of SiC reacted at 1400°C and P(H$_2$O) = 4.0 × 10^{-6} MPa. (A) before reaction, (B) 1 hour, (C) 2 hours, (D) 3 hours, (E) 5 hours, (F) 7 hours.

where K_e is the equilibrium constant and $\Delta G°$ is the standard free energy for the reaction of Eq. (10). Therefore, the corrosion rate of SiC above the transition water vapor pressure is determined by diffusion of the product gases which can be expressed as follows:

$$J(SiC) = h(SiO) P(SiO)/RT$$
$$= \frac{h(SiO)}{P(H_2O) RT} \exp(\Delta S°/R)\exp(-\Delta H°/RT) \quad (12)$$

where h(SiO) is now the effective mass transfer coefficient for the product gas species. According to Eq. (12), the corrosion rate is now inversely proportional to the ambient P(H$_2$O). The corrosion rate was observed experimentally to decrease with P(H$_2$O) as shown in Figure 3. After the transition, the corrosion rate is still a function of the gas velocity since mass transfer coefficient, h(SiO) is similar to that in Eq. (7). The only difference is that the effective diffusion coefficient is now[10]:

$$D = (D_{SiO-H_2} D_{H_2O-H_2})^{1/2} \tag{13}$$

The plot of the corrosion rate versus square root of the gas velocity at high water vapor pressures in Figure 5 is again linear indicating mass transport control. However, in this case, Eq. (12) predicts an exponential temperature dependence with the apparent activation energy being equal to the standard enthalpy for Eq. (10). The temperature dependence of the rate is plotted in Figure 9 and the calculated activation energy is 503 kJ/mol. This is very close to that calculated from thermodynamic data, 532 kJ/mol[7].

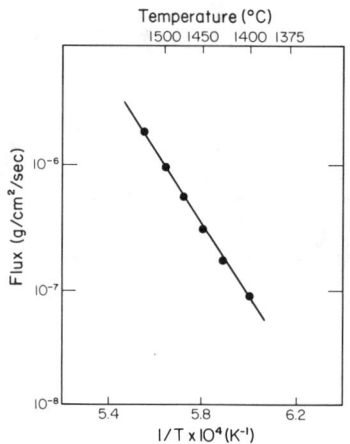

Fig. 9. Corrosion rate as a function of temperature at a water vapor pressure of $P(H_2O) = 1.0 \times 10^{-4}$ MPa.

The dependence of the corrosion rate on the experimental variables, i.e. water vapor pressure, gas velocity, and temperature, at water vapor pressures above the maximum in the corrosion rate, suggests that SiO_2 is present on the surface and its reduction controls the rate of continued active corrosion. Figure 10 shows the development of the surface morphology during corrosion at 1400°C. These micrographs indicate the presence of a "fuzzy" corrosion product, particularly for long corrosion times. Figure 10E shows that this corrosion product dissolves in HF suggesting that it is silica. This microscopic evidence corroborates the presence of SiO_2 at water vapor pressures beyond the maximum in corrosion rate.

CONCLUSIONS

The corrosion of SiC in hydrogen at temperatures between 1400°C and 1527°C in water vapor pressures between 10^{-6} and 10^{-3} MPa occurs by active oxidation of the SiC by the water vapor with the formation of SiO and CO. At low water vapor

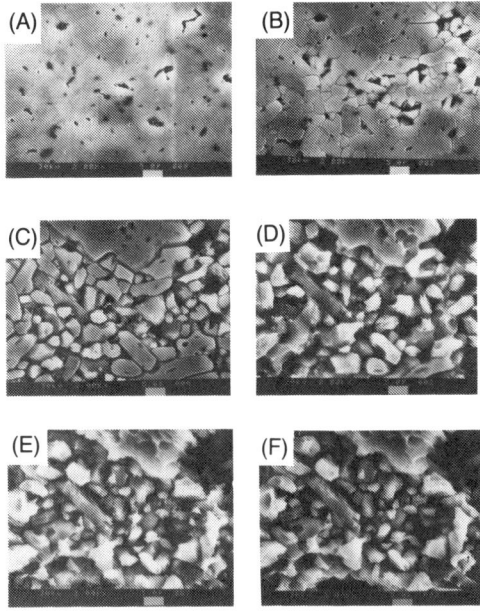

Fig. 10. Evolution of surface morphology of SiC reacted at 1450°C, $P(H_2O) = 1.0 \times 10^{-4}$ MPa. (A) before reaction, (B) 1/2 hour, (C) 1 hour, (D) 2 hours, (E) 3 hours, (F) 3 hours and HF etched.

pressures, transport of H_2O from the gas stream to the surface controls the reaction rate. At high water vapor pressures, SiO_2 is present as discrete particles on the surface and continued active oxidation occurs with the concurrent reduction of this SiO_2.

Acknowledgements

Based on the dissertation submitted by H. Kim for the Ph.D. degree, Department of Ceramic Engineering, The Ohio State University, 1987.

Supported by the National Science Foundation under grant numbers DMR-8410690 and DMR-8607565.

---

*Now with Oak Ridge National Laboratory, Oak Ridge, TN
‡Alpha Silicon carbide, "Hexoloy," Sohio, Niagara Falls, NY
†Model 2000, Cahn Instruments, Cerritos, CA
§Rapid Temp, Bloomfield, NJ
¶Deoxo, Fisher Scientific, Beachwood, OH
**Model 700, Panametrics, Inc., Waltham, MA
††Model ISI-SX-40, International Scientific Instruments, Inc., Santa Clara, CA

References

[1] J. J. Burke, A. E. Gorum and R. N. Katz (eds.), Ceramics for High Performance Applications, Brookhill Publishing Co., 1974.

[2] J. J. Burke, E. N. Lenoe and R. N. Katz (eds.), Ceramics for High Performance Applications II, Brookhill Publishing Co., 1977.

[3] F. L. Riley (ed.), Progress in Nitrogen Ceramics, Martinus Nijhoff Publishers, 1981.

[4] J. A. Costello and R. E. Tressler, "Oxidation Kinetics of Silicon Carbide Crystals and Ceramics: 1, In Dry Oxygen," *J. Amer. Ceram. Soc.*, **69** [9] 674 (1986).

[5] J. W. Hinze and H. C. Graham, "The Active Oxidation of Si and SiC in the Viscous Gas-Flow Regime," *J. Electrochemical Soc.*, **123** [7] 1066 (1976).

[6] T. Bessman, "SOLGASMIX-PV, A Computer Program to Calculate Relationships in Complex Chemical Systems," Report No. PRNL/RM-5775, Oak Ridge National Laboratories, Oak Ridge TN., 1977.

[7] D. R. Stull and H. Prophet, JANAF Thermochemical Tables, National Bureau of Standards, 1971.

[8] W. M. Rohsenow and H. Choi, Heat, Mass and Momentum Transfer, p. 148, Prentice-Hall, 1961.

[9] C. J. Geankoplis, Mass Transport Phenomena, p. 25, Ohio State University, 1972.

[10] H. E. Kim and D. W. Readey, to be published.

[11] H. S. Carslaw and J. C. Jaeger, Conduction of Heat in Solids, p. 215, Oxford University Press, Oxford, 1959.

[12] C. Wagner, "Passivity During the Oxidation of Silicon at Elevated Temperatures," *J. Appl. Phys.*, **29** [9] 1295 (1958).

[13] E. A. Gulbransen and S. A. Jansson, "The High Temperature Oxidation, Reduction, and Volatilization Reactions of Silicon and Silicon Carbide," *Oxid. Metals*, **4** [3] 181 (1972).

[14] J. E. Antill and B. Warburton, "Active to Passive Transition in the Oxidation of SiC," *Corrosion Sci.*, **11** 337 (1971).

Thermal Expansion and Elastic Anisotropies of SiC as Related to Polytype Structure

Z. Li and R. C. Bradt

Dept. of Materials Science and Engineering
University of Washington
Seattle, WA 98195

Abstract

The structural relationships between the polytypes of SiC are such that the anisotropic thermal expansion coefficients (second order tensors) and single crystal elastic constants (fourth order tensors) can be systematically related to the structural stacking layer sequence. In this paper, the concept of the fraction of hexagonal stacking is generally applied to describe the anisotropic thermal expansion coefficients. The single crystal elastic anisotropy for the SiC polytype structures and the temperature dependencies of the anisotropies are also addressed.

Pertinent to transient thermal stresses and to processing related internal micromechanical residual stresses, the concept of an anisotropic thermoelastic stress index is also presented. Its anisotropy is also graphically illustrated for the (3C) and the (6H) SiC polytypes. In addition to its general thermoelastic stress utility, it is also demonstrated to be useful for predicting the most desirable crystal (whisker) growth orientations for SiC whisker incorporation into composite matrices.

Introduction

Association of the principal axial coefficients of thermal expansion and the single crystal elastic constants of crystalline solids is quite logical, for both of these physical properties describe dimensional changes of structural elements and are related through the Gruneisen tensor. In the case of the thermal expansion, dimensional changes are described as affected by changes in temperature, while the elastic constants address dimensional changes with applied stress or pressure. It is only natural to relate those physical properties for engineering applications where the processes involve variations of both temperature and stress, such as the development of thermoelastic stresses. As many applications of SiC are at elevated temperatures and those involve temperature changes, it is appropriate to consider the thermal expansion coefficients, the elastic constants and finally their product, the thermoelastic stress index for SiC at different temperatures in its various structural forms.

As the thermal expansion coefficients are second order (rank) tensors, the principal axial coefficients for the common structural forms of SiC are α_{11} for the cubic structure, α_{11} and α_{33} for the hexagonal polytypes and also an α_{11} and α_{33} for the rhombohedral ones. The single crystal elastic stiffnesses are fourth order (rank) tensors and these require more terms. Included are C_{11}, C_{12}, and C_{44} for the cubic structure, C_{11}, C_{12}, C_{13}, C_{33}, and C_{44} for the hexagonal polytypes and C_{11}, C_{12},

C_{13}, C_{14}, C_{33}, and C_{44} for the rhombohedral ones. These elastic stiffness are easily converted to the elastic compliances, S_{ijkl} for each structure by standard calculations.

The generalized Gruneisen relation can be applied to relate the two sets of physical properties[1] as:

$$\gamma_{ij} = \frac{V}{C_p} \Sigma \alpha_{kl}\, C_{ijkl}, \qquad k,l = 1,2,3, \qquad (1a)$$

or in its inverted form as:

$$\alpha_{ij} = \frac{C_p}{V} \Sigma \gamma_{kl}\, S_{ijkl}, \qquad (1b)$$

where C_p is the heat capacity and V is the molar volume of the crystal. For the cubic structure the general Gruneisen relation relates the thermal expansion coefficient to the elastic compliances as:

$$\alpha_{11} = \frac{C_p}{V} \gamma (S_{11} + 2S_{12}). \qquad (2)$$

For the hexagonal and rhombohedral structures Equation 1b reduces to:

$$\alpha_{11} = \frac{C_p}{V} [\gamma_1 S_{11} + S_{12}) + \gamma_3\, S_{13}], \qquad (3a)$$

and

$$\alpha_{33} = \frac{C_p}{V} [\gamma_1 2S_{13} + \gamma_3\, S_{33}]. \qquad (3b)$$

SiC Structures

Silicon carbide, SiC, exists in many different structures which can be described on the basis of a covalently bonded tetrahedrally coordinated polyhedron, either (SiC_4) or(CSi_4). These tetrahedra form basic layer planes which are successively stacked in different orderly sequences to yield the various polytypes. The stacking layer sequence for each of the numerous SiC polytypes can be conveniently described as a mixture of the simple cubic sequence (ABC), C, and the simple hexagonal one (AB), H. These two closed packing sequences are similar to those for zinc sulphide, ZnS, in the sphalerite (cubic) and wurtzite (hexagonal) crystal structures, respectively[2].

An infinite number of combinations of these C and H stacking layers sequences probably exists, as nearly one hundred different polytypes have already been identified[3]. However, among the many, only five short-period polytypes are considered to be thermodynamically stable. These include the one cubic structure with the space group F$\bar{4}$3m, the (3C), the hexagonal structures with P6mc, (2H), (4H), and (6H) and the rhombohedral structure with R3m, the (15R). The (3C) cubic form is frequently referred to as beta SiC. It is stable at temperatures below about 1600°C. The various hexagonal and rhombohedral polytypes are collectively known as alpha SiC structures and are stable at elevated temperatures, above 2000°C. The layer stacking sequences for the (2H), (3C), (4H), (6H) and (15R) polytypes are (HH), (CCC), (CHCH), (CCHCCH) and (CCHCH)$_3$, respectively.

The non-hexagonal structures, that is the cubic and rhombohedral SiC polytypes can also be described in terms of hexagonal "pseudocells". The two "a"-axes of all of the hexagonal "pseudocells" always lie within the same layer plane, while the "c"-axis is always perpendicular to the layer planes. For the cubic beta structure, the (3C), the hexagonal "pseudocell" "c"-axis is parallel to the cubic $\langle 111 \rangle$ direction and the two "a"-axes are parallel to $\langle 110 \rangle$ directions. For the rhombohedral structure the hexagonal "pseudocell" "c"-axis is parallel to the rhombohedral $\langle 111 \rangle$ direction and the two "a"-axes are parallel to $\langle 1\bar{1}0 \rangle$ directions. This latter hexagonal convention is usually applied in X-ray diffraction technique to describe the rhombohedral crystal indices[4].

THERMAL EXPANSION AND THERMAL EXPANSION ANISOTROPY OF SiC

The principal axial thermal expansion coefficients have been accurately measured for only three of the many different polytypes of SiC, the (3C), (4H), and (6H). For these Structures the α_{ij} have been determined from room temperature to 1000°C by the high temperature X-ray diffraction technique[5-7]. Figure 1 illustrates the thermal expansion coefficients for those three SiC polytypes. It is evident that the thermal expansion coefficient of the (3C) cubic polytype of SiC, α_{11}, is identical to the α_{ij} values within the basal planes of the two hexagonal SiC structures, the (4H) and (6H) polytypes, α_{11}(4H) and α_{11}(6H). The principal axial thermal expansion coefficients of the (4H) and (6H) polytypes perpendicular to their basal planes, α_{33}(4H) and α_{33}(6H) are both lower than α_{11}(3C). These values decrease in the natural order, α_{33}(6H) > α_{33}(4H), an order which is directly related to the structure of these SiC polytypes, specifically the stacking layer sequence along the "c"-axes.

Since all of the SiC polytypes consist of a mixture of the cubic, C, and the hexagonal, H, stacking layer sequences, each polytype can be distinguished by the percentage of its hexagonal stacking layer sequence. Naturally, one could also describe the polytypes on the basis of their percentage of cubic stacking layer sequence. Table 1 lists some of the common SiC polytypes, their various notations and their fraction of hexagonal stacking layer sequence, F^h. A comparison of Table 1 with Figure 1 reveals that the higher is the F^h value of the hexagonal polytype, the lower is the thermal expansion coefficient along the "c"-axis, α_{33}. The relationship of the thermal expansion coefficients perpendicular to the basal plane, α_{33} to F^h for the different SiC polytypes has been expressed as[8]:

$$\alpha_{33}(?H) = \alpha_c (1-F^h) + \alpha_h F^h, \qquad (4)$$

where the α_c is the thermal expansion coefficient of the pure cubic polytype (3C) and α_h is the thermal expansion coefficient of the pure hexagonal polytype (2H)

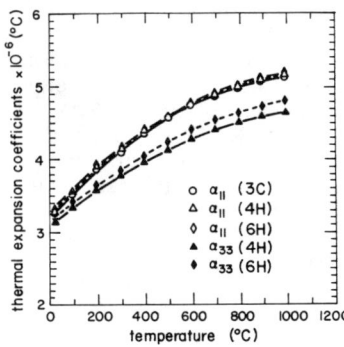

Fig. 1. The principal axial coefficients of thermal expansion for the (3C), (4H) and (6H) SiC polytypes.

Table 1. Notations for the common polytypes of SiC*.

Ramsdell	Wells	Wyckoff	F^h	N
3C β-SiC	a	CCC	0	3
6H SiC II	aaabbb	CCHCCH	0.33	6
15R SiC I	(aaabb)$_3$	(CCHCH)$_3$	0.40	15
4H SiC III	aabb	CHCH	0.50	4
2H SiC	ab	HH	1	2

*F^h is the fraction of hexagonal stacking layers, and N is the number of layers per unit cell.

perpendicular to the basal plane, which can be determined from analysis of the α_{33} of (4H) and (6H) measurements. The average characteristic thermal expansion values are equal to $4.45 \times 10^{-6}/°C$ and $3.63 \times 10^{-6}/°C$ for α_c and α_h, respectively, over the temperature range from room temperature to 1000°C. The F^h quantity is the fraction of hexagonal stacking sequence layers in the polytype structure, which varies from 0 to 1. Since the coefficients of thermal expansion of the rhombohedral polytypes of SiC can be readily expressed in terms of the "a"-axes and "c"-axes of their hexagonal pseudocells, rather than the less familiar rhombohedral ones, they also may be expressed within the same format as those of the hexagonal polytypes on the basis of their appropriate hexagonal "pseudocells".

From the previous analyses of the principal axial coefficients of thermal expansion it is then possible to present the α_{ii} for all of the polytypes in the usual second order polynomial format[8]. The α_{11} are the same for all of the polytypes and can be expressed as:

$$\alpha_{11} = 3.19 \times 10^{-6} + 3.60 \times 10^{-9}T - 1.68 \times 10^{-12}T^2 \ (1/°C). \tag{5a}$$

However, the α_{33} values vary with stacking layer sequence. For the first four hexagonal polytypes the α_{33} values are:

$$\alpha_{33}^{2H} = 2.99 \times 10^{-6} + 1.32 \times 10^{-9}T - 1.87 \times 10^{-13}T^2 \ (1/°C), \tag{5b}$$

$$\alpha_{33}^{4H} = 3.09 \times 10^{-6} + 2.63 \times 10^{-9}T - 1.08 \times 10^{-12}T^2 \ (1/°C), \tag{5c}$$

$$\alpha_{33}^{6H} = 3.18 \times 10^{-6} + 2.48 \times 10^{-9}T - 8.51 \times 10^{-13}T^2 \ (1/°C), \tag{5d}$$

and

$$\alpha_{33}^{8H} = 3.14 \times 10^{-6} + 3.03 \times 10^{-9}T - 1.31 \times 10^{-12}T^2 \ (1/°C). \tag{5e}$$

The α_{11} is identical to the α_c value in Equation 4 and the value for α_{33}^{2H} is that of α_h.

The thermal expansion and the thermal expansion anisotropy of crystals depends on two structural factors[9]: (i) the strengths of the bonds within each of the structural polyhedra, which directly relate to the nearest neighbor effects, and (ii) the angular changes between the polyhedra, or the structural tilting, which may relate to the second, third,... etc, nearest neighbor effects. As illustrated by Megaw[9], the coefficients of thermal expansion for all crystalline structures may be considered to consist of some combination of these two factors.

Hazen and Finger[10] have presented a relationship which was derived on the basis of the strengths of the bonds of crystal polyhedra and which adequately expresses the mean linear thermal expansion coefficients of many different crystal polyhedra between room temperature and 1000°C. That equation is:

$$\bar{\alpha}_{1000°C} = 4.0(4) \frac{N}{S^2 Z_a Z_c} \times 10^{-6}/°C, \tag{6}$$

where $\bar{\alpha}_{1000°C}$ is the mean thermal expansion coefficient between room temperature and 1000°C, N is the structure's coordination number, Z_a and Z_c are the cation and anion valencies, respectively, and S^2 is a factor which is 0.20 for carbides and nitrides. For SiC the N, Z_a and Z_c are each four. Substituting these into Equation 6, yields an $\bar{\alpha}_{1000°C}$ value for the coefficient of thermal expansion of the (CSi_4) or (SiC_4) tetrahedron of about $5 \times 10^{-6}/°C$. This value is in reasonably good agreement with all of the average experimental values expressed by Equations (5a-5e).

This theoretical coefficient of thermal expansion for the SiC tetrahedron is only related to the nearest neighbor atoms. As it is slightly larger than the experimental measurements, this suggests the presence of bond angle and polyhedral tilting contributions, or perhaps even further removed bonding effects which reduce the overall thermal expansion for each of the SiC structures. In addition it should be noted that the differences from the theoretical polyhedral thermal expansion coefficient as expressed by Equation 6 are not the same along the a-axes and c-axes of the various polytypes. This further confirms that it is the differences in the C and the H stacking layer sequences which directly determine the thermal expansion anisotropy of the SiC polytypes.

As previously mentioned, the two basic stacking layer sequences, C and H, of SiC are analogous to the stacking layers of the sphalerite (3C) and wurtzite (2H) structures of ZnS. The crystal energies of these structures are proportional to their Madelung constants, which are 1.638 and 1.641 for the sphalerite and wurtzite structures, respectively[11]. This difference in the Madelung constants for these two structures results from the second nearest neighbor atomic positions along the "c"-axes as the strengths of the bonds along the "a"-axes for these structures are the same, while the strengths of the bonds along the "c"-axis in the wurtzite structure, pure H stacking, are slightly higher than for the sphalerite structure, pure C stacking.

The aforementioned arguments can be translated to the principal axial coefficients of thermal expansion. The α_{11} values, for these two structures should be identical; however, the α_{33} of the wurtzite structure will be lower than the coefficient of thermal expansion perpendicular to the layers in the cubic sphalerite structure. Furthermore, since the thermal expansion of the sphalerite structure is isotropic, the α_{33} coefficient of thermal expansion for the wurtzite structure should then be somewhat lower than the α_{11} value. Thermal expansion coefficient measurements for a number of wurtzite structures, including BeO, AlN and ZnO all confirm that the principal axial coefficients of thermal expansion parallel to the "c"-axes, the α_{33} values, are lower than along the "a"-axes, the α_{11} values. This discussion and their results further substantiate the approach of the derivation of Equation 4. For the α_{33} of the different SiC polytypes, the greater the percentage of the hexagonal stacking layer sequence, the lower the thermal expansion coefficient.

The coefficient of thermal expansion for a single crystal in a given crystallographic direction, α, can be expressed in the form:

$$\alpha = \Sigma \alpha_{ij} \cos^2 \theta_i \qquad i=1,2,3, \qquad (7)$$

where θ_i is the angle between the direction of interest and the principal crystallographic axes. For a cubic crystal, there is only one principal thermal expansion coefficient because of its isotropy. For cubic crystals the α is equal to α_{11} in all crystallographic directions. Figure 2a is a polar plot which illustrates the thermal expansion coefficients of cubic SiC(3C) on the (110) plane as a function of orientation at the different temperatures: RT, 500°C and 1000°C. It is evident that although α is independent of the crystallographic orientation, it is highly temperature dependent. With increasing temperature the α of cubic (3C) SiC increases rapidly from 3.3×10^{-6}/°C at room temperature to 4.6×10^{-6}/°C at 500°C. The $(d\alpha/dT)$ then slightly decreases to yield a thermal expansion coefficient of about 5.1×10^{-6}/°C at 1000°C.

For hexagonal crystals, there are two principal axial thermal expansion coefficients, α_{11} and α_{33}, for which Equation 7 reduces to:

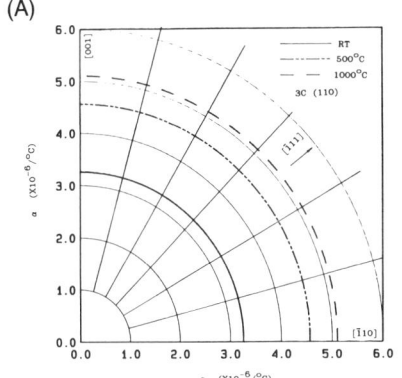 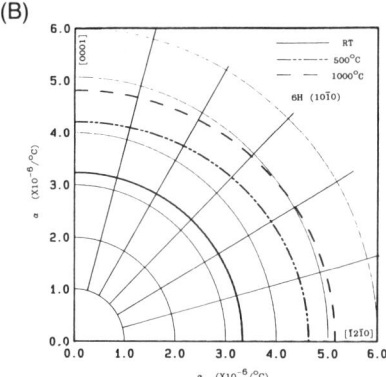

Fig. 2. (A) The coefficients of thermal expansion, α_{ij}, of the cubic (3C) polytype of SiC as a function of orientation on the (110) and (B) the coefficients of thermal expansion, α_{ij}, of the hexagonal (6H) polytype of SiC as a function of orientation on the (10$\bar{1}$0).

$$\alpha = \alpha_{11} + (\alpha_{33} - \alpha_{11})\cos^2\theta, \qquad (8)$$

where θ is the angle between the direction of interest and the hexagonal "c"-axis. Therefore, the thermal expansion for all hexagonal crystals is transversely isotropic with regard to the "c"-axis, but changes for other crystallographic directions as illustrated in Figure 2b. Figure 2b shows the α for the hexagonal (6H) polytype of SiC as a function of orientation on the (10$\bar{1}$0) plane at different temperatures. The profile of α becomes an ellipse due to the thermal expansion anisotropy. The α has a maximum value in the [$\bar{1}$2$\bar{1}$0] direction, α_{11}, and a minimum value in the [0001] direction, α_{33}. Similar to the α for the cubic (3C) polytype of SiC the α of hexagonal SiC(6H) is also highly temperature dependent. The α_{11} and α_{33} are 3.3 × 10^{-6}/°C and 3.2 × 10^{-6}/°C at RT, but increase to 5.2 × 10^{-6}/°C and 4.8 × 10^{-6}/°C at 1000°C, respectively. The $\Delta\alpha$, equal to $(\alpha_{11}-\alpha_{33})$, also increases with increasing temperature from 0.1 × 10^{-6}/°C at 20°C to 0.4 × 10^{-6}/°C at 1000°C.

From both Figures 2a and 2b, it is evident that the effect of temperature on the thermal expansion coefficients of SiC is much greater than the effect of the crystal orientation. However, the single crystal thermal expansion anisotropy of the hexagonal polytypes is by no means insignificant. One definition of the thermal expansion anisotropy for hexagonal crystals is the following expression:

$$"A_t" = (\alpha_{11} - \alpha_{33})/\alpha_{33} = (\alpha_{11}/\alpha_{33}) - 1, \qquad (9)$$

where "A_t" is then considered to be a indication of the thermal expansion anisotropy of the crystal structure.

The thermal expansion anisotropy, "A_t", for the different SiC polytypes is directly related to the temperature and the stacking layer sequence, as might be surmised by the α_{ij} and their temperature dependencies as previously expressed in polynomial form in Equations 5a-e. Figure 3 shows that variation of "A_t" with temperature for four hexagonal polytypes. The solid lines in Figure 3 are "A_t" for both the (4H) and (6H) polytypes as calculated from the experimental measurements, while the dashed lines are the "A_t" vs T curves calculated for the (2H) and (8H) polytypes derived from Equation 4. A series of curves of "A_t" values exists, where the "A_t" of the (2H) polytype is the largest.

Figure 4 depicts the thermal expansion anisotropy, "A_t", versus the fraction hexagonal stacking, F^h, for the different polytypes of SiC at the apparent "A_t" maximum, about 800°C, which is near the reported Debye temperature for SiC, 807°C. It is evident that the more cubic is the stacking layer sequence, the less anisotropic the polytype structure is in its thermal expansion. Patrick[12] has applied similar concepts related to Raman spectroscopy to address the anisotropy of the transverse optical phonon modes ($TO_1 - TO_2$) for different SiC polytypes. Patrick's results are also plotted as the solid line in Figure 4, and reveals the same trend as the one for the maximum in the thermal expansion anisotropy,"A_t".

Thermal expansion anisotropy is related to the internal microstresses in polycrystalline ceramic materials when its effect is combined with that of the elastic anisotropy[13]. For sufficiently large values of "A_t", the spontaneous internal microcracking of polycrystalline ceramics will occur for all grain sizes above a critical grain size, as was demonstrated in the pseudobrookite structure by Kuszyk and Bradt[14]. For the two commonly utilized monolithic polycrystalline ceramics SiC(6H) and Al_2O_3, the "A_t" values between 20°C and 1000°C are 0.06 and 0.15, respectively. Since the thermal expansion anisotropy of SiC is much less than that of Al_2O_3, only about one third, and the elastic constants of the two materials are nearly the same, the magnitudes of the internal residual stresses in the two materials will also compare similarly after cooling from 1000°C and be much less for the SiC.

SINGLE CRYSTAL ELASTIC CONSTANTS OF SIC

The elastic constants of single crystals are fourth order tensors and thus are crystallographically quite different from the thermal expansion coefficients which are only second order tensors. Elastic properties are anisotropic even for cubic crystal structures. Experimental determinations of the single crystal C_{ij} are difficult, consequently the temperature dependencies of the single crystal elastic constants, the dC_{ij}/dT, for SiC have never been measured. However, the single crystal elastic constants, the C_{ij}, have been estimated at elevated temperatures for the (3C) cubic SiC structure from the polycrystalline beta SiC elastic moduli. By applying a tensor transformation method, the C_{ij} at elevated temperatures for the hexagonal SiC structure were also calculated[15,16].

There are three independent elastic constants for cubic single crystals: C_{11}, C_{12} and C_{44}. For the cubic (3C) polytype of SiC, which has $F\overline{4}3m$ symmetry, these three single crystal elastic constants have been reported by Tolpygo[17] to be 352.3, 140.4, and 232.9 GPa, respectively, at room temperature. Since many polycrystalline ceramic materials are macroscopically isotropic and consist of randomly oriented grains or crystals, the description of their elastic properties requires only two independent elastic constants, including either Young's modulus

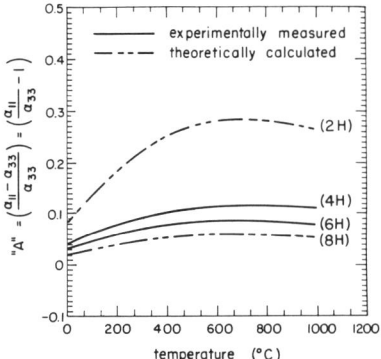

Fig. 3. The anisotropies of the coefficients of thermal expansion, "A", of the hexagonal SiC polytypes as a function of temperature.

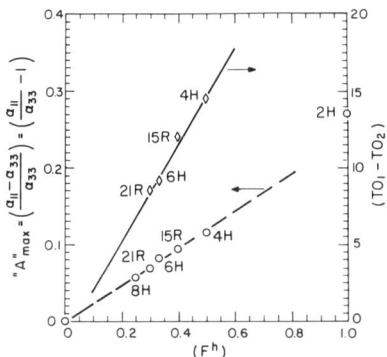

Fig. 4. The anisotropies of the coefficients of thermal expansion, "A", and the transverse optical phonon modes ($TO_1 - TO_2$) of SiC polytypes as a function of hexagonal stacking fraction (F^h).

(E), the shear modulus (G), the bulk modulus (K) or Poisson's ratio (v). Relationships between the single crystal elastic constants and the polycrystalline ceramic elastic moduli can be expressed by the Voigt averages[1] as:

$$E_V = \frac{(C_{11} - C_{12} + 3C_{44})(C_{11} + 2C_{12})}{2C_{11} + 3C_{12} + C_{44}}, \quad (10)$$

Silicon Carbide

$$G_V = (C_{11} - C_{12} + 3C_{44})/5, \tag{11}$$

and

$$K_V = \frac{(C_{11} + 2C_{12})}{3} \tag{12}$$

The temperature dependencies of the elastic constants of single crystal (3C) SiC, the dC_{ij}/dT, have not been measured. However, the elastic moduli of high density beta SiC polycrystalline ceramics have been measured by several different researchers[18,19]. These experimental results can be expressed in the form:

$$E = E_0 + bT, \tag{13}$$

and

$$G = G_0 + b'T, \tag{14}$$

where E_0 and G_0 are the 0°C values. The constants, E_0 and G_0 may be dependent upon the material's microstructure, such as the density, porosity, grain size, etc. and are different for each reported study. However, the values of b and b' are identical and equal to -0.020 and -0.007 GPa/°C, respectively. The calculated value of (dK/dT), utilizing K=f(E,G), is -0.016 GPa/°C.

The temperature dependencies of the polycrystalline elastic moduli which are calculated from the single crystal elastic stiffnesses, the C_{ij}, must be equal to the polycrystalline experimental values. However, since the single crystal elastic constants exist as three independent parameters, it is necessary to develop additional relationships between the temperature dependencies of the single-crystal elastic constants and those of the polycrystalline ceramic elastic moduli. This can be done by initially considering many other cubic crystal structures. The elastic constants, E_V, G_V, C_{11} and C_{44}, for a number of different cubic crystal structure materials[20] have been determined as a function of temperature and are summarized in Table 2.

Table 2. Temperature derivatives of the elastic constants of some cubic crystals.

Crystal	(dC_{11}/dT)	(dE/dT)	(dC_{44}/dT)	(dG/dT)	T (°C)
CaF$_2$	-.039	-.031	-.012	-.012	20-420
LiF	-.091	-.071	-.020	-.022	20-800
CsBr	-.017	-.018	-.009	-.007	20-450
SbIn	-.019	-.028	-.003	-.003	27-327
Al	-.069	-.041	-.015	-.015	20-500
Au	-.035	-.028	-.013	-.011	27-527
Ge	-.014	-.015	-.007	-.007	20-300
Ti	-.022	-.028	-.013	-.011	25-500

These results reveal that most of the values of (dC_{44}/dT) and (dG/dT) for cubic materials are nearly identical and are in fact much smaller than the values of (dE/dT). On this basis, an initial assumption can be made that the temperature dependencies of C_{44} and G, namely (dC_{44}/dT) and (dG/dT) for the (3C) cubic polytype of SiC are also identical.

In combination with this assumption to estimate C_{11} and C_{12} at elevated temperatures, Equations 11 and 12 can be differentiated to yield:

$$\frac{dG}{dT} = \frac{1}{5}\left(\frac{dC_{11}}{dT}\right) - \frac{1}{5}\left(\frac{dC_{12}}{dT}\right) + \frac{3}{5}\left(\frac{dC_{44}}{dT}\right), \tag{15}$$

and

$$\frac{dK}{dT} = \frac{1}{3}\left(\frac{dC_{11}}{dT}\right) + \frac{2}{3}\left(\frac{dC_{12}}{dT}\right), \tag{16}$$

yielding two equations and two unknowns. Both (dC_{11}/dT) and (dC_{12}/dT) can be easily determined utilizing the above equations. Those values are equal to -0.025 and -0.011 GPa/°C, respectively. Figure 5 illustrates the C_{ij} values of cubic (3C) SiC at elevated temperatures. Substituting the above C_{ij} values into Equation 10, the Voigt average for the polycrystalline Young's modulus versus temperature can be determined and then the temperature derivative can be calculated. The (dE/dT) value is -0.020 GPa/°C, which is identical to the experimental polycrystalline values as illustrated in Figure 6. The consistency of this internal check confirms that the above assumptions and calculation procedures are satisfactory for determining the (dC_{ij}/dT) values at elevated temperatures.

The hexagonal crystal structures of SiC have P6mc symmetry and therefore require five independent single crystal elastic constants for complete description. These can be expressed as the elastic stiffnesses (C_{11}, C_{12}, C_{13}, C_{33}, and C_{44}). As previously discussed, the single crystal elastic constants are fourth order tensors and can be expressed in any arbitrary crystallographic direction by[21]:

$$C'_{ijkl} = \beta_{im}\beta_{jn}\beta_{ko}\beta_{lp}C_{mnop}, \tag{17}$$

where the β_{ij} are the directional cosines[‡]. Since a hexagonal "pseudocell" can be specified within any cubic structure, with the hexagonal [0001] direction parallel to the cubic [111], the [1$\bar{2}$10] parallel to the [$\bar{1}$10], and the [$\bar{1}$010] parallel to the [$\bar{1}$$\bar{1}$2], the single crystal elastic constants of the hexagonal structure[§] of SiC can be estimated by utilizing Equation (17) and the transformation matrix:

$$\begin{bmatrix} x'_1 \\ x'_2 \\ x'_3 \end{bmatrix} = \begin{bmatrix} \beta_{11} & \beta_{12} & \beta_{13} \\ \beta_{21} & \beta_{22} & \beta_{23} \\ \beta_{31} & \beta_{32} & \beta_{33} \end{bmatrix} \begin{bmatrix} x_1 \\ x_2 \\ x_3 \end{bmatrix} = \begin{bmatrix} 1/\sqrt{2} & -1/\sqrt{2} & 0 \\ -1/\sqrt{6} & -1/\sqrt{6} & 2/\sqrt{6} \\ 1/\sqrt{3} & 1/\sqrt{3} & 1/\sqrt{3} \end{bmatrix} \begin{bmatrix} x_1 \\ x_2 \\ x_3 \end{bmatrix} \tag{18}$$

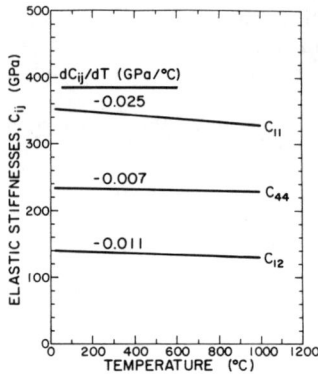

Fig. 5. Single crystal elastic stiffnesses of cubic SiC as a function of temperature.

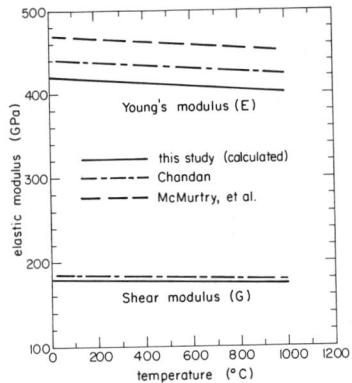

Fig. 6. Young's modulus and shear modulus of cubic polycrystal SiC as a function of temperature.

where the X_1', X_2' and X_3' are the axes of the hexagonal "pseudocell" along the $[\bar{1}10]$, the $[\bar{1}\bar{1}2]$ and the [111] cubic directions, respectively, and X_1, X_2 and X_3 are the axes of the cubic crystal along the [100], the [010] and the [001] directions. Substituting the appropriate β_{ij} values yields the following complete set of relationships between the hexagonal and the cubic single-crystal elastic stiffnesses as:

$$2c_{11}^h = c_{11}^c + c_{12}^c + 2c_{44}^c , \qquad (19a)$$

$$3c_{33}^h = c_{11}^c + 2c_{12}^c + 4c_{44}^c , \qquad (19b)$$

$$6c_{12}^h = c_{11}^c + 5c_{12}^c - 2c_{44}^c , \qquad (19c)$$

$$3c_{13}^h = c_{11}^c + 2c_{12}^c - 2c_{44}^c , \qquad (19d)$$

and

$$3c_{44}^h = c_{11}^c - c_{12}^c + c_{44}^c . \qquad (19e)$$

Equations (19b) and (19e) have also been demonstrated to be equal from measurements of sound velocities in cubic and hexagonal crystals of SiC[22].

Using the Tolpygo C_{ij} values for the cubic (3C) SiC and the above equations, the room temperature single crystal elastic stiffnesses of the hexagonal structure of SiC can be estimated. These calculations are summarized in Table 3 along with the experimental elastic stiffnesses which have been reported for the measurement of a mixed polytype, but predominantly (6H) hexagonal structure specimen of SiC by Arlt and Schodder[23]. Comparison of these results indicates a satisfactory agreement.

Table 3 also lists the experimental and the calculated elastic constants of ZnS, as ZnS also exists in the two similarly related cubic (3C) and hexagonal (2H) polytype structures. The single-crystal elastic constants of both the (3C) cubic structure (sphalerite) of ZnS and the (2H) hexagonal structure (wurtzite) of ZnS have been independently measured by different researchers[24,25]. From Table 3, it is evident that the transformed values are in excellent agreement with the reported experimental values. This further supports the validity of the above transformation procedures.

Similar to the previous discussion of the thermal expansion, which related to different stacking layer sequences, the elastic constants of SiC may also be expected to exhibit stacking layer sequence effects. However, the transformed elastic constants of the hexagonal (2H) ZnS from cubic ZnS(3C), for which F^h is equal to zero, are nearly identical to the experimental hexagonal ZnS(2H) values, for which F^h is equal to 1. This indicates that the stacking layer sequence effects on the elastic constants of ZnS are insignificant and can be similarly assumed negligible for SiC.

The temperature dependencies of the C_{ij} values for the hexagonal SiC polytypes can also be calculated from the set of Equations 19 simply by taking the derivatives with respect to temperature. The results are plotted in Figure 7. Using the elevated temperature C_{ij}, the polycrystalline hexagonal elastic moduli can be calculated and

Table 3. Room temperature single crystal elastic stiffnesses of ZnS and SiC (GPa).

	ZnS				
	C_{11}	C_{33}	C_{12}	C_{13}	C_{44}
Cubic Values[19]	104.6		65.3		46.1
Hexagonal Values[20]	124	140	60	45	286
Transformed Values*	131	140	56	48	285

	SiC				
	C_{11}	C_{33}	C_{12}	C_{13}	C_{44}
Cubic Values[9]	352.3		140.4		232.9
Hexagonal Values[18]	500	564	92		168
Transformed Values*	479	521	98	56	148

* After Equations (19a) to (19e).

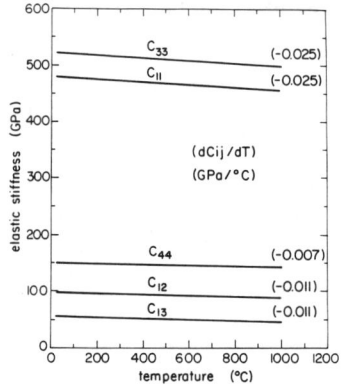

Fig. 7. Single crystal elastic stiffnesses of hexagonal SiC as a function of temperature.

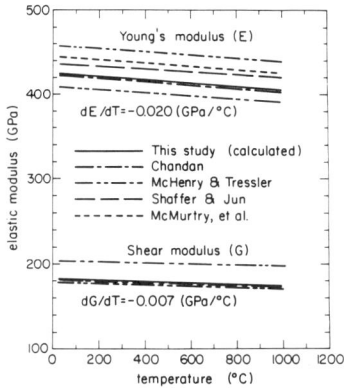

Fig. 8. Polycrystalline elastic moduli of hexagonal SiC as a function of temperature.

these results are compared with published experimental SiC values[18–19,26–27] as illustrated in Figure 8. The results are in good agreement.

As previously discussed, the elastic properties of all single crystals are anisotropic. A convenient method to describe the elastic anisotropy of crystalline solids is to consider the orientation dependence of the various elastic moduli such as Young's modulus, the elastic shear modulus and Poisson's ratio[28]. The Young's modulus is defined the ratio of the normal stress and normal strain in a direction of interest as:

$$E' = \sigma'_{ii}/\epsilon'_{ii} , \qquad (20a)$$

where the prime implies the direction of interest. The elastic shear modulus relates the shear stress and shear strain as:

$$G' = \sigma'_{ij}/\epsilon'_{ij} \qquad i \neq j, \qquad (20b)$$

while Poisson's ratio is defined as:

$$\upsilon' = \epsilon'_{jj}/\epsilon'_{ii} \qquad i \neq j. \qquad (20c)$$

These equations can also readily be shown to exhibit the forms:

$$E = 1/S'_{iiii} , \qquad (21a)$$

$$G = 1/S'_{ijij} , \qquad (21b)$$

and

$$v = -S'_{ijij}/S'_{iiii} ,$$ (21c)

where i,j = 1,2,3 and i≠j. The S'_{ijkl} for the cubic and hexagonal single crystal are listed in the Appendix. The Young's modulus can be specified using only one crystallographic direction and is related to one axis, i. However, the elastic shear modulus, G, and Poisson's ratio, v, are transverse coefficients which require two axes perpendicular to each other, i and j, for definition.

The crystal orientation and temperature dependencies of Young's modulus, the elastic shear modulus and Poisson's ratio for the cubic (3C) SiC polytype on the (110) plane are illustrated in Figures 9a-e. The (110) was chosen because it is the plane which includes the directions with the maximum and minimum values of Young's modulus. Figure 9a depicts the Young's modulus, or E for the cubic (3C) SiC polytype on the (110) plane. The maximum Young's modulus, E, is in the [$\bar{1}$11] direction and is about 510GPa. It is nearly twice as large as the minimum Young's modulus value, which is in the [001] direction, only about 280 GPa. The E in the [$\bar{1}$10] direction is intermediate, about 420 GPa. It is apparent that if the Young's modulus along the length of a single crystal whisker is an important composite design factor, then ⟨111⟩ oriented whiskers should be employed for a maximum elastic modulus and ⟨100⟩ whiskers should be used to achieve the minimum elastic modulus value.

The elastic shear modulus and Poisson's ratio must be graphically represented by a different method than the Young's modulus, because two directions are necessary for their specification. One is the i and j direction rotating together on the (110) plane, which can be expressed as $G_{⟨hkl⟩⟨hkl⟩}$ and $v_{⟨hkl⟩⟨hkl⟩}$ on the (110) plane. The other is to fix the i direction as being perpendicular to the (110) plane and then rotate the j direction within the (110) plane, as $G_{⟨110⟩⟨hkl⟩}$ and $v_{⟨110⟩⟨hkl⟩}$. These two approaches are illustrated in Figure 9b and 9c for the two G possibilities, and in 9d and 9e for the Poisson's ratio cases.

Figure 9b illustrates the elastic shear modulus, $G_{⟨hkl⟩⟨hkl⟩}$ on the (110) plane. The maximum value of the elastic shear modulus is 230 GPa as $G_{[\bar{1}10][001]}$ or $G_{[001][1\bar{1}0]}$, while the minimum value is 45° from [001] direction and only about 120 GPa. The $G_{⟨110⟩⟨hkl⟩}$ is illustrated in Figure 9c. The maximum and minimum values are 230 and 100 GPa for $G_{[\bar{1}10][001]}$ and $G_{[110][\bar{1}10]}$, respectively. The variation of G is from 100 GPa to 230 GPa for the different directions, which further substantiates the large elastic anisotropy that was evident for Young's modulus.

Figure 9d illustrates the Poisson's ratio, $v_{⟨hkl⟩⟨hkl⟩}$, on the (110) plane. The maximum values are 0.44 for $v_{[\bar{1}10][001]}$, which is not equal to the $v_{[001][1\bar{1}0]}$ that is 0.28. The minimum value of Poisson's ratio is 45° from the [001] direction and about 0.04. The $v_{⟨110⟩⟨hkl⟩}$ is illustrated in Figure 9e. The maximum value is about 0.44 for $v_{[110][001]}$, while the minimum value becomes negative about -0.10 for $v_{[110][\bar{1}10]}$. This negative value does not show in Figure 9e. The variation of v is from -0.10 to 0.44 with different directions, highly anisotropic. Figures 9d and 9e also reveal that the Poisson's ratio is independent of temperature while the Young's and shear moduli are only slightly temperature dependent. However, all of the elastic moduli of single crystal cubic (3C) SiC are highly anisotropic and dependent on the crystalline orientation.

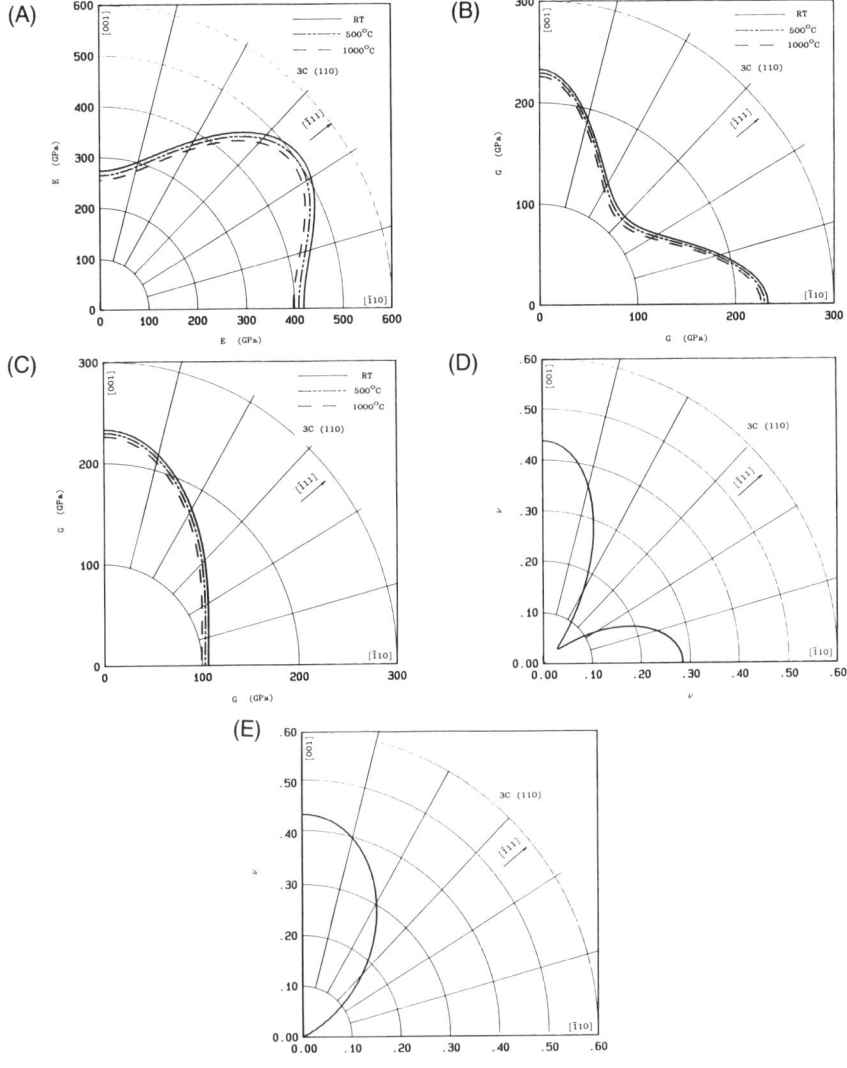

Fig. 9. (A) The Young's modulus, $E\langle hkl\rangle$, of cubic SiC as a function of orientation on the (110), (B) the shear modulus, $G\langle hkl\rangle\langle hkl\rangle$, of cubic SiC as a function of orientation on the (110), (C) the shear modulus, $G\langle 110\rangle\langle hkl\rangle$, of cubic SiC as a function of orientation on the (110), (D) the Poisson's ratio, $\upsilon\langle hkl\rangle\langle hkl\rangle$, of cubic SiC as a function of orientation on the (110), and (E) the Poisson's ratio, $\upsilon\langle 110\rangle\langle hkl\rangle$, of cubic SiC as a function of orientation on the (110).

Silicon Carbide

Since the elastic properties of all hexagonal crystals are transversely isotropic to their "c"-axis, or $\langle 0001 \rangle$ direction, the crystal orientation and temperature dependencies of elastic moduli for the hexagonal SiC on the $(10\bar{1}0)$ plane have been summarized in Figures 10a to 10e by the same method as those of cubic SiC. As previously noted, these elastic anisotropies may be expected to apply to all of the hexagonal polytypes of SiC. Figure 10a illustrates the E of hexagonal SiC on the $(10\bar{1}0)$ plane, as it varies from the "c"-axis, the $[0001]$, to the basal plane, the $[\bar{1}2\bar{1}0]$. The Young's moduli along the $[0001]$ and $[\bar{1}2\bar{1}0]$ of the hexagonal SiC polytypes are identical to those along the $[\bar{1}11]$ and $[\bar{1}10]$ of the cubic SiC polytype as 510 GPa and 420 GPa, respectively. The minimum Young's modulus of about 340 GPa is approximately in the $[\bar{1}2\bar{1}1]$ direction.

The elastic shear modulus, G, and Poisson's ratio, v, are illustrated in Figures 10b–10e. Figure 10b depicts the elastic shear modulus, $G_{\langle hkil \rangle \langle hkil \rangle}$ on the $(10\bar{1}0)$ plane. The minimum value is 130 GPa for $G_{[\bar{1}2\bar{1}0][0001]}$ or $G_{[0001][\bar{1}2\bar{1}0]}$. The maximum value is 45° from the $[0001]$ direction and about 210 GPa. The $G_{\langle 10\bar{1}0 \rangle \langle hkil \rangle}$ is illustrated in Figure 10c. It shows that the $G_{[10\bar{1}0][\bar{1}2\bar{1}0]}$ is slightly larger than the $G_{[10\bar{1}0][0001]}$. The variation of the shear modulus is from 130 GPa to 210 GPa with different directions.

Figure 10d illustrates the Poisson's ratio, $v_{\langle hkil \rangle \langle hkil \rangle}$, in the $(10\bar{1}0)$ plane. The minimum value is about 0.08 for the $v_{[\bar{1}2\bar{1}0][0001]}$ which is not equal to the $v_{[0001][\bar{1}2\bar{1}0]}$ of about 0.10. The maximum value is also 45° from the $[0001]$ direction. The $v_{\langle 10\bar{1}0 \rangle \langle hkil \rangle}$ is illustrated in Figure 10e with a maximum value of 0.26 for the $v_{[10\bar{1}0][\bar{1}2\bar{1}0]}$ and minimum value of 0.08 for the $v_{[10\bar{1}0][0001]}$. The variation of the Poisson's ratio is from 0.08 to 0.26. The temperature dependencies of the elastic moduli for hexagonal SiC are also very small. The elastic moduli for both crystal structures are highly dependent on the crystal orientation but only slightly dependent on the temperature.

The elastic anisotropy of cubic single crystals can be generally evaluated by calculating the Zener Ratio[29], "A_z", which is the ratio of the two shear moduli $G_{\langle 100 \rangle \langle 001 \rangle}$ to $G_{\langle 100 \rangle \langle 110 \rangle}$ as:

$$"A_z" = 2C_{44}/(C_{11} - C_{12}). \tag{22a}$$

In the hexagonal structure, an analogous form for the Zener Ratio, "A_z" is:

$$"A_z" = C_{44}/C_{66} = 2C_{44}/(C_{11} - C_{12}) \tag{22b}$$

The "A_z" are equal to unity for elastically isotropic crystals, whereas values either smaller or greater than unity correspond to the degree of elastic anisotropy possessed by the crystal structure. For cubic SiC the Zener Ratio at room temperature is 2.2 while that for hexagonal SiC is 0.8. This suggests that the elastic anisotropy of cubic SiC is greater than that of hexagonal SiC. The "A_z" is only slightly dependent on temperature and increases less than 5% from room temperature to 1000°C as illustrated in Figure 11.

THERMOELASTIC STRESS INDEX

The previous correlations of the thermal expansion and the elastic anisotropies with SiC polytype structure can be combined to address their associated role in the

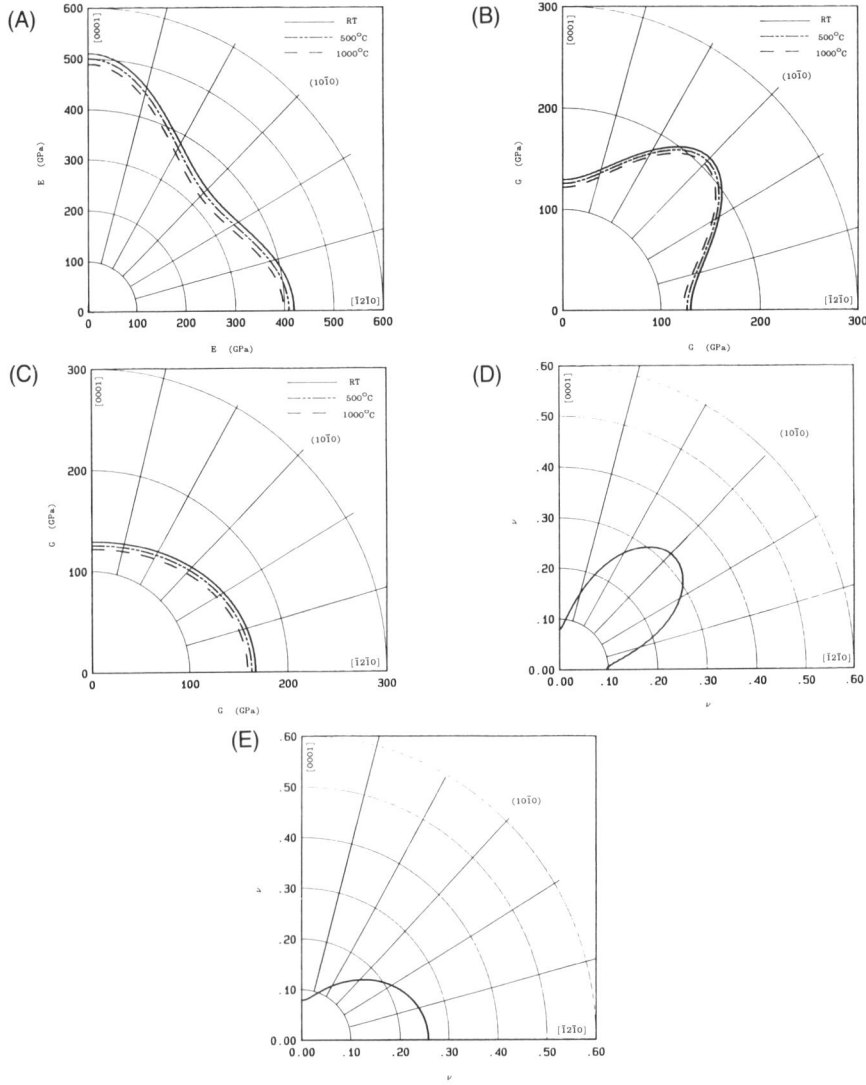

Fig. 10. (*A*) The Young's modulus, $E\langle hkil\rangle$ of hexagonal SiC as a function of orientation on the $(10\bar{1}0)$, (*B*) the shear modulus, $G\langle hkil\rangle\langle hkil\rangle$, of hexagonal SiC as a function of orientation on the $(10\bar{1}0)$, (*C*) the shear modulus, $G\langle 1010\rangle\langle hkil\rangle$, of hexagonal SiC as a function of orientation on the $(10\bar{1}0)$, (*D*) the Poisson's ratio, $\upsilon\langle hkil\rangle\langle hkil\rangle$, of hexagonal SiC as a function of orientation on the $(10\bar{1}0)$, and (*E*) the Poisson's ratio, $\upsilon\langle 1010\rangle\langle hkil\rangle$, of hexagonal SiC as a function of orientation on the $(10\bar{1}0)$.

Silicon Carbide

determination of thermoelastic stresses within SiC single crystals. The concepts advanced in this paper are generally applicable to any single crystal, whether a single-crystal grain within a polycrystalline matrix or a single-crystal reinforcing whisker within the matrix of a composite, either a polymer, glass or metal. For example, as described by Petrovic, et al[30], it is of interest to incorporate SiC whiskers as reinforcement in composites consisting of various crystalline ceramic matrices. When these composites are cooled from their fabrication temperatures, residual internal micromechanical stresses develop as a consequence of the thermal expansion and elastic constant mismatches. While it is beyond the scope of this paper to actually calculate those stresses for all of the different possibilities of matrix materials for their reinforcement by SiC whiskers, it is logical to extend the thermoelastic concepts which have already been discussed relative to single-crystal whiskers of SiC as a familiar illustration.

In the development of the theory of thermoelastic stresses, the strain resulting from thermal expansion, or the restraint thereof, is considered to be equal to an equivalent mechanical strain. For complete restraint in one dimension, the result of that simple equality is the relationship:

$$\sigma_{ts} = \alpha E \Delta T, \qquad (23)$$

where σ_{ts} is the thermoelastic stress, ΔT is the temperature change, and α and E are as previously defined. Although the above equation is without the consideration of the surrounding matrix, which must be incorporated into the composite thermoelastic stress problem for a complete solution, it clearly illustrates the importance of the product (αE), which is referred to as the thermoelastic stress index, of the SiC crystal structure itself. It should be noted by reference to Equation (1a) that the thermoelastic stress index bears some resemblance to the Gruneisen constants.

Having previously addressed the individual α and also the E values for the different SiC polytypes, it is a relatively direct application to consider their product with respect to crystal orientation and temperature. In fact, because neither α nor E are transverse coefficients, the product αE is simply that of α_{ii} and E, or $\bar{S}^{-1}_{iiii}$**. This product can be directly obtained in any crystallographic direction. Figure 12a depicts the thermoelastic stress index for the cubic (3C) polytype of SiC, again on the (110) plane. Comparison of Figure 12a with 9a, reveals that the Young's modulus and the thermoelastic stress index are quite similar in structural anisotropy. This is because of the stress index are quite similar in structural anisotropy. This is because of the thermal expansion of the cubic crystal structure is isotropic. However, the temperature dependence of the thermoelastic stress index is far greater than that for the Young's modulus itself. This is a direct consequence of the increase of the coefficient of thermal expansion of SiC with increasing temperature.

Having established the crystallographic dependence of the thermoelastic stress index, αE, as well as previously considering α and E independently for the (3C) structure, it is possible to address the most desirable single-crystal cubic SiC whisker orientation for composite reinforcement. On the basis of a maximum E criterion, the most desirable whisker orientation is the one with the whisker axis a $\langle 111 \rangle$. Such an oriented whisker may be expected to carry a maximum fraction of the applied load. However, a $\langle 111 \rangle$ whisker also has the largest thermoelastic stress index and in some instances may be expected to develop the largest thermoelastic

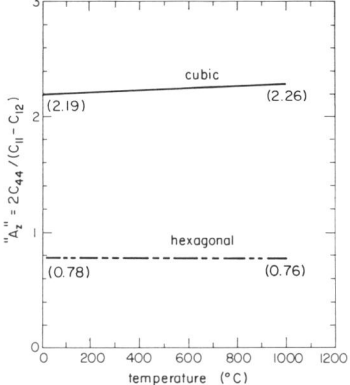

Fig. 11. Elastic anisotropy, Zener ratio of cubic and hexagonal SiC as a function of temperature.

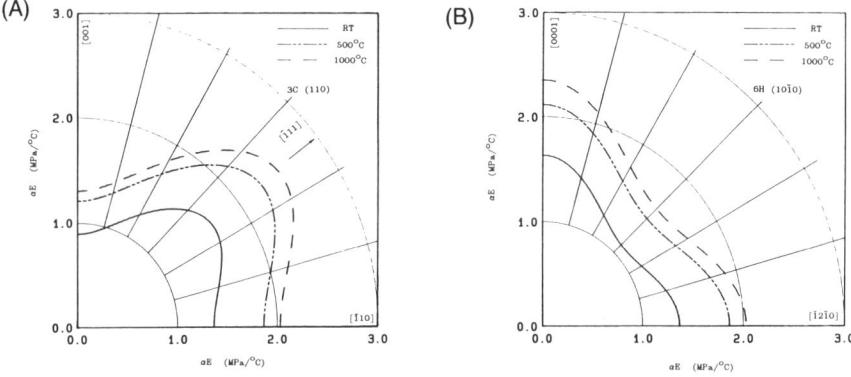

Fig. 12. (*A*) The thermoelastic stress index for cubic SiC(3C) as a function of orientation on the (110), and (*B*) the thermoelastic stress index for hexagonal SiC(6H) as a function of orientation on the (10$\bar{1}$0).

residual stresses on cooling from the fabrication temperatures or during thermal cycling of the composite. A ⟨100⟩ orientation whisker would yield the minimum for both, while a ⟨110⟩ whisker would be a compromise. If the yielding of a metallic matrix to develop a strengthening dislocation substructure and enhance the matrix properties were desired, then ⟨111⟩ oriented whiskers would seem to be the optimal orientation for some composites. Obviously, other criteria can also be applied once the thermoelastic anisotropy of the reinforcing whisker system is established such as depicted in Figure 12a. Naturally, the thermoelastic characteristics of the matrix are also important considerations.

The thermoelastic stress index for the hexagonal (6H) structure of SiC on the $(10\bar{1}0)$ plane is illustrated in Figure 12b. Similar to that for the (3C) cubic polytype, the temperature dependence of the thermoelastic stress index is much greater than that of the Young's modulus alone, as was previously illustrated in Figure 10b. However, because of the thermal expansion anisotropy of the hexagonal structure, the thermoelastic stress index does not exactly parallel the Young's modulus anisotropy as was the situation for the cubic (3C) crystal structure.

It is also of interest to compare the thermoelastic stress indices of the cubic and the hexagonal (6H) SiC as illustrated in Figure 12a and 12b. The maximum values of αE for the cubic (3C) and the hexagonal (6H) structures of SiC are in their $[\bar{1}111]$ and $[0001]$ directions, respectively. At room temperature the maximum αE values for the individual structures are nearly identical at 1.65 MPa/°C. However, at 1000°C the maximum αE value for the (6H) polytype is about 2.35 MPa/°C, slightly smaller than the 2.50 MPa/°C value for the (3C) cubic structure. This decrease is because the "c"-axis thermal expansion coefficient, α_{33} for the (6H) polytype is smaller than α_{11}, as well as the α for the (3C) polytype. The αE along the basal plane for the hexagonal (6H) SiC polytype, $\alpha E \langle \bar{1}2\bar{1}0 \rangle$, is the same as the cubic structure, $\alpha E \langle \bar{1}10 \rangle$, at different temperatures as the thermal expansions and Young's modulus for both materials are identical in those directions.

Figures 13a–13c depict the thermoelastic stress indices for the four common hexagonal polytypes, (2H), (4H), (6H) and (8H) as a function of orientation on their $(10\bar{1}0)$ planes at RT, 500°C and 1000°C. Similar to Figure 3, a series of nearly parallel curves exists. The highest values are those for the (8H) polytype, because the F^h for the (8H) polytype is only 0.25. The lowest values are those for the (2H) polytype, although its F^h is equal to unity, it is the most anisotropic in its thermal expansion. The differences of the thermoelastic indices of the various polytypes at room temperature are only very small, as the thermal expansion coefficients are nearly the same. However with increasing temperature the α also increase and that difference increases as illustrated in Figures 13b and 13c. The αE in the $\langle \bar{1}2\bar{1}0 \rangle$ directions for all of these different hexagonal SiC polytypes are equal, because the α_{11} and $E \langle \bar{1}2\bar{1}0 \rangle$ are identical for the hexagonal polytypes. It is clear that the polytype dependency is not very great for the thermoelastic stress index.

CONCLUSIONS

The elastic and thermal expansion anisotropies and their product, the thermoelastic stress index αE, of the different structural polytypes of SiC have been considered as related to crystal orientation and temperature. From room temperature to 1000°C, the thermal expansion coefficients of SiC are only slightly orientation dependent, but highly temperature dependent, while the elastic moduli are the reverse. The consequence is that the thermoelastic stress index is highly dependent on both the crystal orientation and the temperature. This feature emphasizes the importance of the level of temperature, as well as simply the value of ΔT during thermal cycling of SiC materials, for the thermoelastic stress index directly determines the resulting thermal stress levels which are generated.

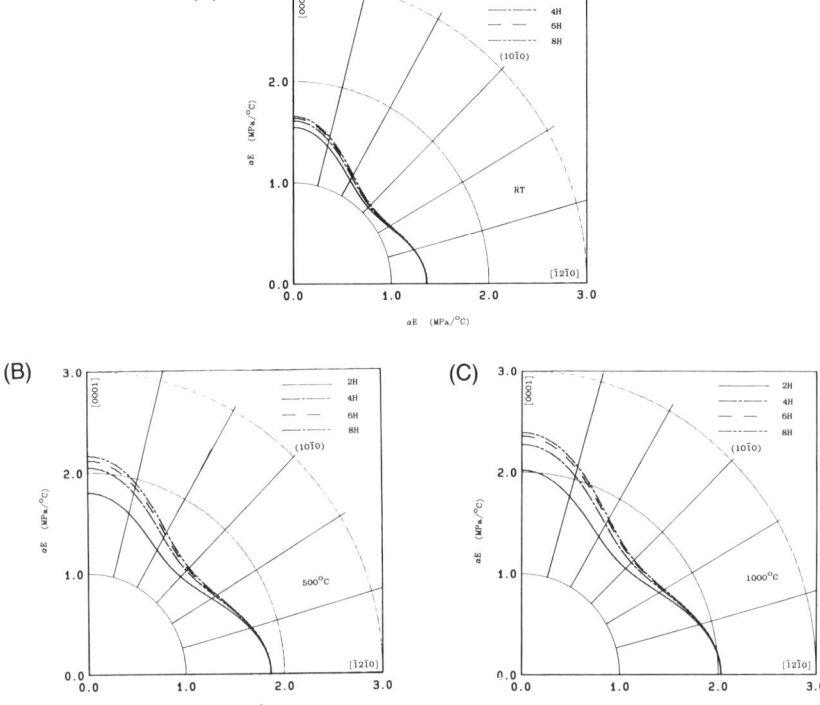

Fig. 13. (A) the thermoelastic stress index for four hexagonal SiC polytypes as a function of orientation on the (10$\bar{1}$0) at RT, (B) the thermoelastic stress index for four hexagonal SiC polytypes as a function of orientation on the (10$\bar{1}$0) at 500°C, and (C) the thermoelastic stress index for four hexagonal SiC polytypes as a function of orientation on the (10$\bar{1}$0) at 1000°C.

APPENDIX

The single crystal elastic constants in an arbitrary direction can be expressed by:

$$S'_{ijkl} = \beta_{im}\beta_{jn}\beta_{ko}\beta_{lp}S_{mnop},$$

where the β_{ij} are the directional cosines with the angle between the i-axis of interest and the j-axis of crystal. For practical application the four-suffix compliances are converted to the two-suffix compliances as:

11--1, 22--2, 33--3, 23--4, 13--5, 12--6.

Silicon Carbide

The Young's modulus, the elastic shear modulus and Poisson's ratio are:

$$E = 1/S'_{11},$$

$$G = 1/S'_{44},$$

and

$$v = -S'_{12}/S'_{11}, \quad \text{respectively.}$$

For cubic crystals:

$$S'_{11} = S_{11} + S_c\,(\beta_{11}^2\beta_{12}^2 + \beta_{12}^2\beta_{13}^2 + \beta_{13}^2\beta_{11}^2),$$

$$S'_{12} = S_{12} + S_c\,(\beta_{11}^2\beta_{21}^2 + \beta_{12}^2\beta_{22}^2 + \beta_{13}^2\beta_{23}^2),$$

$$S'_{44} = S_{44} + 4S_c\,(\beta_{21}^2\beta_{31}^2 + \beta_{22}^2\beta_{32}^2 + \beta_{23}^2\beta_{33}^2),$$

where

$$S_c = S_{11} - S_{12} - (S_{44}/2).$$

For hexagonal crystals:

$$S'_{11} = S_{11}\,(1-\beta_{13}^2)^2 + S_{33}\beta_{13}^4 + (2S_{13} + S_{44})(\beta_{13}^2 - \beta_{13}^4),$$

$$S'_{12} = S_{12} + (S_{11} + S_{33} - S_{44} - 2S_{13})\beta_{13}^2\beta_{23}^2 + (S_{13} - S_{12})(\beta_{13}^2 + \beta_{23}^2),$$

$$S'_{44} = S_{44} + 4\,(S_{11} + S_{33} - S_{44} - 2S_{13})\,\beta_{23}^2\beta_{33}^2 + (2S_{11} - 2S_{12} - S_{44})\,\beta_{13}^2.$$

Acknowledgement

The authors acknowledge the financial support of NASA Grant No. NAGW-199. Discussions with J. A. Salem and M. Taya during preparation of the manuscript were appreciated.

---

†Other averaging schemes are also possible and equally valid for this technique of calculating (dC_{ij}/dT) from polycrystalline data.
‡Usually α_{ij} are used for directional cosines, but in this manuscript that character has already been used for the thermal expansion coefficients.
§In a strict sense, these results are for the trigonal rather than the hexagonal structure, because the $\langle 111 \rangle$ direction of the cubic structure has three fold symmetry. It is reasonable to ignore the C_{14} term when considering the applicability of these elastic constants to hexagonal SiC.
**It is of course possible to also address the thermoelastic shear and Poisson's effects, but they will not be considered in this manuscript.

References

[1] Purdue University, "Thermal Expansion." in *Thermophysical Properties of Matter*, Vol. 13, p. 873-878, edited by Y. S. Touloukian, R. K. Kirly, R. E. Taylor and T. Y. R. Lee, IFI/Plenum, New York, (1970).

[2] R. W. G. Wyckoff, *Crystal Structure*, Vol I, 2nd edition, p.111, Interscience, New York (1963).

[3] N. W. Jepps and T. F. Page, "Polytypic Transformation in Silicon Carbide," p. 259-307 in *Crystal Growth and Characterization of Polytype Structures*, edited by P. Krishna, Pergamon Press, Oxford, (1983).

[4] B. D. Cullity, Elements of X-Ray Diffraction, 2nd edition, Addison-Wesley, Reading, MA, (1978).

[5] Z. Li and R. C. Bradt, "Thermal Expansion of the Cubic (3C) polytype of SiC," *J. Mat. Sci.*, 21, 4366-68, (1986).

[6] Z. Li and R. C. Bradt, "Thermal Expansion of the Hexagonal (4H) Polytype of SiC," *J. Appl. Phys.*, 60, 612-14, (1986).

[7] Z. Li and R. C. Bradt, "Thermal Expansion of the Hexagonal (6H) Polytype of SiC," *J. Amer. Ceram. Soc.*, 69, 863-66, (1986)

[8] Z. Li and R. C. Bradt, "Thermal Expansion and Thermal Expansion Anisotropy of SiC Polytypes," *J. Amer. Ceram. Soc.*, 70, 445-448, (1987).

[9] H. D. Megaw, "Crystal Structures and Thermal Expansion." Mat. Res. Bull., 8, 1007-1018, (1971).

[10] R. M. Hazen and L. M. Finger, Comparative Crystal Chemistry, p.115, Wiley-Interscience, New York, (1984).

[11] W. D. Kingery, H. K. Bowen and D. R. Uhlmann, Introduction to Ceramics, p.43, Wiley-Interscience, New York, (1976).

[12] L. Patrick, "Dependence of Physical Properties on Polytype Structure," Mat. Res. Bull., 4, S129-140, (1969).

[13] V. A. Likhachev, "Microstructural Strains due to Thermal Anisotropy," Soviet Physics—Solid State, 3, 1130-36, (1961).

[14] J. A. Kuszyk and R. C. Bradt, "Influence of Grain Size on Effects of Thermal Expansion Anisotropy in $MgTi_2O_5$," *J. Amer. Ceram. Soc.*, 56, 420-23, (1973).

[15] Z. Li and R. C. Bradt, "The Single Crystal Elastic Constants of Cubic (3C) SiC to 1000°C," Accepted for publication by the J. Mater. Sci.

[16] Z. Li and R. C. Bradt, "The Single Crystal Elastic Constants of Hexagonal SiC to 1000°C," (Submitted for publication by Int. J. High Tech. Ceram.).

[17] K. B. Tolpygo, "Optical, Elastic and Piezoelectric Properties of Ionic and Covalent Crystals with the ZnS Type Lattice," Soviet Physics—Solid State, 2, 2367-76, (1961).

[18] H. C. Chandan, "Elevated Temperature Instrumented Charpy Impact Testing of Commercial and Experimental Silicon Carbides," Ph.D Dissertation, Pennsylvania State Univ. (1980).

[19] C. H. McMurtry, M. R. Kasprzyk and R. G. Naum, "Microstructural Effects in Silicon Carbide," pp. 359 in Silicon Carbide—1973, editors, R. C. Marshall J. W. Faust, Jr and C. E. Ryan, Univ. of South Carolina Press, Columbia, SC. (1974).

[20] G. Simmons and H. Wang, Single Crystal Elastic Constants and Calculated Aggregate Properties: A Handbook, 2nd ed., The M.I.T. Press, Cambridge, (1971).

[21] D. S. Lieberman and S. Zirinsky, "A Simplified Calculation for the Elastic Constants of Arbitrarily Oriented Single Crystals," Acta Cryst. 9, 431-436, (1956).

[22]D. W. Feldman, J. H. Parker, Jr., W. J. Choyke and L. Patrick, "Phonon Dispersion Curves by Raman Scattering in SiC, Polytypes 3C, 4H, 6H, 15R, and 21R," *Phys. Rev.*, **173**, 787–793, (1968).

[23]G. Arlt and G. R. Schodder, "Some Elastic Constants of Silicon Carbide," *J. Acoust. Soc. Amer.*, **37**, 384–385, (1965).

[24]Don Berlincourt, Hans Jaffe, and L. R. Shiozawa, "Electroelastic Properties of the Sulfides, Selenides, and Tellurides of Zinc and Cadmium," *Phys. Rev.*, **129**, 1009–1017, (1963).

[25]C. F. Cline, H. L. Dunegan, and G. W. Henderson, "Elastic Constants of Hexagonal BeO, ZnS, and CdSe," *J. Appl. Phys.*, **38**, 1944–1948, (1967).

[26]K. D. McHenry and R. E. Tressler, "Fracture Toughness and High Temperature Slow Crack Growth in SiC," *J. Amer. Ceram. Soc.*, **63**, 152–156, (1980).

[27]P. T. B. Shaffer and C. K. Jun, "The Elastic Modulus of Dense Polycrystalline Silicon Carbide," Mat. Res. Bull., 7, 63–70, (1972).

[28]J. Turley and G. Sines, "Representation of Elastic Behavior in Cubic Materials for Arbitrary Axes," *J. Appl. Phys.*, **41**, 3722–25, (1970).

[29]D. H. Chung and W. R. Buessem, "The Elastic Anisotropy of Crystals," p. 217 in Anisotropy in Single-Crystal Refractory Compounds, vol. 2 edited by F. W. Vahldiek and S. A. Mersol, Plenum Press, New York, (1968).

[30]J. J. Petrovie, J. V. Milewski, D. L. Rohr and F. D. Gac, "Tensile Mechanical Properties of SiC Whiskers," *J. Mater. Sci.*, **20**, 1167–1177, (1985).

Section IV

Applications

Improved Silicon Carbide for Advanced Heat Engines. Part I: Process Development for Injection Molding

Thomas J. Whalen and Walter Trela

Research Staff
Ford Motor Company
Dearborn, MI 48121

Abstract

Alternate processing methods have been investigated as a means of improving the mechanical properties of injection-molded silicon carbide. Various mixing processes—dry, high-sheer, and fluid—were evaluated along with the morphology and particle size of the starting β-SiC powder. Statistically-designed experiments were used to determine significant effects and interactions of variables in the mixing, injection molding, and binder removal process steps. Improvements in mechanical strength can be correlated with the reduction in flaw size observed in the injection molded green bodies obtained with improved processing methods.

Introduction

The application of silicon carbide for advanced heat engines depends primarily on its mechanical strength and reliability, and the ability to fabricate complex shapes for mass production on an economically sound basis. In February, 1985, a program was begun at Ford Motor Company for the National Aeronautics and Space Administration to develop a high strength-high reliability silicon carbide material which could be mass produced. The advantages and disadvantages of several methods of fabricating SiC were reviewed[1]. Injection molding and pressureless sintering were the processes chosen for this application since these processes are generally considered to be capable of high production rates for complex engines parts at an attractive low cost. The pressureless sintering process involving the addition of carbon and boron to fine silicon carbide powder was selected to provide the dense material required for high strength. It is the purpose of this paper to review the initial baseline work and the process development work on the mixing, molding and dewaxing parts of the program. In Part II of the review, the sintering process development and the microstructural and property relationships will be discussed. The extensive use of statistically-designed experiments[2,3,4] will be evident in both Parts I and II.

Baseline Material Evaluation

A time-line chart of the overall process for making silicon carbide MOR bars or complex shapes is shown in Figure 1. The process controls and the sampling and

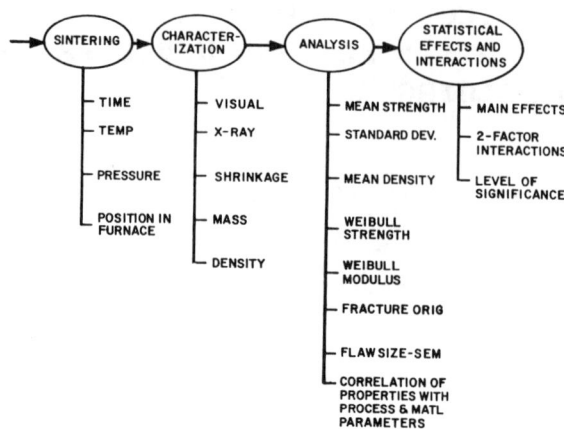

Fig. 1. Process and controls chart for making SiC bars or complex shapes by injection molding.

evaluations are contained on the chart. Statistical process control charts are used in the mixing, molding, dewaxing and sintering processes. The particle size distributions of the primary powders used in the process were determined[5] and the median values of the particle diameter in the volume distribution were 1.05, 2.19, and 7.04 micrometers for SiC, boron and carbon black, respectively. At the beginning of the program the composition of the baseline material was selected from the results of a 2^3 experiment involving the three variables, SiC powder size, carbon source and percentage of solids loading in the molding mix. The choice for the solids loading range was based on previous experience with the injection molding process. A schematic of the experiment is shown in Figure 2.

The initial sintered density results for these compositions when processed through the conventional argon dewax and sintering were disappointing, as shown by the average density in Table I. Subsequent runs, which reduced oxygen contamination by dewaxing and sintering in a vacuum, improved the density to greater than 90% of theoretical. The Table points out an interesting feature of statistically-planned experiments. Although the sintered-density data were disappointing, the three experiments all led to the same conclusions, which were that the finest powder, UF Ibiden (18 m^2/g), was better than the coarser SF Ibiden (13m^2/g), that the carbon addition as carbon black was better than carbon added as resin, and that the higher solids loading (58.5%) was better than the low solids loading (55.5%). These results were used to prepare samples for the baseline study.

Dry mixing of the powders led to poor density (90%) and strength (240 MPa, 35 Ksi) results. Examination of the material showed agglomerates in the mixed powder were above 100 micrometers in diameter. Dry mixing was replaced by high shear mixing in a Haake Mixer, which reduced the agglomerate size to about 80 micrometers and led to the baseline process which was adopted at that time. Statistical process control (SPC) charts were generated for the baseline mixing, molding and sintering processes and these charts were found to be beneficial in

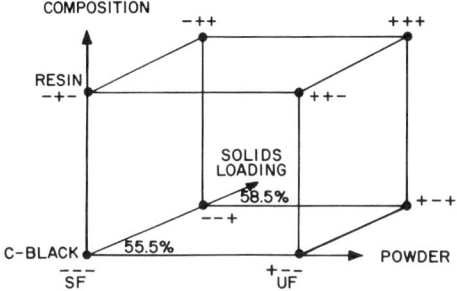

Fig. 2. Schematic of experimental design for baseline composition.

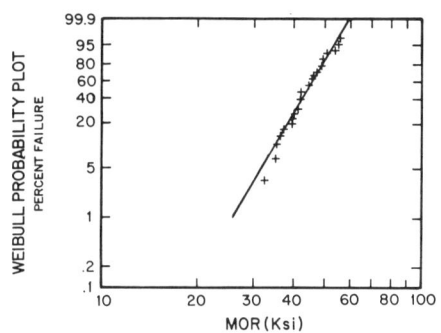

NUMBER OF SAMPLES = 30.00
WEIBULL CHARACTERISTIC VALUE = 45.82
WEIBULL SLOPE = 7.97
DISTRIBUTION MEAN = 43.14
STANDARD DEVIATION = 6.43

Fig. 3. Weibull plot of baseline MOR data.

Table I. Main effects and interactions from two to the third power experiment on sintered density of injection molded bars (% theoretical density).

Identity	Effects on Density		
	Conventional Ar Dewax & Sinter	Two Cycle Vac Dewax & Sinter	One Cycle Vac Dewax & Sinter
Average	72.7 ± 0.1	88.2	90.2
Powder	1.2 ± 0.2	1.8	2.0
Carbon	−5.8 ± 0.2	−4.2	−3.5
Powder-Carbon	−1.6 ± 0.2	−1.8	−1.0
Solids Loading	0.9 ± 0.2	0.8	0.5
Powder-Loading	−0.1 ± 0.2	0.2	1.0
Carbon-Loading	0.2 ± 0.2	0.2	0.5
Powder-Carbon-Loading	0.6 ± 0.2	0.8	0

maintaining and improving process control. The mean strengths on machined bars (9.8 × 4.9 × 51 mm) tested in four-point bending at room temperature, 1000°C and 1400°C were 299 MPa (43.3 Ksi), 285 MPa (41.4 Ksi) and 325 MPa (47.2 Ksi), respectively. Room temperature Weibull data (30 samples) are shown in Figure 3 and mean strengths and standard deviations at high temperatures (6 samples each) are given in Figure 4. The fracture origins in the baseline material were observed by scanning electron microscopy (SEM) to be of three kinds; and iron-rich flaw due to tramp iron from the mixing operation (Figure 5), a boron-rich flaw from poor boron dispersion during mixing (Figure 6) and a sulfur-containing flaw believed to signal a carbon-rich flaw due to the poor dispersion of carbon during mixing (Figure 7). Sulfur was found to be a minor impurity in the carbon black used as a carbon source.

MIXING DEVELOPMENT

Early in the program it was recognized that the mixing process was the key to improved strength of sintered silicon carbide from data on the effect of mixing technique on the strength. Dry mixing provided a mean strength of 230 MPa, high shear mixing gave 299 MPa, and a fluid mixing process yielded 385 MPa when comparable composition and processing were evaluated. A new fluid mixing process[6] was developed which showed considerable promise in the reduction of the agglomerate sizes believed to be controlling the strength. The fluid mixing process is one in which the ingredients are mixed in toluol as the fluid and the molding waxes are introduced as a solution in toluol. Stir drying and pan drying are used to remove the fluid, and a small wax addition is made at the end of the process to reach the correct solids loading. Five duplicate batches were evaluated for reproducibility with statistical process control charts of unwetted agglomerate size and boron agglomerate size. The unwetted agglomerate size was controlled at a mean value of 36 micrometers following mixing and waxing with a three standard deviation range of less than 8 micrometers. The mean boron agglomerate size was controlled below 27 micrometers following drying, mixing and waxing with a three

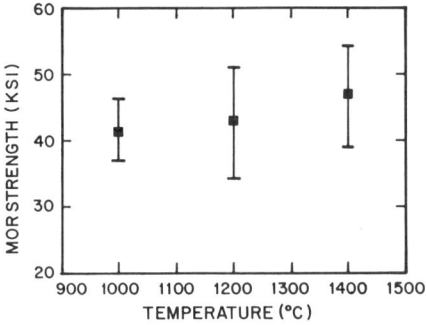

Fig. 4. High temperature MOR strength of baseline material.

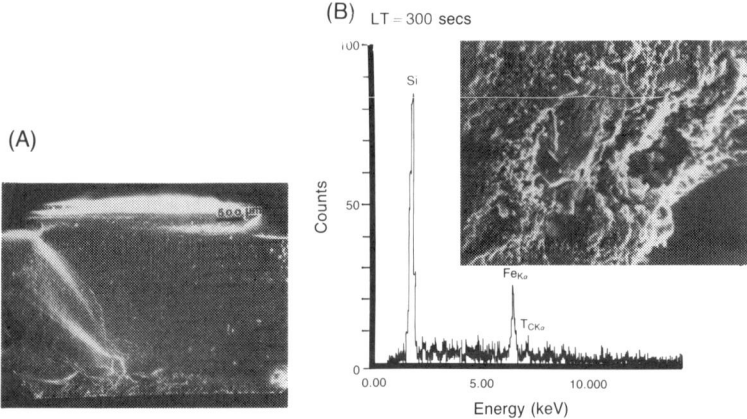

Fig. 5. SEM micrographs showing (A) fracture origin and (B) iron-rich surface defect.

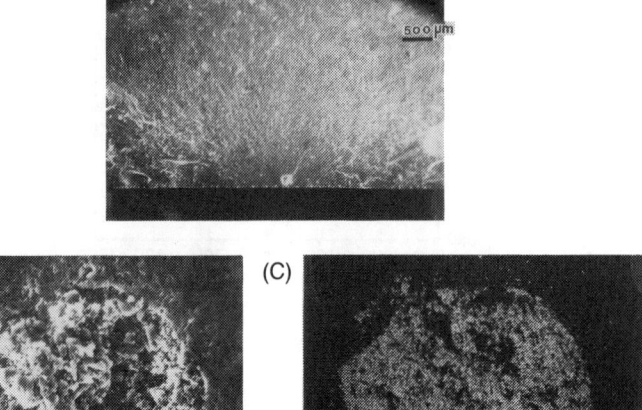

Fig. 6. SEM micrographs of fracture surface showing (A) fracture origin (B) boron-rich flaw and (C) boron map.

standard deviation range below 11 micrometers. From these data and experience in statistical process control analysis, it was concluded that the process was reproducible and under statistical process control.

A matrix experiment of design 2^{4-1} was carried out to improve on the fluid mixing process. Four variable at two levels each, carbon screening (with and without screening), boron screening (with and without screening), ball mill load (10 and 40 balls per charge) and ball mill time (48 and 96 hours) were evaluated. The data generated on agglomerate size are given in Table II in which the negative sign denotes the lower value of the variable and the plus sign represents the higher value. The effects of the variables and an analysis of variance for the unwetted agglomerates are shown in Table III and for boron agglomerates in Table IV. The only significant effect of the four variables is the number of milling balls and milling time, with the number of balls being the more important variable.

Molding Development

In the development of injection molded materials, we have observed the general relationship of flow distance in a spiral mold as a function of surface area of the powders contained in the molding mix. For the silicon based ceramic powders, the data shown in Figure 8 at constant volume percent solids were generated for silicon, silicon nitride and silicon carbide powders. These data have led to the formation of an injection molding processing window, as shown in Figure 9, which guides us

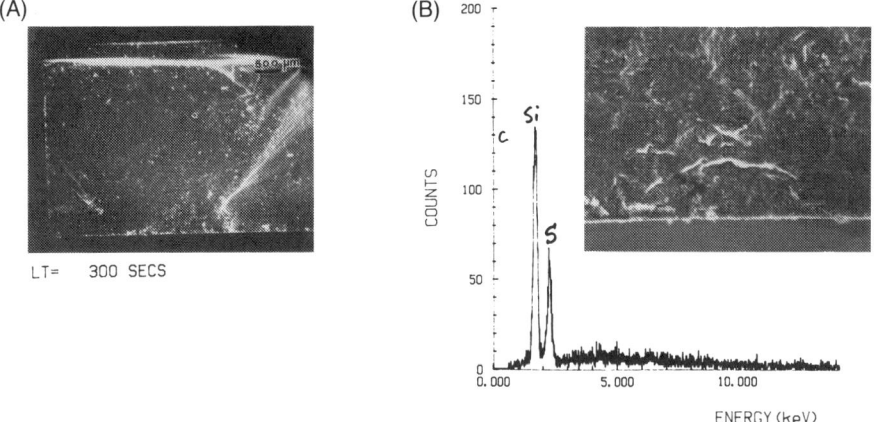

Fig. 7. SEM micrographs of fracture surface showing (A) fracture origin and (B) sulfur-rich flaw.

in the molding of the Ibiden UF and SF powders employed in the program. For the UF powders used in the studies following the baseline characterization, the solids loading (volume percent solids) is maintained between 55 and 60%.

DEWAXING PROCESS DEVELOPMENT

The objective of the dewaxing process is to remove the wax necessary for the injection molding process without causing fracture or chemical change of the material to be sintered. The dewaxing process can be enhanced by a knowledge of the thermal breakdown of the wax binder with increasing temperature. The waxes used in these studies were evaluated by thermogravimetric analysis in both argon and vacuum environments[5]. The critical temperature regions for wax removal in an argon atmosphere was found to be higher by about 100°C than the temperature regions for wax removal in vacuum. The experience gained in dewaxing silicon carbide MOR bars favors vacuum dewaxing on the basis that oxidation of the silicon carbide and carbon powders is more easily controlled and avoided in a vacuum than in an inert gas atmosphere. The dewaxing cycle for MOR bars typically covers four days with a maximum temperature of 300°C.

MODEL TO IMPROVE STRENGTH

A plan to improve the strength by reducing flaw size in the microstructure of sintered silicon carbide was formulated. Data relating maximum agglomerate size to strength in these materials is presented in Figure 10. In the figure one sees the general relationship of the increasing strength with decreasing maximum agglomerate size. When the data for the agglomerate size is transformed to the inverse of the square root of the agglomerate size, a linear relationship is observed with strength. A similar relationship between the inverse square root of crack length

Table II. Mixing study to reduce agglomerate size.

	Experiment No.							
	1	2	3	4	5	6	7	8
	Batch No.							
	J	T	V	U	W	R	S	M
Variables								
C-Screen	−	+	−	+	−	+	−	+
B-Screen	−	−	+	+	−	−	+	+
No. of Balls	−	−	−	−	+	+	+	+
Mill Time	−	+	+	−	+	−	−	+
	Unwetted Agglomerate—Maximum Size (μm)							
Average 5 Samples	36	29	26	38	18	21	21	21
Range 5 Samples	22	6	3	10	0	7	10	7
	Boron Agglomerate—Maximum Size (μm)							
Average 5 Samples	24	21	22	22	10	15	13	8
Range 5 Samples	7	7	13	7	13	5	0	13

Table III. Mixing study to reduce unwetted agglomerate size.

Comparison	Effects and Analysis of Variance[*]			
	Total	Effect	Deg. of F.	Sum of Squares
A (C-Screen)	8	2.0	1	8.0
B (B-Screen)	2	0.5	1	0.5
C (No. of Balls)	−48	−12.0	1	288.0[†]
D (Mill Time)	−22	−5.4	1	60.5[†]
AB,CD	16	4.0	1	32.0
AC,BD	−2	0.5	1	0.5
AD,BC	4	1.0	1	2.0

[*]Variance = 14.4
[†]Highly Significant

Table IV. Mixing study to reduce boron agglomerate size.

Comparison	Effects and Analysis of Variance*			
	Total	Effect	Deg. of F.	Sum of Squares
A (C-Screen)	-3	-0.75	1	1.1
B (B-Screen)	-5	-1.25	1	3.1
C (No. of Balls)	-43	-10.75	1	231.1†
D (Mill Time)	-13	-3.25	1	21.1†
AB,CD	-7	-1.75	1	6.1
AC,BD	3	0.75	1	1.1
AD,BC	-3	-0.75	1	1.1

*Variance = 16
†Highly Significant

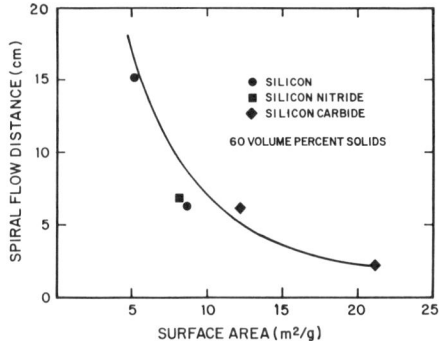

Fig. 8. Injection molding behavior of several silicon based ceramic powders.

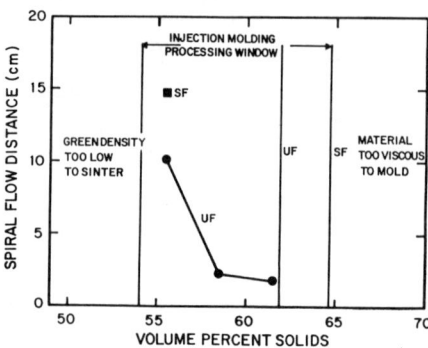

Fig. 9. Injection molding processing window for SiC powders.

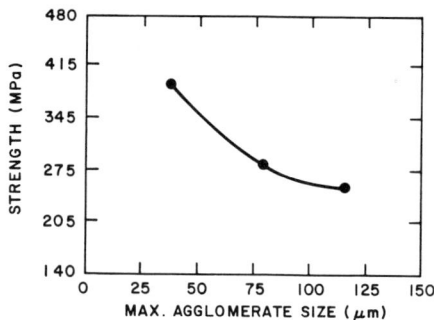

Fig. 10. Mean MOR strengths as a function of maximum agglomerate size and crack length.

and strength in glass was inferred by Griffith[7]. Future work is aimed at reducing agglomerate size in the mixing process by employing a greater milling intensity during the fluid mixing process. The goal of the program is to reach a Weibull characteristic strength of 550 MPa (80 Ksi) and a Weibull modulus of 16. A simple linear model with the transformed agglomerate size data was used to estimate the maximum agglomerate size permitted to reach the program strength goal. A maximum agglomerate size of 16 micrometers was predicted with the simple linear model with least squares parameter estimation.

Summary

A baseline injection-molded silicon carbide was fabricated and evaluated for density and strength. The fracture origins were found to be inclusions and voids believed to be caused by agglomerates formed in the mixing process. A fluid-mixing process was developed which reduced the agglomerate size. Improvements and statistical control of the molding and dewaxing processes were described. A plan to improve strength by reduction in agglomerate size by increasing milling intensity during the fluid mixing process was presented. Part II will discuss the sintering process and strength-processing variable relationships.

Acknowledgement

The authors wish to acknowledge Drs. R. Govila and R. M. Williams (deceased) for mechanical property and microstructural evaluation of these materials, E. Cartwright, B. N. Juterbock and L. V. Reatherford for material preparation and molding, R. Elder and J. R. Baer for dewaxing and sintering, J. Mangels for molding advice and aid and Dr. W. L. Winterbottom for program management. This project was supported in part by NASA Contract # NAS [3-24384].

References

[1] T. J. Whalen, "Processing and Properties of Structural Silicon Carbide," *Ceram. Eng. Sci. Proc.*, 7 [9-10] 1135-1143 (1986).

[2] G. E. P. Box, W. G. Hunter and J. S. Hunter, Statistics For Experiments, John Wiley and Sons, New York, 1978.

[3] C. R. Hicks, Fundamental Concepts in the Design of Experiments, Third Edition, Holt, Rinehart and Winston, New York, 1982.

[4] C. Cochran and G. Cox, Experimental Design, Second Edition, John Wiley and Sons, New York, 1957.

[5] T. J. Whalen and W. L. Winterbottom, "Improved Silicon Carbide for Advanced Heat Engines," NASA Contractor Report 179477, September, 1986.

[6] T. J. Whalen, "Improved Silicon Carbide for Advanced Heat Engines," NASA Contractor Report 180831, October, 1987.

[7] A. A. Griffith, "The Phenomena of Rupture and Flow in Solids," Phil. Trans. Roy. Soc. 221A, 163 (1920); "The Theory of Rupture," Proc. Int. Cong. App. Mech., Delft 1, 55 (1924).

Improved Silicon Carbide for Advanced Heat Engines*. Part II: Pressureless Sintering and Mechanical Properties of Injection Molded Silicon Carbide

Thomas J. Whalen and J. R. Baer

Ford Motor Company
Dearborn, MI 48121

Abstract

The influence on density and strength of pressureless sintering in vacuum and argon environments has been evaluated with injection molded silicon carbide materials. Main effects and two factor interactions of sintering—cycle variables temperature, time, heating rate and atmosphere—were assessed. An improved understanding of the influence of the processing flaws and sintering conditions has been obtained. Strength and density have improved from a baseline level of [299 MPa (43.3 Ksi)] and [94%] of theoretical density to values greater than [483 MPa (70 Ksi)] and [97%.]

Introduction

In Part I, process development work on the mixing, molding and dewaxing processes of the program was reviewed. In Part II, it is the purpose to review the sintering process development and some microstructural and mechanical property relationships. The references[2,3,4] concerning factorial experiments in Part I will also be useful in understanding the details as presented in this Part II.

Initial Argon Sintering Experiment

The baseline material discussed in Part I was sintered in vacuum and the results indicated substantial weight loss from silicon carbide dissociation and a sensitive dependence on sintering temperature and time that was found difficult to control. Sintering in an argon atmosphere appeared to be a potential solution and a simple (2X2) matrix experiment was conducted in which the two variables, sintering temperature and rate of temperature increase, were evaluated at two levels each. Injection molded bars were dewaxed in a vacuum dewaxing cycle and sintered in an argon atmosphere. The density results are shown in Figure 1, Argon Sintering with Two Variables; Rate of Temperature Increase and Sintering Temperature, and it can be seen that an acceptable high density was obtained at 2150°C for 30 minutes at the slower heating rate. The arrow in the Figure indicates the path of greatest ascent for the density data.

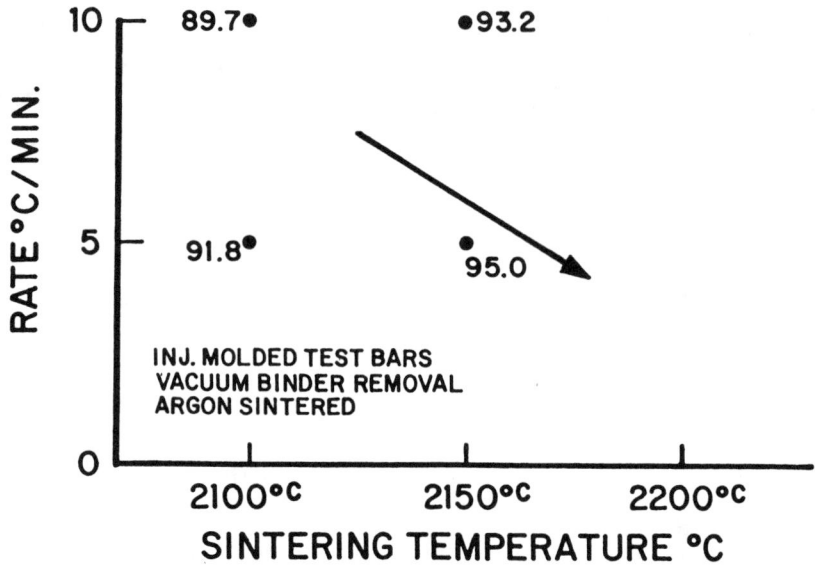

Fig. 1. Argon sintering potential: sintered density vs. sintering.

Large Matrix Experiment of Five Variables to Improve Density and Strength

A one half fraction factorial experiment of a two to the fifth power design was carried out to improve density and strength. The five variables at two levels each were two variations of the fluid mix process, two percentages of solids loading, two dewaxing cycles, two sintering temperatures and two sintering times. A process flow chart for this experiment is shown in Figure 2 which includes the details of the variable values and the process routes. The fluid mixing process (Table I) included two variations in wet ball milling times. The entire matrix was duplicated by sintering in both argon atmosphere, with appropriate sintering times and temperatures, and in vacuum with appropriate vacuum sintering times and temperatures. Eight duplicate injection molded MOR bars were in each group. The mean densities, mean strengths and highest value of strength were treated statistically and the effects were determined as shown in Table II for mean densities of vacuum sintered bars, and in Table III for densities of argon sintered bars. In Table I the average density of all bars was 97% of theoretical density while the average density for argon sintered bars (Table II) was only 91.8%. The effects of mean densities (Table II) for vacuum sintered bars indicate that Process B leads to higher densities than Process A, the higher sintering temperature leads to higher densities than the lower one, and that some interactions between variables are noted. For argon sintered material, the densities were influenced only by the solids loading variable and a strong solids loading process interaction. A comparison of the vacuum sintered and argon sintered MOR averages of mean strengths and highest values and their effects are shown in Table IV. Clearly Process B is superior to Process A for

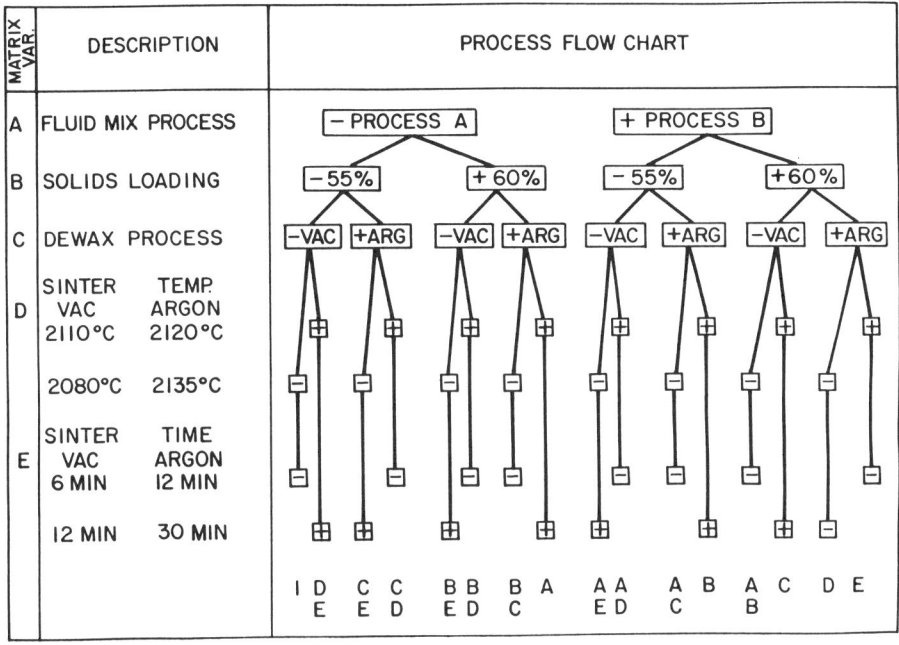

Fig. 2. Process flow chart for a half fraction of 2^5 design.

Table I. Fluid mixing processes.

| | Hours | |
Process Step	Process A	Process B
Wet Ball Mill		
Solids	48	96
Solids + Binders	24	72
Drying		
Stir Dry	14	4
Pan Dry	100	100
Prepare for Molding		
Mix	4	4
Add Wax	4	---

Silicon Carbide 357

Table II. Matrix 1A Mean densities of machined MOR bars yields and effects (%t.d.) vacuum sintered.

Expt. No	Effects	Means Yield	Effects
1	(AVE.)	96.22	(97.01)
2	A	96.39	[.37]+.10*
3	B	95.94	.18
4	AB	97.52	[.58]
5	C	95.91	−.25
6	AC	96.70	.18
7	BC	96.24	.14
8	ABC,DE	97.25	[−.40]
9	D	97.81	[.73]
10	AD	97.06	−.27
11	BD	97.20	.00
12	ABD,CE	97.95	−.08
13	CD	97.17	−.01
14	ACD,BE	97.12	−.08
15	BCD,AE	97.13	−.11
16	E	97.57	.14

*Estimated Standard Error

Variable	−	+
A	Process A	Process B
B	55% Solids	60% Solids
C	Vac Dewax	Argon Dewax
D	2080°C	2110°C
E	6 Min.	12 Min.

Table III. Matrix 1B mean densities of machined MOR bars yields and effects (%t.d.) Argon sintered.

Expt. No.	Effects	Means** Yield	Effects
1	(AVE.)	93.00	(91.84)**
2	A	90.87	0.28 +_ .12*
3	B	91.74	[0.71]
4	AB	92.93	[1.15]
5	C	91.93	-0.12
6	AC	90.82	0.22
7	BC	91.97	0.07
8	ABC,DE	92.45	-0.32
9	D	92.23	-0.24
10	AD	90.41	0.12
11	BD	91.82	0.09
12	ABD,CE	92.59	0.07
13	CD	91.65	0.04
14	ACD,BE	90.99	0.14
15	BCD,AE	91.51	-0.13
16	E	92.57	0.11

*Estimated Standard Error

Variable	−	+
A	Process A	Process B
B	55% Solids	60% Solids
C	Vac Dewax	Argon Dewax
D	2135°C	2120°C
E	30 Min.	60 Min.

** Means of 8 Bars

Table IV. Matrix 1 comparison of vacuum sintered and Argon-sintered mean and highest value strengths of machined MOR bars effects (ksi).

Effects	1A Vacuum-Sintered		1B Argon-Sintered	
	Means	Highest Value	Means	Highest Value
(AVE.)	(57.4)	67.2	52.05	60.0
A	[4.9]+−1.0*	[7.5]+−0.8*	[6.0]+−0.5*	[6.4]+−0.6*
B	−1.3	1.0	0.8	0.9
AB	−1.5	1.1	1.9	0.4
C	3.0	2.2	[−2.9]	−1.2
AC	1.1	3.1	2.1	1.9
BC	2.5	[3.9]	[2.3]	[3.5]
ABC,DE	−0.6	0.1	[2.8]	[3.2]
D	[−3.9]	[−5.3]	−0.6	0.4
AD	−1.6	−2.7	−1.3	−0.2
BD	−1.6	[−4.4]	−0.4	−0.1
ABD,CE	1.3	0.8	−1.3	−1.7
CD	0.5	−2.3	−0.8	−2.8
ACD,BE	−0.4	−0.1	1.1	−0.8
BCD,AE	−1.4	−0.8	0.2	0.8
E	−0.6	1.2	−0.6	−0.3

*Estimated Standard Error

Variable		−	+
A		Process A	Process B
B		55% Solids	60% Solids
C		Vac Dewax	Argon Dewax
D	Vac	2080°C	2110°C
	Vac-Argon	2135°C	2120°C
E	Vac	6 Min.	12 Min.
	Vac-Argon	30 Min.	60 Min.

** Means of 8 Bars

all measures of strength. The other significant effects are specific to vacuum sintering or to argon sintering processes. A summary of the strength data for both normal statistics and Weibull statistics is shown in Table V up to that point in the program. Fractographs from some of these samples were evaluated and the same type of fracture origins were observed as shown in Part I, but the flaw dimensions were diminished which accounted for the increase in mechanical strength. The longer milling time provided in Process B is believed to be the primary process improvement leading to strength improvement.

SINTERING CYCLE IMPROVEMENTS

A fractional factorial experiment of the design, $2^7 - 4$, was performed to evaluate the main effects of seven factors at two levels each in eight experiments. Only the main effects can be estimated, and further experimentation is necessary to elucidate the two factor or higher interactions. The variables and their two levels

Table V. Summary of SIC strength data (room temperature).

		Normal Statistics		Weibull Statistics	
	n	Mean (ksi)	Std. Dev. (ksi)	0 (ksi)	m
BASELINE	30	43.3	5.7	45.8	8.0
FLUID MIXED 5 Batches					
2080°C	22	54.0	8.2	57.1	8.7
2110°C	20	51.7	8.4	55.2	6.7
TASK II MATRIX 1A					
Best Group (4M-60-A-2080)	8	64.5	10.1		
All 4Q	31	61.4	6.6	64.2	10.5
All 4M	32	58.6	10.2	62.9	6.0
MATRIX 1B Vac-Argon Sinter (Density 90-93%)					
Best Group #8 (4M-60-A-2135)-30)	7	60.7	4.5		
All 4Q	31	51.8	11.8	57.1	7.9
All 4M	28	56.3	6.3	59.1	10.1

Table VI. Mean densities of machined MOR bars effects (% t.d.).

Expt. No.	Variables	Mean Yields*	Main Effects
1	(Ave.)	96.3	
2	A	94.0	[1.08] +_.15**
3	B	92.7	−0.23
4	C	94.3	0.38
5	D	93.4	[1.53]
6	E	94.8	[1.43]
7	F	93.5	[1.43]
8	G	97.1	−0.43

*10 Samples
**Estimated Standard Error from Previous Experiments

Variables	−	+	"t" Test Probability
A Fluid Mix Process	"C"	"B"	0.001
B Carbon Source	DeGussa	Carbon Black	0.600
C Argon Ramp Rate	9°C/Min.	5°C/Min.	0.290
D Argon Sinter Temp	2080°C	2120°C	0.000
E Argon Sinter Time	12 Min.	30 Min.	0.000
F Backfill Temp.	1620°C	1760°C	0.000
G Vacuum Hold Time	4 Hr.	2 Hr.	0.200

are listed in Table VI. The first two variables are intended to evaluate a small change in Process B, involving the time of wax introduction, and the source of carbon in the mixes. DeGussa carbon is a source of finely divided carbon in suspension in toluol, whereas carbon black is a powder source that is mechanically mixed during the fluid mixing process. The other five factors are parameters in the argon sintering cycle. The results of this experiment are given in Tables VI, VII and VIII. The average of the densities was above 96% (Table VI) and highly significant effects were observed for variables A, D, E, and F. Process B, 2120°C sintering temperature, 30 min sintering time, and 1760°C backfill temperature were all significantly better than the alternative factors. Student's "t" test probabilities are also listed in Table VI which show significant differences between the two levels of variables A, D, E and F. The mean strengths of MOR bars and the main effects are given in Table VII. Process B, the slower heating rate, 30 min sinter and 1760°C backfill temperature are all favored over the alternative parameters, and the "t" test results support the significance of the effects. Considering the extreme value statistics (highest values) one sees similar results in Table VIII. Process B, 30 min sintering time and 1760°C backfill temperature are all significantly better than the alternative parameter.

Table VII. Mean strengths of machined MOR bars effects (ksi).

Expt. No.	Variables	Mean Yields	*Main Effects
1	(Ave.)	61.0	
2	A	59.1	[3.34] +_ 0.88**
3	B	55.0	-1.01
4	C	53.2	[3.27]
5	D	53.5	-0.76
6	E	63.7	[5.25]
7	F	59.0	[4.95]
8	G	66.0	-0.84

*10 Samples
**Estimated Standard Error

Variables	-	+	"t" Test Probability
A Fluid Mix Process	"C"	"B"	0.08
B Carbon Source	DeGussa	Carbon Black	0.65
C Argon Ramp Rate	9°C/Min.	5°C/Min.	0.07
D Argon Sinter Temp	2080°C	2120°C	0.74
E Argon Sinter Time	12 Min.	30 Min.	0.01
F Backfill Temp.	1620°C	1760°C	0.01
G Vacuum Hold Time	4 Hr.	2 Hr.	0.71

Table VIII. Extreme value statistics, MOR strengths (ksi).

Expt. No.	Variables	Highest Single Strength* (Ksi)	Main Effects (Ksi)
1	(Ave.)	77.0	
2	A	72.0	[3.6] +_ 0.7**
3	B	73.9	-0.3
4	C	64.9	-1.4
5	D	60.5	-1.7
6	E	76.1	[10.6]
7	F	66.3	[4.8]
8	G	79.2	0.3

* From Group of 10 Samples
**Estimated Standard Error from Previous Experiments

Heat Treatment of Silicon Carbide to Improve Strength

We have observed that heat treatment (annealing) of machined silicon carbide test bars improves the strength and, sometimes, reduces the scatter (standard deviation) in the strength data. To quantify these observations, a 2X2 matrix experiment was employed to evaluate two heat treating temperatures, [1200°] and [1400°C] for eighteen hours, and two heat treating environments, air and argon. Weight gains on sintered and machined MOR bars were also measured, and oxide coating thicknesses were calculated (Table IX). As expected, coating thicknesses were greater in air than in argon. The mean strengths and standard deviations of groups of seven or eight bars were measured, as given in Table X. The "t" tests indicated that the [1200°C] heat treatment in both argon and air yielded better results (stronger samples) than the [1400°C] air anneal. When the two temperatures were compared with samples from both argon and air grouped together, the bars from the [1200°C] anneal temperature were significantly better than bars from the higher temperature anneal. These data indicate that strength improvements in the range of 15 to 20% at these strength levels can be realized by heat treatment. It is speculated that the mechanism by which the strength of MOR bars is improved involves the oxidation of the surface to reduce the severity of surface flaws and provides for a greater probability of initiating fracture at internal flaws instead of at surface flaws.

Comparison of Two Silicon Carbide Powder Sources at Two Boron Levels

An experiment was designed to compare two commercially available silicon carbide powders (Ibiden powder from Japan, used exclusively in this program, and Superior Graphite powder from the United States) at each of two boron levels, nominally [0.5%] and [1.0%.] The design of the experiment, a [2X2] factorial, and the density results are given in Table XI. The Ibiden powder sinters to a higher density than the Superior Graphite powder under the conditions of the experiment, and 1.0% boron yields higher density bars than does 0.5% boron. The mean strengths and highest value strengths from groups of eight to ten samples are listed in Table XII along with the effects of the variables. None of the effects is statistically significant in light of the size of the estimated standard error of the experiment. Similar groups of MOR bars were annealed at [1200°C] in air and the strength data were evaluated. The effects of the variables were not significant, although the general strength level tended to be higher than the non annealed group. A comparison was made of these two groups of samples and it was observed that the annealed samples were significantly stronger in several of the groups. When all of the samples are grouped together in non annealed and annealed groups, the annealed samples are significantly stronger, as indicated from the student"t" test.

Summary

Several statistically designed experiments were performed to evaluate process variables which were thought to influence the density and strength of injection molded pressureless sintered silicon carbide. Vacuum sintering and argon sintering were investigated, and an argon sintering cycle was developed which yielded acceptable density and strength. Flaw sizes in the microstructure were shown to be

Table IX. Weight gains on sintered SiC MOR bars after heat treatment for 18 hours.

Atmosphere	Temp °C	Weight Gain (GMS X10000) Bar			Calculated Thickness (μm)
		n	x	σ	
Argon	1200	7	0.6	3.9	0.2
Argon	1400	7	0.4	3.3	0.1
Air	1200	8	3.3	2.1	1.0
Air	1400	8	4.5	2.1	1.4

n = Number of samples

Table X. Strength of heat treated MOR bars.

Atmosphere	Temp °C	n	Strength x (ksi)	σ (ksi)	x % Change*
Argon	1200	7	63.8	6.7	15.6
Argon	1400	8	55.4	11.5	0.4
Air	1200	8	66.4	9.4	20.3
Air	1400	7	57.1	4.1	3.4

*Baseline – 28 MOR Bars – 4M, x=55.2 ksi, σ=9.3 ksi
n = Number of Bars

Table XI. Mean densities of machined MOR bars yields and effects (% t.d.).

Batch	Expt. No.	Effect	Means** Yield	Effects
14A	1	(Ave)	98.4	(96.4)
16A	2	A	94.1	−3.0 ± .1*
4N	3	B	97.4	0.3
15A	4	AB	95.7	1.3

*Estimated Standard Error From Task II – Matrix 1
**Means of 10 Samples

Variables	−	+
A (SiC Source)	Ibiden	Superior Graphite
B (Boron Level)	0.5%	1.0%

Table XII. Mean and highest value strengths of sintered machined MOR bars yields and effects (ksi).

Expt. No.	Effects	Means* Yields	Means* Effects	Highest Values* Yields	Highest Values* Effects
1	(Ave)	66.6	(61.4)	80.6	(75.1)
2	A	49.9	0.0 ± 3.4**	67.9	−3.7 ± 2.9**
3	B	56.3	6.3	73.4	1.8
4	AB	72.9	16.7	78.7	9.0

*Mean and Highest Value of 8 to 10 Samples
**Estimated Standard Error

reduced and strength was increased with improvements in the fluid mixing process. Heat treating (annealing) was found to significantly increase the strength of sintered and machined MOR bars. A second source of silicon carbide powder was found to yield similar strengths as the initial powder source in this program. The general level of strength increased from a mean of [299 MPa(43 Ksi)] for the baseline material to mean values of about [483 MPa (70 Ksi)] for annealed MOR bars.

Acknowledgement

The authors wish to acknowledge Drs. R. Govila and R. M. Williams (deceased) for mechanical property and microstructural evaluation of these materials, E. Cartwright, B. N. Juterbock, L. V. Reatherford, R. Elder and W. Trela for material processing, and Dr. W. L. Winterbottom for program management.

---

*Supported in part by NASA Contract No. [NAS3-24384]

Fabrication of Sintered Alpha-SiC Turbine Engine Components

R. W. Ohnsorg, M. O. Ten Eyck

Standard Oil Engineered Materials Company
Niagara Falls, NY

Abstract

Sintered alpha-SiC turbine engine components for the Advanced Gas Turbine (AGT) Program have been fabricated by Standard Oil Engineered Materials Company almost exclusively using three primary forming procedures—wet-bag isostatic pressing followed by green machining, slip casting, and injection molding. The forming method for each fabricated component is listed and a closed loop fabrication approach is presented which takes into account three general paths utilized in taking a material through to commercial realization.
The AGT 101 transition duct fabricated by isostatic pressing or injection molding, the AGT 100 gasifier turbine scroll formed by slip casting and isopressing, and the AGT 100 gasifier turbine rotor formed by injection molding are discussed in detail with regard to iterative changes throughout the course of development and improvement. Each of these iterative changes has been shown to correspond to one of the three closed loop paths—component development/process refinement, process refinement through performance, or component design through performance.

Introduction

Excellent physical properties and economic considerations make sintered alpha-SiC (SASC) a prime candidate for applications in gas turbine engines where, at temperatures of up to 1371°C (2500°F), strength retention, corrosion, oxidation, creep, and thermal fatigue affect the integrity of the component.
Ceramics offer improved fuel economy due to combustion chamber heat retention as a result of elimination of water cooling in the turbine engine and less dependence upon lubrication.[1] Since SASC weighs about 60% less than metals, less inertial forces in rotating components and improved acceleration will result. Multi-fuel capability will be achieved and reduced emissions result from the higher temperature combustion. In addition, SASC is made from the raw materials sand and coke which are in abundance as opposed to superalloy components containing materials such as cobalt and chromium.
Because of its high temperature strength and stability both Allison and Garrett have elected to include SASC as a prime material in their turbine engine development programs. Standard Oil has a wide range of forming capabilities which include cold pressing, hot pressing, extrusion, dry-bag isopressing, wet-bag isostatic pressing, slip casting, and injection molding.

Forming Processes

The three primary forming procedures used by Standard Oil for the AGT components are wet-bag isostatic pressing followed by green machining, slip casting, and injection molding.

In wet-bag isopressing, a spray-dried free flowing powder containing binders and sintering aids is vibrated into a rubber bag having a rectangular or cylindrical shape. An appropriately sized steel mandrel can be inserted to reduce material usage. Top and bottom rubber closures are incorporated and the assembly is deaired. Hydraulic pressure is subsequently applied through a fluid thereby compressing all outside surfaces of the powder (Fig. 1).[2] This procedure is generally utilized either for simple cylindrical and tubular shapes or for prototypes. The advantages of this process are a uniformly dense green body having excellent fired properties with minimal sintering distortion, low tooling costs, and fast adaptability to design changes. However, fabrication of complex shapes such as turbine rotors is either not possible or extremely time consuming and costly. Low production rates and difficulties in automation are other characteristics for this process.

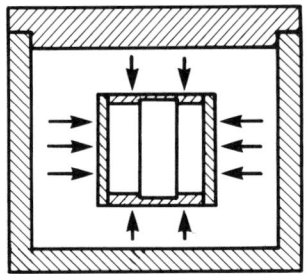

Fig. 1. Wet-bag isopressing schematic.

In slip drain casting a fluid suspension of a sinterable powder containing a deflocculant, binder, and sintering aids is poured into a porous plaster mold. The plaster mold dewaters the slip and the excess slip is drained after a suitable wall thickness is built up. After a suitable time period, the mold halves are separated. Fig. 2 shows the steps in forming a connecting duct, one of the five parts constituting the AGT 100 gasifier turbine scroll. In the case of more complex components, core inserts are removed while the material is still moist but before shrinkage occurs avoiding cracking of the part. Slip drain casting allows undercuts and hollow configurations to be formed which would be impossible using machining or injection molding procedures, and the external configuration of the part is controlled by the plaster mold surfaces. A limitation is that parts need to be of a constant wall thickness. The advantages include net shape or near-net shape forming of complex shapes with constant thickness at low tooling costs. Other

- Description
  - Fluid supension poured into porous plaster mold which dewaters slip; mold inverted and excess slip poured out after suitable wall thickness built up.

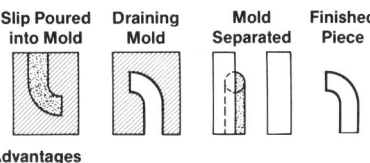

- Advantages
  - Net shape or near-net shape capability
  - Complex shape capability for constant thickness
  - Low tooling costs
- Disadvantages
  - Internal dimensional control limited
  - Low production rates
  - Difficult to automate

Fig. 2.   Slip casting connecting duct.

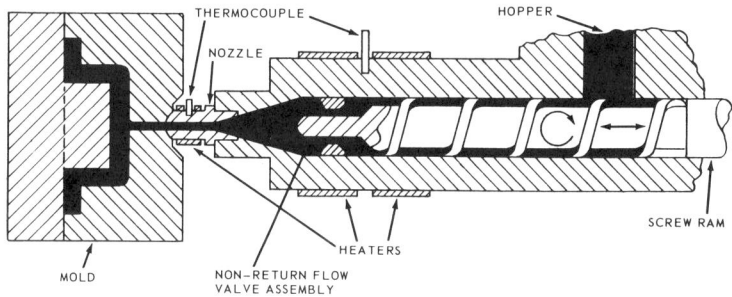

Fig. 3.   Injection molding machine schematic.

disadvantages are internal dimensional control is limited due to slip draining characteristics and production rates are generally low because the process is labor intensive and difficult to automate.

Injection molding is a forming procedure in which a granulated thermoplastic mix is compounded for improved plasticity and homogeneity. At Standard Oil a horizontal reciprocating screw injection molding machine is used to inject the mix into a warm mold. The granulated compound is fed into a heated barrel and replasticized by screw rotation during rearward screw travel thereby placing a predetermined volume of material in front of the screw. Subsequent forward travel of the screw without rotation pushes the thermoplastic through a heated nozzle into a sprue bushing through runners and gates into the part as shown in Fig. 3. In the case of the turbine rotor, the sprue is part of the nose or shaft and no runners or gates are utilized.

The major advantages of injection molding are its net shape or near-net shape capability and its ability to form a wide variety of complex shapes having variable thickness geometry. An important example is the gas turbine rotor where the thickness ranges from 0.76 mm (0.03 in.) for inducer blade tips to over 63.5 mm (2.5 in.) for the central hub portion. In addition, high production rates are possible and the process is easily automated.

Disadvantages include high tooling cost and related lead times as well as complex shrink factor determination and therefore the potential for several tooling revisions. Long binder removal cycles are required for large parts with substantial cross sectional thickness to assure that the resin is properly removed.

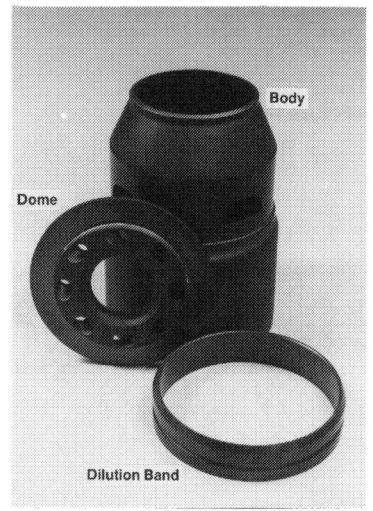

Body — Slip Casting
Dome — Isostatic Pressing
Dilution Band — Isostatic Pressing

Fig. 4. AGT 100 combustor.

AGT Component Fabrication

The AGT 100 components fabricated by Standard Oil can be divided into three basic sections—combustor, gasifier turbine, and power turbine. The combustor (Fig. 4) contains three components, a body having relatively constant wall thickness and a dome and dilution band having symmetrical configuration.

The most critical component in the gasifier section (Fig. 5) is the radial rotor which is a symmetrical one-piece component having variable thickness geometry. The scroll assembly is the only component which not only contains more than one part but utilizes two forming procedures to fabricate its five individual parts.

The vane is relatively small incorporating a critical curvature and is most suitably formed by injection molding. The outer backplate is symmetrical except for the for tabs making it suitable for isopressing followed by two green machining operations—lathe turning and computer numerical control (CNC) machining. The symmetrical inner backplate is smaller than the outer backplate lending itself to isopressing and lathe turning.

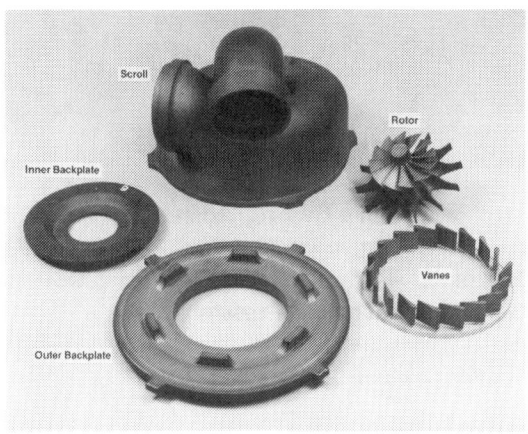

Rotor — Injection Molding
Scroll — Slip Casting/Isostatic Pressing
Vanes — Injection Molding
Outer Backplate — Injection Molding or Isostatic Pressing
Inner Backplate — Isostatic Pressing

Fig. 5. AGT 100 gasifier turbine.

The power section (Fig. 6) uses the same components as but larger than the gasifier side. The same forming procedures therefore apply. Although the power turbine rotor is formed by injection molding, it is fired using a procedure which infiltrates the porous baked part with silicon forming a SiSiC composite. In this way a nearer to net shape part can be molded as opposed to the gasifier rotor which has a linear shrinkage of about 20% by pressureless sintering. The power scroll assembly utilizes only three parts as opposed to five for the original gasifier scroll.

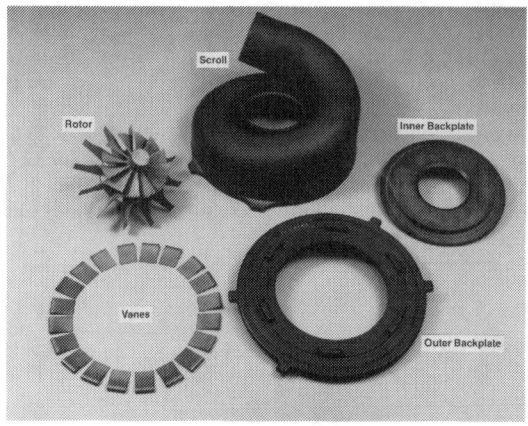

Rotor	— Injection Molding
Scroll	— Slip Casting/Isostatic Pressing
Vanes	— Injection Molding
Outer Backplate	— Injection Molding or Isostatic Pressing
Inner Backplate	— Isostatic Pressing

Fig. 6. AGT 100 power turbine.

The AGT 101 components fabricated by Standard Oil (Fig. 7) with the exception of the stator segments were designed as symmetrical components to aid in fabrication. Additional features such as three protruding tabs on the combustor baffles and three corresponding flats on the transition duct were incorporated for assembly purposes. The transition duct was subsequently modified providing a more complex exterior, and lathe turning plus CNC machining were utilized as well as injection molding in a parallel development program. Injection molding was also the preferred fabrication approach for the turbine shroud largely because of its size and complexity, and the stator segments exclusively because of their complexity.

Closed Loop Fabrication Approach

A closed loop fabrication approach has successfully been used to develop components suitable for engine testing for all the forming procedures utilized.[3] The fabrication approach illustrated in Fig. 8 provides three paths for taking a component through to commercial realization. After general material and process development which includes fabrication and testing of cold pressed bars or plates to generate material properties data, the part designer frequently uses finite element modeling to aid in configuration design. A fabrication procedure which can satisfy

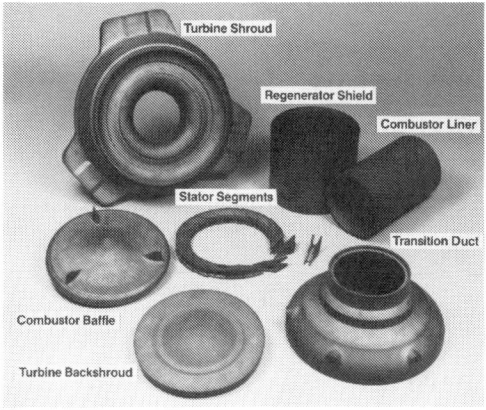

Turbine Shroud	— Injection Molding
Regenerator Shield	— Isostatic Pressing
Combustor Liner	— Extruding or Isostatic Pressing
Transition Duct	— Injection Molding or Isostatic Pressing
Turbine Backshroud	— Isostatic Pressing
Combustor Baffle	— Slip Casting or Injection Molding
Stator Segments	— Injection Molding

Fig. 7. AGT 101 components.

shape and tolerance requirements is selected and components are fabricated. The component development/process refinement loop (solid line) is pursued during the initial stages of component development to improve and optimize the process. In addition, minor component design changes may be incorporated to simplify forming.

Process refinement through component performance is illustrated by a second route within this closed loop fabrication approach (dashed line). Nondestructive evaluation (NDE) acceptable and dimensionally satisfactory components are subjected to rig and/or engine testing. Unacceptable performance may necessitate modification and improvement of fabrication procedures.

Lastly, component or brittle material design through performance is represented by the third loop (dotted line). Unacceptable rig or engine performance caused by localized high stress levels may necessitate a change in component design or material selection. Detailed finite element (zoom) modeling may be required to determine the actual stress level at certain initial design features.

One or more of these loops has successfully been used to fabricate each of the components for the two gas turbine programs. In the following sections three different components made using at least one of the three primary forming procedures are discussed in more detail. The iterative changes based on the closed

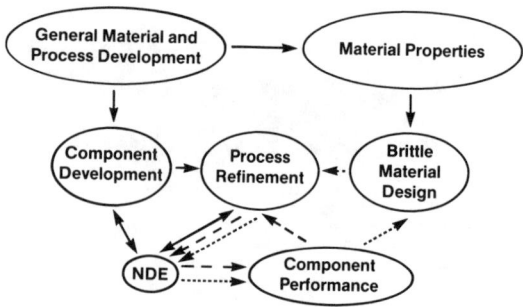

Fig. 8. Closed loop fabrication approach.

loop fabrication approach is illustrated on the AGT 101 isopressed or injection molded transition duct, the AGT 100 isopressed and slip cast gasifier scroll assembly, and the AGT 100 injection molded gasifier turbine rotor.

TRANSITION DUCT

The present transition duct design has integral thermocouple ports and a 180° airflow diverter as shown in Fig. 9. Transition duct development has continued throughout the AGT program on isopressed and since 1984 also on injection molded components. The combustion products enter between the transition duct and combustor baffle and, in order to effect separation, the baffle has three equispaced tabs which initially fit into three corresponding tab details green machined on the inside surface of the original transition duct design as shown in Fig. 10(a). Since both lathe turning and CNC machining were required on the isopressed component, fabrication simplification led to the continuous inside platform. The outer contour was changed opposite the continuous inside platform (b) to minimize thermal stresses due to thickness gradients. A decision to add three thermocouple ports for improved temperature control during testing was accommodated by shrink-fitting tubular inserts (c). Because inadequate joints allowed leakage of gas at relatively low pressures (air at .05 MPa) the ports were brazed using $MoSi_2$. Subsequent failure at low stress levels due to oxidation degradation resulted in a redesign and a modified fabrication approach, providing an example of component design through performance in the closed loop fabrication approach. Integral thermocouple ports and a 180° airflow diverter were incorporated (d), the latter to provide a more uniform air flow within the engine.

Because of the subsequent nonsymmetrical design, extensive CNC green machining was required. As a result, injection molding was pursued in parallel with isopressing and green machining. Early injection molded parts showed nonuniform green density as determined by computed tomography. Cracks developed in the lower molded density regions as shown by both tomography and X-ray on the sintered component. By eliminating the step representing the continuous inside platform and providing a relatively continuous blend as shown in Fig. 11, the injection molding turbulence was reduced, the density gradient was

Fig. 9. Transition duct.

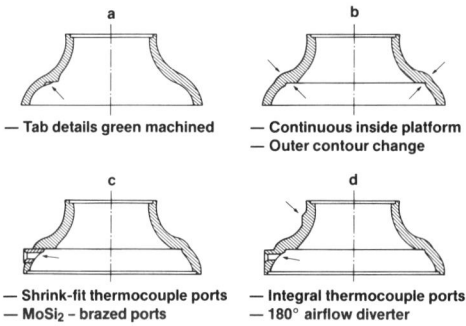

Fig. 10. Transition duct isopressing chronology.

a — Tab details green machined
b — Continuous inside platform
— Outer contour change
c — Shrink-fit thermocouple ports
— MoSi$_2$ – brazed ports
d — Integral thermocouple ports
— 180° airflow diverter

Silicon Carbide

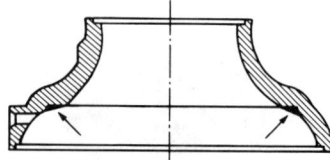

Fig. 11. Transition duct injection molding modification.

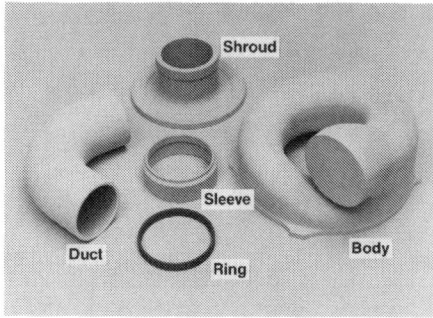

Fig. 12. Five-piece scroll assembly components.

minimized, and the cracking was eliminated. In addition, sprue and gate modifications were incorporated to ensure a more uniform flow of material over the nonsymmetrical configuration. The additional machining required was more than offset at this stage of the development cycle by the higher yields obtained.

Gasifier Turbine Scroll

The gasifier turbine scroll components for the five-piece assembly are shown in Fig. 12.[4] The nonsymmetrical scroll body and connecting duct with constant wall thickness are fabricated by slip casting. The shroud, adapter sleeve, and adapter ring having symmetrical configuration are fabricated by isopressing and lathe turning. The initial assembly procedure utilized a variety of green to sintered and partially sintered to sintered shrinkfitting steps. The mating surfaces of the scroll body and shroud were machined and final assembly was accomplished by joining at high temperature using a $MoSi_2$ braze (Fig. 13). A problem subsequently encountered was failure of the brazed joint in engine tests. Cyclic oxidation testing revealed joint degradation and expansion. Various different high temperature braze materials were investigated and a CrVTi braze developed at Oak Ridge National Laboratories was found to perform best. One of the CrVTi brazed scroll assemblies has passed four 25-hour cycles at 1288°C (2350°F) followed by a 245-kg (540-lb) load test, representing a successful example of process refinement through performance.

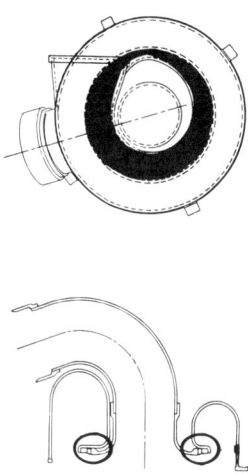

Fig. 13. Scroll assembly braze.

Inadequate braze coverage permitting leakage between hot and cold surfaces led to a design change which consisted primarily of lip removal on the scroll body flange (Fig. 14). The process involved first increasing the diameter of the lip for sintering stability and subsequent machining to match print dimensions. The machined mating surfaces were brazed using CrVTi. The result was stress reduction caused by thermal gradients between components and improved braze coverage largely thought to be due to removal of the restricting flange lip.

Because of alignment and assembly difficulties within the engine, the scroll assembly was converted from a five-piece assembly to a three-piece assembly (Fig. 15). To facilitate this design change the shroud height was increased and the connecting duct was combined with the interturbine coupling to form a single separate component.

Gasifier Turbine Rotor

The gasifier and power turbine rotors are the only rotating AGT ceramic components fabricated by Standard Oil. Program success has largely been measured by satisfactory proof tests and successful rig and engine test results of the gasifier turbine rotor because of its dynamic characteristics and higher temperature requirement. Several development iterations were performed during the program and several hundreds of rotors were fabricated by injection molding.

The original gasifier turbine rotor design (Fig. 16) exhibits a stress distribution at design speed as shown in Fig. 17. The three areas of relatively high stress addressed in subsequent process iterations include the base between the airfoils (a), the base/shaft fillet (b), and the center of the hub (c).

Several molding trials were conducted during this development program, first to determine molding parameters and second to develop a suitable processing

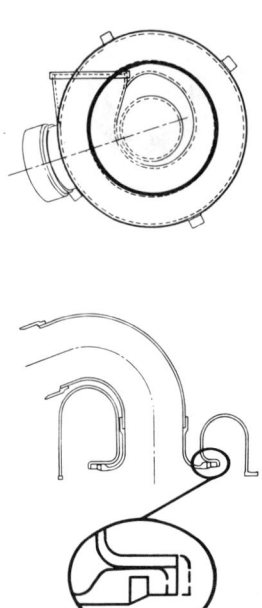

Fig. 14. Scroll body flange modification.

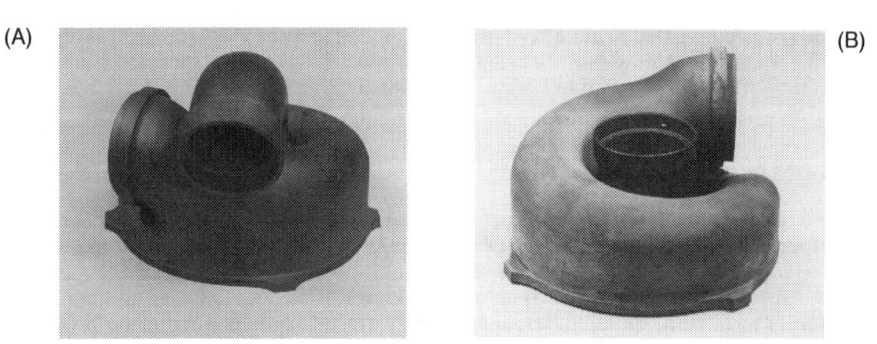

Fig. 15. Five-piece (A) and three-piece (B) scroll assemblies.

Fig. 16. Gasifier turbine rotor.

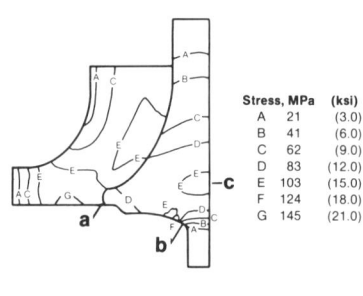

a — Base between airfoils
b — Base/Shaft fillet
c — Center of hub

Fig. 17. Stress distribution at design speed.

sequence. The Group 1 baseline rotors had an average burst speed of 99.3 krpm.[5,6] However, knit lines were detected between the airfoils (Fig. 18). Modification of molding parameters reduced the incidence of knit lines in this area, but the average failure speed dropped to a disappointing 80.5 krpm. Failure analysis by Allison determined that the low spin results could be traced to shallow knit lines on the backface near the shaft fillet. The knit lines had apparently moved as a result of molding condition modification. After removal of a thin layer (about 1.27 mm) through green or final machining the average failure speed increased to 95.1 krpm.

Although successful, this procedure was labor intensive and new measures had to be introduced. The injection molding tool was therefore modified to provide shaft end injection as opposed to nose end injection for improved material integrity in the backface/shaft region (Fig. 19). The series of short shots for both nose-end

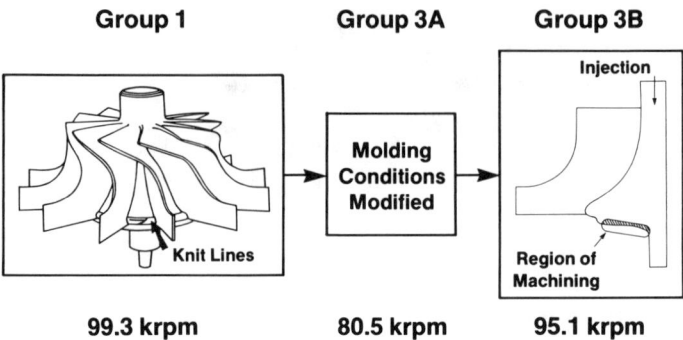

Fig. 18. Rotor iterations.

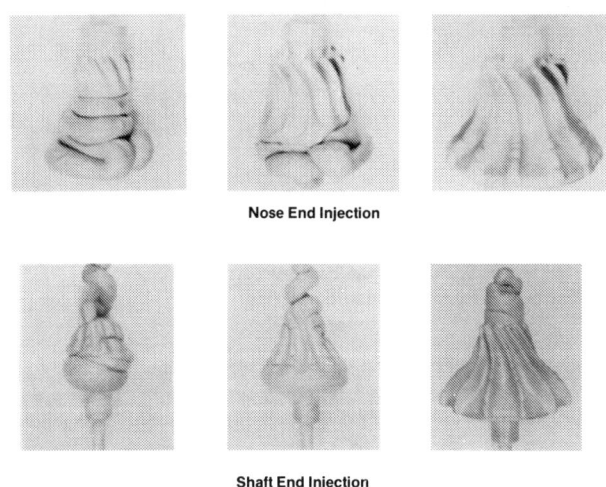

Fig. 19. Injection direction comparison.

and shaft-end injected rotors show crossovers between strands making them susceptible to air entrapment or knit line formation at the rotor base when using nose end injection.

A group of 70 shaft-end injection rotors demonstrated 99.8 krpm average failure speed as shown in Table I. More significantly, the Weibull modulus almost doubled and the overall process yield tripled. Failure of the inducer blades during

Table I. Rotor performance summary.

Rotor Type	No. Spin Tested Burst	No. Spin Tested Proofed	Avg. Failure Speed* (krpm)	Weibull Modulus	Process Yield (%)
Group 1	10	0	99.3	6.9	11
Group 3A	13	3	80.5*	6.6	13
Group 3B	13	17	95.1*	6.5	16
Group 8	40	30	99.8*	12.0	45

*suspended item analysis

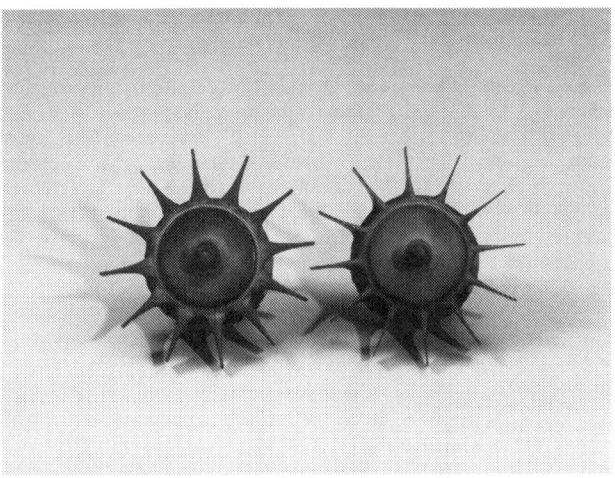

Fig. 20. Original ECR (right) versus thickened airfoil (left).

subsequent engine testing of these Group 8 rotors was attributed to foreign object damage (FOD). In order to improve performance, Allison redesigned and increased the thickness of the inducer portion of the airfoil as shown in Fig. 20. This provided a change in the stress distributions and the maximum stress increased to about 200 MPa (29 ksi) at the base/shaft fillet as opposed to about 152 MPa (22 ksi) at both the underside of the inducer blades and the base/shaft fillet on the original engine configuration rotor (ECR (Fig. 21)). As a result, the spin performance was calculated to decrease by 8.3%. Despite this additional stress increase, a thickened airfoil rotor was recently successfully engine tested at 1079°C (1975°F) and 70% of design speed, clearly a successful example of the closed loop fabrication approach.

Hot isostatic pressing or HIPing was investigated in an attempt to further improve the physical properties.[7,8] It was found that both the twelve rotors HIPed at 138 MPa (20 ksi) and the 33 rotors HIPed at 207 MPa (30 ksi) had an average

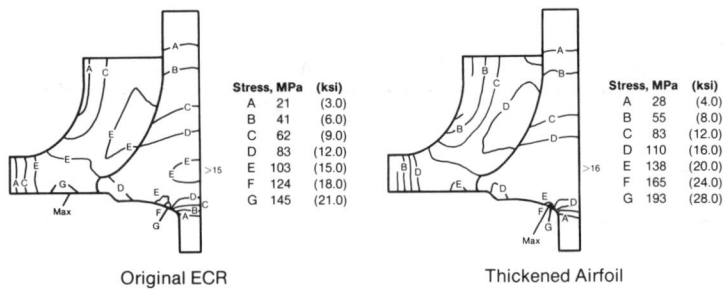

Fig. 21. Stress distribution original ECR versus thickened airfoil.

final density of 3.16 g/cm^3 (98.4% theoretical) even though they exhibited average pre-HIP densities of 3.10 and 3.11 g/cm^3 respectively.

The influence of HIPing on material strength was assessed using test bars cut from rotors. Room temperature flexure strength tests were conducted on 50.8 mm × 6.4 mm × 3.2 mm (2" × 1/4" × 1/8") specimens cut both axially from the center of the hub and radially from the base. Table II shows average flexure strength for rotor test bars tested by both Allison and Standard Oil. A strength increase was observed in all cases resulting in an average overall increase of about 15%.

Table II. Strength comparison sintered versus HIPed rotors.

	Rotors Tested at Allison				
		Axial Direction		Radial Direction	
Condition	No. of Rotors	MOR, MPa, (Ksi)	No. of Bars	MOR, MPa (ksi)	No. of Bars
Sintered	2	325 (47.2)	12	370 (53.6)	28
HIPed (138 MPa)	2	413 (59.9)	12	392 (56.8)	28

	Rotors Tested at Standard Oil				
		Axial Direction		Radial Direction	
Condition	No. of Rotors	MOR, MPa, (Ksi)	No. of Bars	MOR, MPa (ksi)	No. of Bars
Sintered	4	302 (43.8)	18	311 (45.1)	23
HIPed (207 MPa)	2	330 (47.9)	11	370 (53.6)	16

Spin testing of the rotors was carried out in vacuum at room temperature. A total of 23 sintered and 11 HIPed thickened airfoil rotors were either proofed or spun to failure (Fig. 22). Suspended item analysis was used to determine the average burst speed and Weibull modulus for proofed rotors. The Weibull modulus increased for both, but an 8.4% increase in average burst speed (90.6 krpm versus 98.2 krpm) was obtained through HIPing.

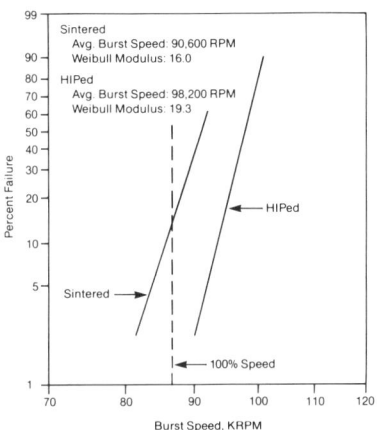

Fig. 22. **Spin results thickened airfoil.**

Computed tomography was used to compare spatial densities of both sintered and HIPed rotors to further aid in rotor analysis. Fig. 23 represents a schematic of the X-ray tomography result for the same rotor pressureless sintered and then HIPed at 2100°C and 207 MPa. The sintered rotor showing a wide variation in density has a low density hub center and a periphery close to theoretical. The HIPed rotor on the other hand shows much less variation and the low density hub center has virtually disappeared. The density gradient changed from about .28 to .10 g/cm^3 and densities in the central hub portion were found to increase by about 10%. This increase in overall densification manifests itself also in an observable volumetric shrinkage of about 1.2%.

As a result of HIPing, bulk density increased from 3.10/3.11 g/cm^3 to 3.16 g/cm^3 average and tomography revealed that density uniformity was greatly improved. The flexural strength of bars cut from rotors increased by 15% and spin performance improved by 8% (at 207 MPa).

SUMMARY

Each of the iterations discussed represents one of the three paths in the closed loop fabrication approach—component development/process refinement through

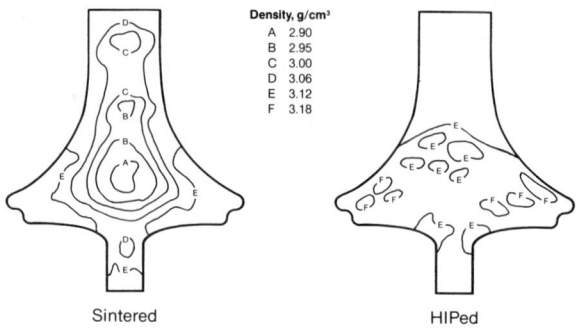

Fig. 23. Computed tomography result.

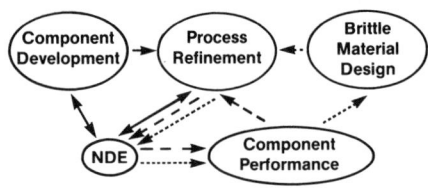

Fig. 24. Closed loop fabrication approach.

performance, and component design through performance. The closed loop fabrication approach is reviewed in Fig. 24 and all of the component iterations are summarized in Table III.

Several additional process iterations have been incorporated throughout the course of the AGT programs and each can be related to one of the three paths discussed. Although this concept is not new, the format presented here enables the part designer and part manufacturer to interact and to determine which path would be most suitable at a given decision point based on economic and performance considerations.

Table III. Summary of component iterations.

AGT 101 Transition Duct

- The inside platform and outer contour was modified.
- Thermocouple parts were added and the 180° airflow diverter was added.
- The inner contour was modified for injection molding.

AGT 100 Gasifier Turbine Scroll

- The braze was changed from $MoSi_2$ to CrVTi.
- The Scroll body flange was modified to remove the lip.
- The assembly was redesigned.

AGT 100 Gasifier Turbine Rotor

- Molding condition were modified to minimize knit lines between the blades.
- Backface machining was incorporated to remove defects near the shaft fillet.
- The injection direction was changed from nose end to shaft end.
- The inducer airfoil was redesigned.
- Hot isostatic pressing was incorporated.

References

[1] R. S. Storm and R. W. Lashway, "Progress in Ceramic Component Fabrication Technology," AIAA-82-1211 (1982).

[2] F. L. Kennard, "Ceramic Component Fabrication," Proceedings of the 13th Automotive Materials Conference, Ann Arbor, 1095-1111 (1985).

[3] J. W. Hinton, J. W. MacBeth, and M. O. Ten Eyck, "Recent Developments in the Fabrication and Testing of Ceramic Engine Components," Proceedings of the Twenty-Second Automotive Technology Development Contractors' Coordination meeting, Detroit, 505-515 (1984).

[4] R. Ohnsorg, M. Ten Eyck, J. Zanghi, and T. Sweeting, "Ceramic Component Fabrication," Proceedings of the Twenty-Third Automotive Technology Development Contractors' Coordination Meeting, Detroit 173-181 (1985).

[5] D. A. Turner and L. E. Groseclose, "Methods for Improving Reliability in Ceramic Turbine Rotors," SAE Paper 851788 (1984).

[6] R. Ohnsorg, M. Ten Eyck, and T. Sweeting, "Development of Injection Molded Rotors for Gas Turbine Applications," ASME 86-GT-45 (1986).

[7] T. J. Whalen, "HIPing of SiC," *Ceramic. Eng. Sci. Proc.*, **5** [5-6] 341-349 (1984).

[8] M. O. Ten Eyck, R. W. Ohnsorg, and L. E. Groseclose, "Hot Isostatic Pressing of Sintered Alpha Silicon Carbide Turbine Components," ASME 87-GT-161 (1987).

Large Silicon Carbide Radiant Tube Production Process

Michael C. Kasprzyk
Vice President—Business Development

Inex Incorporated
South Protection Road
Holland, NY 14080

Abstract

Radiant tubes made of nickel/chrome alloys have long been used in heat treating furnaces where the combustion process must be kept separate from the furnace atmosphere. Unfortunately, these tubes are prone to failure due to high temperature creep, oxidation or melting and furnace owners must plan on their frequent replacement.
    Silicon carbide radiant tubes overcome these problems. Ceramic tubes produced by conventional fabrication methods have, however, proved to be too expensive. Recently invented processing techniques using dry, as purchased, coarse-grained powders to produce reaction bonded silicon carbide tubes are described. This simple process which employs an inductively heated traveling hot zone has potential to dramatically increase process yield and reduce manufacturing costs. Large diameter radiant tubes fabricated in this way will provide furnace operators with a low cost, high-temperature ceramic alternative to metal alloy tubes.

Introduction

Work being done by Inex, Inc. to produce large silicon carbide radiant tubes is presented in this paper. This review consists of a description of the tubes, their use, their current composition and problems which are encountered. The benefits of low cost SiC tubes and the processes necessary to produce high quality, low cost radiant tubes made of SiC are discussed.
    Most heat treat furnaces such as carburizers or carbonitriders used to harden steel alloys and sometimes annealing operations employ an atmosphere controlled furnace. Usually these furnaces separate the combustion process from the atmosphere by the use of a radiant tube (Figure 1). The combustion takes place inside the tube and the heat is radiated out into the controlled atmosphere, hence the name radiant tube. These furnaces currently use alloy tubes made with high percentages of nickel and chrome. These systems generally work well and there are tens of thousands of these burner radiant tubes in use. There are, however, limitations and problems associated with their use.
    The upper limit of their use is controlled by their oxidation and high temperature creep (Figure 2). This in turn, limits the thermal efficiency of the combustion process and the capacity of the furnace. Operating at high temperatures is simply not possible. Secondly, the tubes have poor life, (Figure 3) usually

Fig. 1. Radiant tube.

Ni/Cr Alloy
Operating Temperature Limit

Material	Suggested Maximum Operating Temperature
H A 330	1800° F (982° C)
INCONEL 600	2000° F (1093° C)

Fig. 2. Existing materials.

Fig. 3. Scrapped tubes.

due to high temperature creep and/or burn through. This affects furnace up-time, product quality and manufacturing costs. In addition, high chrome/nickel tubes aren't cheap. In fact, they are very expensive and furnace operators must consider their frequent replacement as part of furnace operating costs.

The solution is simple, replace nickel/chrome alloy tubes with silicon carbide. SiC has excellent thermal conductivity, a higher melting temperature and is not as prone to high temperature creep.

Figure 4 shows the results of some high temperature creep testing done on Inex material compared to Inconel 600, a premium alloy. At 1200°C, the SiC showed no measurable creep while the Inconel 600 showed that creep failure was probable. At 1350°C, the SiC again showed no measurable creep and the alloy actually began to melt (Figure 5).

Several problems could be encountered when using SiC in this application. The brittle nature of ceramics means that extreme care must be employed when handling and installing large thin-walled bodies. Certainly thermal shock is also a concern, especially if the furnace is cycled on/off frequently. These problems can be readily overcome with proper handling during installation and controlled start up/shut down procedures.

There is another more serious problem—SiC radiant tubes have been too costly. In some cases they cost 5 to 10 times as much as metal alloy tubes. The end users simply cannot justify the high cost even when considering longer life, higher thermal efficiency, greater furnace throughput and reduced down-time.

What makes the cost so high? Conventional ceramic forming processes currently used to make large diameter radiant tubes are shown in Figure 6. A manufacturer could slip cast or extrude SiC to produce a green body, then fire it in a furnace to produce reaction bonded SiC or even sinter the material (with the proper licensing, of course). In both processes, a liquid is added, a green body is formed and then the liquid is removed. Anything which can be done to lower the labor content, lower the labor wage rate, or use lower cost energy or fewer facilities to reduce cost, would be beneficial.

Another way to reduce cost would be to improve the yield of the process. It is difficult to improve conventional methods as long as it is necessary to use a process that generates a green body and then requires that body to be handled and further processed. A different solution might be to invent a new process, such as one which doesn't require as much labor, facilities or energy used by conventional methods, or one which dramatically improves the yield.

Figure 6 describes conventional processes. The formation and handling of the green body obviously constitutes a large part of the labor, facilities and equipment costs. It is also during these steps where many problems are encountered.

NEW PROCESS

Inex has explored and recently developed under contract from the Gas Research Institute, Chicago, Illinois, several methods which eliminate the green body processing. In one new method, as purchased free-flowing, dry powders are loaded directly into an induction furnace and fired. Inex has produced small diameter, short tubes up to 3 1/4" in diameter and 18" in length. Equipment used for these small tubes is shown in Figure 7. A small 2.5 KW power supply was used to fabricate these tubes. Larger facilities to enable manufacturing up to 8" in diameter and 9 feet long have been developed—again under a GRI contract. Figure 8 shows this equipment.

¹INCONEL 600   .031 in.
INEX SiC-Si   <.001 in.

¹NOTE: Inconel had considerable scale —
5.5% weight loss

Creep testing done in compression on a "Netsch" testing machine — 2 weeks at 1200°C (2192°F)

Fig. 4. Hi-temp creep.

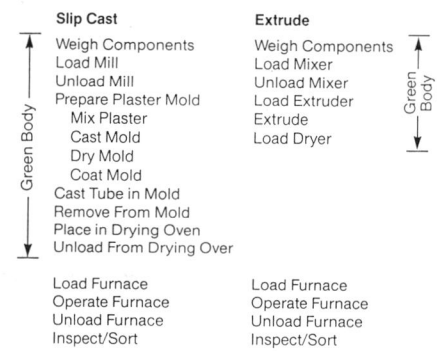

Fig. 5. Creep test pieces.

```
                  Slip Cast              Extrude
     ↑    Weigh Components        Weigh Components    ↑
     │    Load Mill               Load Mixer          │
     │    Unload Mill             Unload Mixer        Green
 Green    Prepare Plaster Mold    Load Extruder       Body
 Body       Mix Plaster           Extrude             │
     │      Cast Mold             Load Dryer          ↓
     │      Dry Mold
     │      Coat Mold
     │    Cast Tube in Mold
     │    Remove From Mold
     │    Place in Drying Oven
     ↓    Unload From Drying Over

          Load Furnace            Load Furnace
          Operate Furnace         Operate Furnace
          Unload Furnace          Unload Furnace
          Inspect/Sort            Inspect/Sort
```

Fig. 6. Conventional processes.

Fig. 7. Prototype equipment.

Fig. 8. Scaled-up equipment.

Coarse-grained SiC powder, particulate silicon, insulation grains and in some cases particulate graphite are loaded into hoppers (Figure 9). These dry powders which must be coarse enough to flow freely by gravity are loaded into a vertical furnace by means of feeder heads (Figures 10 & 11). The inside and outside diameters and therefore, wall thicknesses of the tube are determined by the powder feed mechanism. The length is simply a function of the furnace size and how much powder is loaded. The straightness of the tube is controlled by the vertical furnace tube. The dimensional accuracy, concentricity and roundness are controlled solely

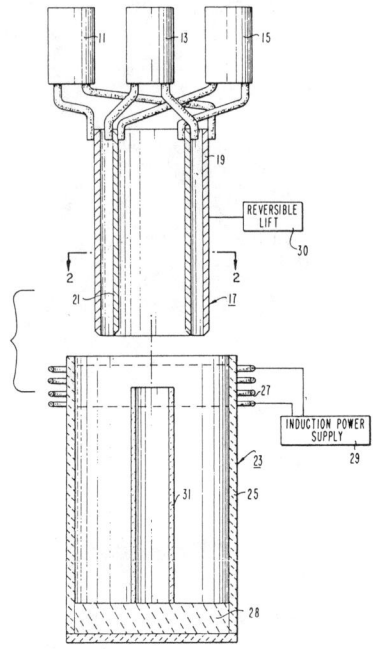

Fig. 9. Process description.

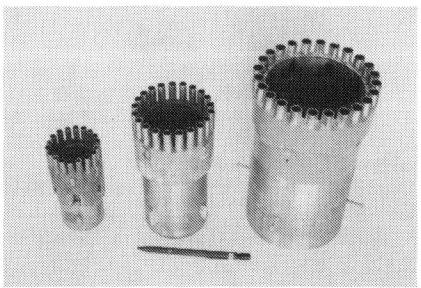

Fig. 10. Powder feed head.

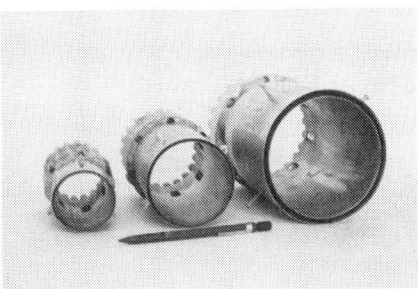

Fig. 11. Powder feed head.

Fig. 12. Moving hot zone.

by the feeder head. Shrinkage related dimensional problems do not exist with this process.

It's ironic that this process, which the Gas Research Institute supported, uses a high frequency electric induction furnace to create a ceramic tube used in gas fired furnaces. One of the keys to this process and to its inherent low cost is the moving hot zone which progressively melts the silicon allowing it to infiltrate the silicon carbide (Figure 12). The reaction is, of course, exothermic which reduces the power requirements. Current designs accommodate both an inert atmosphere or a vacuum. The furnace is allowed to cool and the finished product is removed.

Essentially the process consists of just two steps, powder loading and firing. Both of these steps are easily automated. This new process for which a patent has been applied, along with others under development, should overcome the biggest hurdle to the widespread use of silicon carbide large diameter radiant tubes, which is cost.

Advantages of this process include the use of inexpensive materials. Silicon carbide at about 60 cents per pound and silicon which costs around 90 cents per pound are used. A six and one-half inch O.D. tube, eight feet long has approximately $38 worth of material, not very much for a tube this large. The furnace and energy costs are low because a very short, traveling hot zone is used. Labor costs are low since this is a very simple process and there is no green body. The equipment is simple, inexpensive and flexible.

Different size requirements are easily accommodated. The capacity of the process can economically be increased in small incremental steps. Very little floor space is required.

All elements of cost are minimized by this process. This process is now being used in a pilot plant under GRI contract to determine actual manufacturing costs and to produce tubes for testing. These test pieces will be supplied to the Center for Advanced Materials at Penn State where they will conduct complete physical and thermal material characterization. Real world tests in gas fired heat treat furnaces will also be done as part of this current contract.

SUMMARY

The excellent high-temperature properties of silicon carbide, along with high quality and low cost made possible by new processing methods, could make this product dominate over current high-temperature metal alloy tubes. In fact, it is expected that these SiC tubes will be competitive in price and far superior in performance.

Section V
Fibers and Composites

SiC Whiskers and Platelets

Samuel C. Weaver and Richard D. Nixdorf

American Matrix, Incorporated
118 Sherlake Drive
Knoxville, TN 37922

Gerald Vaughan

University of Tennessee
Knoxville, TN

Abstract

To obtain optimum performance from composite materials there is an advantage to selecting the size, shape, and composition of the reinforcement material to fit the application. American Matrix has developed a process for manufacturing silicon carbide whiskers and platelets which permits control of size and surface morphology. The platelet and whisker products were examined. Preliminary mechanical properties data of platelets in aluminum are discussed.
A product mixture of boron carbide whiskers and platelets has been developed.

Background

Silicon carbide is the ceramic most often used to reinforce composite materials. SiC is chosen for its high strength, high modulus and good oxidation resistance.
To date, most reinforcement by ceramic materials utilizes large diameter SiC fibers or SiC whiskers in the size range of 0.5 micrometers.
To obtain optimum performance from composite materials there is an advantage to selecting the shape and size of the reinforcement material to fit the application.
As we began evaluating the type of products to be developed at American Matrix, Inc. (AMI), we decided to look at a somewhat wider scope of products for use as reinforcement. It is apparent that different material types and shapes will have advantages in different matrices. For instance, silicon carbide whiskers have been particularly effective in toughening Al_2O_3 and Si_3N_4[1-9]. Both silicon-carbide whiskers and silicon-carbide grit have been effective in increasing the modulus of aluminum alloys of 6061-T6 and 2124[10-15]. Particulate, which has been segregated to have an aspect-ratio of 2-3, has been used by DWA to increase the modulus of these alloys. Composite modeling has shown that platelets with higher aspect ratios should show an even further improvement in modulus[16].
Currently, there are no universally accepted models for the materials interactions in various composites which allow them to achieve their desired effects. This absence is pronounced in the case of ceramic materials where extensive efforts are

being made to increase their toughness. In terms of product size and shape, however, we have tentatively selected the following approaches as being optimum.

Ceramics—Discontinuous small fibers (whiskers) and thin, small diameter platelets.

Metals—Discontinuous small whiskers and thin, small diameter platelets.

Polymeric Materials—Large diameter whiskers and large platelets.

Based on a limited amount of data, however, there are competing health and safety aspects to handling small materials (powders and whiskers) particularly in the submicrometer range. Consequently, the industries handling silicon carbide whiskers have been careful to minimize exposure to employees.

The objective of the American Matrix development has been the production of silicon-carbide and boron-carbide materials in both a whisker and a platelet configuration. The goal of the whisker-manufacturing is to make whiskers which are approximately 2 micrometers diameter with aspect ratios greater than 10. The larger diameter whiskers are expected to reduce health risks without sacrificing the effectiveness of the whisker itself. Dr. Terry Tiegs (Oak Ridge National Laboratory) has reported significantly improved toughness and flexural strength data with increasing SiC whisker diameter for whiskers in the size range of 0.2 to 0.7 micrometers[17].

The platelets are being manufactured in three size ranges. The smallest range is approximately 2 micrometers thick with diameters from 25 to 50 micrometers. These are intended for use in ceramics and metals. The intermediate size range is 10 micrometers thick by 250 micrometers diameter; these platelets are for use as reinforcement in polymeric materials. The largest platelets are up to a centimeter in diameter (or larger if possible) for use as a semi-conductor material. Silicon carbide has some attractive semi-conductor properties such as high electron mobility, good radiation resistance, and maintains its semi-conducting properties to 1240°C.

RESULTS

For ease of presentation of the results, the discussion is divided into the following sections:

1. Silicon Carbide Whiskers

2. Boron Carbide Whiskers and Platelets

3. Silicon Carbide Platelets

4. Biological results from exposure to the above products

SILICON CARBIDE WHISKERS

American Matrix has been successful in developing a manufacturing process to make a uniform, reproducible, beta silicon carbide whisker. The manufacturing parameters can be adjusted to make long whiskers having a mean length of

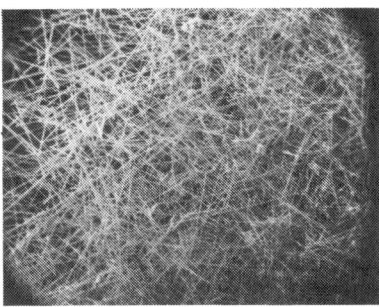

Fig. 1. SEM photomicrographs of SiC whiskers (beta phase) as manufactured by American Matrix, Inc. (magnification 200X).

Fig. 2. SEM photomicrographs of SiC whiskers (beta phase) as manufactured by American Matrix, Inc. (magnification 1000X).

approximately 200 micrometers (Figure 1) or to make short rounded whiskers which have a mean length of about 50 micrometers (Figure 2). Notice that the short whiskers have a much smoother surface with rounded ends while the larger silicon carbide whiskers have a rougher surface combined with ends that are not as rounded.

American Matrix has also been successful in manufacturing whiskers which are primarily alpha phase silicon carbide. X-ray diffraction results showed the fibers to be 67% alpha phase and 33% beta phase. Figures 3 and 4 indicate the highly faceted nature of the alpha phase material. The faceted structure of the alpha phase material is very interesting from an application standpoint. The ridged material should have a higher interlocking effect with the matrix thereby making a more effective load transfer from the matrix to the fiber in both metal and polymeric composites. It may also act like a "rebar" in ceramic materials thereby resulting in

Fig. 3. SEM photomicrographs of SiC whiskers (alpha phase) as manufactured by American Matrix, Inc. (magnification 1000X).

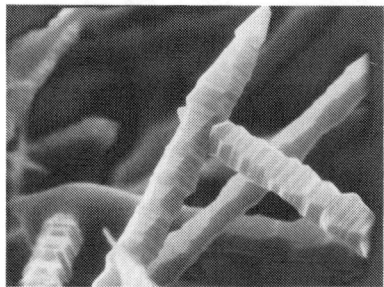

Fig. 4. SEM photomicrographs of SiC whiskers (alpha phase) as manufactured by American Matrix, Inc. (magnification 5000X).

increasing toughness. On the other hand, the intersection of faceted faces may serve as stress concentrators which increase the likelihood of breaking at that point. The final evaluation will have to be judged on the basis of actual experimental results in the different matrices.

The manufacturing procedure for the beta phase material has a very high yield and results in a product which requires very little processing after manufacture. The alpha phase material, however, has a much lower product yield.

BORON CARBIDE WHISKERS

The boron carbide whiskers have a high boron content (in the range of 78%) that indicates they are very close to stoichiometric B_4C. However, some of the results indicated $B_{13}C_2$ and the whiskers are slightly on the boron rich side of B_4C.

Fig. 5. SEM photomicrographs of boron carbide whiskers after separation process to take out platelets and granules (magnification 200X).

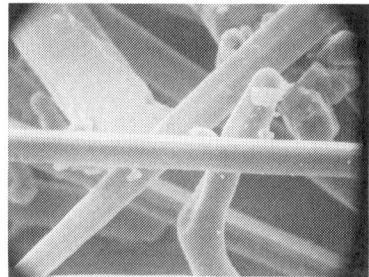

Fig. 6. SEM photomicrographs of boron carbide whiskers after separation. Note smooth, slightly faceted surfaces (magnification 2000X).

The fibers range in diameter from 7-10 micrometers with aspect ratios in the range of 50. Figures 5 and 6 show typical boron carbide whiskers.

Boron carbide platelets have been manufactured as part of a product that results in whiskers, platelets and granules. AMI has recently had success in separating these materials into their individual components. No tests have been performed to date on B_4C platelets only.

Boron carbide is important for several applications. It has an even higher strength to weight ratio than does SiC and is compatible with some matrices that will react with the silicon carbide.

Silicon Carbide Platelets

American Matrix has developed a process for high volume manufacturing of silicon carbide platelets. These platelets range in thickness from 0.5 micrometers to as large as 20 micrometers. By controlling the thickness, the material can be tailored for use in various applications. Figures 7 through 9 show various scanning electron micrographs of the platelets. The platelets are alpha phase material which show the basic configuration of a hexagon shape. In Figure 8, the platelets show three different planes growing out of the same basic crystal structure. Interfacial bonding resulting from different platelets sintering together at the interface is shown in Figure 9.

Fig. 7. SEM photomicrographs of single crystal platelets of alpha phase SiC. Platelets show basic hexagonal configuration (magnification 100X).

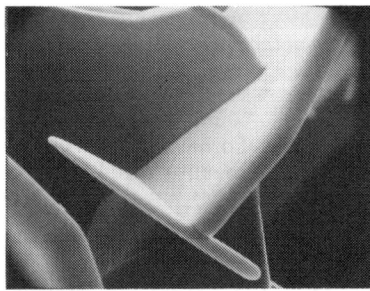

Fig. 8. SEM photomicrographs of single crystal platelets of alpha phase SiC showing three different planes growing out of the same crystal (magnification 1000X).

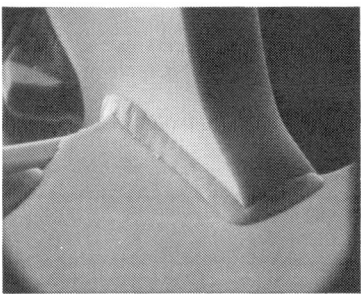

Fig. 9. SEM photomicrographs of single crystal platelets of alpha phase SiC showing two platelets of different orientation which have bonded at the interface (magnification 2000X).

Another characteristic of the SiC platelet material is the very high purity. Most of the material has less than 1000 ppm of impurities while some material has as low as 150 ppm. However, even these very high purity levels are probably not indicative of the true as manufactured material purity, but rather are indicative of impurities which are introduced during further processing. Figure 10 is a picture taken with an optical microscope which shows the light transmission through the platelets, a further indication of their high degree of purity.

Dr. Haldeman (MIT-Lincoln Laboratories) measured the sheet resistance of a large SiC platelet (approximately 1 centimeter diameter). Test results were approximately 8000 ohms/square which suggests a much higher purity level in the crystals than the overall chemical analysis indicates. Dr. Robert Davis (North Carolina State University) made sheet resistance measurements which, while slightly lower, confirmed the measurements of Dr.Haldeman. Haldeman irradiated the SiC platelet with 100 megarads of 1.6 MeV electrons (equivalent to 10 years in space) which resulted in a decrease in resistivity to about 5000 ohms/square.

Mr. Larry Matus (NASA-Lewis) ran some Hall resistivity measurements. The results from these tests for electronic properties are reported in Table 1.

Silicon carbide platelets of three different sizes were fabricated into 6061 aluminum powder metallurgy compacts containing 25% by volume of the silicon carbide platelets. The platelets showed improved properties with each reduction in platelet size. The results for the fabricated composites are tabulated in Table 2.

Because of the improvement in properties of metal matrix composites with reduced platelet diameter, the development of this product was extended to include smaller thicknesses. Platelets as small as 0.5 micrometers in thickness have been manufactured. While test results on this material are not yet available, it is believed that this size of platelet should be particularly effective in reinforcing metal matrices and possibly ceramics.

Preliminary tests of platelets in polymeric matrices indicate the platelet configuration is going to be particularly effective for increasing the modulus. Initial results showed the platelets to mix well and to bond well in polymerics.

Table 1. SiC Platelet electronic data.

- Structure: Alpha Phase
- Sheet Resistance: 8000 ohms/square
- After 100 Megarads of 1.6 MeV Electrons: 5000 ohms/square
- Hall Measurements Using Van der Pauw Measurements Technique:
  Electrical Resistivity: 7.64 ohm-cm
  Hall Mobility: 100
  Carrier Concentration: $8.1 \times 10^{15}/cm^3$

Fig. 10. Transmission light microscopy of Silicon Carbide platelets showing translucent nature of material (magnification 200X).

Table 2. Silicon Carbide platelets in 6061 aluminum*.

Material	Platelet Lot Number	Platelet Diameter	Yield	UTS	Strain	Youngs Modulus
6061	S10	500 Microns	35,700 psi	36,700 psi	0.8%	14.2×10^6 psi
6061	S2R	75 Microns	43,600 psi	47,900 psi	1.9%	15.9×10^6 psi
6061	S34SF	40 Microns	53,000 psi	57,000 psi	1.8-2.0%	16.0×10^6 psi
6061	T6-Wrought	------	40,000 psi	45,000 psi	17.0%	10.3×10^6 psi

*Platelet Loading = 25 Volume %; Platelet Thickness = 10 Microns

Health Effects

Prior to introducing silicon carbide whiskers for commercial sale, American Matrix initiated a study with the University of Tennessee to evaluate the health effects. Several different materials were included in the study in addition to SiC whiskers. The materials being studied are Silicon carbide whiskers (0.5 micrometer diameter, 2 micrometer diameter, and 5 micrometer diameter); silicon carbide platelets (25 micrometer average diameter and 250 micrometer average diameter); boron carbide whiskers (10 micrometer diameter); and control specimens of normal cells and asbestos.

Preliminary data suggests that personnel working with 0.5 micrometer diameter whiskers should handle the material with care to minimize inhalation. The first set of results from these studies should be available for publication the first part of 1988.

Conclusion

A technology has been developed which permits the manufacture of multiple single crystal shapes in more than one composition. The characteristics of single crystal SiC whiskers and platelets, and boron carbide whiskers and platelets are discussed.

References

[1] T. N. Tiegs and B. G. Becher, "Thermal Shock Behavior of an Aluminum-SiC Whisker Composite," *J. Am. Ceram. Soc.*, **70** (5) C-109-C-111 (1987).

[2] B. F. Becher and G. C. Wei, "Toughening Behavior in SiC-Whisker-Reinforced Aluminum," *J. Am. Ceram. Soc.*, **67** (12) C-267-C-269 (1984).

[3] B. F. Becher, T. N. Tiegs, J. C. Ogle, and W. H. Warwick, "Toughening of Ceramics by Whisker Reinforcement," pp. 61-73 in Fracture Mechanics of Ceramics, Vol. 7. Plenum Press, New York 1986.

[4] L. M. Sheppard, "Reliable Ceramics for Heat Engines," Advanced Materials & Processes/Metal Progress; Oct., 1986; pp. 54-66.

[5] D. E. Niesz, "A Visit to Pechiney Ceramics," Ceramic Bulletin, Vol. 65, No. 9, 1986; pp. 1253-1254.

[6] A. J. Klein, "Ceramic-Matrix Composites," Advanced Materials & Processes; Sept, 1986; pp. 26-33.

[7] G. C. Wei and B. F. Becher, "Development of SiC-Whisker-Reinforced Ceramics," Ceramic Bulletin, Vol. 64, No. 2 (1985).

[8] B. F. Becher and T. N. Tiegs, "Toughening Behavior Involving Multiple Mechanisms: Whisker Reinforcement and Zirconia Toughening," *J. Am. Ceram.Soc.*, **70** (9) 651-654 (1987).

[9] N. D. Corgin, G. A. Rossi, P. M. Stephan, "Making Ceramics Tougher," Machine Design, July 23, 1987; pp. 84-89.

[10] Advanced Materials and Processes; Feb. 1988; pg. 29.

[11] W. R. Mohn and D. Vukobratovich, "Engineered Metal Matrix Composites for Precision Optical Systems," Sampe Journal; Jan/Feb 1988.

[12] S. Deramarkar, "Aluminum-Matrix Composites and Reinforced Laminates," Materiaux et Techniques; May-June 1986; pp. 197-200.

[13] DOD Metal Matrix Composites Information Analysis Center Current Highlights; Vol. 7, No.3, Sept. 1987.

[14] W. R. Mohn, "Your MMC Product Will Stay Put," Research & Development, July 1987; pp. 54-68.

[15] C. F. Lewis, "The Exciting Promise of Metal-Matrix Composites," ME, May 1986.

[16] J. Dolowy, Personal Communication; 1984.

[17] T. Tiegs, Personal Communication; April 1987.

Single Phase Alpha-SiC Reinforcements for Composites

Wolfgang D. G. Boecker, Stephen Chwastiak, Frank Frechette
and Sai-Kwing Lau

Standard Oil Engineered Materials Company
Technology Center
Niagara Falls, NY 14302

Abstract

Silicon carbide offers excellent mechanical, physical and chemical properties desired for reinforcement of a wide range of matrices from polymers, glasses, metals to ceramics. Various SiC reinforcements in the form of particulates, fibers and whiskers are of interest in current composite research.

Most of the current commercially available SiC reinforcements, however, will not meet the thermal stability requirements and cost expectations for structural ceramic applications. Recent developments in single phase silicon carbide reinforcements in the form of single crystal platelets with high aspect ratio and continuous polycrystalline filaments are described. The fabrication routes for both SiC platelets and polycrystalline SiC filaments are outlined. When compared to other fabrication routes, both processes offer the potential for low cost materials. Typical microstructures are shown, physical properties are discussed and thermal stability is compared to other SiC reinforcements. Examples of ceramic and metal matrix composites using these reinforcements are also presented.

Introduction

Since Edward Goodrich Acheson's invention[1] in 1892, silicon carbide has been recognized for its excellent chemical, thermal and mechanical properties. This has led to a wide range of applications varying from abrasives and refractories to structural and electronic materials.

Based on its unique combination of properties, SiC has also become one of the most promising reinforcement materials for advanced composite development. However, most of the present SiC reinforcements, including fibers and whiskers, are based on the low temperature cubic structure of SiC, the beta phase. Furthermore, most of the organic precursor derived β-SiC fibers contain major amounts of oxygen while most of the whiskers contain significant amounts of metal impurities. As a result, these reinforcements (perhaps with the exception of the VLS whiskers) have only limited thermal stability. They are thus not totally adequate for fabrication of ceramic (especially non-oxide) composites.

This low thermal stability problem, when compounded with additional high-cost considerations severely impedes the development of high-temperature composite technology. The demand for lower cost reinforcements capable of withstanding fabrication temperatures typical for oxide and nonoxide ceramic composites, e.g., greater than 1500°C and up to 2300°C, has led our research interest to two novel forms of α-SiC: single crystal hexagonal platelets and continuous polycrystalline filaments.

Single Crystal SiC Platelets

SiC has been known to grow into crystals of different shapes and forms such as needles, rods and plates under various growth conditions. Parameters like temperature, atmosphere, impurities or dopants appear to be the critical factors and surface diffusion the key mechanism in controlling the growth of these crystals. Knippenberg[2] has studied such crystal growth phenomena in detail.

The growth of hexagonally shaped plate-like SiC crystals, the subject of the current work, was first discovered by Acheson in a graphite electrode furnace named after him. These crystals are based on the high-temperature alpha structure consisting predominantly of the 6H polytype. They often reach several centimeters in size and are found in the center core region of an Acheson furnace.

Synthesis

While SiC plate-like single crystals, or platelets, are potentially very attractive reinforcements for various matrices, based on their excellent mechanical and thermal properties as well as low cost potential, the Acheson crystals are heavily intergrown. They are also much too large for practical composite applications.

The present work has therefore focused on growing these platelets in ten to hundred micron sizes and in isolated form. High yields form inexpensive raw materials like silica and carbon[3] was another equally important objective. The technical approach taken for this work is based on the use of dopants like boron and aluminum, also well-known as sintering aids for SiC, to effectively promote the formation of these hexagonal crystals.

Platelet yields over 90% have been obtained starting from various raw materials which provide silicon and carbon in stoichiometric amounts. The relationship between platelet characteristics including platelet size and yields and growth conditions like raw materials, critical impurities, temperature, time, atmosphere, as well as nature and amount of growth dopants has been successfully established. Recently, Kistler-DeCoppi and Richarz[4] reported platelet growth from submicron cubic silicon carbide powders using similar growth additives.

A near-perfect single crystal platelet of α-SiC is shown in Figure 1. The maximum dimension of this typical platelet is 400 microns, and its thickness is about 25 microns. This platelet was selected from a product made by heating submicron powder of β-SiC to a temperature greater than 2000°C, at which time the powder transforms to α-SiC. Due to the presence of surface active dopants added, the SiC grows as platelet-shaped crystals. By observation with an optical microscope using transmitted light, these platelets are seen to be translucent as shown in Figure 2.

The scanning electron micrographs of two other platelet samples grown to nominal sizes of 100 microns and 30 microns, respectively, are shown in Figure 3. It can be seen that many of these crystals have one or more dihedral angles of 120°. Although growth steps can be found on some of the faces, the basal planes of these platelets are very smooth. It can also be observed that the finer platelets tend to have rounded corners in contrast to the sharp corners commonly found on the large

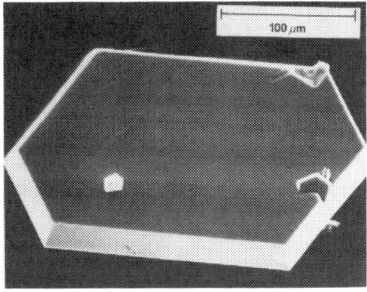

Fig. 1. Scanning electron micrograph of a single α-SiC hexagonal platelet.

Fig. 2. Optical micrograph showing the transparency of α-SiC platelets.

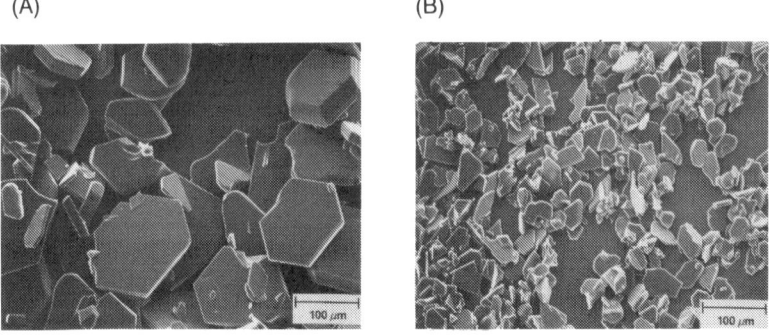

Fig. 3. Scanning electron micrographs of two α-SiC platelets grown to different sizes, (A) average size 100 μm and (B) average size 30 μm.

platelets. However, the aspect ratios of the small platelets remain as high as those of the larger ones.

Properties

The properties of SiC platelets are summarized in Table 1. The size and aspect ratio of the platelets were measured either from SEM photographs, using calibration standards for magnification, or directly under an optical microscope with a high depth-of-field optical system. For example, an average maximum dimension in one typical sample population of about 30 large platelets is 110 ± 9 microns (estimate of error in average based on 95% confidence limits) with a range 60 to 180 microns. A typical average aspect ratio for these large platelets is 13 ± 2, with a range of 3 to 27.

Particularly significant are the high Vickers hardness, 30 GPa, and high elastic modulus, 470 GPa (68 Mpsi), both characteristic of high quality SiC. These properties were obtained by an indentation technique developed by Professor C. Y. Li and co-workers at Cornell University[5]. The apparatus utilizes precise monitoring of position of a diamond tip indentor through a set of capacitance gages, allowing measurement of strains down to nanometer levels. Measurement of the force and strain during elastic recovery after an indentation on the basal plane of a SiC platelet allows calculation of the elastic modulus. The same apparatus is being adapted to measure strength of the platelets. The polytype identifications were made using information from the International Centre for Diffraction Data and described by the Ramsdell notation.

Table 1. Silicon Carbide platelets.

Maximum Dimension, μm	10–500
	Controlled by growth parameters
Aspect Ratio (Maximum Dimension: Thickness)	10–15
Density, g/cm^3	3.21
Elastic Modulus*, GPa (Mpsi)	470 (68)
Hardness (Vicker)*, GPa	30
Crystal Structure	Major: Hexagonal 6H
	Minor: 4H
	Trace: 15R
Morphology	Hexagon, single crystal
Thermal Stability, °C	1600 in Air
	2250 in Nitrogen

*Measurement by Professor C. Y. Li, Cornell University

POLYCRYSTALLINE SiC FILAMENTS

The discovery of pressureless sintering of α-SiC in the early 1970s has opened a new era of application for complex structural components[6]. Advances in ultrafine

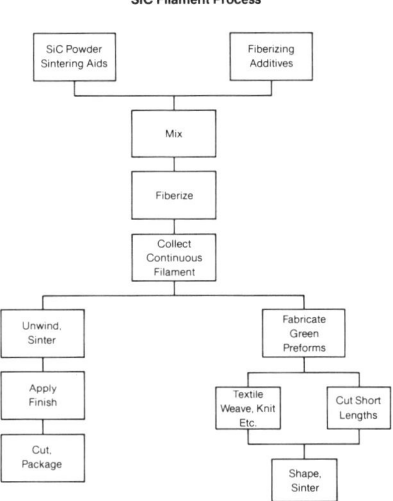

Fig. 4. Flow diagram for processing of continuous α-SiC filaments.

powder development, forming and sintering technologies have led to the feasibility of developing a continuous filament form of α-SiC. These filaments should offer significant economic and performance benefits compared to the fairly slow and complex chemical conversion routes used in other SiC fibers.

Fabrication

The filament forming process developed in the present work is based on conventional fiber spinning techniques. As shown in the schematic flow diagram in Figure 4, fiberizable compositions are prepared by mixing a sinterable SiC powder with fiber forming aids so that a flowable mixture is obtained. The mixture is then extruded through a spinnerette to produce the filament. The diameter of the fiber is determined by the extrusion rate of the mixture and the rotating speed of the collecting drum. Typical fiber diameters range between 75 and 150 microns. Smaller diameter fibers can be produced by making minor changes in the process. Fibers with diameters as small as 20 microns have been produced in laboratory quantities. Preliminary measurements indicate the smaller fibers have higher tensile strengths than the large diameter fibers.

The α-SiC powder has typically a submicron average particle size and is combined with sintering aids well known from literature[7]. Spools of green filament are shown in Figure 5. At this point in the process, the filaments can be sintered using conditions typical for α-SiC.

Fig. 5. Green α-SiC filament packages.

Fig. 6. Woven α-SiC fabric before and after sintering.

The excellent green strengths exhibited by the filaments makes it possible to provide textile type preforms for composites. Continuous green filament yarns have been successfully woven prior to sintering. After sintering the preform can be used to fabricate a composite having the reinforcing elements precisely arranged for optimum performance. An example of such a fabric before and after sintering is shown in Figure 6.

Properties

The properties of the SiC filaments which have thus far been determined are summarized in Table 2. The density was determined by comparing the microstructure as shown in Figure 7 with microstructures of sintered SiC bodies whose densities were known. Filament diameters were measured using a splitting image microscope, the SEM, and image analysis. The tensile strength was calculated by

Table 2. Polycrystalline SiC Filaments.

Density, g/cm^3	$\geq$3.10
Tensile Strength, MPa (kpsi) (gauge length 1/4 inch)	up to 1380 (200)
Elastic Modulus*, GPA (Mpsi)	400 (58)
Diameter, μm	20-150
Crystal Structure	Single Phase Polycrystalline α-SiC Major: Hexagonal 6H Minor: 4H Trace: 15R
Thermal Stability, °C	1600 in Air 2200 in Nitrogen

*Measurement by Professor C. Y. Li, Cornell University

Fig. 7. Optical micrograph of a polished section of α-SiC filament showing detailed microstructure.

determining the break load in tension and measuring the filament cross-sectional area using a Q-10 Image Analyzer (Cambridge Instruments). The strength value shown in the table was obtained on a 150 micron diameter fiber sample.

The elastic modulus measurements were based on stress-strain data using a technique developed by Li et al.[5]. The miniaturized mechanical testing system consists of a load cell and a capacitance gage for monitoring the displacement. The elastic modulus value for the filaments is lower than the one obtained for the platelets and reflects the polycrystalline nature and the sintered density of about 96-98% of theoretical.

High Temperature Stability Study

Experimental

A series of oxidation tests was conducted in order to evaluate the high temperature stability of the SiC polycrystalline filaments and single crystal platelets. In an effort to simulate conditions generally encountered during the fabrication of oxide matrix composites, the temperature chosen for these experiments was 1550°C in static air, and the oxidation time was 5 hours.

For comparison purposes, several commercial state-of-the-art SiC reinforcements including Nicalon[*] ("Standard Grade"), and Tyranno[†] (TRN-M801) fibers were also oxidized in the same experiments.

To conduct the oxidation experiments, small bundles of the various SiC fibers were cut into 7 to 10 cm long sections and placed in alumina or SiC crucibles. The amount of sample tested for each type of fiber was usually less than 1 g. The fiber and platelet samples were gradually heated to 1550°C and held for five hours in a stagnant air environment. Several oxidation tests at a higher temperature of 1600°C were also conducted.

Post-Test Analysis

Post-test analysis including weight change measurements, optical, and SEM examination was conducted on each reinforcement sample after the 1550°C oxidation test. Preliminary results indicated that the weight change data, with the exception of the polycrystalline SiC fibers, were usually inconsistent as reproducibility from test to test was difficult to obtain. It was observed that the polycrystalline SiC fibers and SiC fibers and SiC platelets generally gained weight during the oxidation while the Nicalon and Tyranno fibers tended to lose weight, revealing the existence of volatile species. Most of the important post-test information was actually derived from the SEM examinations shown in Figures 8-11.

Figures 8a and 8b depict the Tyranno SiC fiber strand before and after the oxidation test, respectively. It can be seen that severe oxidation and cracking of the fiber surface was encountered. A significant degree of swelling has occurred. In addition, a molten oxide surface layer was often observed on the Tyranno fiber bundle as well.

The oxidation behavior of Nicalon fibers was similar to that of the Tyranno fibers in that severe oxidation with surface cracking and spalling was encountered. However, a unique feature observed on the oxidized Nicalon fibers was the formation of numerous glassy bubbles as shown in Figure 9. The size of these bubbles was generally in the range of 50-100 microns and they seemed to have grown from the surface of the Nicalon fibers due to the formation of gaseous species such as SiO and CO during oxidation. In addition, as observed on the Tyranno fibers, the oxidized Nicalon fibers were extremely brittle and friable. This kind of oxidation accompanied by scale spalling and fiber fracturing was probably the main reason why reproducible weight change data could not be obtained for the commercial SiC fibers.

The SEM pictures of the polycrystalline SiC fibers prepared in the study before and after the oxidation test are shown in Figures 10a and 10b. It can be easily seen that excellent integrity of the fibers has been maintained. While a thin oxide layer was formed on the fiber surface, there was no evidence of cracking or spalling.

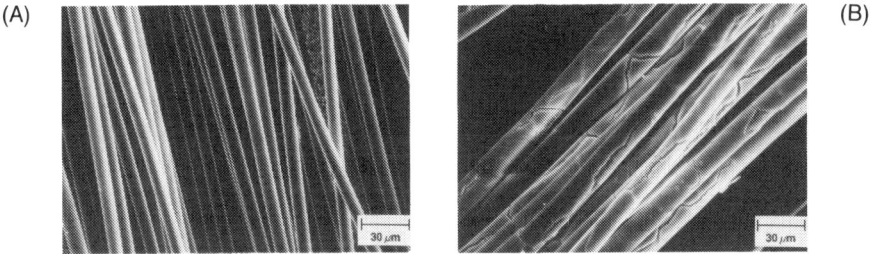

Fig. 8. Scanning electron micrograph of Tyranno SiC fibers (TRN-M801), (A) before and (B) after oxidation at 1550°C in air for five hours.

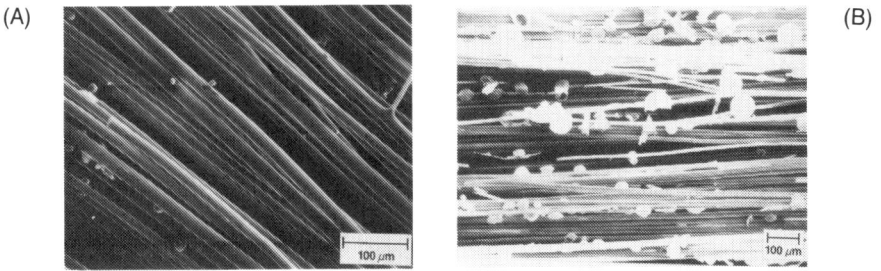

Fig. 9. Scanning electron micrograph of Nicalon SiC fibers (A) before and (B) after oxidation at 1550°C in air for five hours.

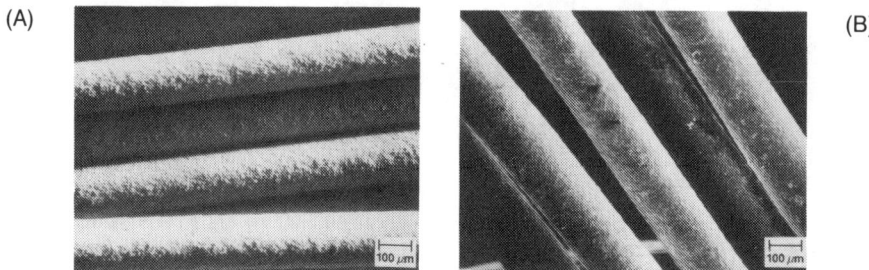

Fig. 10. Scanning electron micrograph of α-SiC filaments from the present work (A) before and (B) after oxidation in air at 1500°C for five hours.

Fig. 11. Scanning electron micrograph of α-SiC platelets (A) before and (B) after oxidation in air at 1550°C for 5 hours.

Furthermore, in contrast to the commercial SiC fibers described earlier, the weight change data obtained on the polycrystalline SiC fibers were very reproducible. The deviation from run to run was within 5 percent. The average weight gain was 1.35×10^{-4} g/cm^2. This was equivalent to the formation of a 1.7 micron thick SiO$_2$ layer after the five-hour oxidation. This high stability was not surprising because the oxidation resistance of the polycrystalline SiC fibers could be expected to be similar to that of monolithic sintered SiC.

The oxidation resistance of the SiC single crystal platelets was also found to be excellent. SEM pictures shown in Figs. 11a and 11b revealed that besides the rounding of the platelet corners, virtually no difference could be found before and after oxidation.

It can therefore be concluded that the two single phase SiC reinforcements prepared in this study possess far superior high temperature stability than any of the state-of-the-art commercial SiC fibers. Limited experiments in air at 1600°C and in inert gas environments at temperature as high as 2250°C have also been conducted. Preliminary results again revealed that the SiC platelets and polycrystalline filaments are exceptionally stable even under such conditions.

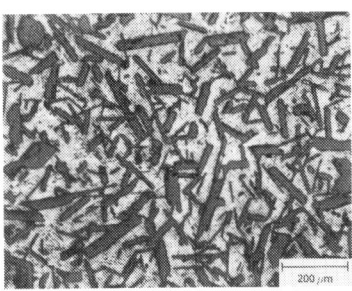

Fig. 12. Optical micrograph of squeeze casted aluminum alloy (E-332) matrix composite containing 40 Vol % SiC platelets.

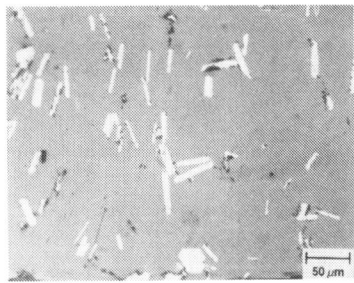

Fig. 13. Optical micrograph of hot-pressed Al_2O_3 matrix composite with 7 Vol % SiC platelets.

EXAMPLES OF COMPOSITES CONTAINING SiC PLATELETS AND FILAMENTS

The described forms of silicon carbide reinforcements have been incorporated into a variety of matrices. An example of a metal matrix composite with 40 vol % platelets is shown in Figure 12. The composite was fabricated by squeeze casting of aluminum alloy E 332.

SiC platelets in an oxide ceramic matrix are seen in Figure 13 where fine alumina powder was hot-pressed with 7 vol % of platelets into a dense composite.

The high temperature stability of platelets is demonstrated in a pressureless sintered α-SiC matrix composite with 20 vol % platelet loading and sintered without pressure at 2150°C with a final density of about 95% of theoretical. Figure 14

Silicon Carbide

Fig. 14. Fracture surface of sintered α-SiC matrix composite with 20 Vol % SiC platelets.

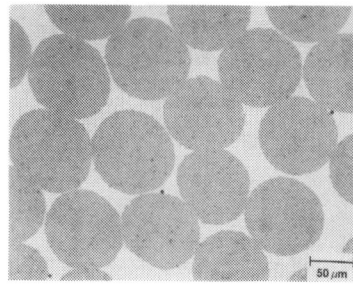

Fig. 15. Optical micrograph of silicon matrix composite with 75 Vol % SiC polycrystalline filaments.

shows a fracture surface of this sample with evidence of transgranular fracture through several platelets.

The polycrystalline SiC filaments have been studied as continuous reinforcements in a silicon matrix. The example in Figure 15 shows a cross section of a silicon/SiC filament composite with a fiber loading of over 75 vol % produced by liquid infiltration technique.

A series of other matrices are currently being investigated using both types of reinforcements for advanced composites to increase fracture toughness as well as strength.

Summary

Pilot processes for fabricating two novel forms of α-SiC reinforcements have been developed. The first produces hexagonal single crystal platelets with basal plane dimensions ranging from 10 to 500 microns and an aspect ratio of 10 to 15. The second one provides continuous polycrystalline filaments having a diameter of approximately 20 to 150 microns and a density of over 96% of theoretical.

These two newly developed reinforcements share several important common features: they are both fabricated from low cost raw materials using inexpensive processing steps. Both have very high modulus of elasticity in the range of 55 to 68 Mpsi (379 to 470 GPa). The most unique feature of these two reinforcements is their chemical stability at high temperatures. Experiments conducted in this study have demonstrated that the α-SiC platelets and polycrystalline continuous filaments are stable up to 1600°C in air and 2250°C in inert gas, far superior than most of the commercially available β-SiC based fibers. These properties make the fine α-SiC reinforcements developed in this study uniquely suitable for high-temperature composite applications.

In a parallel composite development program, the single platelets and polycrystalline continuous fibers developed have both been incorporated into various matrices including oxide and nonoxide ceramics, glasses, metals and polymer using processing techniques including hot pressing, pressureless sintering, reaction bonding and squeeze casting. Property evaluations of these composites are in progress.

Acknowledgments

The authors wish to thank M. Srinivasan, S. Seshadri and W. Binnie for their analytical support and C. McMurtry, R. Maier and S. Calandra for assistance in the composite fabrication.

---

[*]Nippon Carbon Co., Japan.
[†]Ube Industrials Ltd., Japan.

References

[1] E. G. Acheson, "Production of Artificial Crystalline Carbonaceous Materials," U.S. Patent 492,767, February 28, 1893.

[2] W. F. Knippenberg, "Growth Phenomena in Silicon Carbide," *Philips Res. Repts.* **18**, No. 3, 161–274, 1963.

[3] W. D. G. Boecker, S. Chwastiak, T. M. Korzekwa and S. K. Lau, "Hexagonal Silicon Carbide Platelets and Preforms and Methods of Making and Using Same," U.S. Patent 4,756,895, July 12, 1988.

[4] P. A. Kistler-De Coppi and W. Richarz, "Phase Transformations and Grain Growth in Silicon Carbide Powders," *Int. J. High Techn. Ceramics* **2**, 99–113, 1986.

[5] D. Stone, W. LaFontaine, S. Ruoff, S. P. Mannula, B. Yost and Che-Yu Li, Materials Res. Soc., Symposium Proc., 127–131, 1986.

[6] J. A. Coppola, L. N. Hailey and C. H. McMurtry, "Process for Producing Sintered Silicon Carbide Ceramic Body," U.S. Patent 4,174,667, November 7, 1978.

[7] W. D. G. Boecker and H. Hausner, "The Influence of Boron and Carbon Additions on the Microstructure of Sintered Alpha Silicon Carbide," *Powder Met. Int.*, **10**, No. 2, 87–89, 1978.

Silicon Carbide Fibers from Methylpolysilane Polymers

J. Lipowitz, G. LeGrow, T. Lim, N. Langley

Dow Corning Corporation
Midland, MI 48640

Abstract

A broad range of ceramic compositions, centered about silicon carbide, can be produced from polymethylsilane polymers. This family of polymers is prepared by redistribution of methylchlorodisilanes. Quarternary phosphonium salt catalysis leads to a methylchloropolysilane intermediate which is derivatized with organomagnesium (Grignard) reagents. This derivatization replaces the residual Si-Cl ligands with aliphatic or aromatic organic substituents, thus producing the family of polyorganosilane polymers.

These polymers can be melt spun, cured in the solid state, and pyrolyzed to produce ceramic fibers. Chemical and physical properties of the silicon carbide-like ceramic fibers produced are described. Methods used to characterize these ceramic fibers includes elemental analysis, density, X-ray diffraction, infrared spectroscopy, ^{29}Si and ^{13}C solid state, magic angle spinning, nuclear magnetic resonance, scanning Auger depth profiling, porosity measurements by X-ray scattering, and thermal stability measurements to 1600°C.

Introduction

Ceramics can be prepared by pyrolysis of various organosilicon polymers. The advantages of this method of ceramics formation include: 1) the ability to prepare ceramic shapes which are difficult to achieve by conventional powder processing techniques (e.g., fibers, films), 2) lower temperature processing than conventional technologies, 3) the ability to achieve very high purity because the reagents used to prepare polymers can be purified by well-established methods (e.g., distillation, recrystallization), 4) the ability to vary ceramic composition by varying polymer composition, and 5) the ability to prepare metastable ceramic structures[1]. This technology has been discussed by a number of authors[2-10].

The goal of the present work was to develop a family of stable, melt-spinnable polymers which can be rapidly cured in the solid state and polymerized to ceramic fibers with compositions which are stoichiometric silicon carbide or which are carbon-rich or silicon-rich silicon carbide. An additional goal was to characterize these ceramic fibers in a similar manner to earlier characterization of ceramic fibers derived from pyrolysis of polymers[1]. Silicon carbide-like compositions are desirable because they are likely to be the most thermally stable compositions, in both inert and oxidative atmospheres, in the Si-C-N compositional system. Silicon carbide-like compositions are also expected to have a higher elastic modulus than silicon nitride-like compositions based on comparative moduli of dense, polycrystalline silicon carbide (E approximately 60×10^6 psi, 415 GPa) and dense, polycrystalline silicon nitride (E approximately 44×10^6 psi, 310 GPa)[11].

Methylpolysilanes can be prepared by phosphonium salt catalyzed redistribution of mixed methylchlorodisilanes. The preparation method has been described[12-14]. The methylchlorodisilanes are a high boiling component of the reaction product of silicon with methyl chloride, used industrially to prepare methylchlorosilanes (direct process).

The methylchloropolysilane intermediate is then reacted with alkyl-magnesium halides or phenylmagnesium halides (Grignard reagent) to replace the reactive Si-Cl groups with more stable Si-R groups (R = alkyl or phenyl). The methylpolysilane polymer (MPS) can then be melt spun, crosslinked (cured) in the solid state and pyrolyzed in an inert atmosphere at 1000 to 1400°C to produce a MPS-derived ceramic fiber. These ceramic fibers will be referred to as MPS ceramic fiber. By choice of the alkyl Grignard reagent used (generally methyl) and by the ratio of alkyl to phenyl Grignard reagent used, ceramic fibers ranging from carbon-rich through stoichiometric Si-C to silicon-rich can be produced. Increasing phenyl group content in the polymer leads to increasing carbon content in the ceramic product. Figure 1 shows an outline of the process.

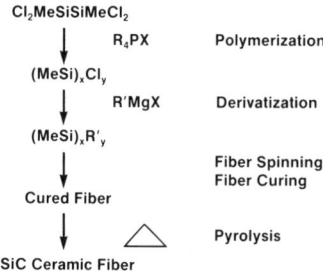

Fig. 1. Processing steps to obtain ceramic fiber from MPS polymer.

Results and Discussion

Composition

Table 1 shows the range of elemental analyses which have been obtained in the ceramic fibers by control of the polymer composition. Analysis methods have been described as have many of the characterization methods used in this work[1]. Stoichiometric silicon carbide is 70 wt% silicon and 30 wt% carbon. Small amounts of oxygen are inadvertently introduced during the multiple processing and handling steps even though an inert atmosphere is used during processing. Traces of nitrogen are introduced into the ceramic if pyrolysis is performed in a nitrogen atmosphere.

A rule-of-mixtures composition for one MPS ceramic fiber sample which was characterized by a variety of methods is shown in Table 2. This sample contained

Table 1. Elemental analysis of ceramic fiber derived from methylpolysilane polymers.

Wt %	Range
Si	52–75
C	24–47
N	0–0.1
O	0.5–6.0
H	0–0.1

Table 2. Elemental composition of ceramic fibers derived from a methylpolysilane polymer.

SiO_2	12 Wt%	0.10 Mole
SiC	81 Wt%	1.00 Mole
C	7 Wt%	0.28 Equiv.

Measured ρ_o = 2.62 g/cm^3 (Gradient Column Method)
Calculated ρ_t = 2.92 g/cm^3
Calculated Porosity Fraction = $1-(\rho_o/\rho_t)$ = 0.10

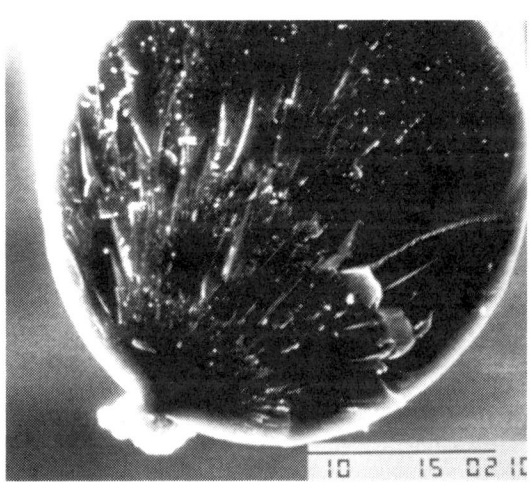

Fig. 2. SEM fractograph of fiber from MPS polymer showing critical surface flaw, 1.2 GPa tensile strength, 29 micron diameter, 25 mm gauge length (10 µm bar), one end of a matched pair.

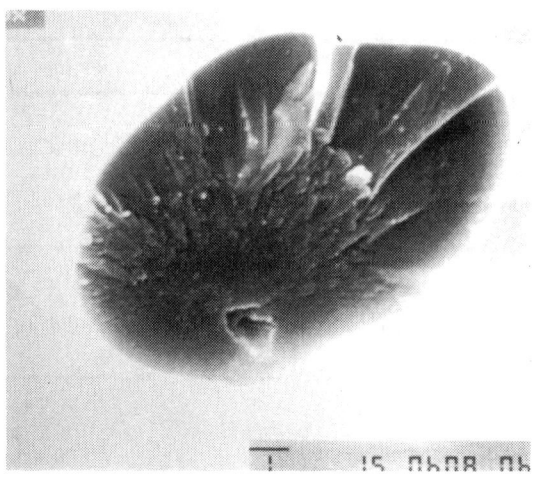

Fig. 3. SEM fractography of fiber from MPS polymer showing critical internal flaw, 2.1 GPa tensile strength, 9.6 micron diameter, 25 mm gauge length (1 μm bar), one end of a matched pair.

59.1 wt% Si, 29.3 wt% C, 6.28 wt% O and 0.09 wt% N. Unfortunately the oxygen content was considerably higher than the typical 1 or 2 wt%. The small amount of N was ignored in the rule-of-mixture calculation and in the calculation of theoretical density, ρ_t[1]. Note that the calculated volume fraction of porosity is 0.10. This is a minimum value since a gradient column density rather than bulk density was used to calculate porosity. The ceramic fibers prepared from polymeric precursors always appear to contain some porosity.[1] The nature of the porosity will be discussed in a later section.

Mechanical Properties and Fractography

Typical recent tensile strength and elastic modulus properties measured for these fibers are $\sigma = 240 \times 10^3$ psi (1.7 GPa) and $E = 30 \times 10^6$ psi (210 GPa). Fiber diameters are typically 10 to 20 microns. The fiber test method has been published[15].

Figures 2 and 3 show typical fractographs of low and higher strength MPS fibers. Primary fracture ends were retained for SEM analysis by capturing them in a water soluble grease which was subsequently washed away[16]. The fractographs show the classic features of brittle tensile failure of glassy materials; a fracture mirror surrounding the critical flaw which is in turn surrounded by a region of mist and hackle. Ceramic fibers prepared from polymer precursors generally show these features[17]. As shown in Figures 2 and 3, weak fibers tend to break at surface flaws. Strong fibers tend to break at internal flaws. As an indication of potential fiber properties, individual fibers of tensile strength approaching 2.8 GPa (400 × 10³ psi) with elastic modulus of 240 GPa (35 × 10⁶ psi) have been obtained.

Thermal Stability

Figure 4 shows thermogravimetric analysis (TGA) traces obtained in flowing helium for a commercial Si-C-O ceramic fiber[*] (CGN) and for a MPS ceramic fiber[†], respectively.[†]

Samples were heated at 5 C/min to 1400°C, held 5 hours, heated to 1570°C and held an additional 2 hours. The MPS sample lost only 3 % weight whereas the CGN sample lost 18% weight. The weight loss is largely determined by oxygen content as weight is primarily lost by carbothermic reduction of the SiOSi bonds in the ceramic to carbon monoxide, as shown by the following model reaction:

$$SiO_2 + 3C \rightarrow SiC + 2CO$$

The CGN sample contained 11 wt% oxygen and the MPS sample contained only 2 wt% oxygen. Small amounts of SiO may also volatilize from the sample.

Figures 5 and 6 are SEM images comparing degradation of the CGN and MPS fibers after 2 hour exposure at 1500°C in flowing argon (4 cm/sec flow rate). Note that the MPS fiber shows much less surface and interior degradation to large silicon carbide grains than the CGN fiber, again due to its lower oxygen content.

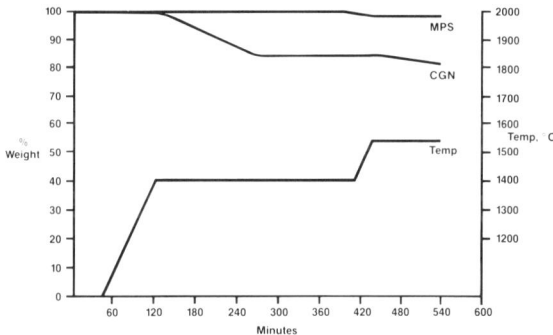

Fig. 4. Thermogravimetric analysis trace in Helium of CGN and MPS ceramic fibers.

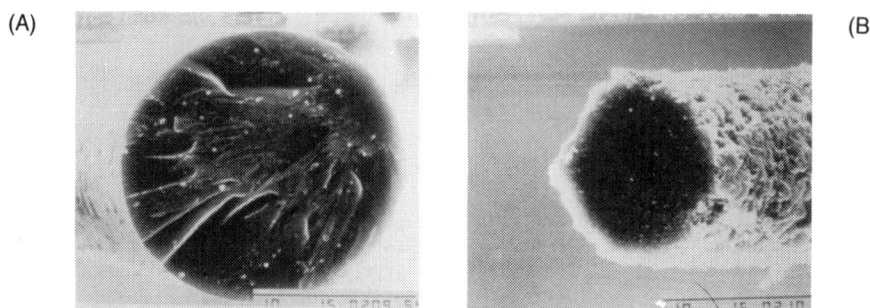

Fig. 5. SEM of CGN ceramic fiber (A) before and (B) after thermal aging at 1500°C for 2 hours in Argon (10 μm bar).

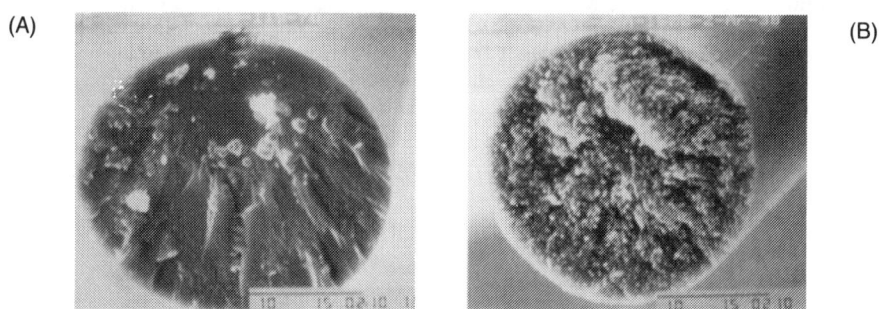

Fig. 6. SEM of ceramic fiber derived from MPS polymer (A) before and (B) after thermal aging at 1500°C for 2 hours in Argon (10 μm bar).

Chemical Structure

Figure 7 shows an infrared spectrum of MPS ceramic fiber, run as a KBr pellet[1]. The predominant band is due to Si-C bonds at 840 cm^{-1}. A weak shoulder at 1080 cm^{-1} is due to Si-O bonds. Other weak bands near 2950 cm^{-1} are due to a small amount of residual C-H bonds. The continuous increase in absorption towards the visible may be attributed to black-body absorbance by excess C.

Figures 8 and 9 show solid state, magic angle sample spinning nuclear magnetic resonance (MAS NMR) spectra of powdered MPS ceramic fiber. Figure 8 is a ^{29}Si MAS NMR spectrum obtained at 6.4 Tesla, with 4 KHz sample spinning. A total of 248 pulses were acquired in the FT-NMR mode with 2 minute delays between pulses to permit T_1 relaxation. The peak at $\delta = -14$ ppm, referenced to external TMS, closely corresponds to literature[20] chemical shifts for β-SiC (-18.3 ppm) and

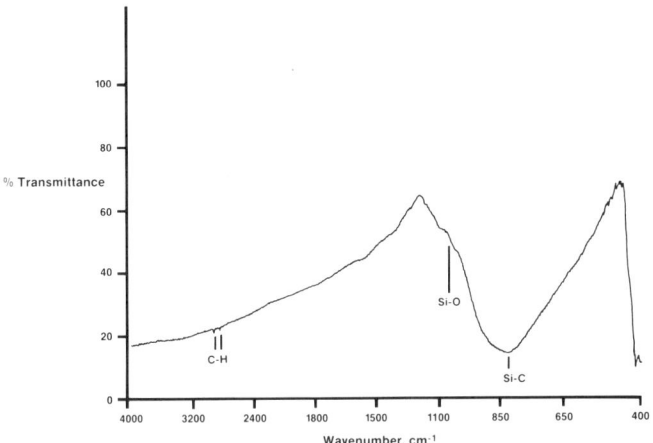

Fig. 7. Infrared spectrum of ceramic fiber derived from MPS polymer, run as a KBr pellet.

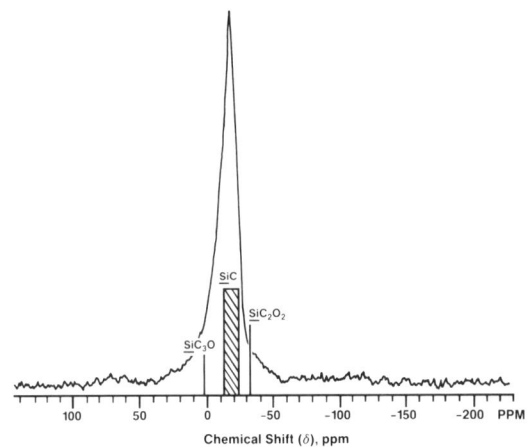

Fig. 8. ^{29}Si MAS NMR spectrum of ceramic fiber derived from MPS polymer.

Silicon Carbide

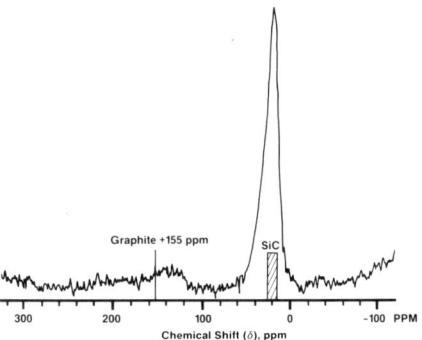

Fig. 9. ^{13}C MAS NMR spectrum of ceramic fiber derived from MPS polymer.

6H SiC (-13.9, -20.2 and -24.5 ppm, 3 equal intensity signals). Some silicon oxycarbide species appear to be present as shown by signal intensity at approximately +1 ppm ($SiOC_3$ species) and -30 ppm (SiO_2C_2 species)[1].

Figure 9 shows the 13_c MAS NMR spectrum of powdered MPS ceramic fiber obtained using the same instrument. A BN rotor was machined for sample spinning (3.8 KHz) to eliminate interfering 13_c signals from the usual polymeric rotors. A total of 282 pulses were acquired in the FT-NMR mode with 2 minute delays between pulses to permit T_l relaxation. The major peak at δ = +21 ppm corresponds closely to the published chemical shift of silicon carbide of unspecified polytype at δ = +21 ppm[21].

Broad tails of the peak may correspond to resonance of C-Si-O species as observed by ^{29}Si MAS NMR. A weak, broad signal at +140 ppm can be assigned to excess carbon as it corresponds closely to the reported[21] chemical shift of graphite at +155 ppm (broad). NMR data thus support the IR spectroscopic evidence that the sample consists primarily of a silicon carbide-like structure. NMR data also support the rule of mixtures calculation indicating that excess carbon is present.

A scanning Auger depth profile§, Figure 10, confirms the predominantly SiC-like structure. However, the near surface region of the fiber is quite rich in oxygen, suggesting a silicon oxycarbide structure.

X-Ray Diffraction

X-Ray diffraction (Figure 11) shows a broad peak at 2θ = 36 degrees for fiber exposed to 1200°C in argon, suggesting that nanocrystalline silicon carbide is present. This is confirmed by additional broad peaks at 2θ = 60 and 72°. A crystallite size of 2 nm is estimated from broadening of the [111] peak at 2θ = 36°. Darkfield TEM suggests a somewhat larger crystallite size of 3 to 4 nm[22]. The tungsten and aluminum peaks are standards used to calibrate wt% crystallinity[23].

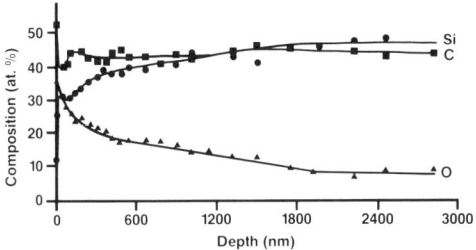

Fig. 10. Scanning Auger depth profile of ceramic fiber derived from MPS polymer.

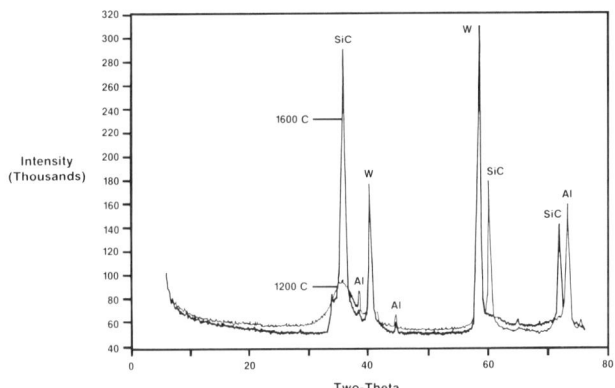

Fig. 11. X-ray diffractograms of ceramic fiber derived from MPS polymer after exposure to 1200°C and 1600°C.

The sample exposed to 1200°C contained 50 wt% microcrystalline SiC. On exposure to argon at 1600°C, the SiC peaks sharpen considerably indicating an increase in crystallite size to 50 nm. Total crystalline content increases to 86 wt%, with 60 wt% β-SiC and 26 wt% α-SiC present.

Porosity from X-Ray Scattering

In Figure 12, it can be seen that low angle scattering intensity ($2\theta < 30°$) is considerably greater for the sample exposed to 1200°C than for the sample exposed

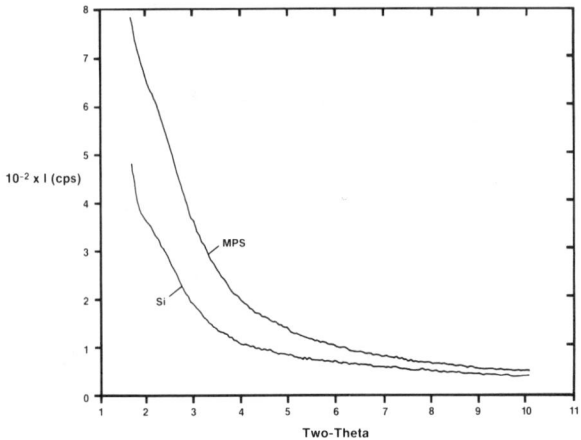

Fig. 12. X-ray scattering curves for ceramic fiber derived from MPS polymer and from a highly crystalline silicon sample.

to 1600°C. Sample weight and count time were the same. More precise X-ray scattering data was obtained[¶] for this sample in comparison with a highly crystalline, fully dense silicon sample having minimal low angle scattering (Figure 12). It is apparent that the MPS sample exhibits considerably more scattering in the $2\theta = 2$ to 10-degree region than does the Si sample.

Scattering intensity was corrected after equating scattering at $2\theta = 9.5°$ for the two samples. From a Guinier plot (corrected intensity vs. $S**2$), an apparent radius of gyration, R_g, for density inhomogeneties in the MPS sample of 0.89 nm is obtained[24]. This corresponds to a spherical scatterer radius of 1.0 nm. From unpublished x-ray scattering work with partially microcrystalline CGN and amorphous HPZ fibers[25], covering abroad range of scattering angles ($2\theta = 0.1$ to 20°), the scattering intensity can be attributed to porosity on a nanometer size range in MPS fiber. The pore fraction of 0.10 calculated from gradient column density vs. calculated rule-of-mixtures density thus appears to consist primarily of nanometer-size porosity. Some surface-connected porosity in this size range is evident from the difference between gradient column density ($\rho_o = 2.336$ g/cm³, using a mixture of tetrachloroethylene and tribromomethane) and a density by He pycnometry of 2.70 g/cm³, as compared to a calculated density of 2.92g/cm³.

Most of this x-ray scattering intensity is lost on heating the sample to 1600°C, implying loss of nanoscale porosity. Further characterization of this nanoporosity is underway.

SUMMARY

Methylpolysilane polymers can be processed into strong, high modulus ceramic fibers. Efforts to increase properties are underway. Ceramic composition is controllable from silicon-rich through stoichiometric silicon carbide to carbon-rich.

These ceramic fibers, when low in oxygen content, show the lowest weight loss when heated to 1570°C in an inert atmosphere of any polymer-derived ceramic fiber studied to date. They also show the least surface degradation by SEM when heated to 1500°C in an inert atmosphere of any polymer-derived ceramic fiber studied to date.

For one sample containing 6 wt% oxygen, which is higher than typical, the ceramic structure appears to consist of 50 wt% nanocrystalline SiC having 2 nm crystallite size in a continuous, amorphous silicon oxycarbide phase. Ceramic fibers containing 2% oxygen contain 80 wt% microcrystalline SiC. The glassy silicon oxycarbide phase contains nanometer size porosity which is largely removed on heating to 1600°C. A small quantity of excess carbon, 7 wt%, is present and probably is in the form of microcrystalline, turbostratic pyrolytic carbon as is found in other polymer-derived ceramic fibers[1].

ACKNOWLEDGEMENTS

Work was performed under DARPA Contract No. F33615-83-C-5006, administered by the Air Force, Wright Aeronautical Labs. We acknowledge the support and encouragement of K. Rhyne (DARPA) and A. P. Katz (AFWAL). We also thank C. T. Li for the SEM images, A. Zangvil for the scanning Auger data and R. Miller for the X-Ray scattering data.

---

*NICALON Ceramic Grade Fiber, Nippon Carbon Co., Ltd., Tokyo, Japan.
†This MPS ceramic fiber contained 2 wt% O.
‡Thermal Analyzer Model 429 equipped with 1600°C furnace, Netzsch Incorporated, Exton, PA.
§A. Zangvil, University of Illinois, using a Model 395 Scanning Auger Microscope, Perkin-Elmer Corp., Physical Electronics Division, Eden Prairie, MN.
¶A wide angle Phillips diffractometer using CuKα radiation, equipped with a curved graphite crystal monochomator and 1/8 degree slits, was operated in air at 40 KV and 30 mA in a step scanning mode, R. Miller, Michigan Molecular Institute.

References

[1] J. Lipowitz, H. A. Freeman, R. T. Chen, and E. R. Prack, "Composition and Structure of Ceramic Fibers Prepared from Polymer Precursors," *Adv. Ceram. Mater.*, 2 (2), 121-28 (1987).

[2] K. J. Wynne and R. W. Rice, "Ceramics via Polymer Pyrolysis," *Ann. Rev. Mater. Sci.*, 14, 297-334 (1984).

[3] R. W. Rice, "Ceramics from Polymer Pyrolysis, Opportunities and Needs—A Materials Perspective," *Am. Ceram. Soc. Bull.*, 62 [8] 889-92 (1983).

[4] S. Yajima, "Special Heat-Resisting Materials from Organometallic Precursors," *ibid*, 62 [8] 893-98, 903 (1983).

[5] R. West, L. D. David, P. I. Durovich, H. Yu, and R. Sinclair, "Polysilastyrene: Phenylmethylsilane-Dimethylsilane Copolymers as Precursors to Silicon Carbide," *ibid.*, 62 [8] 899-903 (1983).

[6] R. Wills, R. A. Markle, and S. P. Mukherjee, "Siloxanes, Silanes, and Silazane in the Preparation of Ceramics and Glasses," *ibid.*, 62 [8] 904-11, 915 (1983).

[7] C. L. Schilling, Jr., J. P. Wesson, and T. C. Williams, "Polycarbosilane Precursors for Silicon Carbide," *ibid.*, 62 [8] 912-15 (1983).

[8] R. H. Baney, "Some Organometallic Routes to Ceramics," Ch. 20 in Ultrastructure Processing of Ceramics, Glasses and Composites. Edited by L. L. Hench and D. R. Ulrich. John Wiley and Sons, NY, 1984.

[9] R. H. Baney, "Molecular Design and Evaluation of Preceramic Polymer for High Temperature Structural Ceramics," presented at Division of Industrial and Engineering Chemistry, I. Structural, Electronic, and Refractory Ceramics, ACS Symposium on Emerging Materials Technologies, September, 1985.

[10] T. Mah, M. G. Mendiratta, A. P. Katz and K. S. Mazdiyasni, "Recent Developments in Fiber-Reinforced High Temperature Ceramic Composites," *Am. Ceram. Soc. Bull.*, 66 (2), 304-8 (1987).

[11] D. W. Richerson, "Modern Ceramic Engineering," M. Dekker, New York, p. 72 (1982).

[12] R. H. Baney and J. H. Gaul, U.S. Patent 4,298,559 (1981).

[13] R. H. Baney and J. H. Gaul, U.S. Patent 4,310,651 (1982).

[14] R. H. Baney, J. H. Gaul, and T. K. Hilty, "Methylchloropolysilanes and Derivatives Prepared from the Redistribution of Methylchlorodisilanes," *Organometallics*, 2, 859-864 (1983).

[15] C. T. Li and N. R. Langley, "Improvement in Fiber Testing of High-Modulus Single-Filament Materials," *J. Am. Ceram. Soc.*, 68 (8), C-202-C-204 (1985).

[16] Private communication, C.T. Li.

[17] M. Jaffe and L. C. Sawyer, "Strength Limiting Features of Polymer Derived Ceramic Fibers," Proc. of the 3rd International Conf. on Ultra Structure Processing of Ceramics, Glasses and Composites," Los Angeles, 1987, in press.

[18] Spectra were obtained by G. L. Turner (Ref. 19), Spectral Data Services, Inc., Champaign, IL.

[19] G. L. Turner, R. J. Kirkpatrick, S. H. Risbud and E. Oldfield, "Multinuclear Magic-Angle Sample-Spinning Nuclear Magnetic Resonance Spectroscopic Studies of Crystalline and Amorphous Ceramic Materials, *Am. Ceram. Soc. Bull.*, 66 (4) 656-631 (1987).

[20] G. R. Finlay, J. S. Hartman, M. F. Richardson, and B. L.Williams, "^{29}Si and ^{13}C Magic Angle Spinning NMR Spectra of Silicon Carbide Polymorphs," *J. Chem. Soc., Chem. Commun.*, 159 (1985).

[21] D. T. Haworth and C. A. Wilkie, "The Solid State ^{13}C-NMR Spectra of Some Carbides," *J. Inorg. Nucl. Chem.*, **40**, 1689-90 (1978).

[22] Y. W. Chang, A. Zangvil and J. Lipowitz, "Characterization of Si-C-N-O Fibers by Analytical STEM and Scanning Auger Techniques," these proceedings.

[23] L. K. Frevel and W. C. Roth, "Semimicro Assay of Crystalline Phases by X-Ray Powder Diffractometry," *Anal. Chem.*, **54**, 677 (1982).

[24] L. E. Alexander, "X-Ray Diffraction Methods in Polymer Science," Krieger Pub. Co., Huntington, NY, 1979, Chapt. 5.

[25] G. E. LeGrow, T. F. Lim, J. Lipowitz and R. S. Reaoch, "Ceramics from Hydridopolysilazane," *Am. Ceram. Soc. Bull.*, **66** (2) 363-67 (1987).

Characterization of Si, C, N, O Fibers by Analytical STEM and Scanning Auger Techniques

Yeu-Wen Chang and Avigdor Zangvil

Materials Research Laboratory and
Department of Materials Science and Engineering
University of Illinois at Urbana-Champaign
Urbana, IL 61801

Jonathan Lipowitz

Dow Corning Corporation
Midland, MI 48686

Abstract

Several kinds of continuous Si, C, N, O ceramic fibers from organic precursors have been characterized microstructurally and microchemically. Quantitative microchemical analysis of fibers at and below the surface was accomplished in the scanning Auger microprobe by in situ ion sputtering and Auger analysis. Fibers with known compositions were used as standards for the determination of the elemental Auger sensitivities. Composition vs. depth profiles showed the existence of surface layers of various depths and compositions. A solvent-desized Nicalon fiber was found to have an ~ 50 nm carbon-rich surface layer and a Si, N, C, O fiber (from HPZ precursor) had an ~ 0.65 μm deep SiO_2-rich surface layer. TEM analysis showed that the HPZ fiber is amorphous and the Si, C, O (MPS) fiber is microcrystalline, with a crystallite size of about 3 ~ 4 nm. Windowless and ultra-thin window energy dispersive x-ray spectrometry yielded semiquantitative results for the light elements. Granular defects depleted of nitrogen and containing carbon and oxygen were observed in the interior of HPZ fibers.

Introduction

Continuous ceramic fiber reinforced ceramic composites are promising materials for high temperature structural applications. The strong ceramic fibers provide the higher strength and toughness of the composites through various energy dissipation processes during crack propagation. The microstructure, microchemistry, mechanical properties and thermal stability of fibers are important parameters to decide whether or not the composites can be successfully used in a high temperature environment. The characterization also provides information which will help develop appropriate processes to produce fibers for high temperature applications. Many investigators[1-6] have characterized Nicalon fibers by transmission electron microscopy (TEM), scanning electron microscopy (SEM), scanning Auger microprobe (SAM), X-ray diffraction (XRD), and mechanical tests. Some other techniques have also been used to analyze the chemical bonding structure of fibers[7].

In this study, the ceramic fibers were characterized microstructurally and microchemically. Microstructural analysis was accomplished by selected area electron diffraction and TEM/STEM with an ultra-thin window (UTW) energy dispersive x-ray spectrometer (EDS). Scanning Auger microprobe was used for the study on the microchemistry of the fibers. SEM and SAM with UTWEDS were used to analyze the fracture surfaces of the fibers.

Experimental Procedure

The ceramic fibers characterized in this study include three kinds of organosilicon polymer derived fibers—HPZ (hydridopolysilazane)[7-8], MPS (methylpolysilane)[9-10], and Nicalon[11-12]. Both MPS and Nicalon fibers are Si, C, O materials, with the MPS fiber being much lower in oxygen content, and HPZ fibers are Si, N, C, O materials.

The composition depth profiling was accomplished by using in situ ion sputtering and surface analysis in the scanning Auger microprobe* which is equipped with an argon ion sputtering gun. The fibers for this analysis were pressed onto and wrapped with an indium foil. It has been reported[13-14] that curved surfaces eroded non-uniformly during ion sputtering due to the variation of the angle of ion incidence on the surface. It means that the rate of sputtering at different positions on the curved surface is variable. Therefore, the position for sputtering and analysis was fixed at the middle point of the fiber, that would have the same incidence angle as the flat SiO_2/Si sample which was used to calibrate the rate of sputtering.

Fibers fractured in tension were mounted in a specially built holder for SEM and SAM studies. SEM was used to study fractography and search for the defect on the fracture surface. Since the EDS with Be window attached on the SEM could not detect the light elements, such as carbon, nitrogen and oxygen, the samples were moved to the SAM to analyze the defects on the fracture surface by an ultra-thin window EDS.†

Microstructural analysis was carried out by TEM/STEM.‡ An ultra-thin window EDS§ attached on this instrument was used to yield semiquantitative results in the microchemical study. The samples for TEM were prepared by mounting the fibers on copper grids, and ion milling along the direction of the fibers. Another method, used for HPZ fibers which have low electrical conductivity, was to embed fiber fragments in a dental amalgam and then employ conventional polishing, disc cutting, dimpling and ion milling.

Results and Discussion

SAM analysis

Scanning Auger microprobe can be used to analyze the composition of the top 2-20 atomic layers at the surface of materials. Insulating materials may cause a charging problem which will affect the analysis. Limited charging only causes peak shifts to higher energy, but quantitative analysis is still possible. However, serious charging will deflect the electron beam, and analysis becomes completely impossible. Approaches to overcome the surface charging problem have been suggested[15-16]. In this study, the fibers were embedded in indium foil, and the beam voltage and current were reduced as low as 3 KV and 0.01 μA, respectively. Serious charging did not occur in most polymer-derived fibers. Fibers that caused serious charging were coated with a thin layer of gold before sputtering to avoid charging.

The relative elemental sensitivity factors were obtained from spectra of several HPZ fibers, of which the bulk compositions were known. These sensitivity factors were also applicable to MPS and Nicalon fibers. The silicon peak could occur at two different energy levels, 76 eV and 92 eV[17], which have different sensitivity factors. The peak from SiO_2 is located at 76 eV. HPZ fibers, which have a SiO_2-rich surface layer, exhibited a split silicon peak, which should be taken into account with different sensitivity factors for quantitative analysis.

Composition depth profiles were obtained by alternately sputtering and obtaining an Auger spectrum. The automatic depth profiling routine was not used because it relies upon unchanging peak positions. In some fiber materials, the Auger peaks, and in particular, the silicon peak did change positions.

Figure 1 shows the composition depth profile of standard grade Nicalon (SGN), which had been desized with an organic solvent. The fiber had an ~ 50 nm carbon rich surface layer, which may be partly due to an incomplete removal of the organic sizing. The composition within the fiber also shows excess carbon to exist. This is deduced from a rule-of-mixtures calculation, assuming that oxygen occurs as SiO_2, nitrogen as Si_3N_4 and carbon as SiC or C.

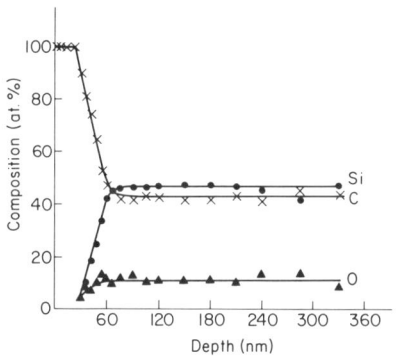

Fig. 1. Composition depth profile of Nicalon (SGN) fiber. Fiber has an ~ 50 nm carbon rich surface layer.

Figure 2 shows the composition depth profile of a Si, N, C, O (HPZ) fiber. It had an ~ 0.65 μm SiO_2-rich surface layer. The spectrum shown in Fig.3(a) was taken from the SiO_2 rich region. The silicon peak was split into two peaks, which means that the material contains mixtures of Si-O (the lower energy peak) and Si-C, Si-N, or Si-Si (the higher energy peak) bonds. A typical spectrum of HPZ fiber

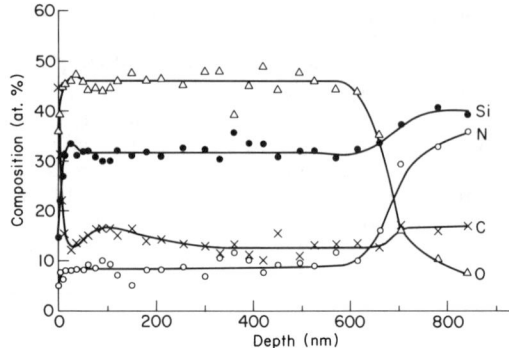

Fig. 2. Composition depth profile of HPZ fiber, showing an ~ 0.65μm deep SiO$_2$-rich surface layer.

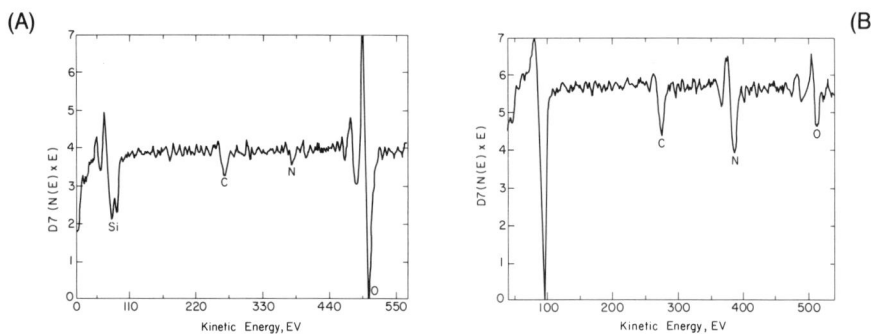

Fig. 3. (*A*) shows a split Si peak spectrum that was taken from the SiO$_2$-rich surface layer of the fiber analyzed in Fig. 2. (*B*) is a typical spectrum of HPZ fiber.

is shown in Fig. 3(b). The bulk composition still shows that the excess carbon exists based on a rule-of-mixtures calculation, but the total amount of carbon is low.

Figure 4 shows the composition depth profile of a Si, N, C, O fiber which was enriched with carbon. The carbon concentration increased through the whole depth of the fiber. The surface also shows a thin layer of SiO$_2$ with C.

The composition depth profile of a Si, C, O (MPS) fiber is shown in Fig.5. The bulk composition is similar to Nicalon fiber except that the oxygen content is about 6 at.% compared with 10 ~ 15 at.% for Nicalon. However, the microchemistry of

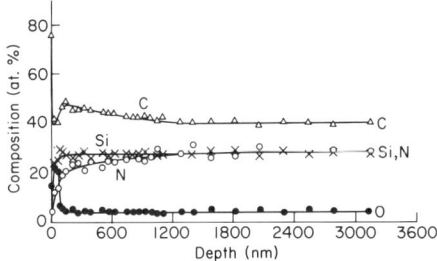

Fig. 4. Composition depth profile of carbon enriched HPZ fiber.

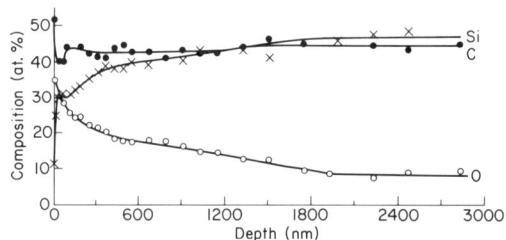

Fig. 5. Composition depth profile of MPS fiber.

the surface layer is different. The oxygen decreased gradually inward to about 1.8 μm depth and a high concentration carbon layer is not present.

TEM/STEM Analysis

Figure 6 shows a typical TEM micrograph and selected area electron diffraction (SAD) pattern of an HPZ fiber, which indicates that the fiber is amorphous. This is in agreement with previous XRD results[7]. A few small defects were found in the HPZ fiber shown in Fig.7(a). The thin window EDS in Fig. 7(b) shows the defects have higher oxygen content and perhaps somewhat higher carbon than the regular fiber material. The windowless EDS spectra of regular HPZ fiber are shown in Fig. 8. It shows a higher intensity of light elements in a thinner area. This confirmed that the x-ray intensity from light elements is sensitive to the thickness, as would be expected from absorption calculations.

Figure 9(a) shows a dark field micrograph of the MPS fiber. The bright spots were contributed by the first ring (111) of the SAD pattern (Fig. 9(b)), which shows a β-SiC phase. The size of the crystallites is about 3-4 nm, which compares well with results from x-ray diffraction line broadening[10].

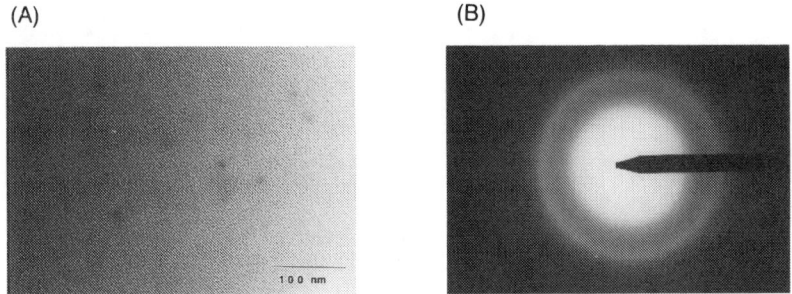

Fig. 6. (A) is a bright field TEM micrograph of HPZ fiber; SAD pattern (B) indicates the material is amorphous.

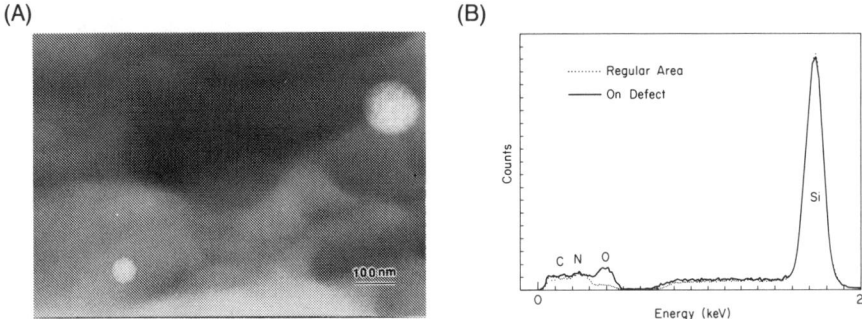

Fig. 7. (A) shows defects in the HPZ fiber; (B) shows ultra-thin window EDS spectra of HPZ at and near defect.

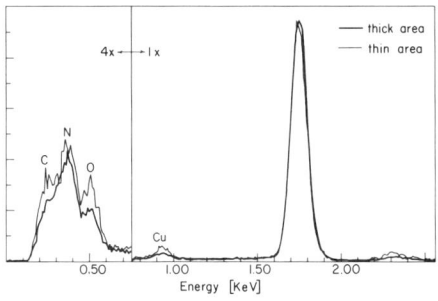

Fig. 8. Windowless EDS spectra of HPZ fiber, showing the effect of the thickness on the analysis.

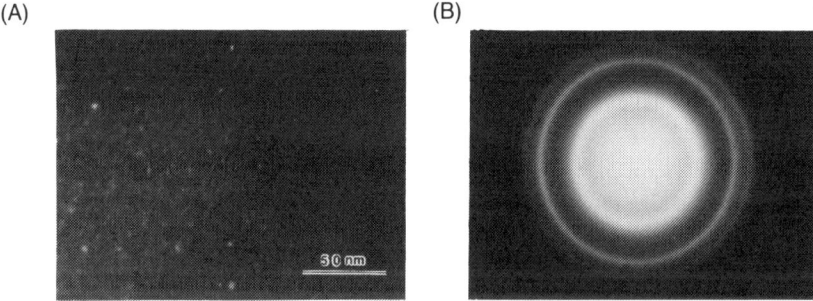

Fig. 9. (*A*) is a dark-field electron micrograph of MPS fiber. (*B*) is a SAD pattern and indicates crystallites are β-SiC.

FRACTURE ANALYSIS

HPZ fibers with low and medium strength were selected for fracture analysis. In many cases, the failure initiated from surface defects which were induced by spinning or mechanical damage[8]. Few failure-initiating defects were found at the interior of the fiber. Figure 10 shows a granular defect at the locus of the failure of an HPZ fiber. Ultra thin window EDS analysis at the defect showed that the defect was depleted of nitrogen relative to the adjacent fiber material.

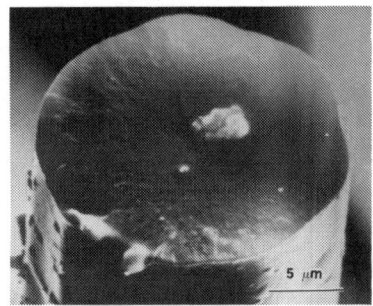

Fig. 10. SEM secondary electron image of fracture in HPZ, showing a granular defect.

Summary

1. The microchemical characterization of fibers at and below the surface can be accomplished by in situ sputtering and SAM analysis. Nicalon fiber which has been solvent-desized has an ~ 50 nm carbon rich surface layer and HPZ fiber has an ~ 0.65 μm SiO_2-rich surface layer.

2. TEM micrographs and SAD patterns show that HPZ fiber is amorphous and the MPS fiber is microcrystalline. Windowless and ultra-thin window EDS can yield semiquantitative results for light elements.

3. Granular defects depleted of nitrogen and containing carbon and oxygen were observed in the interior of HPZ fibers.

Acknowledgments

The microanalytical work was performed at the Center for Microanalysis of Materials of the Materials Research Laboratory at the University of Illinois, which is supported by the U.S. Department of Energy, Division of Materials Sciences under contract No. DE-AC02-76ER01198. The authors thank the staff of the Center, and Ms. N. Finnegan in particular, for their continued assistance. We are also grateful to M. L. Trautman, DMD, for his kind assistance. The study was supported by DARPA through the Dow Corning Corporation under contract No. F33615-83-C-5006.

---

*Model 595, Physical Electronics Industries, Eden Prairie, MN.
‡Kevex corporation, Foster City, CA.
†Model EM420, Philips, Eindhoven, The Netherlands.
§EDAX International, Prairie View, IL.

References

[1] T. Mah, N. L. Hecht, D. E. McCullum, J. R. Hoenigman, H. M.Kim, A. P. Katz, and H. A. Lipsitt, "Thermal Stability of SiC Fibres (Nicalon)," *J. Mater. Sci.*, **19** (1984) pp. 1191-1201.

[2] G. Simon and A. R. Bunsell, "Mechanical and Structural Characterization of the Nicalon Silicon Carbide Fibre," ibid, pp. 3649-3557.

[3] G. Simon and A. R. Bunsell, "Creep Behavior and Structural Characterization at High Temperatures of Nicalon SiC Fibers," ibid, pp. 3658-3670.

[4] L. C. Sawyer, R. Arons, F. Haimbach, M. Jaffe and K. D. Rappaport, "Characterization of Nicalon: Strength, Structure and Fractography," Ceramic Eng. and Sci. Proc., July-August (1985), pp. 567-587.

[5] T. J. Clark, M. Jaffe, J. Rabe and N. R. Langley, "Thermal Stability Characterization of SiC Ceramic Fibers: I. Mechanical Property and Chemical Structure Effects," ibid, July-August (1986), pp. 901-913.

[6] L. C. Sawyer, R. T. Chen, F. Haimbach IV, P. J. Harget, E. R. Prack, and M. Jaffe, "Thermal Stability Characterization of SiC Ceramic Fibers: II. Fractography and Structure," ibid, pp. 914-931.

[7] J. Lipowitz, H. A. Freeman, R. T. Chen and E. R. Prack, "Composition and Structure of Ceramic Fibers Prepared from Polymer Precursors," *Advanced Ceramic Materials*, **2** [2] (1987) pp. 121-128.

[8] G. E. Legrow, T. F. Lim, J. Lipowitz, and R. S Reaoch, "Ceramics from Hydridopolysilazane," *Am. Ceram. Soc. Bull.*, **66** [2], (1987) pp. 363-368.

[9] T. Mah, M. G. Mendirata, A. P. Katz and K. S. Mazdiyasni, "Recent Developments in Fiber-Reinforced High Temperature Ceramic Composites," ibid. pp. 304-307.

[10] J. Lipowitz. N. R. Langley, G. E. LeGrow and T. F. Lim, "SiC Fibers from Methylpolysilane Polymers," this book.

[11] S. Yajima, Y. Hasegawa, J. Hayashi, and M. Iimura, "Synthesis of Continuous Silicon Carbide Fiber with High Tensile Strength and High Young's Modulus, Part I. Synthesis of Polycarbosilane as Precursor," *J. Mat. Sci.*, **13** (1978) pp. 2569-2576.

[12] Y. Hasegawa, M. Iimura and S. Yajima, "Synthesis of Continuous Silicon Carbide Fiber, Part II. Conversion of Polycarbosilane Fibre into Silicon Carbide Fibres," ibid., **15** (1980) pp. 720-728.

[13] A. D. G. Stewart and M. W. Thompson, "Microtopography of Surfaces Eroded by Ion-Bombardment," ibid., **4** (1969) pp. 55-60.

[14] B. B. Meckel, T. Nenadovic, B. Perovic, A. Vlahov, "Experimental Investigation of the Sputter-Topographic Evaluation of a Cylindrical Surface," ibid, **10** (1975) pp. 1188-1193.

[15] B. Goldstein and D. E. Carlson, "Determination of the Composition of Glass Surfaces by Auger Spectroscopy," *J. Am. Ceram. Soc.*, **55** (1972) pp. 51-52.

[16] W. Brue, H. J. Dudek and G. Ziegler, "Application of AES and XPS for Microstructural Characterization of Dense Si_3N_4 and Oxynitride Glasses," British Ceram. Proc. No. 34, August (1984), pp. 32-34.

[17] L. E. Davis, N. C. MacDonald, P. W. Palmberg, G. E. Riach and R. E. Weber, "Handbook of Auger Electron Spectroscopy," 2nd Ed. (1976), Perkin-Elmer Corp., Eden Prairie, MN 55343.

Section VI
Electronic Applications

CRYSTAL GROWTH OF SiC FOR ELECTRONIC APPLICATIONS

LAWRENCE G. MATUS AND J. ANTHONY POWELL

NASA Lewis Research Center
Cleveland, OH 44135

ABSTRACT

The development of SiC as a useful semiconductor for high temperature and other applications has long been hindered by the lack of a suitable crystal growth process. Recent progress in the epitaxial growth of single crystal SiC films on inexpensive single crystal silicon wafers has been encouraging. In the last year, much has been learned regarding qualities of these films pertinent to semiconductor applications. Examples are the high level of compensation in unintentionally doped films and the prevalence of antiphase disorder in all films grown on (001) Si. Specific progress will be described in the improvement of film morphology and the elimination of antiphase disorder through the use of Si wafers whose orientation is slightly off axis from (001).

INTRODUCTION

The NASA Lewis Research Center currently has a program to develop electronic devices capable of operation at high temperatures for extended periods of time. This program supports a major element of the Center's mission: to perform basic and developmental research aimed at improving aeropropulsion systems. The need for high temperature electronic devices in research and development of advanced aircraft engines and operational engine applications such as engine monitoring and control has already been described[1].

In addition to aeronautics, there are many other areas that would benefit from the existence of high temperature electronic devices. Space applications include power electronics for the space station and other satellites. Since power electronics require radiators that dissipate waste energy, electronic devices that operate at higher temperatures would allow a reduction in radiator size. Other applications on Earth include deep-well drilling instrumentation, power electronics, and nuclear reactor instrumentation and control.

The desired operating temperature for some of these applications approaches 600°C, which is well beyond the capability of currently available semiconductor devices. It is clear that a new semiconductor material will need to be developed to meet the high temperature needs of the applications mentioned above. This paper will describe progress made at the NASA Lewis Research Center in the development of silicon carbide (SiC) as a high temperature semiconductor material.

Comparison of Materials

In comparing potential candidate materials for high temperature semiconductor devices, SiC stands out not only because of its excellent high temperature electronic properties but also because it is a very stable ceramic material up to temperatures of 1800°C. Six materials were selected for comparison and some pertinent properties of these materials are assessed in Table I. There certainly are other materials that could have been included but the six selected are either commercially available or offer much potential for future development.

The first material listed in Table I is the most commonly used semiconductor, silicon; it provides a basis of comparison for the other materials. The next two in the list, gallium arsenide and gallium phosphide, are representatives of the many III-V compounds and their alloys. Two silicon carbide crystalline forms are included because they are both potentially useful high temperature semiconductors but have significantly different electronic and optical properties. Diamond is listed because of its long-term potential.

The maximum operating temperature for a semiconductor is determined by the forbidden energy bandgap. The temperature limit is reached when the number of intrinsic carriers, thermally excited across the energy gap, approaches the number of extrinsic carriers. This temperature (when expressed as the absolute temperature) is roughly proportional to the energy bandgap. In Table I, a maximum temperature of 300°C was assumed for silicon and based on the ratio of bandgaps, the operating temperature was calculated for the other materials. Diamond is a special case in that a phase change around 1100°C determines its maximum temperature. These temperatures should be used only as a starting point since other factors also affect the maximum temperature, such as the type of device and the length of service required for a given application.

Other material properties of major importance for the operation of a semiconductor at high temperatures are the physical stability, carrier mobility, thermal conductivity, and breakdown field. The thermal conductivity is of particular importance if much power is to be dissipated in a device.

In addition to the properties listed in Table I, there are other considerations in selecting a semiconductor material. For example, the technology base for silicon is so far ahead of the other materials that there must be compelling reasons for choosing any other material. In the case of the Lewis program for developing electronics for advanced turbine engines, the target operating temperature was set at 600°C. This ruled out silicon. Gallium arsenide and gallium phosphide are also commercially available, and gallium phosphide meets the Lewis temperature goal. However, the stability of this material (and the other III-V compounds) was not considered sufficient for reliable operation for long periods at 600°C.

A major disadvantage of the remaining three materials in Table I has been the lack of reproducible processes for growing the large-area single-crystal substrates that are needed for commercial production of devices. This is still a problem with diamond, but considerable progress has been made recently on the crystal growth of SiC. The remaining part of this paper will describe this recent progress.

Silicon Carbide Crystal Growth

Until recently, there was no process whereby single crystals of SiC with sufficient size, purity, and perfection could be grown reproducibly. SiC does not melt at any reasonable temperature and pressure conditions so this rules out the

Table I. Comparison of electronic materials.

Material	Bandgap, (EV)	Maximum Operating Temperature °C	Physical Stability	Carrier Mobility	Thermal Conductivity	Breakdown Field
Silicon	1.1	300	Good	Good	Good	Good
Gallium Arsenide	1.4	460	Fair	Excellent	Fair	Good
Gallium Phosphide	2.2	875	Fair	Fair	Fair	Good
Beta Silicon Carbide	2.3	925	Excellent	Fair	Very Good	Excellent
6H Silicon Carbide	2.9	1240	Excellent	Fair	Very Good	Excellent
Diamond	5.5	1100	Very Good	Good(?)	Excellent	Excellent

growth-from-melt technique commonly used to obtain other semiconductor single crystals. Historically, vapor phase growth processes have proven to be the most successful method for producing SiC crystals. Early research was done on SiC crystals that were a by-product of the industrial process for making sandpaper grit and abrasives. In the industrial process, SiC is formed at 2500°C by the reaction of silica and coke. At these temperatures, gas pockets form within the SiC reaction product. The SiC sublimes and then condenses on the walls of pockets located at cooler parts of the reaction product. Occasionally, isolated SiC crystals are produced within these pockets during the production process. The larger and better crystals are selected for research purposes.

In 1955, Lely[2] developed a laboratory version of the industrial sublimation process and was able to produce rather pure SiC crystals. Encouraged by the Lely process, NASA Lewis and other laboratories pursued the development of SiC semiconductor devices during the sixties. Unfortunately, by the early seventies the Lely process and other processes had not matured to the point where high-quality large-area crystals could be grown reproducibly. Since crystal substrates are crucial to device fabrication, interest in SiC waned, and from 1973 to 1980, there was very little effort in the U.S. on SiC. However, research did continue in Japan and in Europe during this period.

In 1980, because of the increased need for high temperature electronics in advanced turbine engines, NASA Lewis again embarked on a high temperature electronics program. The emphasis again has been on developing SiC.

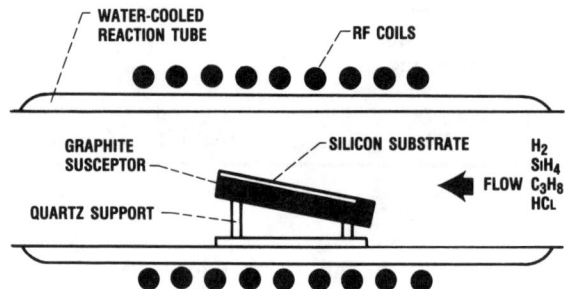

Fig. 1. Schematic diagram of reaction chamber for the cubic SiC crystal growth process.

The problem regarding the crystal growth of SiC is rooted in the fact that SiC crystals can take on many different structural forms called polytypes. In most cases, SiC crystals grown by sublimation techniques are a mixture of different polytypes. Since each SiC polytype has its own electronic properties, sublimation grown SiC crystals usually contain heterojunctions and possses unpredictable electronic properties. To favor growth of a single polytype of SiC, epitaxial growth on a host crystal substrate from gases containing silicon and carbon was hypothesized. The host crystal imparts its crystalline regularity to the thin growing layer. Since silicon is available in perfect, large, and low-cost crystals, many attempts were made on the heteroepitaxial growth of SiC on Si. These efforts were largely unsuccessful because of the large lattice mismatch that exists between Si and SiC (e.g. the SiC lattice is 20% smaller than the Si lattice).

Large area heteroeptiaxial growth of cubic SiC on Si was finally achieved at the NASA Lewis Research Center by using a chemical vapor deposition (CVD) process[3]. Crystal growth takes place at atmospheric pressure in a fairly conventional horizontal CVD system. A complete system description is given in Powell, et al[4]. The growth chamber is shown schematically in Figure 1.

The quartz reaction tube is water-cooled and coaxial with an RF coil that inductively heats a rectangular graphite susceptor. The graphite susceptor, which is coated with SiC to minimize the introduction of contaminants from the graphite, supports the Si substrates on which SiC is epitaxial grown. Temperature is measured and controlled by an automatic optical pyrometer.

The process steps are shown in Figure 2. Hydrogen, purified by passage through a palladium alloy diffusion cell, is the carrier gas and flows continuously during the entire process. The sources of Si and C are silane (3% SiH_4 in H_2) and propane (3% C_3H_8 in H_2) respectively. To provide a good surface for the initial SiC growth, HCl gas is added for two minutes with the Si substrate at 1200°C to chemically etch 2 μm of the Si surface. The essential step in the growth process is the rapid temperature ramp from near room temperature to the growth temperature in the presence of a hydrocarbon. When near room temperature is reached after the HCl etch step, C_3H_8 is added to the carrier gas. After allowing an additional 30 seconds for flow equilibration, the temperature of the susceptor and substrate is ramped to 1360°C in about 30 seconds. The C_3H_8 is left on for an additional 90

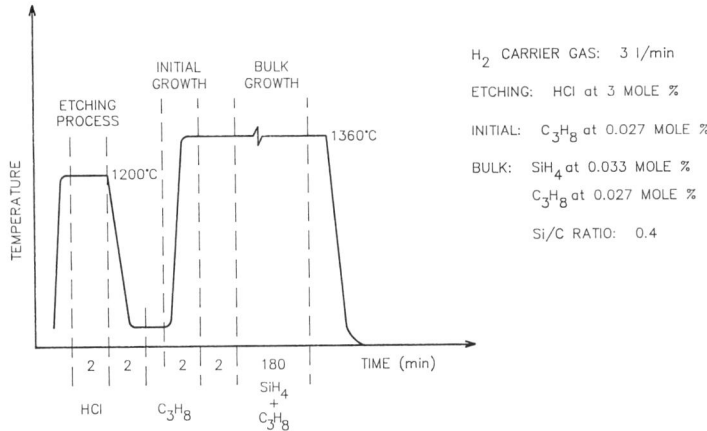

Fig. 2. Growth schedule for producing a single-crystal cubic SiC film on Si.

seconds and then turned off. During the 2 minute period (ramp plus 90 seconds), a single crystal cubic SiC film about 20 nm thick is produced on the Si substrate. After a two-minute purge with the H2 carrier gas, SiH_4 and C_3H_8 are added to initiate the final step, the bulk growth of cubic SiC to the desired thickness. During this time, the cubic SiC layer grows at a rate of 3 to 4 μm/hr.

EVALUATION OF SiC FILMS

Initially, the transition from the Si substrate to the SiC epitaxial layer was thought to occur by means of a thin buffer layer or transition layer of the order of 20 nm thick[3,5]. However, high resolution transmission electron microscopy (TEM) has demonstrated that the SiC/Si interface is abrupt with no transition region[6,7]. The cubic SiC films do contain a large density of defects that include interfacial twins, stacking faults, and antiphase disorder[6-8]. The defect density in the films is greatest near the SiC/Si interface and decreases with distance away from the interface[9]. It is generally felt that the high defect density in the epitaxial cubic SiC films results from the thermal expansion and lattice mismatch between SiC and Si.

A particular type of lattice defect, called antiphase boundaries (APB's), was reported to be present in all SiC films grown on Si at NASA Lewis[8]. The APB's can be decorated by chemical etching, sputter etching, wet oxidation or cubic SiC growth in the presence of diborane[10] and have the appearance of irregular shaped boundaries as shown in Figure 3. APB's form in the initial stages of growth on the Si substrate when SiC islands of opposite phase nucleate and grow together. Across the APB, the chemical bonding is between like atoms (i.e., Si-Si or C-C), instead of the normal Si-C bond between neighboring atoms. Early theories suggest that APB's will form on substrates with steps on the atomic scale. If the height of the steps is an odd number of atoms, APB's form (Figure 4a). If the height is an even number of atoms, no APB's form (Figure 4b).

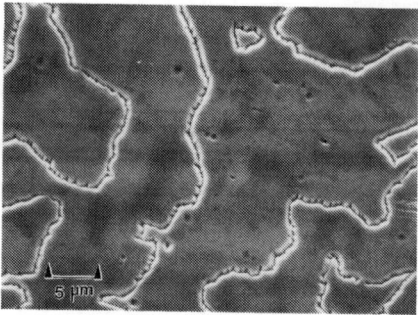

Fig. 3. SEM micrograph of "boron decorated" APB's (grooves in surface) in a cubic SiC film grown using an on-axis (001) Si substrate. Film consists of a 1-μm-thick boron-doped layer on a 10-μm-thick undoped layer.

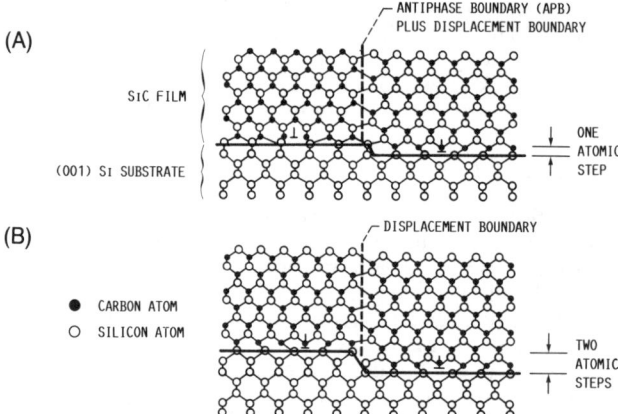

Fig. 4. Model of SiC/Si interface; (A) Substrates with odd number atomic steps result in APB formation. (B) Substrates with even number atomic steps result in APB-free growth.

Normally, in epitaxial growth on Si, the surface of the Si substrates is oriented precisely parallel to an atomic plane, e.g. the (001) plane. It has been found in the growth of gallium arsenide (GaAs) on Si, that APB's are eliminated by orienting the substrate slightly off-axis from the (001) plane[11]. The reason for this success is as follows: It has been shown that the atomic step height on off-axis Si substrates is an even number of atoms[12]. Hence, as predicted by the model illustrated in Figure

4, the APB's are eliminated. This technique was applied to SiC grown at NASA Lewis with the following results: for Si substrates that were tilted 2° to 4° from the (001) plane, all APB's were eliminated from the subsequent SiC films grown. In addition, the resultant SiC films were smoother by a factor of 2 to 3 than the films grown on on-axis substrates. Although not yet proven, APB's may account for some of the degraded performance of SiC devices.

Electrical characterization to determine electrical properties of the cubic SiC films is an important and necessary evaluation step if high quality SiC films are to be achieved. Room temperature Hall measurements were made on cubic SiC films grown at NASA Lewis[13] using the van der Pauw configuration[14]. Ohmic electrical contacts consisted of sputtered tantalum followed by sputtered gold. The Hall measurements indicated that unintentionally doped films were always n-type. It was suspected that the n-type character of the films was caused by the nitrogen atom acting as a donor impurity. In order to perform a detailed analysis of the charge carrier concentration, Hall measurements were made on three cubic SiC films over the temperature range 50-300K. Experimental results for the films studied are shown as points in Figure 5. The data was analyzed using the appropriate relation

$$\frac{n(n+N_A)}{N_D-N_A-n} = \frac{N_c}{S} \exp\left(\frac{-E_D}{kT}\right)$$

which is accurate for a single level and nondegenerate carriers. The measured carrier concentration is n, the donor impurity concentration and the acceptor impurity concentration are N_D and N_A, respectively. E_D is the activation energy of the donor level, N_c, s and k are constants and T is the temperature. The fitting of the charge carrier concentration as a function of temperature, n(T), was obtained by a linearized least-squares fit with E_D, N_A, N_D as variables. This three-parameter fit yielded the solid line curves in Figure 5, and the values given in Table II. The first point evident from the tabulation is that the films are highly compensated with N_A/N_D 0.90. The second is that the value of E_D is consistent with the idea that the donor impurity is nitrogen[13]. The identity of the acceptor impurity is not so clear. Crystal defects such as vacancies, interstitial atoms, antisite atoms, stacking faults, or dislocations may be acting as acceptor impurities which leads to a reduction of conduction band electrons by recombination.

It is believed that all present day cubic SiC films grown at NASA Lewis and elsewhere are compensated. The significance of compensation is the degradation of device quality. Compensated semiconductors have many ionized impurities (positive for donors and negative for acceptors) embedded in the crystal lattice and these ions serve as scattering centers for the moving charge carrier. This increases the impurity scattering and reduces the total mobility of the charge carrier compared to uncompensated material with the same density of free carriers.

Concluding Remarks

The development of semiconductor materials does not occur overnight. First silicon and then gallium arsenide for example, have come to the marketplace after years in the laboratory and countless dollars spent for development costs. The

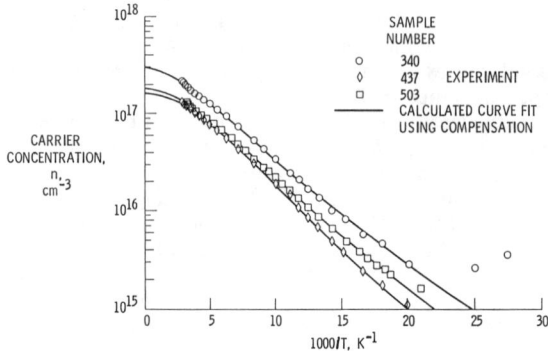

Fig. 5. Carrier concentration n vs 1000/T for three cubic SiC samples. The points are experimental results while the solid lines are calculated fits using a compensation model.

Table II. Parameters obtained in a linearized least-squares fit using a compensation model.

Sample	N_D 10^{18} cm^{-3}	N_A 10^{18} cm^{-3}	N_A/N_D	E_D MeV
340	2.79	2.50	0.90	13.2
437	1.76	1.60	0.91	17.2
503	2.15	1.97	0.92	14.6

history of SiC as a high temperature semiconductor has been one of high expectations followed by disappointment. Recent advances in the crystal growth of SiC and the increased knowledge of the bulk material properties of the grown SiC are cause for renewed enthusiasm. SiC now appears ready to emerge as a useful semiconductor material.

References

[1] W. C. Nieberding and J. A. Powell, "IEEE Transactions on Industrial Electronics," Vol. IE-29, No. 2, May 1982 p. 103.

[2] J. A. Lely "Darstellung von Einkristallen von Silizium carbid und Beherrschung von Art und Menge der eingebauten Verunreinigungen," Ber. Dt. Kerm. Gas, Vol. 32, pp. 229-231, 1955.

[3] S. Nishino, J. A. Powell, and H. A. Will, Appl. Phys. Lett. 42, 460 (1983).

[4] J. A. Powell, L. G. Matus, and M. A. Kuczmarski, *J. Electrochem. Soc.*, 134, 1558 (1987).

[5] A. Addamiano and J. A. Sprague, Appl. Phys. Lett. 44, 525 (1984).

[6] C. M. Chorey, P. Pirouz, J. A. Powell, and T. E. Mitchell, in Semiconductor-Based Heterojunctions: Interfacial Structure and Stability, edited by M. L. Green et al., (the Metallurgical Society, Inc., Warrendale, PA, 1986) pp. 115-125.

[7] S. R. Nutt, D. J. Smith, H. J. Kim, and R. F. Davis, Appl. Phys. Lett. 50, 203 (1987).

[8] P. Pirouz, C. M. Chorey, and J. A. Powell, Appl. Phys. Lett. 50, 203 (1987).

[9] J. Ryu, H. J. Kim, and R. F. Davis, Appl. Phys. Lett. 47, 850 (1985).

[10] J. A. Powell, L. G. Matus, and M. A. Kuczmarski, Appl. Phys. Lett. 51, 823 (1987).

[11] M. Kawabe and T. Veda, Jap. *J. Appl. Phys.*, 25, L285 (1986).

[12] R. Kaplan, *Surf. Sci.*, 93, 145 (1980).

[13] B. Segall, S. A. Alterovitz, E. J. Haugland, and L. G. Matus, Appl. Phys. Lett. 49, 584 (1986).

[14] L. J. van der Pauw, Philips Res. Rep., 13, 1 (1958).

Chemical Vapor Deposition, In Situ Doping and MESFET Performance of Beta-SiC Thin Films

Hyeong Joon Kim, Hua-Shuang Kong, John A. Edmond,
Jeffrey T. Glass* and Robert F. Davis[†]

Department of Materials Science and Engineering
North Carolina State University
Raleigh, NC 27695-7907

Abstract

High purity β-SiC films have been epitaxially grown on Si (100) and α-SiC (0001) at 1633K-1823K and 0.1 MPa using SiH_4 and C_2H_4 carried in H_2. Films produced on Si (100) have also been doped with B or Al (p-type) or P or N (n-type) during CVD deposition. Films grown on Si contain substantial concentrations of line and planar defects, especially in the interface region; those produced on α-SiC contain relatively few defects. The ratios of ionized dopant concentration to total dopant concentration for N, P, B and Al are 0.1, 0.2, 0.002 and 0.01, respectively. The solubility limits of N, P, and B at 1633K were determined to be $\sim 2 \times 10^{20}$, 1×10^{18}, and 8×10^{18} cm^{-3}, respectively; that of Al exceeds 2×10^{19} cm^{-3}. Metal-semiconductor field effect transistors (MESFETs) and related contacts have also been investigated. Saturation of the drain currents was achieved at room temperature. The I-V characteristics, measured from 298K to 623K, indicated that these MESFETs performed reasonably well throughout this temperature range.

Introduction and Background

Silicon carbide is the only compound species that exists in the solid state in the Si-C system, but it can occur in many polytype structures. The lone cubic polytype crystallizes in the zinc blende structure and is denoted β-SiC. The ~ 170 known additional hexagonal and rhombohedral polytypes are collectively referred to as α-SiC. The electron Hall mobility of high-purity undoped β-SiC has been postulated from theoretical calculations to be greater than that of the alpha forms over the temperature range of 300-1000K because of the smaller amount of phonon scattering in the cubic material[1]. Although this result has served as one catalyst for the current international interest in β-SiC, this material also possesses a unique combination of additional properties including a high melting point (3103K at 30 atm)[2], high thermal conductivity (3.9 W/cm·deg)[3], wide band gap (2.2 eV at 300K)[4], high breakdown electric field (2.5×10^6 V/cm)[5], high saturated drift velocity (2×10^7 cm/s)[6] and small dielectric constant (9.7)[7]. As such, β-SiC has been theoretically shown[8] to be superior to Si, GaAs or InP using either Johnson's[9] or Keyes'[10] figure of merit for high-frequency, high speed and high power transistor applications.

The high thermal conductivity and breakdown field also indicate that the integration of devices made from β-SiC can be achieved with high densities. Two additional reasons for the renewed interest in β-SiC are the significant advances in the growth of monocrystalline thin films of this material by chemical vapor deposition (CVD) and the ability to dope this material with n- and p-type dopants during growth or via ion implantation. As such, devices from this material have now become a reality.

Monocrystalline Si has been universally adopted as the current substrate of choice for the growth of the β-SiC thin films because of the availability of the former in well characterized and reproducible forms of controlled purity. Unfortunately, the $\sim$ 8% and 20% mismatches in the coefficients of thermal expansion and lattice parameters, respectively, of these two materials did not previously allow the thickness of the films to exceed $\sim 2 \times 10^{-6}$ m without considerable microcracking on cooling. However, amelioration of these differences via the initial reaction of the Si (100) surface with a C-containing gas followed by the successful epitaxial deposition of relatively thick (up to 30 μm), crack-free β-SiC films on this converted layer using individual C- and Si-containing gases has been reported by Nishino et al.[11] and subsequently by Suzuki and co-workers[12], Addamiano and Klein[13], Sasaki et al.[14] and Liaw and Davis[15]. This two-step process is described in the following section.

Beta-SiC has also been previously grown on monocrystalline α-SiC substrates in the temperature range of 1773K-1973K[16,17]. However, the growth conditions were not optimized and the interface between the SiC epilayer and the SiC substrates was not investigated.

A limited number of studies have been conducted to investigate the incorporation via CVD techniques of electronically active impurities in β-SiC. Bartlett et al.[18], Long et al.[19], and Nishino et al.[20] produced p-type β-SiC by adding B_2H_6 or $AlCl_3$ to the gaseous reactants during the growth of this normally n-type (unintentionally doped) material. In addition, von Muench and Pettenpaul[21] doped polycrystalline β-SiC p-type by flowing H_2 through trimethylaluminum (TMA) during crystal growth via the van Arkel process.

Good contacts are essential for successful device fabrication in any semiconductor. Several metallic elements (e.g., W, Mo, Cr, Ni) and alloys (e.g., Au-Ta, Au-Ta-Al, W-Mo, Cr-Ni, Cu-Ti, Al-Si) have been developed as ohmic contacts for α-SiC[22-26]; however, no information on the specific resistivities of the contacts was reported. For the majority of the contact materials noted above, annealing temperatures exceeding 2000K were required to achieve ohmic character. This temperature treatment often results in deep penetration of the deposited contact material which can cause electrical shorting of devices. Moreover, the success of these contacts on hexagonal, α-SiC does not guarantee that they will be suitable for the cubic, beta form, which is of primary interest in the current study. In the case of β-SiC as-deposited and annealed Al have been shown[27] to be ohmic on n-type and p-type material, respectively. Again, no values of the contact resistivities were reported. Finally, in no case has the dependence of the contact resistivity on operating temperature been determined for ohmic electrical contacts on any polytype of SiC.

The fabrication of MESFETs in SiC has also been previously attempted by several investigators. Muench and his coworkers[28] reported MESFET fabrication on bulk crystals of α-SiC in 1977. Saturation of drain currents was achieved and the maximum transconductance reported was 1.75 mS/mm. Device research involving β-SiC[29-30] has also been made possible by the aforenoted advances in the growth of these films on Si (100) substrates[11-15]. Unfortunately, in both of these device studies, saturation of drain currents was not achieved. A possible cause of this was believed to be the leakage current through the β-SiC p-n junction between the active n layer and the buried p layer[29-30]. Other possible causes include the bypass leakage current from the drain to the source, the dense defects near the Si/β-SiC interface[31] and the leakage current at the surface of the β-SiC film. It should be noted that none of these studies reported MESFET characteristics at elevated temperatures.

In the following sections the procedures and results of (i) growth of β-SiC on Si (100) and α-SiC (0001), including theoretical calculations of the compositions and the amounts of condensed product gas phases which are predicted to occur under equilibrium conditions at various Si/(Si+C) ratios, (ii) observation of defect structures in the SiC thin films, (iii) in situ doping of the films, (iv) the contact resistivities of some of the commonly used as well as newly developed materials for contacts on β-SiC and (v) MESFET fabrication and characterization.

EXPERIMENTAL PROCEDURES

Chemical equilibrium calculations involve the computation of the composition of a system, subjected to certain constraints, which contains the minimum free energy. The constraints in CVD systems are the preservation of the masses of each element present, constant temperature, and constant total pressure. White et al.[32] initially developed a computational technique which involves the minimization of the summation of the free energies of all the species present in a given system. Erikson[33] has extended the method to include systems containing more than one condensed phase, including solid solutions, and developed a companion program (SOLGASMIX-PV) for performing the calculations. This program has been used in the present research to calculate the number of moles of the condensed phases which would deposit and to determine the types and amounts of the product gases which would be present under equilibrium conditions. In these calculations, the elements of Si, C and H were the principal species, since SiH_4 and C_2H_4 were the reactant gases, and H_2 was the carrier gas. The total amount of Si and C was maintained at 1 mole; thus, Si/(Si+C) ratio varied from 0 to 1.0. The Si substrate was assumed to be inert although, in reality, it has a moderate vapor pressure near its melting point and could, therefore, be a small part of the overall reaction if a high growth temperature is experimentally employed, especially under low-pressure conditions. The thermodynamic values for all species used in these calculations were taken from the JANAF tables[34]. The $H_2/(SiH_4+C_2H_4)$ mole ratio and the total pressure were fixed at 1000 and 1 atm, respectively. The temperature was chosen to be 1630K, which is essentially that used in the growth of β-SiC on Si substrates.

Epitaxial films of β-SiC have primarily been grown in this research on chemically converted surfaces of high resistivity (ρ = 5000 - 8000 Ωcm), p-type monocrystalline Si (100) substrates by CVD using the high-purity gases of silane (SiH_4) and ethylene (C_2H_4) entrained in the purified (Pd/Ag cell) carrier gas of H_2. For the chemical conversion process in our reactor, an experimentally determined optimum amount of 0.30 mole % of C_2H_4 (1 sccm) in flowing H_2 (3000 sccm) was introduced into the cold wall, barrel-type reaction chamber at room temperature. The substrates (and the SiC-coated graphite susceptor on which they rest) were immediately and rapidly heated to 1633K over a total period of 150 s to produce a very thin chemically converted layer of monocrystalline β-SiC[35]. The subsequent

CVD growth of β-SiC (100) films on this layer was achieved by establishing SiH_4 and C_2H_4 flow rates in the 3000-sccm H_2 flow at a temperature of 1633K and 101.3 kPa total pressure. Growth rates were typically 2 μm/h. For a complete review of this process and the results of analyses of the converted layers, the reader is referred to Ref. 35.

The considerable mismatch in the lattice parameters of Si (100) and β-SiC (100) noted above resulted in the formation of numerous microtwins, intrinsic stacking faults and antiphase boundaries (APB) (see discussion below). As such, investigations concerned with growth on off-axis Si (100) and on the (0001) plane of α-SiC have been conducted. The former approach has been used extensively by researchers in the GaAs community in their analogous attempt to grow large diameter, monocrystalline APB-free films of this material on Si (100) (for a review of this research, see various chapters in Ref. 36). Moreover, Shibahara et al.[37] have recently shown that APBs can be successfully eliminated in β-SiC using off-axis Si (100) with the [100] inclined 2° toward [011]. In our research both 2° and 4° inclinations of the Si [100] toward [011] were investigated; the resultant β-SiC films were ~5 μm thick after a growth period of 7.2 ks.

In an attempt to eliminate all of the various defects simultaneously, growth on the Si (0001) and C (0001) faces of Acheson-derived 6H α-SiC substrates has been studied within the temperature range 1683K-1823K at 101.3 kPa total pressure. Following evacuation of the growth chamber to 1.33 mPa, flowing H_2 (3000 sccm) was admitted into the chamber, the substrates heated at the temperature of growth for 600 s followed by the introduction of SiH_4 and C_2H_4. The SiH_4/C_2H_4 flow rate ratio was 2; the total pressure was 101.3 kPa.

In order to achieve p- and n-type layers in the β-SiC films grown on Si (100) substrates, the p-and n-type dopants of Al and B, and N and P, respectively, were incorporated into the films directly during growth. This was accomplished by introducing $Al(CH_3)_3$, B_2H_6, N_2 (or NH_3) and PH_3 directly into the primary gas stream. The atomic concentrations of these dopants were measured as a function of depth using secondary ion mass spectrometry (SIMS);[†] corresponding carrier concentrations were determined via differential capacitance-voltage measurements[§] using a Hg-probe.

For the evaluation of electrical contacts to β-SiC films on Si substrates, the as-grown films were first polished with 0.1 μm diamond paste to remove the 100-200 nm surface roughness and to eliminate possible surface defects introduced during the final stages of growth. This procedure removed approximately 300 nm of material from the SiC surface. After polishing, the samples were immersed in H_2SO_4 for five minutes at 433K-453K, rinsed in deionized (DI) water, soaked in a 1:1 mixture of NH_4OH and H_2O_2 for five minutes at 333K-343K, rinsed in DI water, etched in buffered HF for two minutes and rinsed in DI water at least five times. Immediately after cleaning, the samples were oxidized at 1473K in dry O_2 for 1.5 hours in order to remove polishing-induced surface damage and residual contaminants. During oxidation, approximately 100 nm of SiO_2 formed which corresponds to the removal of approximately 40 nm of SiC. This oxide was then removed by a brief immersion in HF.

Following the cleaning procedure described above, samples, other than those on which contacts of $TaSi_2$ were to be placed, were mounted in a vacuum evaporator which was subsequently evacuated to a base pressure of 1×10^{-6} Torr. The n-type ohmic contacts of Al, Ni or 97 a/o Au-3 a/o Ta and Cr, and the p-type ohmic contacts of Al or 91 a/o Au-2 a/o Ta - 7 a/o Al were individually deposited by thermal evaporation from a W-boat. Tantalum silicide ($TaSi_2$), another n-type ohmic contact material, was deposited via rf sputtering of a $TaSi_2$ target. The thickness of each of these contacts was approximately 250 nm. An additional p-type ohmic contact was prepared by depositing 50 nm of Al by thermal evaporation, followed by sputtering a 250 nm convering layer of $TaSi_2$. Following deposition, circular contacts with an area of 3.14×10^{-4} cm^2 were delineated on all samples using photolithographic techniques.

Ohmic contacts to β-SiC samples were annealed at various temperatures (see "Results and Discussion") in vacuum (10^{-6}-10^{-5} Torr) or, in the case of Al, a 0.3 Torr flowing Ar atmosphere to obtain electrically stable contacts. Current-voltage (I-V) characteristics were then measured using an HP 4145A semiconductor parameter analyzer and a probe station equipped with a hot stage open to air. Specific contact resistivities were determined by both the three contact[38] and extrapolation[39] method. Following the annealing procedure above, the same instrumentation with a hot stage attachment to the probe station was utilized to measure the change in contact resistivity values with varying operation temperature.

MESFETs were produced in a 600 nm thick, undoped n-type β-SiC film epitaxially deposited on a 7 μm thick buried Al-doped p-type β-SiC layer previously grown on a p-type Si substrate. The latter layer was used to (i) confine the current to a thin n-type active layer and (ii) move this active layer away from the region of highest defect density which extended ~3 μm into the β-SiC film from the Si/SiC interface. The carrier concentration of the n- and p-type layers were 5×10^{16} cm^{-3} and 3×10^{17} cm^{-3}, respectively. A three level mask set employing a concentric ring geometry wherein the gate pattern completely enclosed the center, 100 μm (drain) contact was used. The gate length and the source-to-drain distance were 3.5 μm and 10.5 μm, respectively. Sputtered $TaSi_2$ was used as the source and drain ohmic contacts. These contacts were annealed at 1173K for 300 s in vacuum to minimize contact resistance. Thermally evaporated Au was used as the gate rectifying contact. An HP 4145A semiconductor parameter analyzer was used to obtain drain current-drain voltage (I_D-V_D) data.

RESULTS AND DISCUSSION

Thermodynamic Predictions

The equilibrium number of moles of the two condensed phases of β-SiC and Si (normalized to a maximum value of 1.0), calculated from the SOLGASMIX-PV program as a function of the Si/(Si + C) ratio, is shown in Fig. 1(A). It is not surprising that the amount of β-SiC has a maximum at Si/(Si + C) = 0.5, since the total input amount of Si and C was maintained constant, that is, at 1 mole. No free C or free Si is formed at ratios less than 0.5. However, in the range of Si/(Si + C) > 0.5, free Si forms with the β-SiC in an amount of which increases linearly with the increase in the Si/(Si + C) ratio. It should be noted that the calculations also show that a small amount of free Si forms even at Si/(Si + C) = 0.5. Thus, from this viewpoint, the proper ratio for the growth of β-SiC should be in the carbon-excess region.

Although the product gas phase contains Si, SiH, SiH_4, C_2H_2, C_2H_4, CH_3, and CH_4, the calculations show that the predominant Si- and C-containing species to be SiH_4 (the same as the reactant gas) and CH_4, respectively. The variations in the total partial pressures of all the Si-containing and all the C-containing product gases as a function of the Si/(Si + C) ratio are shown in Fig. 1(B). Each of the totals of the partial pressures of both the Si- and C-containing gases varies markedly in

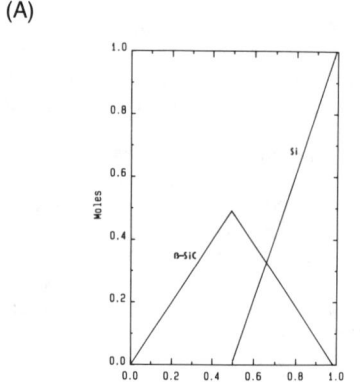 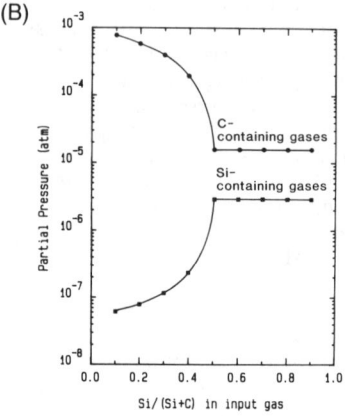

Fig. 1. (A) The amounts of the condensed phases of β-SiC and Si, as a function of the Si/(Si+C) ratio in the reaction gas stream as predicted from thermodynamic calculations using the "SOLGASMIX-PV" program. (B) The total pressures of all the Si and C-containing gases (expressed as partial pressure of a total gas mixture which also contains H_2) as a function of the Si/(Si+C) ratio in the reaction gas stream. The values noted in the graph were obtained from thermodynamic calculations using the "SOLGASMIX-PV" program.

the Si/(Si + C) range < 0.5 but is constant at 0.5 and at all values of the ratio greater than 0.5. At Si/(Si +C) = 0.5, the total pressure of all the C-containing gases (it is essentially all CH_4) reaches a minimum equilibrium pressure; the total pressure of all the Si-containing gases reaches a maximum equilibrium pressure as would be expected, since the amounts of the Si- and C-containing gases are increasing and decreasing, respectively, from Si/(Si + C) = 0.1.

The constancy in the partial pressure of the C- and Si-containing product gases above Si/(Si + C) = 0.5 occurs primarily as a result of the increasing formation of Si. This combined with the formation of SiC and the decrease in the mole fraction of C_2H_4 entering the gas stream act in concert to maintain the constant partial pressures. At Si/(Si + C) = 1.0, all the SiH_4 is predicted to have reacted to form Si; thus, no Si- or C-containing product gases exist at this point.

As noted above, the results of these thermodynamic calculations indicate that the optimum Si/(Si + C) ratio for the formation of the single phase of β-SiC under the stated conditions should be slightly less than 0.5. Furthermore, it may be reasoned that the fastest growth rate of the β-SiC would occur at approximately this ratio as a consequence of the formation of the largest amount of this material (although the thermodynamics tells one nothing regarding the kinetics of growth). However, these results must be compared with carefully conducted experiments to determine their approximation to results obtained under the nonequilibrium conditions extant in the CVD reactor.

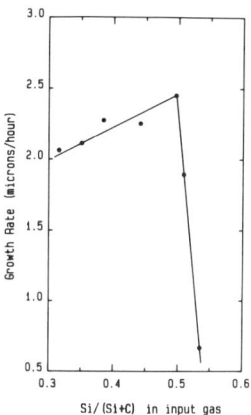

Fig. 2. The experimentally determined growth rate of the β-SiC films as a function of the Si/(Si+C) ratio in the reaction gas stream. The total pressure, H_2 flow rate, temperature, and elapsed time for all depositions was 0.1 MPa, 3000 sccm, 1633K, and 7.2×10^3 s, respectively.

Effect of Si/(Si+C) Ratio on Deposition Rate and Microstructure

The growth rate of the β-SiC films is shown in Fig. 2 as a function of the Si/(Si+C) ratio. The growth rate increases with the increase in the Si/(Si+C) ratio up to Si/(Si+C) = 0.496. As the Si/(Si+C) ratio was decreased below 0.5, black inclusions appeared in the β-SiC films and increased in size and density as the amount of C in the gas phase increased. Essentially all the inclusions possessed quasi-hexagonal faceting as shown by SEM. SIMS and analytical SEM were employed to chemically analyze the particles. The results derived from both instruments indicated essentially no differences in the concentrations of Si and C in the inclusions and in the β-SiC matrix. From these results, it may be concluded that the inclusions are SiC particles.

Figure 2 also shows that the growth rate is rapidly reduced beyond Si/(Si+C) = 0.5. The exact reason for this phenomenon is not known; however, numerous spherical particles occurred in the β-SiC film growth at Si/(Si+C) = 0.535 which were not observed at lower values of this ratio. We believe these particles to be free Si which the thermodynamic calculations predict should be produced in this range. It is also believed that they are produced by homogeneous nucleation in the gas phase. Moreover, their residence time in the gas phase may allow them to serve as sites for the additional heterogeneous nucleation and growth of Si which would normally have been used in the reaction with C_2H_4 to form SiC. Furthermore, if this process is enhanced by the increase in the amount of SiH_4 in the input gas stream, this could explain the observed rapid drop in the growth rate of the films.

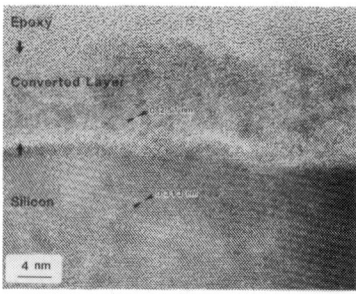

Fig. 3. High-resolution XTEM micrograph of converted layer on the Si substrate prior to CVD deposition (Courtesy of S. Nutt, Brown University, Providence, RI).

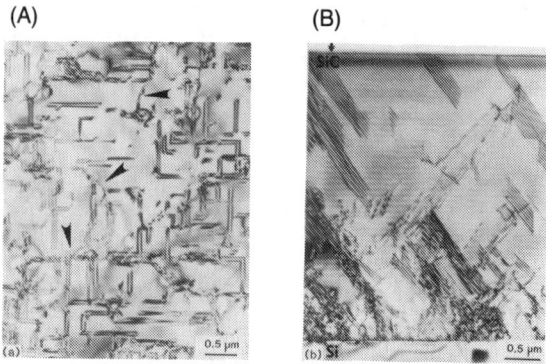

Fig. 4. TEM micrographs showing the general microstructure of β-SiC films: (A) $\langle 100 \rangle$ plan view showing stacking faults and APBs (arrows); (B) $\langle 110 \rangle$ cross-sectional view showing interfacial strain contrast, dislocations, and stacking faults.

Defect Structures in β-SiC Films

The β-SiC layer produced by chemical conversion of the Si (100) surface was monocrystalline and microscopically rough, varying in thickness from 5–12 nm (Fig. 3). It contained a high density of planar defects that were primarily {111} microtwins and intrinsic stacking faults. In addition, localized regions of the converted layer exhibited disorder.

Plan-view and cross-sectional transmission electron micrographs of a β-SiC thin film are shown in Figs. 4(A) and 4(B), respectively. The micrographs show planar defects on {111} planes that intersect at 90° angles in the $\langle 100 \rangle$ projection [Fig.

4(B)] and at 70°32′ angles in the ⟨110⟩ projection [Fig. 4(B)]. As in the converted layer, these defects were identified as microtwins and intrinsic stacking faults. The density of defects is higher at the interface; it decreases over a distance of 3-4 μm from the interface and becomes approximately constant to the surface even for 20 μm thick films.

There are additional defects in Fig. 4(A) that appear as bands of mottled contrast that extend from the Si-SiC interface to the growth surface (see arrows). These defects have recently been identified as antiphase boundaries (APBs)[40]. The unusual appearance of the APBs may be caused by their interaction with additional defects present in the films. To eliminate these defects, Si substrates with the [100] inclined 2° and 4° toward [011] were used. It has been found that the APBs were eliminated except in the edge regions of the films grown on the 2° off axis Si. A plan-view micrograph of a β-SiC thin films grown on a 4° off axis Si (100) substrate is shown in Fig. 5. It can be seen that the APBs observed in Fig. 4(A) are no longer present.

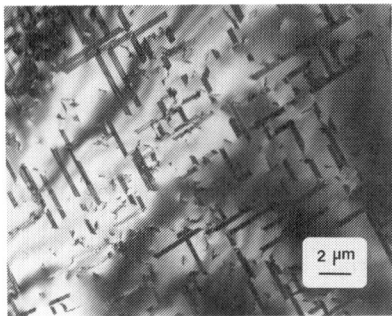

Fig. 5. Plan view TEM micrograph showing absence of APBs in the β-SiC film grown on 4° off-axis Si (100), however, stacking faults are still present.

The growth of the films on α-SiC substrates has been conducted on both the C (0001) and the Si (0001) faces of the latter material. The resulting film was β-SiC (111) at all temperatures studied. However, the surface of the film grown on (0001) was very smooth and reflective whereas that grown on (0001) was relatively rough and unsuitable for device fabrication. In contrast to films grown on Si substrates, no defects were observed in these films when examined by cross sectional transmission electron microscopy (XTEM). In fact, high resolution XTEM shows an abrupt and coherent β-SiC/α-SiC interface (Fig. 6). A single atomic layer runs completely across the interface which indicates that the growth direction of the β-SiC films was exactly [111] in the region (for the growth of β-SiC on α-SiC C face). However, examination in plan view revealed the presence of double positioning (DP) boundaries. In these defects, the material enclosed by the boundaries is oriented 60° to the surrounding matrix around ⟨111⟩. Very recent research has employed α-SiC crystals with the [0001] inclined 3° toward [1120] heated to 1773K for 600 s. The DP boundaries were eliminated by this procedure

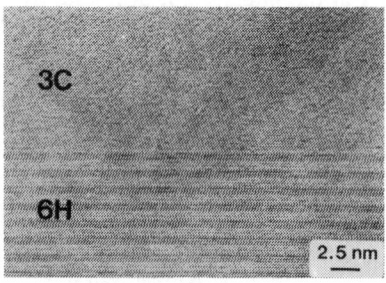

Fig. 6. High-resolution XTEM micrograph of the β-SiC/6H-SiC interface (Courtesy of S. Nutt, Brown University, Providence, RI).

and a mirror smooth film was produced. However, the resulting film was the 6H α-SiC polytype. Additional characterization is now ongoing. As a consequence of the very recent nature of this study utilizing α-SiC substrates, the results of most of the remaining research reported below are concerned with β-SiC grown on Si.

Impurity Incorporation During Deposition

Figure 7 shows the results of the experimental SIMS measurements of atomic dopant concentration as a function of the partial pressure of the dopant source gas for Al, P, B and N. The linear character of both the curve for Al and the sections of the graphs for P, B and N at lower pressures is predicted from considerations of Henry's law. Moreover these curves are similar to those for dopant incorporation into monocrystalline Si.

Solubility data from direct measurements such as diffusion of lattice parameter studies are not available in the literature for the various dopants for the temperature of 1633K used for the CVD growth. However, the change in slope of the atomic concentration curves for P, B and N as well as the changes in the surface character and the x-ray Laue patterns of these heavily doped samples relative to the undoped (or lightly doped) materials indicate the onset of polycrystallinity. These changes are believed to be triggered by the introduction of the various dopants in excess of their solubility limits. Furthermore, the grain boundaries can act as sinks for excess dopant and thus allow continued incorporation of these species at levels significantly higher than allowed by the lattice. The maximum in the Al concentration could not be determined because of increased gas phase nucleation and the resultant occurrence of poor films at high TMA input.

Each plot in Fig. 7 also reveals a major difference between the atomic concentration and the carrier concentration for each dopant. The reasons for this include the measured deep energy levels for the p-type dopants of Al (0.24 eV) and B (0.735 eV) as well as the possibility of compensation from unintentionally introduced n-type dopants (e.g., N). Other possibilities include (i) compensation from line or point defects and/or trapping of impurities at the dislocations and

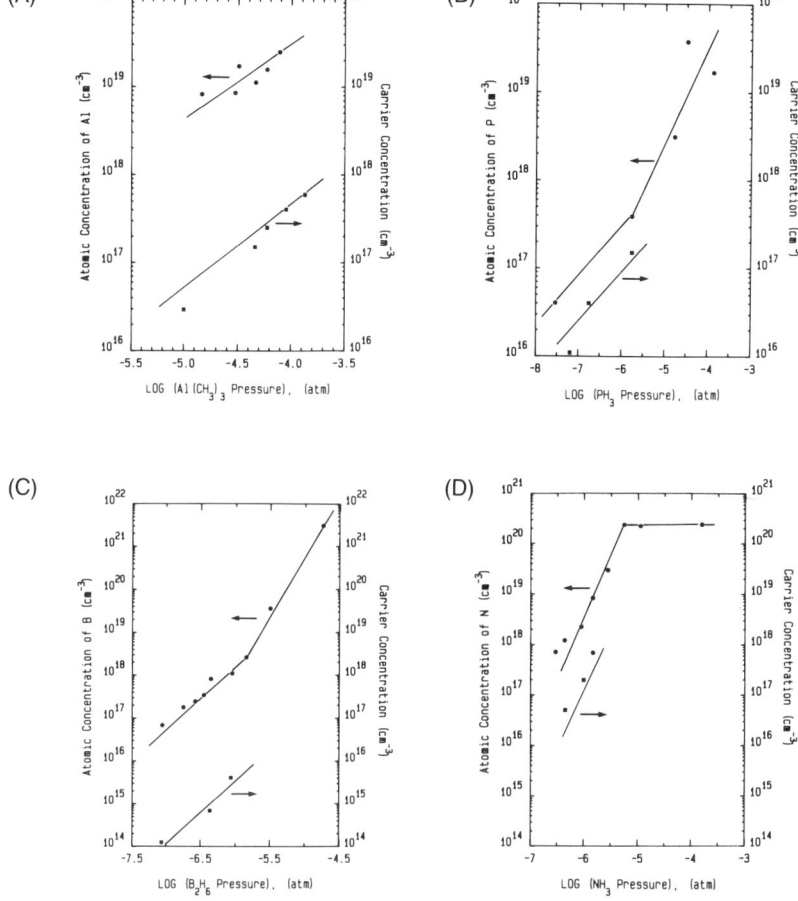

Fig. 7. Atomic and carrier concentrations of (A) Al, (B) P, (C) B, and (D) N as a function of the partial pressure in the CVD chamber of the respective dopant gases of $Al(CH_3)_3$, PH_3, B_2H_6, and NH_3. Aluminum and B are p-type dopants while P and N are n-type dopants in β-SiC.

stacking faults in the material, (ii) dopant-Si and/or dopant-C interaction[41], and (iii) location in nonelectrically active interstitial sites (especially plausible for B). Thus a portion of each of the dopants is either not ionized or located on nonelectrically active sites or forms complexes with Si or C. Combinations of these events are also probable.

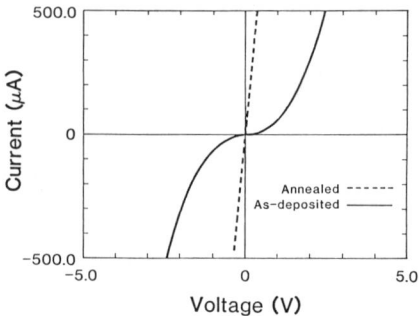

Fig. 8. Room temperature I-V plot showing ohmic contact properties of rf-sputtered $TaSi_2$ on n-type β-SiC as deposited and annealed at 1123K-300 s in vacuum.

Ohmic Contacts

n-type β-SiC: Of the contacts made to n-type β-SiC in this research, as-deposited Ni, Au-Ta, $TaSi_2$ and Cr exhibited nonlinear I-V characteristics; however, the resistance to current flow in either voltage direction was small, as seen in the representative curve for as-deposited $TaSi_2$ shown in Fig. 8.

Specific contact resistance values for n-type β-SiC (carrier concentration ~ 5 × 10^{16} cm^{-3}) are listed in Table I. Prior to each measurement, the Ni, Cr and Au-Ta contacts were independently heated for 300 s to 1523K; the $TaSi_2$ contact was similarly heated to 1123K. The contact resistivity value for Al is for as-deposited material. The $TaSi_2$ contact had a slightly lower resistivity than the other contacts. Furthermore, it was observed that its contact resistance was not significantly changed with annealing temperatures above 1123K (the maximum annealing temperature for this contact material was 1473K). Thus, reproducible and moderately low resistance ohmic contacts may be obtained on n-type β-SiC by rf sputtering of $TaSi_2$ coupled with thermal annealing. As such, this material will be the principal focus of the following discussion regarding ohmic contacts on β-SiC.

The nonlinear I-V characteristics previously noted and shown in Fig. 8 for the as-deposited $TaSi_2$ changed upon annealing at 1123K for 300 s to become linear. The same effect was observed when heating the Ni, Cr and Au-Ta contacts to 1523K. Thus, thermal annealing of these contact materials on SiC was required to achieve operational ohmic character and to minimize contact resistance.

In contrast to the above results, Al (a common p-type dopant in SiC) evaporated onto n-type β-SiC showed ohmic behavior in the as-deposited state. However, upon heating to 1173K for 180-300 s, this contact became rectifying. Daimon et al[27], have recently published similar results showing that the rectifying properties are caused by the diffusion of the Al into the β-SiC and the resultant formation of an alloyed p-n junction.

The operation temperature dependence of the contact resistance for $TaSi_2$ is given in Table II. Heating this combination of materials to as high as 573K did not significantly alter the contact resistance. However, upon heating this sample to

Table I. Contact resistivities of Al, Ni, Cr, Au-Ta and TaSi$_2$ on n-type β-SiC having a carrier concentration of 5×10^{16} cm^{-3}.

Materials	Contact Resistivity (Ω-cm^2)
Al	1.6×10^{-1}
Ni	1.4×10^{-1}
Cr	7.0×10^{-2}
Au-Ta	3.0×10^{-2}
TaSi$_2$	2.0×10^{-2}

Table II. Change in contact resistivity with temperature for TaSi$_2$ on n-type β-SiC.

Temperature (K)	Contact Resistivity (Ω-cm^2)
298	2.0×10^{-2}
373	2.1×10^{-2}
473	2.2×10^{-2}
573	1.8×10^{-2}
673	9.6×10^{-3}

673K, this value decreased and the ohmic character of the contacts improved. This is illustrated in the I-V plot shown in Fig. 9. A decrease in contact resistance with increasing temperature is expected, since the donor concentration in the material is increasing in this temperature range[42]. However, it is not presently understood why a decrease is not observed when heating from 298K up to 573K. It is important to note that the process of decreasing and increasing contact resistivity with temperature was reversible through many heating and cooling cycles. This discounts the possibility of any additional solid state reactions between the contact material and semiconductor beyond which may have occurred during the initial high temperature annealing.

p-type β-SiC: Aluminum, a 91 a/o Au - 2 a/o Ta - 7 a/o Al alloy and the special Al/TaSi$_2$ alloy described previously were examined for ohmic behavior on p-type β-SiC (Al doped, carrier concentration ~ 1×10^{16} cm^{-3}). Table III summarizes the specific contact resistance values obtained for these three materials. Even after heating at 1473K for 1800 s, the Au-Ta-Al alloy contact showed nonlinear I-V characteristics and possessed a high specific contact resistance (~ 4.7×10^{-1} Ω-cm^2). Thus it was unsuitable as an ohmic contact. By contrast, contact resistance for the Al/TaSi$_2$ combination was measured to be 2.0×10^{-1} Ω-cm^2 after a 1473K, 1800 s anneal. (This contact exhibited very poor, non linear behavior after heating at 1073K for 1800 s). Thus, although linear I-V behavior was observed, the resistance of this contact was rather high.

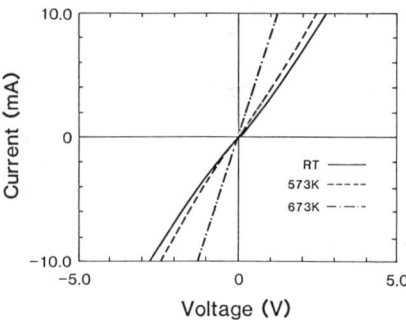

Fig. 9. Operation temperature dependence on ohmic contact properties of rf-sputtered $TaSi_2$ on n-type β-SiC; room temperature, 573K and 673K results.

Table III. Contact resistivity of Al, Au-Ta-Al and Al-$TaSi_2$ on p-type β-SiC ($p \approx 1 \times 10^{16}$ cm^{-3}).

Material	Contact Resistivity (Ω-cm^2)
Au-Ta-Al	4.7×10^{-1}
Al-$TaSi_2$	2.0×10^{-1}
Al	3.1×10^{-2}

Annealed Al was also utilized as an ohmic contact on p-type β-SiC. As-deposited this contact had a very high resistance. After heating to 1150K for 180 s, the contact was ohmic with a resistivity of 3.1×10^{-2} Ω - cm^2. This is illustrated in the I-V plot shown in Fig. 10. It is speculated that annealing at 1150K results in a very shallow p+ region under the contact area created by Al diffusion into the p-type β-SiC. Further annealing to temperatures as high as 1273K did not result in any measurable decrease in contact resistivity from the 1150K value.

The temperature dependence of contact resistivity for Al on p-type β-SiC is summarized in Table IV. In every instance, an increase in measurement temperature resulted in a decrease in the contact resistivity. This phenomenon was also reversible, i.e., cooling from 673K to 298K resulted in an increase of the contact resistivity from 2.2×10^{-3} to 3.1×10^{-2} Ω - cm^2. Figure 11 shows the linearity of these contacts at the various measurement temperatures. No change was observed in the shape or slope of any of these curves (or those shown for $TaSi_2$ in Figure 9) with time (for periods up to eight hours). This demonstrates the usefulness of this contact material on p-type β-SiC at elevated temperature.

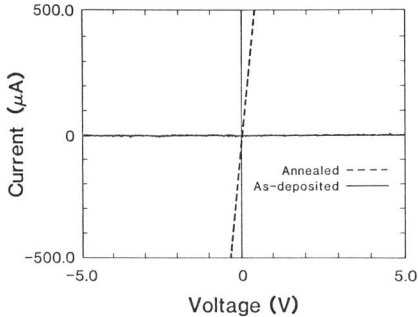

Fig 10. Room temperature I-V plot showing ohmic contact properties of evaporated Al on p-type β-SiC; as-deposited and annealed at 1150K for 180 s in .3 T flowing Ar.

Table IV. Change in contact resistivity with temperature for Al on p-type β-SiC.

Temperature (K)	Contact Resistivity (Ω-cm^2)
298	3.1×10^{-2}
373	1.4×10^{-2}
473	6.3×10^{-3}
573	3.3×10^{-3}
673	2.2×10^{-3}

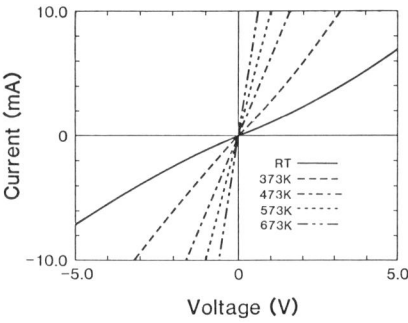

Fig. 11. Operation temperature dependence on ohmic contact properties of evaporated Al on p-type β-SiC; room temperature, 373K, 473K, 573K and 673K results.

Silicon Carbide

MESFETs

Typical room temperature drain current vs drain voltage (I_D-V_D) characteristics of the MESFETs are shown in Fig. 12. The gate voltage (V_G) was varied from 0.6 V to -1.5 V in -0.3 V steps. It can be seen that very good drain current saturation is achieved as the drain voltage increases. The maximum transconductance in the saturated region for this device was 0.64 mS/mm, however, a maximum transconductance of 2.1mS/mm was measured on other devices in this sample. The threshold voltage was -1.6 V, although after subtracting the leakage current, it is reduced to -1.4 V. It was observed that for V_G < -2 V the drain current was almost independent of the gate voltage, and thus the device could not be fully turned off. For example, for the device in Fig. 12, the drain current was 2 μA at a drain voltage of 4 V and a gate voltage < -2.5 V. This indicates that there is some leakage current between the gate and drain. This leakage may be caused by the p-n junction underneath the thin n layer, by the leakage current between the gate and source and/or by the defects in the β-SiC film. As previously reported,[31,43] the defect density, including the antiphase domain boundaries in the bulk of the film, is very high. Research is underway to fabricate MESFETs in films grown on α-SiC, since it has been shown that these films do not appear to contain any antiphase domain boundaries and much fewer line and planar defects than films grown on Si (100) substrates[44]. This will allow the determination of the role of these defects with regard to leakage current in the film.

MESFETs were also examined at temperatures as high as 623K; a limit imposed by the maximum temperature of the experimental arrangement rather than the devices. Figure 13 shows examples of these measurements on the same device used to determine the room temperature data of Fig. 12. In these measurements the drain voltage was applied from 0 to 10 volts in order to more clearly illustrate the dependence of the I_D-V_D characteristics on temperature. Inset in these figures are the I-V characteristics of the gate-drain diode at the various temperatures. It can be seen that as the temperature was increased, the MESFET drain current achieved less saturation and the gate-drain diode (see inset) at reverse bias yielded more leakage current. At room temperature, this diode leakage current was 5 μA at 8.5 V reverse bias, whereas, at 623K it increased to approximately 70 μA. This was probably caused by an increase in the generation current in the depletion region as temperature was increased. These generated carriers also contribute to the drain-to-source leakage current. The maximum transconductance of this device decreased approximately 21%, as the temperature was increased to 623K. This was expected because electron mobility decreases as temperature increases due to the enhanced lattice scattering[45].

CONCLUSIONS

Reproducible, high purity β-SiC thin films have been epitaxially grown on Si (100) and α-SiC (0001) substrates. However, defects generated from the β-SiC/Si interface and the double positioning boundaries existing in the β-SiC on α-SiC are still remaining problems to be solved. Nevertheless these thin films have allowed substantial progress in the in situ doping, contact development and MESFET fabrication in β-SiC. The contact investigation showed that both $TaSi_2$ and Al are stable ohmic contacts to operating temperatures as high as 673K. MESFETs were operated reasonably well from room temperature to 623K.

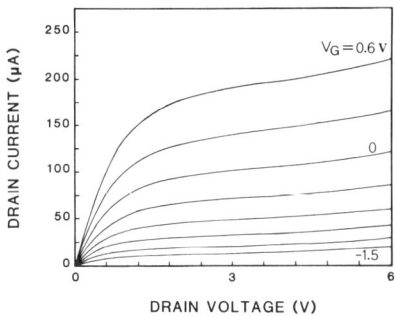

Fig. 12. Drain current-voltage characteristics at room temperature of a MESFET with a gate length of 3.5 μm at room temperature and fabricated in a β-SiC (100) monocrystalline film deposited on a Si (100) substrate.

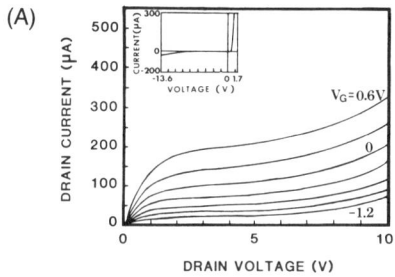

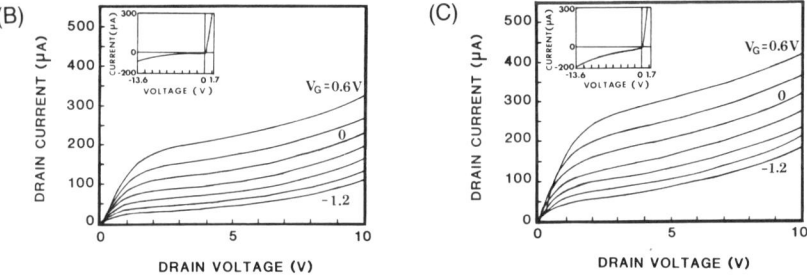

Fig. 13. Drain current-voltage characteristics of the MESFET used to obtain the data in Figure 12. (A) Room temperature, (B) 473K and (C) 623K. The current-voltage characteristics of the gate-drain Schottky diode are included as insets at each temperature.

Silicon Carbide 473

Acknowledgements

The authors gratefully acknowledge support of this program by the Office of Naval Research under contract N00014-82-K-0182 P005. Appreciation is also expressed to C.H. Carter, Jr. (NCSU) and S. Nutt (Brown University) for the TEM micrographs and many helpful discussions.

---

*On assignment from Kobe Development Corporation, Research Triangle Park, NC 27709.
‡Member, American Ceramic Society.
†CAMECA TMS-3f ion microprobe.
§LEI Model 2019 Miller Feedback Profiler.

REFERENCES

[1] P. Das and D. K. Ferry, "Hot Electron Microwave Conductivity of Wide Bandgap Semiconductors," *Solid-State Electron* **19** [10] 851-855 (1976).

[2] R. I. Skace and G. A. Slack, "The Si-C and Ge-C Phase Diagrams," in Silicon Carbide, A High-Temperature Semiconductor, edited by J. R. O'Connor and J. Smiltens pp. 24-28 (Pergamon, New York, 1960).

[3] E. A. Bergemeister, W. von Muench and Pettenpaul, E., "Thermal Conductivity and Electrical Properties of 6H Silicon Carbide," *J. Appl. Phys.* **50** [9] 5790-5794 (1979).

[4] H. P. Philipp and E. A. Taft, "Intrinsic Optical Absorption in Single Crystal Silicon Carbides," in Ref. #2, pp. 366-370.

[5] W. von Muench and I. Pfaffender, "Breakdown Field in Vapor-grown Silicon Carbide p-n Junctions," *J. Appl. Phys.* **48** [11] 4831-4833 (1977).

[6] W. von Muench and E. Pettenpaul, "Saturated Electron Drift Velocity in 6H Silicon Carbide," *J. Appl. Phys.* **48** [11] 4823-4825 (1977).

[7] S. Nishino, Y. Hazuki, H. Matsunami and T. Tanaka, "Chemical Vapor Deposition of Single Crystalline β-SiC Films on Silicon Substrate with Sputtered SiC Intermediate Layer," *J. Electrochem. Soc.* **127** [12] 2674-2680 (1980).

[8] J. D. Parsons, R. F. Bunshah and O. M. Stafsudd, "Unlocking the Potential of Beta Silicon Carbide," *Solid State Technol.* **28** [11] 133-139 (1985).

[9] E. O. Johnson, "Physical Limitations on Frequency and Power Parameters of Transistors," *RCA Rev.* **26** [2] 163-177 (1965).

[10] R. W. Keyes, "Figure of Merit for Semiconductors for High-Speed Switches," *Proc. IEEE* **60** [2] 225 (1972).

[11] S. Nishino, J. A. Powell and H. A. Will, "Production of Large-area Single-crystal Wafers of Cubic SiC for Semiconductor Devices," *Appl. Phys. Lett.* **42** [5] 460-462 (1983).

[12] A. Suzuki, K. Furukawa, Y. Higashigaki, S. Harada, S. Nakajima and T. Inoguchi, "Epitaxial Growth of β-SiC Single Crystals by Successive Two-step CVD," *J. Cryst. Growth* **70** [1-2] 287-290 (1984).

[13] A. Addamiano and P. H. Klein, "Chemically-formed Buffer Layers for Growth of Cubic Silicon Carbide on Silicon Single Crystals," *J. Cryst. Growth* **70** [1-2] 291-294 (1984).

[14] K. Sasaki, E. Sakuma, S. Misana, S. Yoshida and S. Gonda, "High-temperature Electrical Properties of 3C-SiC Epitaxial Layers Grown by Chemical Vapor Deposition," *Appl. Phys. Lett.* **45** [1] 72-73 (1984).

[15] P. H. Liaw and R. F. Davis, "Epitaxial Growth and Characterization of β-SiC Thin Films," *J. Electrochem. Soc.* **132** [3] 642-648 (1985).

[16] P. Rai-Choudhury and N. P. Formigoni, "β-Silicon Carbide Films," *J. Electrochem. Soc.* **116** [10] 1440-1443 (1969).

[17] I. Berman, C. E. Ryan, R. C. Marshall and A. Littler, "The Influence of Annealing on Thin Films of Beta SiC," AFCRL report #72 - 0737, Dec. 19, 1972.

[18] R. W. Bartlett and R. A. Muller, "Epitaxial Growth of β-Silicon Carbide," *Mater. Res. Bull.* **4** S341-S354 (1969).

[19] N. N. Long, D. S. Nedzvetski, N. K. Prokofera and M. B. Reifman, "Line Spectrum of Donor-Acceptor Pairs in β-SiC Crystals Doped with Aluminum," *Opt. i. Spektroskopiya* **29** [4] 388-390 (1970).

[20] S. Nishino, H. Suhara and H. Matsunami, "Reproducible Preparation of Cubic-SiC Single Crystals by Chemical Vapor Deposition," p. 317-320 in extended

Abstracts of the 15th Conference on Solid State Devices and Materials, Tokyo, Aug. 30, 1983.

[21]W. von Muench and E. Pettenpaul, "Preparation of Pure and Doped Silicon Carbide by Pyrolysis of Silicon Compounds," *J. Electrochem. Soc.* **125** [2] 294-299 (1978).

[22]R. N. Hall, "Electrical Contacts to Silicon Carbide," *J. Appl. Phys.* **29** [6] 914-917 (1958).

[23]H. J. van Daal, W. F. Knippenberg and A. Huizing, "Silicon Carbide Semiconductor Device," U.S. Patent 3047439, July 31, 1962.

[24]W. V. Muench, "Silicon Carbide Technology for Blue-Emitting Diodes," **6** [4] 449-463 (1977).

[25]A. Addamiano, "Semiconductor Crystals of Silicon Carbide with Improved Chromium-containing Electrical Contacts," U.S. patent 3510733, May 5, 1970.

[26]T. S. Shier, "Ohmic Contacts to Silicon Carbide," *J. Appl. Phys.* **41** [2] 771-773 (1970).

[27]H. Daimon, M. Yamanaka, E. Sakuma, S. Misawa and S. Yoshida, "Annealing Effects on Al and Al-Si Contacts with 3C-SiC," *Jap. J. Appl. Phys.* **25** [7] L592-L594 (1986).

[28]W. von Muench, P. Hoeck and E. Pettenpaul, "Silicon Carbide Field-Effect and Bipolar Transistors," pp. 337-339, Technical Digest of 1977 International Electronic Device Meeting Institute of the Electrical and Electronic Engineers, New York, 1977.

[29]S. Yoshida, H. Daimon, M. Yamanaka, E. Sakuma, S. Misawa and K. Endo, "Schottky-barrier Field-Effect Transistors of 3C-SiC," *J. Appl. Phys.* **60** [8] 2989-2991 (1986).

[30]G. Kelner, S. Binari, K. Sleger and H.S. Kong, "β-SiC MESFETs," presented at the Third National Review Meeting on Growth and Characterization of SiC and its Employment in Semiconductor Applications, North Carolina State University, No. 1986.

[31]C. H. Carter, Jr., J. A. Edmond, J. W. Palmour, J. Ryu, H. J. Kim and R. F. Davis, "Cross-sectional Transmission Electron Microscopy of Defects in Beta Silicon Carbide Thin Films," pp. 593-598 in Microscopic Identification of Electronic Defects in Semiconductors, Materials Res. Soc. Symp. Proc., Vol. 46 edited by N. M. Johnson, S. G. Bishop and G. Watkins. North Holland, Amsterdam, 1985.

[32]W. B. White, W. M. Johnson, and G. B. Dantzig, "Chemical Equilibrium in Complex Mixtures," *J. Chem. Phys.* **28** [5] 751-755 (1958).

[33]G. Eriksson, "Thermodynamic Studies of High Temperature Equilibrium," *Chemica Scripta* **8** [3] 100-103 (1975).

[34]JANAF Thermodynamic Tables, 2nd ed., edited by D. R. Stull and H. Prophet (National Bureau of Standards, USA, 1971).

[35]H. J. Kim and R. F. Davis, "Physical and Chemical Nature of Films Formed on Si (100) Surface Subjected to C_2H_4 at Elevated Temperatures," submitted to J. Electrochem. Soc.

[36]*Heteroepitaxy on Silicon*, Materials Research Society Symposia Proceedings Vol. 67, edited by J. C. C. Fan and J. M. Poate, Materials Research Society, Pittsburgh, PA, 1986.

[37]K. Shibahara, S. Nishino, and H. Matsunami, "Antiphase-domain-free Growth of Cubic SiC on Si (100)," *Appl. Phys. Lett.* **50** [26] 1888-1890 (1987).

[38]H. H. Berger, "Models for Contacts to Planar Devices," *Solid State Electron.* **15** [2] 145-150 (1972).

[39] B. L. Sharma, "Ohmic Contacts to III-V Compound," pp. 1-3 in "Semiconductors and Semimetals," Vol. 15, Edited by Willardson and A. C. Beer, Academic Press, New York, 1981.

[40] P. Pirouz, C. M. Chorey and J. A. Powell, "Antiphase Boundaries Epitaxially Grown β-SiC," *Appl. Phys. Lett.* **50** [4] 221-223 (1987).

[41] H. Kong (private communications).

[42] J. S. Ryu, Ph.D. Thesis, "Ion Implantation Annealing Characterization and Device Development in β-Silicon Carbide Single Crystalline Thin Films," pp. 192-194, North Carolina State University, Raleigh, NC (1986).

[43] C. H. Carter, R. F. Davis and S. R. Nutt, "Transmission Electron Microscopy of Process-induced Defects in β-SiC Thin Films," *J. Mater. Res.* **1** [6] 811-819 (1986).

[44] H. S. Kong, J. T. Glass and R. F. Davis, "Epitaxial Growth of β-SiC Thin Films on 6H α-SiC Substrates via Chemical Vapor Deposition," *Appl. Phys. Lett.* **49** [17] 1074-1078 (1986).

[45] J. S. Ryu, in Ref. 42, pp. 137-138.

Structural Characterization of Ion Implanted Beta-SiC Thin Films

J. A. Edmond and R. F. Davis

Department of Materials Science and Engineering
North Carolina State University
Raleigh, NC 27695-7907

S. P. Withrow

Sold State Division
Oak Ridge National Laboratory
Oak Ridge, TN 37831

Abstract

Thin films of β-SiC (100) grown on Si, have been implanted with Al, P, Si and Si plus C. The temperature dependence of implantation-induced damage was studied between 77K and 1023K using transmission electron microscopy and Rutherford backscattering/channeling. Amorphous layers resulting from implantation of any of the above species recrystallized via solid-phase-epitaxy during post-implantation annealing between 1973K and 2073K. As such, this latter process necessitated the prior removal of the Si substrate. However, the resultant layers were very defective. Increasing the sample temperature to 1023K during implantation allowed in situ dynamic annealing of the samples and therefore, greatly reduced the residual damage to the lattice. No additional annealing was required for structural recovery.

Introduction

Cubic (zincblende structure) β-SiC is the principal and perennial candidate material for electronic devices for operating under one or more of the conditions of high temperature, high power or high speed. The resurgence in interest in this material has been driven by (1) the increased need for devices operable under the aforenoted conditions, (2) the fact that the recent advances in Si-based thin film deposition and device fabrication technologies can, in most cases, be directly translated to SiC, and (3) the drive to achieve very large numbers of integrated devices per unit area. As the planar density of devices increases, the thermal load and management become major problems. However, the high thermal conductivity of SiC (5 W/cm-deg—approximately that of W) will allow the necessary dissipation of heat and thus the desired large scale device integration without degradation in device performance.

An unstated but critical factor underlying some of the above comments is that the operation of most electronic devices require that the material be amenable to doping with *both* n (donor)—and p (acceptor)—type dopants. This has been accomplished in SiC but not in GaN or C (diamond)—the two rival materials for devices for employment under severe conditions.

Two common methods of doping SiC have been in situ incorporation from the gas phase during growth and from diffusion sources at high temperatures. An alternative method of controllably introducing impurities, and one more suitable for device fabrication in SiC, is ion implantation.

In the process of ion implantation, nuclear scattering events occur inside the host material during the deceleration of an implanted atom and result in regions containing lattice damage. As the dose (ions/cm^2) of the implanted specie is increased, the near-surface region of the implanted material becomes progressively damaged; atomic disorder and eventual amorphization of the structure occurs. Early work by Hart et al.[1] utilized Rutherford Backscattering/channeling (RBS/C) techniques in order to study both disorder production in monocrystalline 6H α-SiC by ion implantation of Sb and N at room temperature and the subsequent structural recovery of the lattice during thermal annealing. They showed that the amorphous regions reordered considerably upon annealing at 923K. However, the extent of the residual line and planar defects present after annealing was not determined.

More recently, cross-sectional transmission electron microscopy (TEM) has been used to show that regrowth of amorphous layers produced via ion implantation invariably occurs by solid-phase epitaxy (SPE) from existing crystalline surfaces[2]. However, the results of this research have also shown that the quality of these SPE regrown layers in compound semiconductors is generally very poor. The two major problems are that nonstoichiometry results during implantation and that dissociation of the constituent elements of the compound semiconductor usually occurs at different temperatures during thermal annealing. Finally, the need for regrowth may be ameliorated or even eliminated by implantation at an elevated temperature sufficient to cause dynamic annealing of the damage.

In the present research, ion implanted layers were produced in monocrystalline β-SiC thin films at liquid nitrogen (LN), room (RT) and high temperatures (T ≥ 673K) and characterized using cross-sectional TEM and/or RBS/C. These results as well as those determined from samples subjected to post implantation annealing will be presented.

Experimental Procedures

Thin films of β-SiC were epitaxially grown in-house on silicon (100) wafers via chemical vapor deposition[3]. Prior to implantation each sample was mechanically polished, oxidized, and etched in HF in order to obtain a clean, undamaged and smooth surface. After mounting in high vacuum, samples were implanted at LN, RT or high temperature (T ≥ 673K) with either Al, P, Si or Si and C. The first two species were implanted in order to dope β-SiC p-type and n-type, respectively; the latter two were used for preamorphization prior to subsequent dopant introduction at concentrations below which amorphization occurs. All implants were made at an incident angle 7° off normal to avoid channeling effects. A summary of implant species and conditions is given in Table I.

Following implantation under LN or RT conditions, samples to be characterized in TEM were sectioned in half. One section was used to observe the as-implanted damage; the other was used in annealing experiments. (Samples implanted at high temperature were not sectioned or annealed.) Before annealing, films were separated from the Si substrate on which they were grown by etching in an (1:1) HF:HNO$_3$ solution until the Si was dissolved. The thickness of each film was

Table I. Summary of ion implantation conditions.

Figure No.	Ion Species	Energies(keV)	Doses(cm^{-2})	*Peak Atomic Conc. (cm^{-3})	Implant Temp (K)
1	^{27}Al$^+$	130	1×10^{15}	1×10^{20}	77
2	^{31}P$^+$	110,220	6,10×10^{14}	1×10^{20}	77
3(B)	^{28}Si$^+$	120,160,320	2.3,3.2,5.1×10^{14}	3×10^{19}	77
(C)	(3(B)+)^{12}C$^+$	50,67,141	2.7,3.2,4.8×10^{14}	3×10^{19}	77
4	^{27}Al$^+$	110,190	6,9×10^{14}	1×10^{20}	298
5	^{31}P$^+$	110	7.7×10^{13}	1×10^{19}	298
6(A)	^{31}P$^+$	110	1×10^{15}	1×10^{20}	298
(B)	^{31}P$^+$	110	1×10^{15}	1×10^{20}	77
8(A)	^{27}Al$^+$	130	4×10^{14}	4×10^{19}	1023
(B)	^{27}Al$^+$	130	4×10^{14}	4×10^{19}	77

*Peak concentration values determined by LSS calculation

approximately 15 microns. Upon loading a sample, the annealing chamber was evacuated to a pressure of 10^{-5} Torr and backfilled with Ar to a pressure ~1 atmosphere. Argon was used in order to help reduce the rate of Si loss during the heating of SiC. Sample halves were annealed at 1973K-2073K for a period of 300 s. Cross-sectional TEM samples of these implanted materials were prepared for observation, as discussed in Ref. 4.

Backscattering spectra as well as cross-section TEM results were also obtained on β-SiC implanted with Al at RT, 623K, 823K and 1023K to study the effect of in situ annealing during implantation. The RBS/C analyses were performed using 2.0 MeV ^{4}He$^+$ ions incident along the [110] axial direction.

RESULTS AND DISCUSSION

LN Implantation of Al and P

Figure 1 shows cross-sectional TEM micrographs of a sample implanted under LN conditions with Al to a peak atomic concentration of 10^{20} Al/cm^3. This concentration is typical for forming a p-type layer in our material. During implantation, the surface became amorphous to a depth of ~ 0.20 microns, as shown in Fig. 1(A). Thermal annealing of this sample at 1973K caused the layer to recrystallize. As shown in Fig. 1(B), a line of defects, commonly referred to as straggling ion damage, resulted near the original lower amorphous-crystalline (a-c) interface. Additionally, a broad band of polycrystalline SiC formed within the central region of the original amorphous layer. This band is bounded by single crystal material with the same crystallographic orientation as the bulk crystal. Three

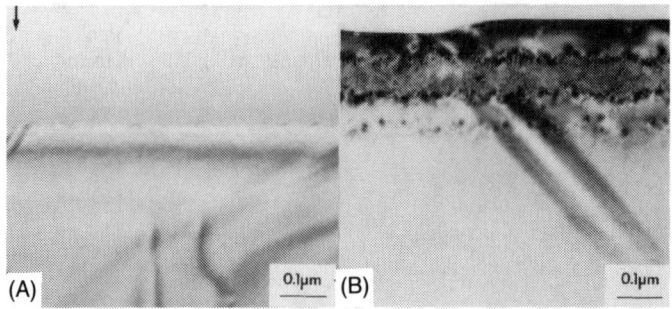

Fig. 1. Cross-sectional TEM micrographs showing the near-surface region of a sample which has been implanted at LN temperature with Al to a peak concentration of 10^{20} Al/cm^3. (A) As-implanted, (B) annealed at 1973K for 300 s.

growth mechanisms by which the observed microstructures could have resulted are described below.

There are two possibilities if one assumes the layer to be amorphous to the very surface. Firstly, the high free energy associated with a surface could result in heterogeneous nucleation and growth from that surface during annealing. Likewise, SPE from the lower a-c interface would occur. The convergence of these two growth fronts could result in a polycrystalline layer centered about the implant peak, as was experimentally observed. Secondly, SPE could occur from the lower a-c interface only. In this case, single crystal growth could initially occur, become textured polycrystalline in the region of the implant peak where the highest concentration of Al exists, and return to single crystal as the Al concentration decreases toward the surface.

The third possible growth mechanism assumes that a thin crystalline cap, not visible at the magnification used to obtain Fig. 1(A), remains following this implant. Under these conditions, SPE could occur from both the lower and upper a-c interfaces during annealing which would again lead to convergence of the moving growth fronts within the region of maximum Al concentration.

Figure 2 illustrates the regrowth properties a P double implant with a peak concentration of 10^{20} P/cm^3. An ~ 0.5 micron amorphous layer was created similar in appearance to that shown in Fig. 1(A). The annealing temperature was 1973K. A polycrystalline layer also resulted, in this case, after ~ 0.1 micron of single crystal regrowth from the lower a-c interface. Straggling ion damage was also observed in the a-c region. The solubility limit of this element in our films during growth at 1660K has recently been determined to be ~ 10^{18} P/cm^3 [5]. Although the annealing temperature of 1973K exceeds the growth temperature, it is believed that the P became supersaturated at the growth front and subsequently formed precipitates which caused the onset of polycrystallinity.

Fig. 2. Cross-sectional TEM micrograph of a sample which has been implanted at LN temperature with P to a peak concentration of 10^{20} P/cm^3 and subsequently annealed at 1973K for 300 s. The near-surface region appears rough as a result of polycrystalline regrowth.

LN Implantation of Si and Si plus C

As seen in Figs. 1(B) and 2, straggling ion damage usually occurred at the original a-c interface following annealing. This damage may cause high leakage currents in diodes, for example, since the damage is near the electrical junction. An increasingly popular method of circumventing this problem is to conduct a high-energy implant of an electrically neutral species prior to implanting the dopant at a lower energy in order to create a deep amorphous layer. As a result, the defects that form after annealing will be relatively far away from the electrical junction. This method also allows complete activation of the dopant for low dose implants and produces sharper dopant profiles, since it prevents channeling of the dopant ion. In an attempt to achieve these effects in β-SiC, films from our research were implanted with self-ions.

The cross-sectional TEM micrographs in Fig. 3 directly compare the structural regrowth properties of implanted and amorphized layers created using Si and Si plus C. Figure 3(A) shows the amorphous layer which was formed by triple implantation of Si to a peak concentration of 3×10^{19} Si/cm^3. The a-c interface is located ~ 0.40 micron below the sample surface. After annealing at 1973K, the layer regrew epitaxially without the severe faulting noted above. However, a high concentration of precipitates and/or loops formed throughout the regrown bulk. In an attempt to eliminate these defects, a triple C implant was superimposed on the triple Si implant thus simulating implantation of SiC into SiC. The projected range peaks were matched (1:1) Si to C using LSS theory[6] in order to obtain the correct stoichiometry. Figure 3(C) shows a cross-sectional TEM micrograph of the regrown layer previously implanted and amorphized with SiC at a peak concentration of 3×10^{19} for both species. The annealing conditions were the same as described for the sample shown in Fig. 3(B). Quite clearly, implanting Si plus C did not structurally improve the quality of the region layer. In fact, microfaulting and microtwinning occurred upon regrowth from ~ 0.14 micron to the sample surface.

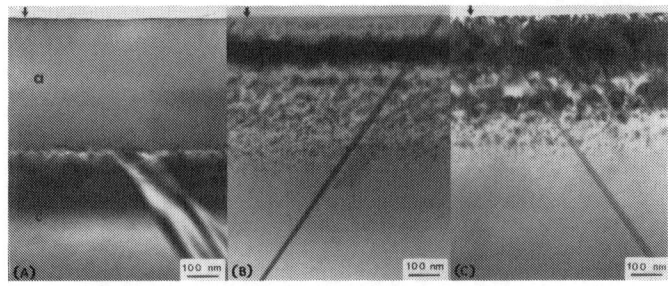

Fig. 3. Cross-sectional TEM micrographs comparing the regrowth properties of samples implanted under LN conditions with equal atom concentrations (3 × 10^{19} cm^{-3}) of Si (B, center) and Si + C (C, right). The as-implanted amorphous layer is also shown in (A). Samples were annealed at 1973K for 300 s.

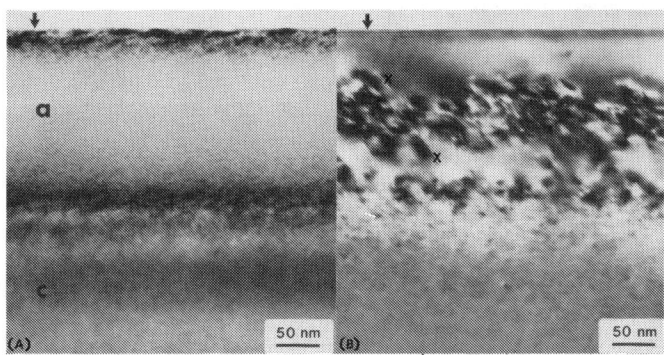

Fig. 4. Cross-sectional TEM micrographs showing the near-surface region of a sample which has been implanted at RT with Al to a peak concentration of 10^{20} Al/cm^3. (A) As-implanted, (B) annealed at 2073K for 300 s. X denotes microfaulted regions.

RT Implantation of Al and P

Figure 4 shows cross-sectional TEM micrographs of β-SiC double implanted with Al at RT to a peak concentration of 10^{20} Al/cm^3. A buried amorphous layer having a crystalline cap of ~ 0.01 micron resulted after implantation (Fig. 4(A)). This upper, and the lower a-c interface located at a depth of ~ 0.17 micron, were very diffuse as a result of implanting at RT. After annealing at 2073K, the amorphous layer regrew by SPE. However, a high concentration of defects was observed (Fig. 4(B)). Precipitates and/or dislocation loops formed where the lower a-c interface was initially located. A broad band of defects (0.05-0.11 micron) resulted where the two a-c interfaces converged during SPE regrowth. Unlike the previous Al implant, however, this region was single crystal with precipitates and some microfaulting (see areas labeled X in Fig. 4(B)). Additionally, approximately the first 0.05 micron below the surface proved to be virtually free of the planar and line defects.

In order to prevent amorphous layer formation during P implantation, and subsequently, polycrystalline regrowth, (as was described in Section (1)), a lower dose RT implant was implemented. The peak atomic concentration was 10^{19} P/cm^3. Evidence of implant damage appeared as a band of dark contrast parallel to the surface in the as-implanted material (Fig. 5(A)). This dark band was not amorphous since stacking faults were visible in this area. After annealing at 2073K, the band of dark contrast was replaced by defect clusters (Fig. 5(B)). No second-phase diffraction spots were observed in selected area diffraction patterns generated from the P-implanted region.

The effect of the implant temperature on the extent of amorphization is illustrated in Fig. 6. Both micrographs are of films that were implanted with P to a peak concentration of 10^{20} P/cm^3, but the sample in Fig. 6(A) was implanted at RT while the sample in Fig. 6(B) was maintained under LN conditions. The RT sample had a ~ 0.01 micron crystalline surface, an amorphous depth of ~ 0.13 micron, and visible damage to ~ 0.23 micron. The LN implant created a thinner crystalline cap (~ 0.005 micron), an amorphous depth of ~ 0.17 micron, and visible damage to ~ 0.35 micron. The same effect of similar magnitude has been observed in Si wafers implanted at RT and LN[7]. This similarity was unexpected, as the sublimation temperature of SiC is > 3073K, while that of Si is 1679K. These results led the authors to the notion that heating β-SiC above RT during implantation may result in in situ annealing of lattice damage. The results of the high temperature (HT) implantation experiments are given below.

HT Implantation of Al

The effect of sample heating during implantation on damage production was investigated using Al implants produced at RT, 623K, 823K and 1023K. Figure 7 compares RBS/C spectra for samples implanted at these four temperatures to a peak concentration of 7×10^{19} Al/cm^3. As shown in this figure, a buried amorphous layer was created within the near-surface region after RT implantation. Heating similar samples to 623K and higher temperatures during implantation progressively reduced the induced damage. The spectra from the 1023K implants nearly coincided with that of the virgin aligned. This latter spectrum is taken from an unimplanted region adjacent to the implanted area. Therefore, the crystallinity of the lattice was virtually recovered during HT implantation, within the sensitivity of RBS/C. However, this technique is not as sensitive to line and planar defects as is TEM. For that reason, two β-SiC samples, implanted under like conditions except temperature, were analyzed in TEM.

Figures 8(A) and 8(B) illustrate the damage produced during implantation of Al in β-SiC at 1023K and 77K, respectively. As shown in Fig. 8(A), implantation at 1023K resulted in neither visible lattice damage nor precipitates. The implant under LN conditions resulted in the creation of a buried amorphous layer. Upon annealing the latter sample to 1973K, residual defects similar to those obtained in Fig. 4(B) were observed. Annealing at this high temperature required Si removal;

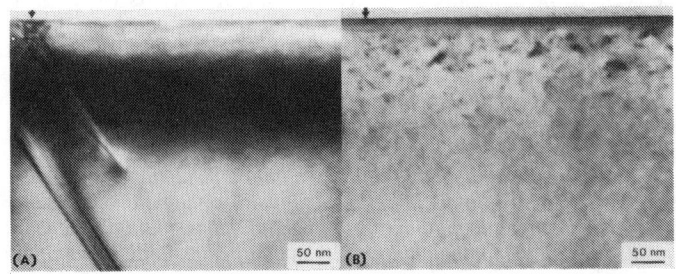

Fig. 5. Cross-sectional TEM micrographs showing the near-surface region of a sample implanted at RT with P to a peak concentration of 10^{19} P/cm^3. (A) As-implanted, (B) annealed at 2073K for 300 s.

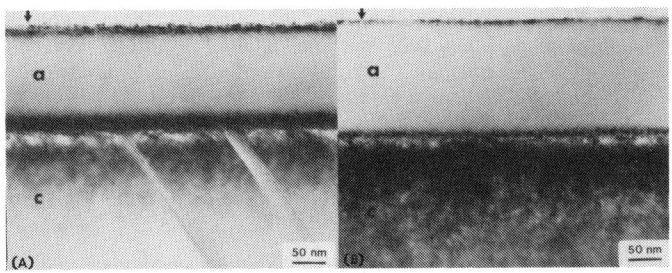

Fig. 6. Cross-sectional TEM micrographs comparing the near-surface damages created by implantation of P to a peak concentration of 10^{20} P/cm^3 under (A) RT and (B) LN conditions.

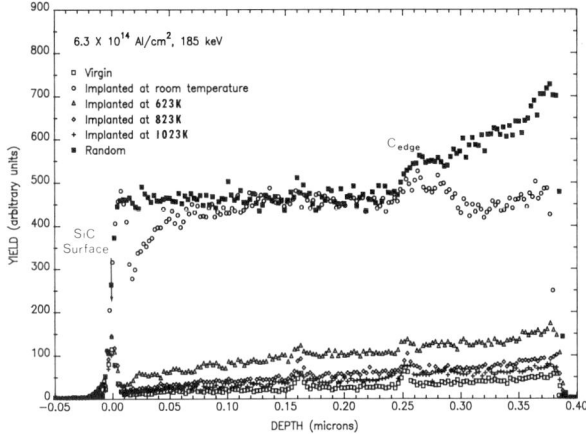

Fig. 7. 2.0 MeV He RBS/C spectra for Al-implanted β-SiC showing the decrease in lattice damage along [110] with an increase in implant temperature.

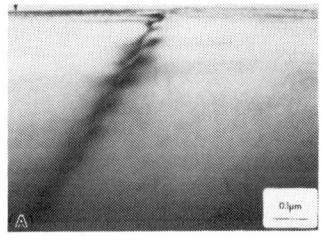 

Fig. 8. Cross-sectional TEM micrographs comparing the near-surface damage created by implantation of Al to a peak concentration of 4×10^{19} Al/cm^3 at (A) 1023K and (B) LN temperature.

Silicon Carbide 487

whereas, annealing to 1023K during implantation did not. This is yet another advantage to high temperature implantation in this material.

Summary and Conclusions

Ion implantation of Al, P, Si and Si plus C into single crystal β-SiC thin films was performed. The temperature dependence of implantation-induced damage was studied between 77K and 1023K. It was determined that amorphous layers produced in β-SiC regrew via SPE in the temperature range of 1973K-2073K. In all cases, however, the quality of regrown regions were very poor; various combinations of precipitates, dislocation loops, microfaults and polycrystalline regions often resulted. It was shown that heating the sample as low as RT during implantation, reduced the width of the amorphous layer that resulted from LN implantation. Finally, it was shown that implantation at higher temperature (623K-1023K) resulted in structural recovery of a layer during ion bombardment. This allows one to introduce electrically active dopants into β-SiC controllably without damaging the crystal lattice and just as importantly, doing so without having to remove the silicon substrates on which they are grown.

Acknowledgement

The authors gratefully acknowledge the support of this program by the Office of Naval Research under contract N00014-82-K-0182 P005 and to the ONR Fellowship program for support of one of the authors (Edmond). Work at Oak Ridge National Laboratory was supported by the Division of Materials Sciences, U.S. Department of Energy under contract DE-AC05-840R21400 with Martin Marietta Energy Systems, Inc. Appreciation is also expressed to C. H. Carter, Jr. for assistance in the preparation and interpretation of the TEM micrographs.

References

[1] R. R. Hart, H. L. Dunlap and O. J. Marsh, "Disorder Production in SiC by Ion Bombardment," Rad. Effects, 9 [5] 261-266 (1971).

[2] J. Narayan, "Interface Structures During Solid-Phase-Epitaxial Growth of Ion Implanted Semiconductors and a Crystallization Model," 53 [12] 8607-8614 (1982).

[3] H. P. Liaw and R. F. Davis, "Epitaxial Growth and Characterization of β-SiC Thin Films," 132 [3] 642-648 (1985).

[4] C. H. Carter, Jr., J. A. Edmond, J. W. Palmour, J. Ryu, H. J. Kim and R. F. Davis, "Cross-Sectional Transmission Electron Microscopy of Defects in β-SiC Thin Films," pp. 485-491 in Microscopic Identification of Electronic Defects in Semiconductors. Edited by N. M. Johnson, S. G. Bishop and G. Watkins. North-Holland, Amsterdam, 1985.

[5] H. J. Kim and R. F. Davis, "Theoretical and Empirical Studies of Impurity Incorporation into β-SiC Thin Films During Epitaxial Growth," J. Electrochem. Soc., 133 [11] 2350-2357 (1986).

[6] J. F. Gibbons, W. S. Johnson and S. W. Mylroie, Projected Range Statistics: Semiconductors and Related Materials, 2nd Edition. Dowden, Hutchinson and Ross, Inc., Stroudsburg, PA, 1975.

[7] W. P. Maszara, G. A. Rozgonyi, L. Simpson and J. J. Wortman, "Temperature Dependent Amorphization of Silicon During Self-Implantation"; pp. 381-387 in Beam-Solid Interactions and Phase Transformations. Edited by H. Kurz, G. L. Olson and J. M. Poate. Elsevier, New York, 1986.

Effects of Cathode Materials and Gas Species on the Surface Characteristics of Dry Etched Monocrystalline Beta-SiC Thin Films

John W. Palmour and R. F. Davis

Department of Materials Science and Engineering - Box 7907
North Carolina State University
Raleigh, NC 27695-7907

P. Astell-Burt and P. Blackborow

Plasma Technology, Inc.
145 Sydney Street
Cambridge, MA 02139

Abstract

Monocrystalline β-SiC (100) thin films were dry etched in CF_4 and NF_3 by reactive ion etching (RIE) and in SF_6 by plasma etching. The effects of cathode materials and gas species on the etched surface of SiC were determined by Scanning Electron Microscopy (SEM) and Auger Electron Spectroscopy (AES). Reactive ion etching in NF_3 or CF_4 utilizing a stainless steel or anodized Al cathode created a roughened surface, which was contaminated with traces of Fe of Al_2O_3, respectively. The roughness was caused by micromasking of the SiC by material sputtered from the cathode. By contrast, a carbon cathode yielded a very smooth, and much cleaner etched surface. Plasma etching of SiC in SF_6 resulted in a crystallographically roughened surface, but the AES spectra indicated almost no contamination and very little fluorine or native oxide. The optimal condition for etching SiC was determined to be RIE in NF_3 on a carbon cathode.

Introduction

Beta-SiC has been investigated for potential use in high frequency, high power and high temperature devices, because of its wide band gap, high saturated electron drift velocity, and high electron mobility[1,2]. In order to electrically characterize this material via *pn* junction mesa diode measurements, as well as fabricate devices such as impact-avalanche transit-time (IMPATT) diodes and bipolar junction transistors, a method of selective, controllable etching is needed. However, SiC is an extremely inert material that can only be conventionally etched by molten salts or Cl_2 or H_2 gases at high temperatures[3]. Therefore, dry etching techniques utilizing fluorinated gases have been investigated.

Sputter deposited films, presumably amorphous (α-SiC), have received the most attention in previous dry etching experiments on this material. It has been reported that XeF_2 vapor combined with Ar^+ bombardment from an ion gun readily etches α-SiC, while the use of XeF_2 vapor alone does not[4]. The major Si-containing product from this reaction was found to be SiF_4, while the major C-containing product was proposed to be either CF_4 or CF_2, although it could not be identified. Reactive ion etching of α-SiC has also been performed in $CF_4 + O_2$, $SF_6 + He$, and Ar gases[5] as well as NF_3[6]. Dry etching experiments on single crystal β-SiC thin films have used CF_4 and $CF_4 + O_2$ mixtures in a variety of modes, including reactive ion beam etching[7], reactive ion etching[8], and plasma etching[9]. The use of SF_6 in plasma etching of monocrystalline β-SiC thin films has also been recently reported[10].

For the case of RIE of SiC in CF_4, a spiked surface was reported, along with the presence of Fe, which was presumably sputtered from the stainless steel cathode onto the etched surface[8]. These conditions are undesirable for the development of SiC devices, since high quality ohmic contacts are to be made on the etched surface. As such, a smooth, chemically clean etch is imperative. The purpose of this research has been to determine the physical and chemical characteristics of dry etched SiC surfaces and the effects of cathode materials, gas selection, and process configuration.

EXPERIMENTAL

The β-SiC thin films used in these experiments were epitaxially grown on Si (100) substrates using a two-step process involving a carburized buffer layer followed by chemical vapor deposition of SiC[11]. These samples were polished with 0.1 μm diamond paste, oxidized in dry O_2 at 1473 K for 90 min., etched in HF for 5 min., and rinsed in deionized water. The samples to be used for SEM were then masked with evaporated Al that was patterned using standard photolithographic techniques; the AES samples remained unmasked.

The reactive ion etching experiments were performed in a parallel plate reactor (Plasmalab RIE 80) with a 28.0 cm diameter Al anode and a 17.0 cm diameter anodized Al cathode with a plate separation of 5.0 cm. Either the anodized Al cathode or a C coverplate were used to support the samples, which were kept at a temperature of 308 K. The samples etched with a stainless steel cathode were from a previous study with a different RIE system[8]. A 13.56 MHz rf supply powered the cathode at densities of 0.440 to 0.548 W/cm^2. The chamber was evacuated to a pressure of 5×10^{-5} Torr and the gas (CF_4 or NF_3) was introduced at 25 sccm, maintaining a pressure of 40 mTorr.

The plasma etching experiments were performed in a parallel plate system (Plasmalab DP 80) having upper and lower Al electrode diameters of 28.0 cm and 24.0 cm, respectively, and a plate separation of 3.0 cm. The upper plate was powered by a 13.56 MHz rf supply at a density of 0.325 W/cm^2. A base pressure of 5×10^{-5} Torr was attained and SF_6 was introduced at 30 sccm, giving a constant pressure of 100 mTorr.

RESULTS AND DISCUSSION

A typical Auger spectrum for unetched β-SiC, shown in Fig. 1(a), reveals the presence of native oxide on the SiC surface. The oxide, typically 20 Å thick[12] is characterized by the suppressed Si peak (relative to that obtained from an Ar^+

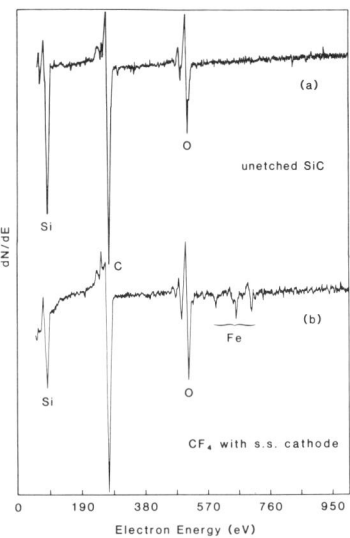

Fig. 1. Auger spectra obtained at 3 kV and 0.5 μA for (a) an unetched SiC surface, and (b) a SiC surface after RIE in CF_4 (40 mT, 0.548 W/cm^2, 10 min.) on a stainless steel cathode.

sputtered surface[8]) in conjunction with the O peak. The nature of the carbon bonding can be determined via the relative heights of the three differentiated fine structure peaks on the low energy side of the C KLL peak. A carbidic C line shape can be seen in Fig. 1(a), as evidenced by the higher peak height of the third, or highest energy, C fine structure peak. By contrast, if the second, or middle C fine structure peak were highest, the C bonding would be graphitic.

After reactive ion etching in CF_4 on a stainless steel cathode, the β-SiC surface exhibited a graphitic C lineshape, as indicated by the dominant second C fine structure peak in the AES spectrum in Fig. 1(b). This is probably due to the preferential etching by F of the Si from the matrix, leaving the excess C on the surface to graphitize.

As noted earlier, reactive ion etching of β-SiC in CF_4 with a stainless steel cathode left Fe on the surface of the SiC, as signified by the three small AES peaks from 590 to 700 eV in Fig. 1(b). No large amount of F was detected, although a very small F peak would have been overlapped by the Fe peak at 651 eV. Also, a very rough spiked surface was formed during this process, as shown in Fig. 2. This spiked structure is strongly believed to be caused by the unintentional formation of micromasks via the deposition of stainless steel from the cathode. This micromask effect disallowed etching under the steel particles. Micromasking has also been reported for dry etching of single crystal Si[10].

When the same conditions were used in a different RIE chamber containing an anodized aluminum cathode, the surface roughening was greatly reduced, as shown

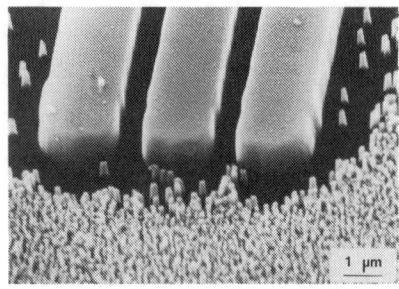

Fig. 2. Spiked structure on SiC after RIE in CF_4 (40 mT, 0.548 W/cm^2) on a stainless steel cathode. Spikes are due to Fe micromasking during etching.

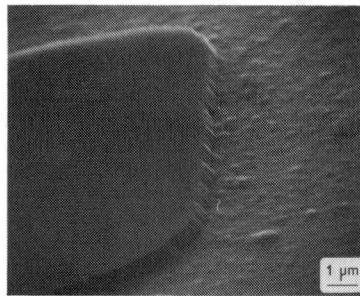

Fig. 3. Etched surface of SiC after RIE in CF_4 (40 mT, 0.440 W/cm^2) on an anodized Al cathode.

in Fig. 3. The small amount of roughness which did result may be due to micromasking, to a lesser degree, by Al_2O_3 sputtered from the cathode. The AES spectrum in Fig. 4(a) shows the presence of strong Al LMM peaks at 38 and 53 eV, which are characteristic of Al_2O_3, as well as the presence of an unusually strong O peak. This spectrum also shows a high F content on the surface. The C lineshape is not graphitic and is only slightly carbidic, appearing to be more like free C. These two facts indicate that fluorocarbon polymerization on the SiC surface occurred during etching. This polymer was then decomposed by the electron beam during Auger analysis, resulting in a "free C" peak and an F peak.

Because of the polymerization problem commonly associated with etching in pure CF_4 and the discovery that F is the chief reactant for both Si and C[4,8], another fluorinated gas was investigated. The most desirable gas was NF_3 because (1) it is more efficiently broken into free F radicals, and (2) all of the possible by-products

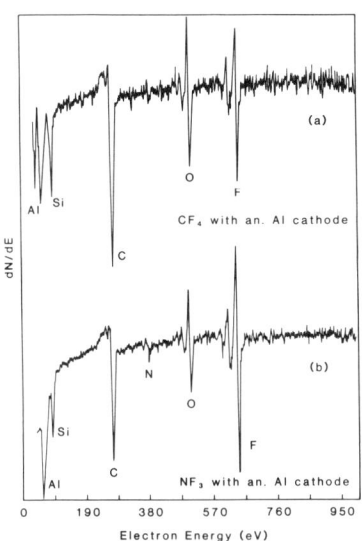

Fig. 4. Auger spectra obtained at 2 kV and 0.5 μA for SiC after RIE in (a) CF_4 on an anodized Al cathode, and (b) NF_3 on an anodized Al cathode, at 40 mT and 0.440 W/cm^2 for 10 min.

of ionization are volatile[14]. Indeed, in this study NF_3 yielded etch rates as fast as 211 nm/min, the highest etch rates reported to date for the dry etching of SiC.

However, when the same etching conditions noted above were used with NF_3, AES again showed a very strong F peak, as shown in Fig. 4(b), and the C peak appeared to again be neither carbidic nor graphitic. Secondly, the presence of Al_2O_3 is again strongly apparent, and a very small nitrogen KLL peak is present at 383 eV. It is assumed that in both CF_4 and NF_3, the very non-reactive anodized Al cathode does not react with the F species in the chamber. This allows these species to accumulate and polymerize with the exposed C on the SiC surface. Again, the fluorocarbon polymer is decomposed by the electron beam during analysis, explaining the "free C" lineshape.

The SiC surface shown in Fig. 5 was very rough after this etch, greatly exceeding the roughness present after etching in CF_4 with an anodized Al cathode. Therefore, the RIE process is again causing micromasking of the SiC and by the same process of depositing Al_2O_3 from the cathode as noted above for CF_4. The increased roughness of the NF_3 etch, as compared with the CF_4 etch, is most likely caused by the increase of the SiC etch rate by the NF_3 without a concomitant increase in the etch rate of the micromasking material.

In order to eliminate the deposition of Al_2O_3 and fluorocarbons onto the etched surface, a C cathode coverplate was placed in the RIE chamber. The advantages of using a C cathode are that it has a very low sputter yield and it is reactive with F, thus decreasing the possibility of micromasking and F species accumulation. The

Fig. 5. Roughened etched surface of SiC after RIE in NF_3 (40 mT, 0.440 W/cm^2) on an anodized Al cathode.

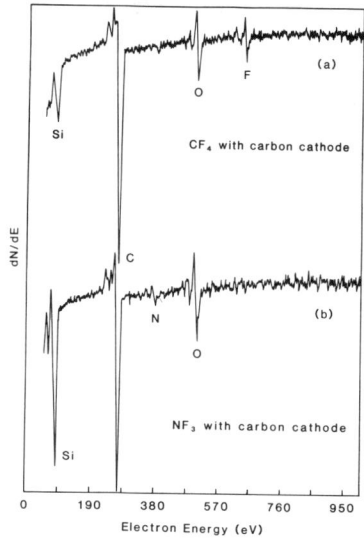

Fig. 6. Auger spectra obtained at 2 kV and 0.5 µA for SiC after RIE in (a) CF_4 on a carbon cathode, and (b) NF_3 on a carbon cathode, at 40 mT and 0.440 W/cm^2 for 10 min.

AES spectrum in Fig. 6(a) shows a β-SiC surface that was reactive ion etched in pure CF_4 using a C cathode coverplate. The F peak is now greatly reduced, indicating only a small amount of fluorocarbon polymerization, and there is not a detectable Al signal. The C KLL peak at 269 eV has increased markedly relative to the Si LVV peak; it also has a graphitic lineshape. Whether the graphitic nature of the C peak is due to preferential etching of Si from the surface, with subsequent graphitization of the remaining C, or material being deposited from the graphite coverplate, is unclear. However, if C is being deposited from the cathode, it is not acting as a micromask as evidenced by the very smooth etched surface seen in Fig. 7.

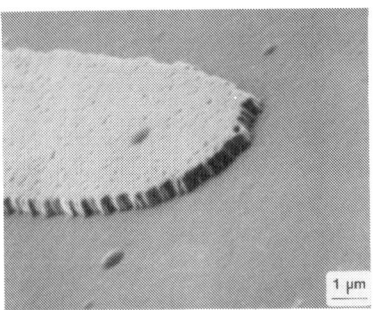

Fig. 7. Smooth etched surface of SiC after RIE in CF_4 (40 mT, 0.440 W/cm^2) on a carbon cathode.

The same C cathode coverplate system was used for RIE in pure NF_3. The AES spectrum in Fig. 6(b) shows a very clean surface, with no F present, and is almost identical to that of the unetched surface seen in Fig. 1(a), the only difference being the small N peak present after etching. There is no significant increase in the C signal relative to that of the unetched surface, and the C lineshape is carbidic. Reactive ion etching in NF_3 with the C cathode coverplate also produced a very smooth surface, shown in Fig. 8, comparable to that of the unetched material. Thus, the best conditions for RIE were obtained with NF_3 on a C cathode.

Plasma etching of β-SiC in SF_6 on an Al anode resulted in good etch rates (up to 50 nm/min.) but substantial roughening of the etched surface with time. The surface roughness after a 10 minute plasma etch in SF_6 showed a few small spikes among indistinguishable roughness features. However, after etching for 60 min, these features grew into large four-sided pyramidal structures, as seen in Fig. 9(a). These pyramidal structures were all oriented in the same direction, shown in Fig. 9(b), indicating a crystallographic etching phenomenon. While Kelner, et al.[10], did report surface roughness after plasma etching in SF_6, crystallographic spiking was not reported. However, the power densities used in this study, 0.325 W/cm^2, are much higher than the 0.12 W/cm^2 that Kelner, et al., used. These larger power densities and resulting depth of etch could account for the difference; but this has not yet been determined. These spikes should not be caused by micromasking

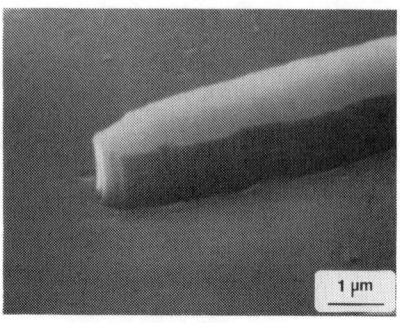

Fig. 8. Smooth etched surface of SiC after RIE in NF_3 (40 mT, 0.440 W/cm^2) on a carbon cathode.

 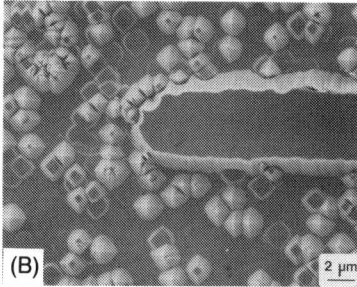

Fig. 9. Spiked surface of SiC after plasma etching in SF_6 for 60 min. at 0.2 Torr and 0.325 W/cm^2 at (A) 70° tilt and (B) 0° tilt. Pyramids are crystallographically faceted.

because there is very little directional ion bombardment, i.e. little sputtering, taking place during plasma etching, as opposed to reactive ion etching.

The absence of micromasking is further evidenced by the very chemically clean surface, shown by the AES spectrum in Fig. 10, yielded by this process. No traces of Al from the electrodes were found. Surprisingly, the weaker O peak and the stronger Si signal in this spectrum indicates less native oxide than that of an unetched sample. Moreover, this spectrum approaches that of an in situ sputter cleaned film. The C lineshape is carbidic. There is possibly a small sulfur LMM peak at 151 eV and a small F peak at 651 eV. The reason for the smaller amount of O on the surface cannot be explained at present, but it is possible that S and F are passivating the Si from oxidation and are subsequently volatilized by the electron beam during Auger analysis. The reason for the crystallographic spiking is unclear at present.

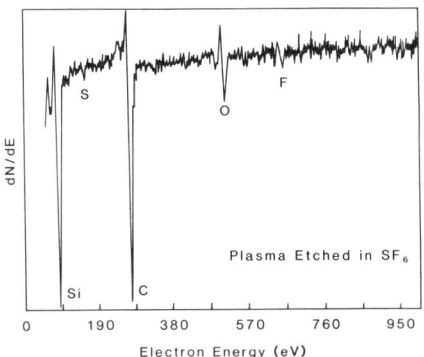

Fig. 10. Auger spectrum obtained at 2 kV and 0.5 µA for SiC after plasma etching in SF_6 at 0.2 Torr and 0.325 W/cm² for 10 min. Note that there is less native oxide present than there was on unetched SiC.

CONCLUSIONS

For reactive ion etching of β-SiC, the choice of cathode material plays a major role in the chemical and physical characteristics of the etched surface. Stainless steel and anodized Al cathodes were found to cause micromasking of the SiC during etching, as evidenced by the detection of these elements on the etched surface and the presence of surface roughness. The degree of roughness over a given amount of time presumably depends on the choice of fluorinated gas, with the faster etching NF_3 gas causing more roughness than CF_4 employed under the same conditions. The use of a non-reactive anodized Al cathode caused accumulation of F in both gases used, allowing it to polymerize with the C on the SiC surface. The use of a C cathode coverplate ameliorated the micromasking problem and left a smooth and chemically clean surface, with little or no fluorocarbon polymerization. Plasma etching in SF_6 caused crystallographically spiked formations that grew with etching time, but left a very clean surface that actually had less native oxide than unetched β-SiC. Because high quality ohmic contacts require both a smooth and chemically clean surface, it has been determined that the optimal configuration for dry etching SiC is RIE in NF_3 using a C cathode.

ACKNOWLEDGMENTS

The authors express their appreciation to W. Hale, T. M. Wallet and K. B. Bhasin for their participation in some of these experiments, and M. Mantini and B. Williams for their assistance in the Auger and SEM analysis, respectively. This work was partially sponsored by the Office of Naval Research under Contract No. N00014-82-K-0182P0005.

References

[1] R. W. Keyes, "Silicon Carbide from the Perspective of Physical Limits on Semiconductor Devices," pp. 534-41 in *Silicon Carbide—1973*. Edited by R. C. Marshall, J. W. Faust, Jr., and C. E. Ryan. University of South Carolina Press, Columbia, SC, 1974.

[2] C. E. Ryan, "International Conference on SiC-1973: Closing Remarks," pp. 651-53 in *Silicon Carbide—1973*. Edited by R. C. Marshall, J. W. Faust, Jr., and C. E. Ryan. University of South Carolina Press, Columbia, SC, 1974.

[3] J. W. Faust, Jr., "The Etching of Silicon Carbide," pp. 403-19 in *Silicon Carbide—A High Temperature Semiconductor: 1959*. Edited by J. R. O'Connor and J. Smiltens. Pergamon Press, New York, NY, 1960.

[4] H. F. Winters, "Etch Products from the Reaction of XeF_2 with SiO_2, Si_3N_4, SiC, and Si in the Presence of Ion Bombardment," *J. Vac. Sci. Technol. B*, 1 [4] 927-31 (1980).

[5] J. Sugiura, W. J. Lu, K. C. Cadien, and A. J. Steckl, "Reactive Ion Etching of SiC Thin Films Using Fluorinated Gases," *J. Vac. Sci., Technol. B*, 4 [1] 349-54 (1986).

[6] W. J. Lu, A. J. Steckl, T. P. Chow, and W. Katz, "Thermal Oxidation of Sputtered Silicon Carbide Thin Films," *J. Electrochem. Soc.*, 131 [8] 1907-1913 (1984).

[7] S. Matsui, S. Mizuki, T. Yamato, H. Aritome, and S. Namba, "Reactive Ion-Beam Etching of Silicon Carbide," *Jpn. J. Appl. Phys.*, 20 [1] L38-L40 (1981).

[8] J. W. Palmour, R. F. Davis, T. M. Wallet, and K. B. Bhasin, "Dry Etching of β-SiC in CF_4 and $CF_4 + O_2$ Mixtures," *J. Vac. Sci. Technol. A*, 4 [3] 590-93 (1986).

[9] S. Dohmae, K. Shibahara, S. Nishino, and H. Matsunami, "Plasma Etching of CVD Grown Cubic SiC Single Crystals," *Jpn. J. Appl. Phys.*, 24 [11] L873-L875 (1985).

[10] G. Kelner, S. C. Binari, and P. H. Klein, "Plasma Etching of β-SiC," *J. Electrochem. Soc.*, 134 [1] 253-54 (1987).

[11] P. Liaw and R. F Davis, "Epitaxial Growth and Characterization of Beta Silicon Carbide Thin Films," *J. Electrochem. Soc.*, 132, 642-48 (1985).

[12] M. Rahaman, Y. Boiteux, and L. DeJonghe, "Surface Characterization of Silicon Nitride and Silicon Carbide Powders," *Am. Ceram. Soc. Bull.*, 65 [8] 1171-76 (1986).

[13] D. N. K. Wang, (private communication).

[14] D. H. Bower, "Planar Plasma Etching of Polysilicon Using CCl_4 and NF_3," *J. Electrochem. Soc.*, 129 [4] 795-799 (1982).

Microstructure and Thermoelectric Energy Conversion in Porous SiC Ceramics

Kunihito Koumoto, Mitsuhide Shimohigoshi, Shunji Takeda, and Hiroaki Yanagida

Department of Industrial Chemistry
University of Tokyo
7-3-1 Hongo, Bunkyo-ku
Tokyo, 113, Japan

Abstract

Porous silicon carbide ceramics, with roughly 60% relative density, obtained by sintering pressed powder compacts at 2000°C for 3h in atmospheres of N_2 or Ar, have been found to show high figures of merit for thermoelectric energy conversion in the temperature range 400-1100°C. Photomicrographs showing the representative microstructures (SEM and TEM) for α- and β-SiC, fired in each atmosphere, are presented. Figures showing the temperature dependence of electrical and thermal conductivity, Seebeck coefficient, and figure of merit are included. Potential applications of this material for high temperature energy conversion, utilizing the high thermal and oxidation resistance, are discussed.

Introduction

When a temperature differential exists between the ends of a solid sample, a thermoelectric force is generated, which is well known as the Seebeck effect. Utilizing this phenomenon, conversion of thermal energy to electric energy, namely thermoelectric power generation, is possible in principle.

The maximum efficiency of energy conversion, η, of a thermoelectric couple is expressed as follows[1]:

$$\eta = \frac{T_h - T_c}{T_h} \cdot \frac{M - 1}{M + (T_c/T_h)} \quad (1)$$

where T_h and T_c are the absolute temperatures of the hot and cold ends of a thermocouple, respectively. M is given by:

$$M = (1 + Z_c T)^{0.5} \quad (2)$$

where T is the average temperature, and Zc is the so-called figure of merit, which is a function of electrical conductivities, S1, S2, Seebeck coefficients, A1, A2, and thermal conductivities, k1, k2 of the respective legs of the thermocouple, as follows:

$$Zc = \frac{(A1 - A2)^2}{[(k1/S1)^{0.5} + (k2/S2)^{0.5}]^2} \quad (3)$$

It is seen from Equations (1) - (3) that the larger the figure of merit and the higher the temperature, the higher the efficiency of energy conversion will be. Hence for the purpose of materials design for thermoelectric energy conversion, it is convenient to define a figure of merit for an individual material as:

$$Zc = S A^2/k \quad (4)$$

For practical applications, materials having high electrical conductivity, large Seebeck coefficient, and low thermal conductivity, that are chemically, thermally, and mechanically stable, would be desirable. Based on these requirements, porous SiC ceramics were chosen for study, to evaluate their potential as thermoelectric semiconductor materials.

Experimental Procedure

Porous SiC ceramics with about 60% relative density were fabricated by sintering isostatically pressed powder compacts, with no additives, at 2000°C for 3h in N_2 or Ar atmospheres. Both α- and β-SiC powders were used as starting materials. Electrical conductivity was measured at 400-1100°C in an argon atmosphere using a D.C. four-probe technique, with Pt wires for electrodes. Thermal conductivity was measured at 400-1100°C in vacuum by a laser flash method. Thermoelectromotive force was measured at 400-1100°C in Ar using two Pt-Pt 13% Rh thermocouples embedded in drilled holes in each end of the specimen. The temperature difference between the specimen ends was controlled at 2-10K by flowing cold air through an alumina tube, placed near one end of the test specimen. The relationship between temperature difference and thermo-electromotive force was always linear, and the Seebeck coefficient was calculated from the slope of the line. X-ray diffraction analysis, using Ni- filtered CuK-α radiation, was done on the sintered samples. Fracture surfaces were studied by scanning electron microscope (SEM). The sintered porous samples were also observed with transmission electron microscope (TEM); the samples were prepared by embedding in resin, followed by cutting, polishing, and ion-thinning by an Ar ion beam.

Results

Microstructure

SEM micrographs of the porous SiC samples are shown in Figure 1. The average grain sizes were about 3 microns for α-SiC and about 0.2 microns for β-SiC.

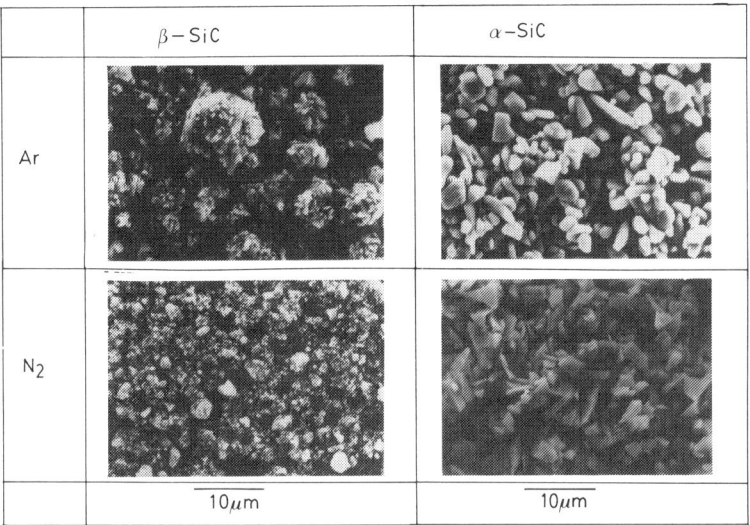

Fig. 1. SEM micrographs showing the representative microstructure of the porous α- and β-SiC samples after firing in N_2 and Ar.

It is noted that grains of Ar-sintered SiC have sharp edges, while those of N_2-sintered SiC become rounded during sintering. X-ray diffraction data shown in Figure 2 indicate that there was very little change in the phases during sintering. The α-SiC consists of the 6H polytype. TEM micrographs are shown in Figure 3. It can be seen that stacking faults exist in almost every grain, for each condition.

Electrical Conductivity

The dependence of electrical conductivity on temperature is shown in Figure 4. β-SiC samples sintered in N_2 and Ar showed n-type conduction as indicated by the sign of the Seebeck coefficient. Nitrogen appears to behave as a donor to increase the conductivity, which was supported by chemical analysis, as 0.33 wt.% was present in specimens sintered in N_2. α-SiC samples sintered in Ar showed p-type conduction, while those sintered in N_2 showed n-type conduction. Temperature dependence of electrical conductivity for α-SiC ceramics, shown in Figure 4, indicates the compensation of acceptors and donors taking place during sintering in a nitrogen atmosphere. The nitrogen content after sintering in N_2 was about 0.15 wt.%, while the nitrogen content was 0.06 wt.% for the specimen sintered in Ar.

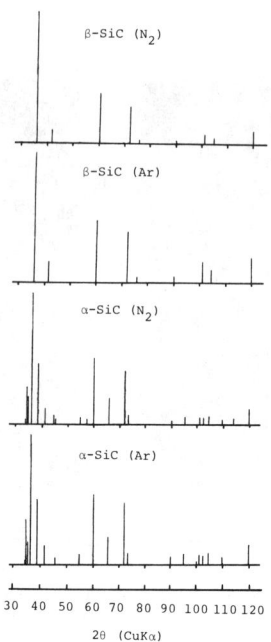

Fig. 2. X-ray diffraction results for α- and β-SiC samples after firing in N_2 and Ar.

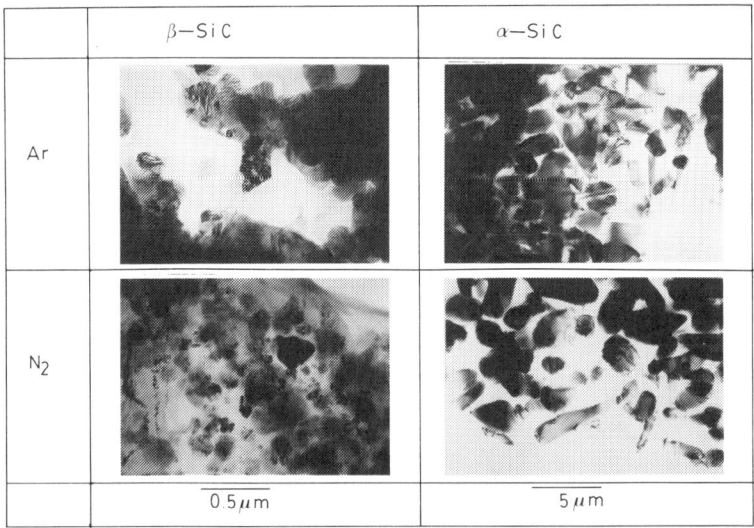

Fig. 3. TEM micrograph showing the representative microstructure of the porous α- and β-SiC samples after firing in N_2 and Ar.

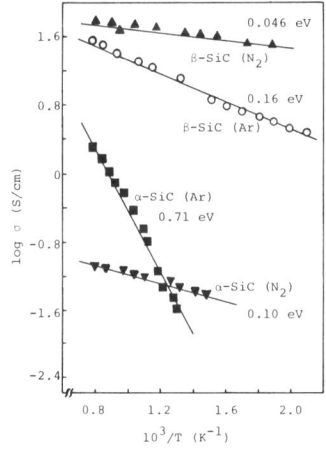

Fig. 4. Plot showing the dependence of electrical conductivity on firing temperature, for porous α- and β-SiC samples fired in N_2 and Ar.

Thermal Conductivity

Due to the porous nature of the samples, the thermal conductivity was almost independent of temperature as shown in Figure 5. The thermal conductivity was less than roughly 10% of the values reported for dense SiC ceramics[2,3], while electrical conductivity was found to be comparable to or higher than for single crystals[4-6]. This relationship indicates a great advantage for increasing the energy conversion efficiency.

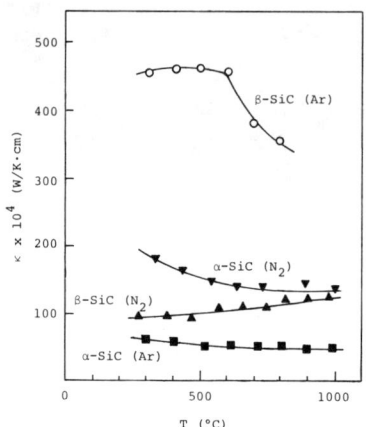

Fig. 5. Plot showing the dependence of thermal conductivity on firing temperature, for the porous α- and β-SiC samples fired in N_2 and Ar.

Seebeck Coefficient

As shown in Figure 6, the β-SiC samples sintered in N_2 and Ar, and the α-SiC samples sintered in N_2 had a negative Seebeck coefficient, while the α-SiC samples sintered in Ar had a positive Seebeck coefficient. The absolute values for the Seebeck coefficients increased with increasing temperature in all cases, which is advantageous for increasing the energy conversion efficiency at high temperature.

Figure of Merit

The calculated figures of merit for porous SiC ceramics are shown in Figure 7. The values are smaller by one or two orders of magnitude than those for other semiconductor materials[1,7] like $GeTe$-Bi_2Te_3, PbTe, $FeSi_2$ and boron carbide. However, these other materials more easily decompose, melt, or oxidize at temperatures above 1000°C. Since the figure of merit for SiC ceramics increases with increasing

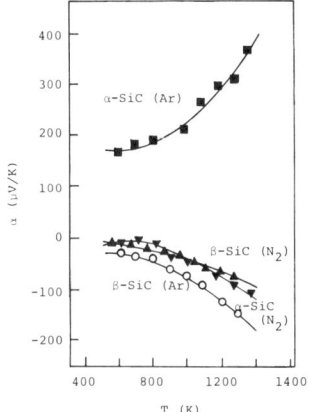

Fig. 6. Plot showing the dependence of Seebeck coefficient on firing temperature, for the porous α- and β-SiC samples fired in N_2 and Ar.

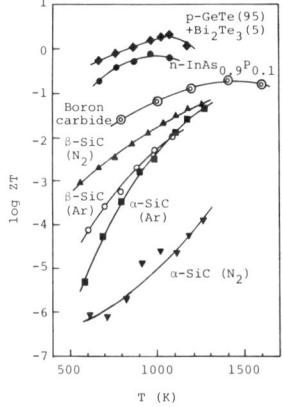

Fig. 7. Plot showing the dependence of the figure of merit (calculated) on firing temperature, both for the porous α- and β-SiC samples fired in N_2 and Ar, and selected other semiconductor materials.

Silicon Carbide 507

temperature even above 1000°C, this material seems to offer high potential for high-temperature thermoelectric energy conversion.

DISCUSSION

The temperature dependence of Seebeck coefficient found in this study for porous SiC ceramics cannot be explained by simple semiconductor theory, considering only that the electron diffusion generates the thermoelectric force under the temperature gradient. Phonon contribution to the Seebeck effect, which is called a phonon-drag effect, has been found for metals and compound semiconductors at low temperatures[7]. The phonon-drag effect is associated with the interaction between current carriers and phonons leading to an apparent increase in the absolute Seebeck coefficient. The temperature dependence of the Seebeck coefficient (shown in Figure 6) is postulated to indicate a phonon-drag effect occurring in porous SiC ceramics.

According to Parrott,[8] the phonon-drag Seebeck coefficient can be expressed as follows:

$$A = A_o + A_p \qquad A_p = \frac{F_p}{F_{ep}} \frac{ms^2 2}{eT} \qquad (5)$$

$$\frac{1}{F_p} = \frac{1}{F_{pp}} + \frac{1}{F_{pb}} + \frac{1}{F_{pe}} \qquad (6)$$

where A_o is the normal Seebeck coefficient without a phonon-drag effect, s the mean velocity of phonons, e the electronic charge, T the absolute temperature and F is the relaxation time, respectively. Subscripts for relaxation time F represent the collision types; for example, F_{ep} means the relaxation time of electron scattering by phonons. Subscript letters e, p, and b denote electron, phonon, and boundary, respectively. If phonon scattering by boundaries (stacking faults, dislocations, etc.) is predominant, relaxation time for the responsible phonon scattering would be expressed as follows:

$$\frac{1}{F_p} = \frac{1}{F_{pb}} = \frac{s}{L^*} \qquad (7)$$

where L^* is the calculated thermal diffusivity. This assumption is the only one which explains the abnormal temperature dependence of Seebeck coefficient (shown in Figure 6). Then A_p can be written as follows, assuming current carriers (electrons or holes) are predominantly scattered by phonons:

$$A_p = \frac{L^* s}{u_o T} \qquad (8)$$

where u_o is the carrier mobility, and its temperature dependence is expressed by the following equation.

$$u_0 = C T^{2-n} \quad (9)$$

where C is a constant and the value n depends on the phonon mode. Carrier scattering by acoustic and optical phonons would give n values of 1.5 and 2.5 respectively.

Using these equations, combined with conventional semiconductor theory, the observed temperature dependence of Seebeck coefficient was simulated by curve-fitting analysis. The solid lines in Figure 6 represent the best fit obtained by the least squares method. This analysis gave information about the carrier-scattering process and thermal diffusivity which comes from the phonons responsible for the phonon-drag effect. The results of the curve-fitting analysis are shown in Table 1.

The calculated n values are about 1.5 in all cases, which indicates the carrier electrons are predominantly scattered by acoustic phonons. L^*s is the calculated thermal diffusivity and Ls is the measured thermal diffusivity. For α-SiC these two thermal diffusivity values are similar, but for β-SiC, they are quite different.

Table 1. Comparison of calculated and observed n values and thermal diffusivity results.

	n	L^*s (m^2/s)	$\bar{L}$ s (m^2/s)
β-SiC in Ar	1.7	948 × 10^{-6}	12.8 × 10^{-6}
β-SiC in N$_2$	1.5	159 × 10^{-6}	1.92 × 10^{-6}
β-SiC in N$_2$	1.5	3.68 × 10^{-6}	4.44 × 10^{-6}

These data may suggest that the phonon modes responsible for a phonon-drag effect and thermal conduction are similar for α-SiC, but different for β-SiC.

It is necessary for L*s to be increased to increase the absolute value of Seebeck coefficient, and hence the figure of merit for thermoelectric energy conversion. Since L* is the mean free path of acoustic phonons, and is directly related to the microstructure, the microstructural features must be considered for purposes of optimizing the energy conversion. The TEM observations (Figure 3), which showed that the porous SiC ceramics had many lattice defects in grains with different sizes, may indicate that the average grain size should be increased and lattice defects should be eliminated to increase L*, as microstructural variation would also alter both electrical conductivity and thermal conductivity of the material, it would be necessary to design and the optimum microstructure to give a maximum figure of merit for energy conversion.

SiC ceramics do seem to offer good potential for high-temperature thermoelectric energy conversion. The ideal microstructure and the method for its control in SiC ceramics, have not yet been established, so further study is needed.

References

[1] C. Wood and D. Emin, Mat. Res. Soc. Symp. Proc. 24 (1984) 199.
[2] G. A. Slack, *J. Appl. Phys.*, **35** (1964) 3460.
[3] MCIC Report, "Engineering Property Data on Selected Ceramics," Vol. 2, Metals and Ceramics Information Center, Columbus, OH (1979) p. 5.2.3-2.
[4] H. H. van Daal, Philips Res. Rep. Suppl. 3 (1965) 70.
[5] H. Kang and R. E. Hilborn Jr., "Silicon Carbide—1973," University of South Carolina Press, Columbia (1974) p. 493.
[6] K. Sasaki, E. Sakuma, S. Misawa, S. Yoshida and S. Gonda, *Appl. Phys. Lett.*, **45** (1984) 72.
[7] Y. Suge (ed.), "Thermoelectric Semiconductors," Maki Shoten, Tokyo (1966) p. 295-355.
[8] J. E. Parrott, Proc. Phys. Soc. B70 (1957) 540.

Index

Abrasion testing, 11
Ackerman, J. L., 187
Advanced Gas Turbine Program, 367
Alloys, 63, 275
Alpha SiC, 253, 105, 113, 137, 175, 201, 215, 241, 253, 367, 407
Alumina, 227
Aluminum, 113, 397
  nitride, 65, 227
Analytical STEM techniques, 435
Annealing, 479
Antiphase disorder, 447
Applications, 215
  electronic, 151, 447, 457, 479, 491
  engines, 153, 275, 343, 355, 367
  heat exchangers, 275, 289
  high-temperature, 201, 227, 241, 447
  radiant tubes, 387
Argon sintering, 355
Astell-Burt, P., 491
Attrition grinding, 125, 137
Auger techniques, 435

Baer, J. R., 355
Baumgartner, H. Robert, 3
Beryllium oxide, 76, 83
Beta-SiC, 3, 35, 93, 113, 157, 343, 457, 479, 491
Binder removal, 157, 343
  plasticizers, 187
Bishop, Bruce A., 157
Blackborow, P., 491
Boecker, Wolfgang D. G., 407
Boron, 3, 113, 125, 364
  carbide, 397
  nitride, 75
Bostedt, Eva, 175
Bowen, H. Kent, 157
Bradt, R. C., 241, 313
Bruner, Susan L., 227

Carbon, 125, 157, 222
Carbothermal reduction, 227
Carlsson, Roger, 175
Carlstrom, Elis, 175
Carter, C. H., Jr., 253
Cathode materials, 491
Cawley, J. D., 47
Ceramic
  component formation, 367, 387
  compounds, 63, 83
  fibers, 407, 421, 435
Chang, Yeu-Wen, 83, 435
Characterization, 3, 35, 93, 265, 421, 435, 479
Chemical vapor deposition, 457
Chevron-notch test, 241
Chia, K. Y., 215
Chlorination, 275, 289
Chwastiak, Stephen, 407
Closed-loop fabrication, 372
Composite materials, 397, 407, 421, 435
Consolidation techniques, 201
Core-shell structure, 35
Corrosion behavior, 11, 275, 289, 301
Crack growth, 241
Creep properties, 253
Crystal
  elastic constants, 335
  growth, 447
Cutler, Raymond A., 17

Davis, R. F., 253, 457, 479, 491
Deformation behavior, 253
Densification, 3, 201
Deposition rate, 463
Dewaxing process, 349
Dispersants, 137
Dispersions, 175
  nonaqueous, 157
Dry-bag isopressing, 367
Dry-etched monocrystalline films, 491
Dutta, Sunil, 201

Edmond, J. A., 457, 479
Elastic anisotropies, 313
Electronic applications, 151, 447, 457, 479, 491
Electrothermal furnace, 93
Ellingson, W. A., 187
Engines, 153, 275, 343, 355, 367
Etching, 491
Exothermic reactions, 17
Eyck, M. O. Ten, 367

Fatigue testing, 10
Fibers, 407, 421, 435
Filaments, 407
Film
  growth, 447, 457, 479
  morphology, 447
Flexural strength, 215
Flow-through furnace processing, 35
Forming process, 368
Fox, Dennis S., 275
Fracture
  analysis, 441
  toughness, 8, 125, 215, 227, 241
Frechette, Frank, 407
Free carbon, 3, 105
Furnace processing, 35, 93, 389

Garrido, L., 187
Gas phase reactions, 35
Gasifier turbine rotor, 377
Gasifier turbine scroll, 376
Ghosh, Asish, 241
Glass encapsulation, 201
Glass, Jeffrey T., 457
Goldberger, W. M., 93
Grain
  growth, 175, 215
  size, 220
Green
  compacts, 157
  state ceramics, 187
Grinding
  attrition, 125, 137
  Ultrasonic Impact, 149
Gronemeyer, S., 187

Hangas, Jon, 113
Heat exchangers, 275, 289
Hexagonal stacking, 313
High-temperature
  applications, 201, 227, 241, 447
  deformation, 253
Hot
  isostatic pressing (HIP), 201
  pressing, 63, 83, 125
Hoyer, Jesse L., 137
HPZ fibers, 435
HSC Process, 93
Hurford, Andrew C., 227
Hydrogen, low dew-point, 301
Hydridopolysilazane, 436

I-V characteristics, 457
Imaging, 187
Implantation, 479
In situ doping, 457
Indirect free carbon method, 108
Injection moulding, 343, 355, 367
Ion
  etching, 491
  implanted films, 479
Ip, S. Y., 289
Isopressing, 157, 367
Isostatic pressing, 367

Jackson, T. Barrett, 227
Jacobson, Nathan S., 275
Jenkins, Michael G., 241

Kasprzyk, Michael C., 387
Kim, Hyeong Joon, 457
Kim, Hyoun-Ee, 301
Knoch, H., 105
Kobayashi, Albert S., 241
Kong, Hua-Shuang, 457
Koumoto, Kunihito, 501
Kreidler, Eric R., 275
Kristoffersson, Annika, 175

Lane, J. E., 253
Langley, N., 421
Lau, Sai-Kwing, 407
LeGrow, G., 421
Li, Z., 313
Lim, T., 421
Lipowitz, Jonathan, 421
Liquid-phase sintering, 227
Long, W. D., 105

Machining techniques, 149
Maeda, Kunihiro, 113
Marra, John E., 275
Marasperse N-22, 137
Matrix composites, 397, 407
Matus, Lawrence G., 447
McNallan, M. J., 289
Mechanical properties, 215
Medical imaging, 187
Metal semiconductor field effect transistors (MESFETs), 457
Metallothermal reduction, 17
Methylpolysilane, 436
  polymers, 421
Microchemical analysis, 435
Microstructural analysis, 435
Microstructure, 63, 83, 201, 215, 463, 501
Mixing process, 343
Modeling
  carburization, 59
  reaction mechanism, 21
  strength and flaw size, 349
Moore, David O., 149
Morse, R., 93

Nicalon fibers, 435
Nitrogen, 83
Nixdorf, Richard D., 397
Nonaqueous dispersions, 157
Nuclear magnetic resonance, 187

Ohmic contacts, 468
Ohnsorg, R. W., 367
OLOA™ 1200, 157
Oxidation, 275, 289, 301

Palmour, John W., 491
Persson, Michael, 175
Phase stability, 63, 83, 113
Plasma
  arc method, 3
  etching, 491
  SiC, 3
Platelets, 397, 407
Polycrystalline filaments, 407
Polymers, 421
Polyorganosilane polymers, 421
Polyphenylene, 175
Polytype structure, 253, 313
Polytypic lamellar formation, 113
Porosity, 215
Porous ceramics, 501
Powders
  alpha Si-C, 175
  beta Si-C, 157
  classified, 157, 175
  coated, 175
  gas phase-synthesized, 3, 35
  hot-pressed, 125
  plasma-synthesized, 3
  submicron, 17
  ultrafine, 93, 125, 137
Powell, J. Anthony, 447
Pressureless sintering, 3, 105, 157, 215, 355
Pyrolysis, 35, 421

Quantitative analysis, 105

Radiant tubes, 387
Reaction
  bonding, 253
  mechanism, 21
Readey, Dennis W., 35, 301
Reed, A. K., 93
Reinforcement materials, 397, 407, 421, 435
Reliability, 201
Rigtrup, Kevin M., 17
Rossing, Barry R., 3
Rotors, turbine, 377
Ruh, Robert, 63, 83

SAM analysis, 436
Scanning Auger techniques, 435
Schreiner, M. E., 289
Schwetz, K. A., 105
Scrolls, turbine, 376
Semiconductors, 447, 457, 479
Seshadri, S. G., 215
Sheek, J. G., 47
Shimohigoshi, Mitsuhide, 501
Shinozaki, Samuel S., 113
SiC-Al_2OC systems, 227
SiC-Al_2OC-AlN system, 73
SiC-AlN system, 65
SiC-BeO system, 76, 83
SiC-BN system, 75
Silicon oxychloride compounds, 275
Single-crystal
    films, 447, 457
    platelets, 407
    silicon, 47
Single-phase reinforcement, 407
Sinter-HIPing, 201
Sintering, 175, 253
    additives, 113
    liquid phase, 227
    pressureless, 3, 105, 157, 215, 355
Size classification, 159, 175
Slip casting, 175, 367
Slurry pressing, 201
Small-bore imaging, 187
Soeta, Atsuko, 113
Solid-state imaging, 187
Srinivasan, M., 215
STEM techniques, 435
Strength, 201, 215, 227, 241
Structural ceramics, 187
Submicron SiC, 17
Suspension processing, 157
Synthesis
    gas phase, 35
    HSC Process, 93
    plasma, 3
    submicron, 17

Takeda, Shunnji, 501
TEM/STEM analysis, 439
Temperature, and deformation, 253
Tetramethylsilane, 35
Thermal
    carburization, 47
    decomposition, 35
    expansion, 313
Thermoelastic stress index, 313
Thermoelectric energy conversion, 501
Thin films, 457, 479, 491
Titanium, 113
Transistor applications, 457
Transition duct, 374
Trela, Walter, 343
Turbines, 367
Turbomilling, 125, 137

Ultrafine powders, 93, 125, 137
Ultrasonic Impact Grinding, 149

Vacuum sintering, 355
Vapor
    depositing, 253
    phase transport, 47
Vaughan, Gerald, 397

Weaver, Samuel C., 397
Wet-bag isostatic pressing, 367
Whalen, Thomas J., 343, 355
Whiskers, 313, 397
White, Ken W., 241
Withrow, S. P., 479
Wittmer, Dale E.,Sr., 125
Wu, Huann-Der, 35

Yanagida, Hiroaki, 501

Zangvil, Avigdor, 63, 83, 435